Climate-Smart and Resilient Food Systems and Security

Mohamed Behnassi · Abdulmalek A. Al-Shaikh ·
Riaz Hussain Qureshi · Mirza Barjees Baig ·
Turki Khalufa A. Faraj
Editors

Climate-Smart and Resilient Food Systems and Security

Springer

Editors
Mohamed Behnassi
Economic Social and Environmental
Council (CESE)
Rabat, Morocco

Center for Environment, Human Security
and Governance (CERES)
Agadir, Morocco

Riaz Hussain Qureshi
University of Agriculture
Faisalabad, Pakistan

Turki Khalufa A. Faraj
Prince Sultan Institute for Environmental
Water and Desert Research
King Saud University
Riyadh, Saudi Arabia

Abdulmalek A. Al-Shaikh
Prince Sultan Institute for Environmental
Water and Desert Research
King Saud University
Riyadh, Saudi Arabia

Mirza Barjees Baig
Prince Sultan Institute for Environmental,
Water and Desert Research
King Saud University
Riyadh, Saudi Arabia

ISBN 978-3-031-65967-6 ISBN 978-3-031-65968-3 (eBook)
https://doi.org/10.1007/978-3-031-65968-3

This Springer imprint is published by the registered company Springer Nature Switzerland AG
The registered company address is: Gewerbestrasse 11, 6330 Cham, Switzerland

Foreword by Dr. Ismahane Elouafi

The climate crisis is threatening our ability to nourish the world. Globally, more than 780 million people are affected by hunger, a number that will grow larger for every degree the planet warms. Seventy-six percent of the world's population gets most of its daily nutrients from plants, but more frequent, devastating droughts and floods are destroying staple crops.

Deeply affected by the climate crisis, agrifood systems can also be part of the solution. Today, the world has an opportunity to reimagine how it sustains itself—an opportunity the global community needs to take. By transforming agrifood systems for greater resilience, sustainability, and inclusivity, we can improve our ability to nourish both people and planet.

When humanity has faced its greatest challenges, innovation has provided the solutions. The progress the world witnessed over the past 400 years, including longer lifespans and better quality of life, is directly linked to scientific advancement. Now, science—from discovery to development—holds the key to tackling climate change and transforming fragile agrifood systems. By prioritizing innovation, encouraging technology transfers between countries, and building cross-sectoral partnerships, the global community can develop climate-smart food systems that can help feed the world while also protecting its natural resources.

Already, science is leading the way towards progress. Climate-smart crop varieties that can withstand droughts, floods, salinity, and extreme temperatures, for instance, are helping to protect farmers from climate emergencies. More sustainable farming approaches, including zero-tillage practices, are cutting farm-related greenhouse gas emissions and increasing agriculture's capacity for carbon sequestration, relieving pressure on the atmosphere. New technologies, like artificial intelligence, are also helping improve farmers' lives and livelihoods by giving them access to the information and data that they need.

But the world is currently underinvesting in these solutions. When science is translated into scalable innovations, its potential for impact massively grows; now, funders, governments, and partners across different sectors need to continue investing in these tools to realize their potential.

climate change, ensuring access to water resources in an equitable and sustainable manner. We need to acknowledge that transforming food systems will require IWRM and efficient water use in both irrigated and rainfed agriculture. We must emphasize that water resources management needs to be integrated at all levels, inclusive of all stakeholders, and coherent across all sectors of agriculture, forestry, fisheries and aquaculture, land and soil, environment, biodiversity, climate change, and One Health, among others. The stress about growing competition for water and degradation of its quality calls for improved governance to ensure sustainable and equitable water allocation, integrated management, and efficient use. Last but not least, we need to seize the international momentum to take bold and innovative initiatives regarding surface and groundwater management in agrifood systems to support governance of water.

Our agenda and guiding our work for next years rests on our vision towards climate-resilient agriculture to transform agrifood systems to be more efficient, more inclusive, more resilient, and more sustainable, accelerating the achievement of the 2030 Agenda for Sustainable Development, responding to the worldwide challenge of assuring water and food security, tackling the impacts of the climate crisis, while aiming to address a broad range of interlinked challenges, including biodiversity loss, desertification, land, and environmental degradation, the need for accessible, affordable renewable energy, and food and water security.

Sustainable and resilient food systems provide a variety of solutions that respond to the climate crisis and water security challenges and contribute to restoring degraded natural and managed ecosystems. I believe that holistic, integrated actions, springing from coordinated and collective efforts, will make our food systems a central part of the global solutions for climate change and water availability, as well as for our planet to be a better place to live, ensuring food security and a more equitable future for all. We must look beyond only water security, food production, or climate change by considering crops and livestock, forests, fisheries and aquaculture and related value chains, livelihoods, biodiversity, and ecosystems in a holistic manner, as well as embracing the indispensable role of women, youth, and Indigenous Peoples, as essential agents of change.

Climate change is a global challenge that requires comprehensive and cross-sectoral action. Such action needs to be taken in full consideration of international goals and agreements. To accelerate its contribution towards the 2030 Agenda, it presents the role of integrated water management and food systems as part of the solution to climate change. The urgency to act on climate change impacts on water and food security has never been clearer, as indicated by the latest reports of the Intergovernmental Panel on Climate Change. We must learn from good practices to promote transformative adaptation and mitigation policies, plans, and actions. Action in a coherent manner, as appropriate, in accordance with and dependent on national contexts and capacities through food systems and integrated water management, which offer considerable potential to maximize co-benefits of adaptation and mitigation.

The water–food–climate nexus asks for urgent actions at global, regional, country, and local levels and across sectors. We must help countries, especially developing countries, implement their country-driven commitments and plans, including Nationally Determined Contributions (NDCs), National Adaptation Plans (NAPs), National Pathways for transforming food systems, ensuring that a diverse range of stakeholders—all levels of governments, youth, business and investors, civil society, frontline communities, Indigenous Peoples, and others—are fully enabled, empowered, and involved in decision-making processes around specific solutions that must be scaled up this decade, build resilience, and mobilize finance at scale.

This book is not only timely but also necessary. Not only because it is fully addressing the water–food–climate nexus in a future orientation but also because it is addressing emerging solutions to the nexus challenges. The science-based approach combined with empirical research give it more impetus. It invites us all to follow the recommendations to guide fundamental changes at all levels, especially at national levels. It gives us inspiration to act, to act now!

I conclude with the words of our famous leaders Nelson Mandela: "It always seems impossible, until it is done," and Jacques Yve Cousteau: "It takes generosity to discover the whole through others. If you realize you are only a violin, you can open yourself up to the world by playing your role in the concert."

Dr. Hans Hoogeveen
Independent Chairperson of the FAO Council
Co-chair of the Champions 12.3
Co-chair of the Alliance for Climate Smart Agriculture

Dr. Hans Hoogeveen was elected as Independent Chairperson of the Council of FAO in June 2021. FAO Members unanimously re-elected him in July 2023 for a second, 2-year term. He is the Ambassador and former Permanent Representative of the Netherlands to FAO, WFP, and IFAD (2016–2021). He is duly acknowledged as a consummate professional and a man of food, agriculture, and natural resources to the utmost degree. With unwavering focus on the stimulation and promotion of sustainable food systems and sustainable agriculture globally, food security and safety, he has served in numerous international functions and boards. Board membership includes Yale University Governance, Environment and Market Initiative. He is a strong supporter of multilateralism, focusing on consensus building. As Chairperson of the 7th Session of the United Nations Forum on Forests Bureau, his leadership resulted in an international instrument deemed a landmark achievement by the UN Secretary-General following 15 years of encumbered international negotiations—a highly noteworthy accomplishment within the forum, which earned him a UN award. He is one of the founding members and current co-Chair of the Global Alliance for Climate Smart Agriculture (GACSA). In addition, he is a founding member and current co-Chair of Champions 12.3, an international coalition of eminent commissioners and CEOs of companies, international organizations, and NGOs, which focus on a 50% worldwide reduction of food losses and waste by 2030. Having worked in almost all UN capitals, including New York, Nairobi, and Geneva, his broad and well-sustained international network is highly regarded. Among his many initiatives is a long list of international events, most of which have been organized by himself. Furthermore, he has experience in a variety of UN bodies other than FAO.

Dr. Hoogeveen chaired numerous bodies such as in the context of FAO (Chairperson of the Program Committee for four years), IFAD (Governing Body), and WFP. Early in his career, he became well acquainted with various aspects of the policy field: The Brussels component, political current affairs, and other party negotiations. He came to regard it as the most integral policy domain, impressing upon him its crucial implications for all living entities—humans, animals, and plants. Fully dedicated to contributing to a high goal in areas such as Food Security and the elimination of poverty, he feels strongly that activity through joint effort in a multilateral field ultimately leads to the achievement of certain pivotal objectives. He has authored a book on Sustainable Development Diplomacy as well as a number of articles on sustainable development and natural resources policies. His academic posts include Professor of Practice in Natural Resource Policy at the Fletcher School of Law, visiting Professor Sustainable Development Diplomacy at Wageningen University and Research, the Netherlands, and Lead Lecturer in European Policy. He holds a degree in Law obtained at Leiden University, the Netherlands, a master's degree in Public Management, and a doctorate from the Fletcher School of Law and Diplomacy, Boston, and Wageningen University, the Netherlands. He has received the following awards: Commander of Gregorius (Holy See) and Commander de l'Orde Grand-Ducal de la Couronne de Chene (Luxembourg).

Acknowledgements

It is with great pleasure that I share the editorship of this volume with my esteemed colleagues: Dr. Abdulmalek A. Al-Shaikh, Prof. Dr. Riaz Hussain Qureshi, Prof. Dr. Mirza Barjees Baig, and Dr. Turki Khalufa A. Faraj. Their expertise and dedication were invaluable throughout the entire publishing process.

While the editors oversaw the publication, the true credit for this volume's content lies with the chapter authors. Their insightful contributions underwent a rigorous peer-review and proofreading process, ensuring the highest quality content. Their cooperation, responsiveness, and commitment throughout this journey were truly commendable. Any shortcomings, however, remain the responsibility of the editors.

The exceptional quality of this volume is also a testament to the invaluable input and dedication of our reviewers. We, the editors, are deeply grateful for their willingness to share their time and expertise. Their critical analyses and expertise significantly enhanced the readability and overall quality of the chapters. On behalf of my co-editors, I would like to express our sincere gratitude to the following reviewers:

- Dr. Umair Riaz, Scientific Officer, Soil and Water Testing Laboratory for Research, Bahawalpur, Agriculture Department, Government of Punjab, Pakistan
- Dr. Nandita Baruah, Assistant Professor Department of Soil Science, BNCA, Assam Agricultural University, Jorhat, India
- Dr. Renu Gupta, Associate Professor Division of Soil Science and Agricultural Chemistry, Faculty of Agriculture, SKUAST-Jammu, India
- Prof. Sanjay-Swami, Ph.D. (Soil Science and Agril. Chemistry), School of Natural Resource Management, College of Post Graduate Studies in Agricultural Sciences, (Central Agricultural University), Meghalaya, India
- Prof. Dr. Nima Norouzi, Amirkabir University of Technology, Tehran, Iran
- Prof. Dr. Lorenzo Traverso, Economy, Engineering, Society and Business Department, University of Tuscia, Viterbo, Italy
- Prof. Dr. Nguyen Bao Quoc, Research Institute for Biotechnology and Environment, Nong Lam University, Ho Chi Minh City, Vietnam

- Dr. Asif Sardar, National Center of Industrial Biotechnology, PMAS-Arid Agriculture University, Rawalpindi, Pakistan
- Prof. Dr. Maqsood Anwar, Dean Faculty of Sciences, PMAS-Arid Agriculture University, Rawalpindi, Pakistan
- Prof. Dr. Rafiq Islam, Soil, Water, and Bioenergy Resources, Ohio State University, South Centers, USA
- Prof. Dr. Özlem Korkut, Atatürk Üniversity, Faculty of Engineering, Erzurum, Turkey
- Dr. Muhammad Abid, State Key Joint Laboratory of Environment Simulation and Pollution Control, School of Environment, Tsinghua University, Beijing, China
- Dr. Asim Jilani, Center of Nanotechnology, King Abdulaziz University, Jeddah, Saudi Arabia
- Dr. Muhammad Rizwan, University of Chinese Academy of Sciences, Beijing, China
- Dr. Muhammad Aamir Shakoor, Department of Agricultural Engineering, Bahauddin Zakariya University (BZU), Multan, Pakistan
- Dr. Ijaz Ahmad, Director, Remote Sensing Agricultural Wing, SUPARCO, Islamabad, Pakistan
- Dr. Amer Masood, Department of Space Science, University of the Punjab, Lahore, Pakistan
- Dr. Syed Muhammad Hasan Raza, Department of Space Science, University of the Punjab, Lahore, Pakistan
- Dr. Omar Jouzdan, Expert—Soil and Water Sciences—Visiting Researcher, Lancaster University, UK
- Prof. Dr. M. M. Abd El-Azeim, Vice Dean, Faculty of Agriculture, Department of Soil and Water Sciences, Minia University, Egypt
- Prof. Dr. Muhammad Arshad, King Abdulaziz University, Jeddah, Saudi Arabia
- Dr. Ammari Abdelhadi, The National Higher School of Hydraulics, Blida, Algeria
- Prof. Dr. Rafiq Islam, Program Director—Soil, Water and Bioenergy Resources, Ohio Sustainability Institute, College of Food, Agriculture and Environmental Sciences (CFAES) The Ohio State University, USA
- Prof. Dr. Özlem Korkut, Faculty of Engineering, Atatürk University, Erzurum, Turkey
- Dr. Khondker Iftekhar Iqbal. Faculty of Arts and Social Sciences, Universiti Brunei Darussalam, Brunei Darussalam
- Prof. Dr. Syed Serajul Islam, Department of Political Science, Lakehead University, Orillia, ON, Canada
- Prof. Dr. Shaikh M. Kais, Department of Sociology, University of Rajshahi, Rajshahi, Bangladesh
- Dr. Abed A. Alataway, Prince Sultan Institute for Environmental, Water, and Desert Research, King Saud University, Riyadh, Saudi Arabia
- Dr. Turki K. Alasiri, Prince Sultan Institute for Environmental, Water, and Desert Research, King Saud University, Riyadh, Saudi Arabia
- Engr. Ali Wafa Abu-Risheh, Prince Sultan Institute for Environmental, Water, and Desert Research, King Saud University, Riyadh, Saudi Arabia

- Dr. Abdulmalek A. Al-Shaikh, Director, Prince Sultan Institute for Environmental, Water, and Desert Research, King Saud University, Riyadh, Saudi Arabia
- Dr. Waqar Akram. Research Scholar, SAFE, Department of Management Sciences, University of Waikato, Hamilton, New Zealand
- Prof. Dr. Shahbaz Khan, Ph.D., D.Sc. Director of UNESCO MS Regional Office for East Asia and Rep to China, DPRK, Japan, Mongolia, and the Republic of Korea
- Prof. Dr. Gary Straquadine, Associate Vice President for Technical Education at Utah State University Eastern, Utah, USA
- Dr. Michael R. Reed, Professor Emeritus, University of Kentucky, United States
- Dr. Muazamma Batool, Regional Agricultural Research Institute, Bhawalpur, Pakistan
- Dr. Asif Sajjad, Associate Professor, Department of Entomology, Islamia University Bahawalpur, Pakistan
- Dr. Anderson Kehbila, Stockholm Environment Institute, Nairobi, Kenya
- Kevin Enongene, FOKABS INC, Ottawa, Canada
- Prof. Dr. Mohammad Suhail, Centre of Applied Remote Sensing and GIS Applications, Samarkand State University, Samarkand Uzbekistan

About the Publishing Institutions

Center for Environment, Human Security and Governance (CERES), Morocco

CERES, formerly the North-South Center for Social Sciences (NRCS), emerged in 2015 as an independent, non-profit research institute built upon the expertise of Moroccan researchers. Its global ambition is to be a leading think tank and a reference point for positive change, especially in the Global South. Since its founding, CERES has fostered a robust network encompassing researchers, policymakers, students, practitioners, and journalists across diverse fields. This network has propelled CERES members to contribute to prestigious assessments like Ipbes, Medec, and EuroMeSco. The Center's expertise is further recognized by its membership in the MedThink 5+5, a group shaping research and policy agendas in the Mediterranean. CERES not only provides expert knowledge to various institutions but also organizes international conferences and capacity-building workshops. Their extensive publications, including books, scientific papers, and policy studies, enjoy global recognition and distribution. The Center focuses on emerging research areas within the human–environment–climate nexus, employing multidimensional, multiscale, interdisciplinary, and policymaking perspectives. Through these initiatives, CERES aims to deliver cutting-edge expertise, advance scientific understanding and application, and ultimately bridge the gap between science and policy for impactful solutions.

Contents

About the Editors

Dr. Mohamed Behnassi is currently a Senior Environmentalist Expert at the Economic Social & Environmental Council (CESE), Rabat, Morocco. From 2004 to 2022, he served as a Professor, Head of the Public Law Department, and Director of the Research Laboratory for Territorial Governance, Human Security and Sustainability (LAGOS), Faculty of Law, Economics, and Social Sciences of Agadir, Ibn Zohr University. He holds a Ph.D. in International Environmental Law and Governance (2003), an M.Sc. in Political Sciences and International Relations (1997), and a B.A. in Administration (1995) from the Faculty of Law, Economics, and Social Sciences, Hassan II University of Casablanca. He obtained a Diploma in International Environmental Law and Diplomacy from the University of Eastern Finland and UNEP, 2015, and is also an Alumnus of both the International Visitors Leadership Program of the Department of State (2011, USA), and the Near East South Asia Center for Strategic Studies (NESA), National Defense University, Washington, DC (2022–24, USA). He is currently the Founding Director of the Center for Environment, Human Security and Governance (CERES). He was appointed in 2019 as an Expert Evaluator for the National Center for Scientific and Technical Research (CNRST/Morocco) and selected twice (2019–2024) as an Assessment Scoping Expert and a Review Editor by the Intergovernmental Science-Policy Platform on Biodiversity and Ecosystem Services (IPBES—Thematic assessment of the interlinkages among biodiversity, water, food and health). He was a Lead Author (LA) of the Human Security

Chapter of 1st MEDECC Assessment Report (MAR1): Climate and Environmental Change in the Mediterranean Basin—Current Situation and Risks for the Future (2021) and Coordinating Lead Author (CLA) of the MEDECC Special Report on "the Nexus between Water, Energy, Food and Ecosystems." In 2024, he was nominated to the UNFCCC Roster of Experts (Paris Agreement Article 6 Technical Review Team) and as a Member of the PRIMA Foundation Scientific Evaluation Committee (SEC) to evaluate proposals submitted to the Thematic Area "Leveraging urban and local food systems for sustainable food systems transformation."

Dr Behnassi is a Senior Researcher in the areas of International Law and Politics of Environment and Human Security. In addition to his participation in more than 70 international and national conferences, he has published 66 scientific papers and 22 books in these areas, including recent ones: *The Water, Climate, and Food Nexus: Linkages, Challenges and Emerging Solutions* (Springer, 2024); *Food Security and Climate-Smart Food Systems—Building Resilience for the Global South* (Springer, 2022); and *The Climate-Conflict-Migration Nexus from a Human Security Perspective* (Springer, 2022). He serves as a reviewer for many global publishers (such as Routledge and Springer) and scientific journals with high-impact factor. He has organized many international conferences covering the above research areas, managed many research and expertise projects for national and international institutions, and is regularly requested to provide scientific expertise nationally and internationally. Other professional activities include social compliance auditing and consultancy by monitoring human rights at work and the sustainability of the global supply chain.

Dr. Abdulmalek A. Al-Shaikh is the Director of the Prince Sultan Institute for Environmental, Water, and Desert Research at the King Saud University (KSU) in Riyadh, Saudi Arabia. He earned his Ph.D. in Arid Land Studies from the University of Arizona, USA in 1983. He also serves as the General Secretary of the Prince Sultan bin Abdulaziz International Prize for Water (PSIPW). The prize is awarded bi-annually at the United Nations Headquarters. Being the Director and Research Chair, he oversees the Research Programs at the Institute. He is the team leader of many research projects concerning the environment, desertification, climate change, and water harvesting in Saudi Arabia including the King Fahd Project for Water Harvesting and Storage in Saudi Arabia, Prince Sultan Project for Villages and Hamlets Rehabilitation in Saudi Arabia, the Space Images Atlas of Saudi Arabia and the Environmental Atlas of Saudi Arabia. Being the Chairman of the international conferences on water resources and arid environments, he successfully organized them at the King Saud University. He has brought world-famous water scientists to Saudi Arabia and published nine proceedings of organized international conferences. He has published extensively in journals of international repute.

Prof. Dr. Riaz Hussain Qureshi is a Fellow of the Pakistan Academy of Sciences, Professor Emeritus and former Vice Chancellor of the University of Agriculture Faisalabad, and former Advisor (Quality Assurance and Learning Innovation/HRD) of the Higher Education Commission of Pakistan. He obtained Ph.D. from the University of Wales, and M.Sc. and B.Sc. with distinction from the University of Agriculture, Faisalabad. He has been President of the Soil Science Society of Pakistan, Member of the Steering Committee of the International Irrigation Management Institute (IWMI), Member of the Research Advisory Committees of PARC and NIBGE, and National Coordinator of Saline Agri. Project for eight years. He supervised 13 Ph.D. and 74 M.Sc. (Hons) students and authored over 200 research papers (including 1 in Nature) and reports. He was the Chief Editor Pakistan Journal of Agricultural Sciences for eight years and was In-charge of many international projects from the UK, Australia, the USA, the Philippines and Japan. He is a pioneer in introducing Saline Agriculture Technology in Pakistan. His co-authored book on Saline Agriculture for Irrigated Land in Pakistan has been published by the Australian Centre for International Agriculture Research. The Report on Water Security Issues of Agriculture in Pakistan by Prof. Qureshi and Muhammad Ashraf, Chairman, PCRWR was published by Pakistan Academy of Sciences, Islamabad in 2019. He received many honors and awards, including Sir William Roberts shield and scholarships for M.Sc. and Ph.D. studies, the Borlaug Award, the Pak Academy of Sciences Best Scientist of the Year Award, and the Pakistan Civil award Pride of Performance.

Dr. Mirza Barjees Baig is a professor at the Prince Sultan Institute for Environmental, Water, and Desert Research, King Saud University, Saudi Arabia. He earned his M.S. degree in International Agricultural Extension in 1992 from Utah State University, Logan, Utah, USA, and was placed on the "Roll of Honor." He completed his Ph.D. in Extension for Natural Resource Management from the University of Idaho, USA, and was honored with the "1995 Outstanding Graduate Student Award." He has published extensively on the issues associated with natural resources in national and international journals. He has also made oral presentations about agriculture, natural resources, and the role of extension education at various international conferences. He serves on the editorial boards of many international journals and is a member of many international professional organizations. He has produced six books on climate change, food security, and climate-smart agriculture. Food waste, water management, degradation of natural resources, the deteriorating environment, and their relationship with society and the community are his areas of interest. He has attempted to develop strategies to conserve natural resources, promote the environment, and develop sustainable communities. He started his scientific career in 1983 as a researcher at the Pakistan Agricultural Research Council, Islamabad, Pakistan. He served at the University of Guelph, Ontario, Canada, as a Special Graduate Faculty member from 2000 to 2005. He served as a foreign professor at the Allama Iqbal Open University (AIOU), Pakistan, through the Higher Education Commission from 2005 to 2009.

Dr. Turki Khalufa A. Faraj is an Associate Professor at the College of Food and Agriculture Sciences and an Environmental Consultant in Prince Sultan Institute for Environmental Water and Desert Research, King Saud University, Kingdom of Saudi Arabia. He is the Supervisor in the Mentoring Track at the Distinguished and Talented Students Program, and the Vice-Rectorate for Educational and Academic Affairs. He has a Ph.D. in Environmental Science, Public Health, Risk Assessment and Epidemiology, from the University of East Anglia, UK. He earned his master's degree in environmental science, Environmental Health, from the Faculty of Meteorology, Environment and Arid Land Agriculture, King Abdul-Aziz University, Jeddah, KSA. He completed his bachelor degree program in Environmental Science, at the Faculty of Meteorology, Environment and Arid Land Agriculture, King Abdul Aziz University, Jeddah, KSA. He is the member of many scientific projects on "water and Environment." He has published dozens of research articles in highly reputed scientific journals. He has represented his institute and the university at numerous meetings, workshops, symposiums, and conferences at the national and international levels. He has translated many prestigious books from English to Arabic language for the use of the civil society and the scientific community. He is a member of the many scientific bodies/fora at the national and international levels.

Abbreviations and Acronyms

AEZ	Agri-Export Zone
AF	Agroforestry
AI	Agricultural Instructors
ANACIM	Agency for Civil Aviation and Meteorology
ANN	Artificial Neural Networks
BNRCC	Building Nigeria's Response to Climate Change
CA	Conservation Agriculture
CAMF	Carbon Accounting Model for Forest
CCAFS	CGIAR Research Program on Climate Change, Agriculture and Food Security
CGIAR	The Consortium of International Agricultural Research Centers
CSA	Climate-Smart Agriculture
CSAPs	Climate-Smart Agricultural Practices
CT	Conventional Tillage
DEA	Data Envelopment Analysis
DT	Drone Technology
FAO	Food and Agriculture Organization
FDA	Agricultural Development Fund
FEFO	First-Expired-Initially-Out
FFP	Farmers Fertilization Practices
FIVIMS	Food and Nutrition Security Conceptual Framework
GHG	Greenhouse Gas
GLOFs	Glacial Lake Outburst Floods
GNSS	Satellite-Based Positioning Systems
GPR	Ground Penetrating Radar
GPS	Global Positioning Systems
GWP	Global Warming Potential
HVAC	Heating Ventilation and Air Conditioning
ICARDA	International Center for Agricultural Research in the Dry Areas
ICS	Irrigation Control System
ICT	Information and Communication Technologies

IEA	International Ergonomics Association
IGP	Indo-Gangetic Plains
ILK	Indigenous and Local Knowledge
ILO	International Labor Organization
IoT	Internet of Things
IPCC	Intergovernmental Panel on Climate Change
IWRM	Integrated water resources management
JICA	Japan International Cooperation Agency
LDN	Land Degradation Neutrality
LIFDCs	Low-Income Food-Deficit Countries
LPG	Liquefied Petroleum Gases
LUCF	Land Use Change and Forestry
MENA	Middle East and North Africa
ML	Machine Learning
MSDs	Musculoskeletal Disorders
MSW	Municipal Solid Waste
NASA	National Aeronautics and Space Administration
NbS	Nature-Based Solutions
NCCP	Pakistan's National Climate Change Policy
NDCs	Nationally Determined Contributions
NDVI	Normalized Difference Vegetation Index
NEH	North Eastern Hill Region
NER	North Eastern Region
NEST	Nigerian Environmental Study/Action Team
OF	Organic Farming
PA	Precision Agriculture
PAT	Precision Agricultural Technologies
PCCA	Pakistan Climate Change Act
PCRWR	Pakistan Council of Research in Water Resources
PLF	Precision Livestock Farming
PPPs	Public-Private Partnerships
RDF	Recommended Fertilizers Doses
RFID	Radio Frequency Identification
RNN	Recurrent Neural Networks
SBM	Slack-Based Model
SDGs	Sustainable Development Goals
SFCS	Smart Field Cultivation Server
SIM	Second Inter-Monsoon
SLM	Sustainable Land Management
SOC	Soil Organic Carbon
SSNM	Site-Specific Nutrient Management
SWM	South-West Monsoon
TR	Technology Readiness
UAVs	Unmanned Aerial Vehicles
VIA	Vulnerability, Impacts, and Adaptation

VRA	Variable-Rate Application
VRT	Variable-Rate Technology
WBCSD	World Business Council for Sustainable Development and Output Insights
WEF	Water, Energy, and Food
WFE	Water-Food-Environment
WFP	World Food Program
WMO	World Meteorological Organization
WSNs	Wireless Sensor Networks

List of Figures

Bundling Climate Smart Agriculture Systems in Morocco: Insights, Advances and Perspectives in a Changing Climate

Integrating Climate-Smart Agriculture and the Water, Energy, and Food Nexus Approach to Address Sustainability Challenges Under a Changing Climate

Food and Water Security Through Climate-Smart Agriculture in Asia and Africa

Energy-Economy-Environment Nexus: Moving Toward an Innovative Framework for Food-Secure Future

Climate Change and Innovative Ecological Interventions Through Climate-Smart Agriculture in the Cholistan Desert of Pakistan

Enhancing Food Security Through Climate-Smart Agriculture in Kenya

Ergonomics for Sustainable Agriculture and Food Security

Contribution of Precision Agriculture to Drought and Food Security

Climate Change Impacts on Agricultural Sector in Sri Lanka: Necessity of Precision Agricultural Technologies

Climate Change Challenges and Its Preparedness Toward Agriculture Using Climate-Smart Agriculture in Potohar Plateau of Pakistan

Climate-Smart Water Management Technologies in Pakistan

Applications of Smart Agriculture in Irrigation Water Management in Developing Countries: The Way Forward

Mitigation of the Effects of Climate Change on Agriculture Through the Adoption of Precision Agriculture Technologies

Effects of Climate Change and Food Security on Sustainable Agricultural Development and Potential Solutions: Evidence from China

Can Seasonal Climate Forecasts Improve Household Welfare? Insights from Rural Senegal

Contribution of Renewable Energy Resources for Food Production and Food Security

List of Tables

Bundling Climate Smart Agriculture Systems in Morocco: Insights, Advances and Perspectives in a Changing Climate

Energy-Economy-Environment Nexus: Moving Toward an Innovative Framework for Food-Secure Future

Climate Change and Innovative Ecological Interventions Through Climate-Smart Agriculture in the Cholistan Desert of Pakistan

Enhancing Food Security Through Climate-Smart Agriculture in Kenya

Ergonomics for Sustainable Agriculture and Food Security

Climate Change Challenges and Its Preparedness Toward Agriculture Using Climate-Smart Agriculture in Potohar Plateau of Pakistan

Applications of Smart Agriculture in Irrigation Water Management in Developing Countries: The Way Forward

Trending Smart Agricultural Techniques in Postharvest Handling to Mitigate Climate Change Impacts in Sri Lankan Context

Mitigation of the Effects of Climate Change on Agriculture Through the Adoption of Precision Agriculture Technologies

Effects of Climate Change and Food Security on Sustainable Agricultural Development and Potential Solutions: Evidence from China

Can Seasonal Climate Forecasts Improve Household Welfare? Insights from Rural Senegal

Climate-Smart and Resilient Food Systems and Security: Introduction

Mohamed Behnassi, Riaz Hussain Qureshi, Abdulmalek A. Al-Shaikh, Mirza Barjees Baig, and Turki Khalufa A. Faraj

Abstract The demand for food to feed an estimated world population of 9.7 billion by 2050 is expected to increase manifold. This increase in food production is often associated with the expansion of agriculture along with increased use of primary inputs, i.e., land, energy, and water resources. As of today, the world's agricultural systems have major environmental and climate impacts, including deforestation, biodiversity loss, unsustainable land and water use, and increased pollution and greenhouse gas emissions. Many of the impacts of climate change are already affecting the natural base of food systems, leading to reduced production and increased losses in resource base, thereby undermining the resilience and food security of vulnerable communities. Reversing these trends will require major changes in farming practices. Climate-smart agriculture (CSA) is promoted as a promising approach to change and reorient food systems within the new environmental, climatic, technological, and socio-economic settings. Current scientific evidence demonstrates

M. Behnassi (✉)
Economic Social and Environmental Council (CESE), Rabat, Morocco
e-mail: behnassi@gmail.com

Center for Environment, Human Security and Governance (CERES), Agadir, Morocco

R. H. Qureshi
University of Agriculture, Faisalabad, Pakistan
e-mail: rqureshi@alum.hec.gov.pk

A. A. Al-Shaikh
Prince Sultan Institute for Environmental, Water and Desert Research, King Saud University, Riyadh, Kingdom of Saudi Arabia
e-mail: aasheikh@ksu.edu.sa

M. B. Baig
Prince Sultan Institute for Environmental, Water and Desert Research, King Saud University, Riyadh, Saudi Arabia
e-mail: mbbaig@ksu.edu

T. K. A. Faraj
College of Food and Agriculture Sciences and Environmental Consultant, Prince Sultan Institute for Environmental, Water and Desert Research, King Saud University, Riyadh, Kingdom of Saudi Arabia
e-mail: turkialasiri@yahoo.com; talasiri@ksu.edu.sa

M. Behnassi et al. (eds.), *Climate-Smart and Resilient Food Systems and Security*,
https://doi.org/10.1007/978-3-031-65968-3_1

the applicability and scalability of CSA practices on a wider scale. The volume *Climate-Smart and Resilient Food Systems and Security* addresses the above issues in some detail. This introductory chapter underscores the scope, content, and objectives of the volume, which contributes theoretically and empirically to the scholarly debate on CSA. The volume also highlights opportunities to accelerate the transformation of existing food systems in terms of climate adaptation, mitigation, and sustainability, thus leading to food security and resilience. Structured in this way, the volume has implications for both research, innovation, and policy making.

Keywords Food systems · Agriculture · Climate risks · CSA practices · Resilience · Nexus · Food security · Renewable energies

1 Background, Scope, and Objectives

Our planet is currently under severe stress due to many interlinked and complex challenges, particularly the unprecedented increase in global population and its demands, resource depletion and scarcity, ecosystem degradation, and excessive waste production leading to global warming. Among these challenges, climate change has become a threat multiplier, threatening our socio-economic systems. Climate change has widespread negative impacts on vital sectors (such as agriculture and fisheries) and resources (such as water and land), that support human security and development. These impacts, which include increasing temperatures, shifting precipitation patterns, more severe and frequent extreme weather events, sea-level rise, increase in the occurrence of weeds, pests, and diseases, and loss of biodiversity and ecosystem services, are increasingly undermining food production systems and affecting food, water, and health security.

Global food demand is expected to increase tremendously to feed a projected global population of 9.7 billion people by 2050, while the climate-related impacts are currently leading to complete crop failures, reduction in arable land, and increase in drought and pests. They are also increasing the vulnerability of human societies, particularly the resource-poor communities, by reducing their resilience to food and nutrition instability. Even if global warming is limited to 1.5 degrees Celsius, the vulnerable and disadvantaged regions and communities are expected to suffer from climate-induced human insecurity problems. In the Global South, these inter-linkages are already conspicuous and mostly direct, especially where the impacts of climate change are compounded with poverty, social vulnerability, resource scarcity, and governance deficit. Therefore, due to climate change and its impact resulting in resource scarcity, advances and achievements made in food production and security are being gradually lost in many regions worldwide, with many implications such as conflicts, social instability, and displacement.

According to many recent studies and assessments, the new climate risks to food production systems (i.e., cropping, livestock, and fisheries) are expected to increase in the coming decades, particularly in vulnerable countries where adaptive capacity is weak. Impacts on agriculture threaten both food security and the pivotal role of agriculture in rural livelihoods and the broad-based development. Climate changes not only threaten crops but also the postharvest life of agricultural products, thus further worsening the food and nutrition situation of small farm householders. Increased postharvest losses also contribute to the increase in global warming. According to the World Bank, if food waste were a country, it would be the third highest emitter of GHGs in the world (World Bank, 2023).

Traditionally, the increase in food production has been linked to agricultural expansion and increased use of chemical inputs, land, energy, and water resources (70% of freshwater). Today, the global agri-food system has major environmental footprints, including deforestation, biodiversity loss, unsustainable land management, soil degradation, and water pollution from high nutrient loading and various agrochemicals from farm emits. It is reported that one-third of all GHG emissions (especially methane emissions) are from agriculture, which are increasing in developing countries—a worrying trend that must be reversed (World Bank, 2023).

In these settings, it is important to recall that food systems are shaped by human-interlinked value-added activities and decisions involved in the production, aggregation, processing, distribution, consumption, and disposal of food products that originate from agriculture. To make such systems sustainable and resilient, appropriate agriculture practices and management can play a key role in coping with the challenges induced by climate change. In addition, amongst the major GHG emission sources, agriculture, forestry, and other land use (AFOLU) sector uniquely comprises both emission sources and sinks.

To reverse the above trends, major shifts in agricultural practices are needed in order to reduce or remove their negative footprint while improving the lives of those still locked in food insecurity and poverty and to prevent the loss of gains already achieved. Traditionally, many of the concerned challenges, though interconnected, have been addressed singly, at times reducing one problem while exacerbating others. Currently, there is a need for a shifting paradigm that simultaneously examines interactions and trade-offs and assesses synergies and achievements, especially in the context of Sustainable Developments Goals (SDGs). Such a paradigm should incite key stakeholders to work in an integrated way and not in silos. The implications of post COVID-19 crisis and other recurring disasters including the ongoing Russian-Ukraine war for energy and food security provide an opportunity to trigger such a shift. These crises are sending clear signals that the global value chains are being tested for their resilience in meeting critical needs that are affected by transport and labor restrictions, the volatility of energy and agricultural input markets, and climate risks. These crises are a powerful wakeup call for current input-intensive food systems to change, learn from past inaccuracies, accept the need for a shift from former inappropriate development paths, assess where practices went wrong, and chart a strategy that makes agriculture profitable, competitive, resilient, sustainable, and inclusive.

In this perspective, Climate-Smart Agriculture (CSA) has been advocated as an integrated approach to address many interlinked challenges such as food security and environmental and climatic changes. CSA is an integrated approach to managing landscapes—cropland, livestock, forests, and fisheries—that addresses the interlinked challenges of food security and climate change. This approach helps transform and reorient food systems under the new environmental, climatic, technological, and socio-economic realities. CSA is a set of agricultural practices and technologies that simultaneously boost productivity, improve resilience, reduce GHG emissions, and thus enhance the achievement of sustainable national food and nutrition security and SDGs. More specifically, with the expectation of overcoming the adverse impact caused by climate change on food security and agriculture, the Food and Agriculture Organization (FAO) has been emphasizing the importance of adopting CSA since 2010.

Although it is built on existing agricultural knowledge, technologies, and sustainability principles, CSA is distinct in several ways. First, it has an explicit focus on addressing climate change in the agri-food system. Second, CSA systematically considers the synergies and tradeoffs that exist between productivity, adaptation, and mitigation. Third, CSA encompasses a range of practices and technologies that are tailored to specific agro-ecological conditions and socio-economic contexts including the adoption of climate-resilient crop varieties, conservation agriculture techniques, agroforestry, precision farming, water management strategies, and improved livestock management (World Bank, 2023).

Moreover, CSA strategies include land reclamation, such as reclaiming the desert lands into productive ecosystems, and conservation of existing natural resources, i.e., land, water, and biodiversity, along with possible alternatives to improve the livelihood of the local communities. CSA practices are labor conserving and, as such, they help famers and workers to pursue other off-farm jobs that can generate additional incomes for households, therefore supporting livelihood security. Also, the introduction of robotics, drone, satellite technology, and information technology would offer a vital strategy for food security and resilient food systems. Genetic techniques can, in turn, increase drought tolerance and heat resistance in crop plants for their better adaptation to future climate. Agroforestry is also an integral part of CSA that involves the growing of forest trees and raising agricultural crops and livestock on a particular piece of land. This practice is seen as a traditional farming system and common land management approach in irrigated ecosystems where trees are grown in combination with agricultural crops, therefore storing more carbon and maintaining resilience. In line with this concept, a model was developed and tested in Pakistan where salt-affected lands and saline water were used for improving the economic conditions of farmers of degraded lands by growing salt-tolerant species of crops, forest trees, and bushes for livestock, fisheries, and apiculture (Qureshi & Barrett-Lennard, 1998).

Adaptation, which leads to resilience, means altering our behavior, systems and, in some cases, ways of life to protect households, economies, and the environment in which we live from the impacts of climate change. In the context of CSA, adaptation comprises three pillars, i.e., (1) *productivity* which aims to sustainably increase the agricultural productivity and incomes from crops and livestock without negatively impacting the environment or over-consuming scarce resources; (2) *reduction in farmers' exposure* to short-term risks while also strengthening their resilience by building their capacity to adapt and prosper in the face of shocks and longer term stresses; and (3) *mitigation,* wherever and whenever possible, by climate change adaptations to reduce and/or remove GHG emissions. The more we reduce emissions now, the easier and less costly it will be to adapt to inevitable changes.

Even though the implementation of CSA approaches is mostly a win–win situation given its various benefits, the adoption remains, however, quite low in many regions. Moreover, funding for CSA is marginal and needs to be increased to align available finance with the relevance of the sector. Despite causing one third of global greenhouse gas emissions, agri-food systems currently receive only 4% of climate finance and only a fifth of this goes to smallholders. Current financial flows need to be realigned to support a sustainable agri-food system transformation (World Bank, 2023).

The research work carried out so far on CSA has a worldwide relevance and a potential to benefit many communities and countries and the existing scientific evidence does show a wider applicability and scalability in this context. Within this framework, publishing the *Climate-Smart and Resilient Food Systems and Security* volume is crucial and timely. It contributes to the scientific debate about the above-mentioned issues while highlighting the opportunities to accelerate the transformation of current food systems within the perspective of climate resilience, sustainability, and food and nutrition security. Most chapters of this volume are mostly based on specific case studies, especially from the Global South (Morocco, Nigeria, Senegal, China, India, Pakistan, Sri Lanka) and Poland. They present various perspectives and lessons from the field—especially in vulnerable and resource-constrained countries—on the most pertinent issues related to the resilience and sustainability of food systems and security. The volume, which provides policy-oriented inputs and recommendations to guide change processes at multiple scales, will have long-term implications for research, innovation, and policy design worldwide.

2 Contents

In Chap. 2, "*Transforming Agri-Food Systems Through Climate-Smart Practices: Ending Hidden Costs, Maximizing Benefits*", Behnassi et al. delve into the paradox of the current agri-food system. While it generates significant economic output (over $10 trillion annually), its hidden inefficiencies create environmental and social costs exceeding $12.7 trillion. These hidden costs, rooted in labor issues, pollution, and excessive resource use, contribute to climate change, environmental degradation, and social inequalities. Wealthier nations are the primary culprits, while poorer nations bear the brunt of the consequences. The chapter emphasizes the urgent need to transform this system by addressing hidden costs and adopting climate-smart practices, paving the way for a more sustainable and equitable future.

In Chap. 3, *Climate-Smart Agriculture (CSA) Practices for Sustainable Intensification in Major Agri-Food Systems of South Asia*, Jat et al. focus on the case of Indo-Gangetic Plains (IGP) of South Asia, which are known for major cereal-based agri-food systems sustaining food requirements of 50% population in the region. However, due to the continuous cultivation of monotonous rice–wheat (RW) cropping system, the system's sustainability and capacity to go on meeting the growing demands is seriously threatened, especially in the era of climate risks. To reverse such scenarios, the authors advocate CSA practices as essential to design a new generation of cereal systems with the potential to reduce farmers' climate vulnerability, who are mostly smallholders. By reference to on-station and off-station trials in South Asia, the authors demonstrate that CSA-based modules had a potential to increase productivity and profitability in major cereal-based agri-food systems while reducing GHG emissions. Additionally, CSA-based production systems are found to be more resilient by improving soil quality and saving irrigation water, with slightly higher yields compared with conventional tillage-based systems. The findings showed a wider practical applicability and scalability, which may pave the way paradigm for scaling up cereal systems in South Asia in near the future.

In Chap. 4, *Climate-Smart Agriculture: Potential Role in Carbon Sequestration and to Address Climate Change under Irrigated Agro-Ecosystems*, Komal et al. assert that the survival and existence of ecosystems and their associated living organisms are highly threatened by climate change, while agriculture happens to be the major contributor to the atmospheric concentration of carbon dioxide. Therefore, reducing carbon emissions and sequestrating it into the agricultural landscape (soil, trees, crops) is the most effective, cheaper eco-friendly approach. Indeed, soil carbon sequestration is the most efficient tool to enrich soil organic carbon pools, which depend on improved agricultural practices. Based on this, the authors have presented the analysis of the current knowledge about soil organic carbon (SOC) sequestration processes and their management in agroforestry systems, while agroforestry (AF) is an integral part of CSA that involves the growing of forest trees and raising agricultural crops and livestock on a particular piece of land. The findings reveal that CSA has great potential to lockup carbon and reduce carbon emissions to the atmosphere.

Marbet et al. focus on the effects of climate change in Morocco in Chap. 5, *Bundling Climate Smart Agriculture Systems in Morocco: Insights, Advances and Perspectives in a Changing Climate.* They reported that food and nutrition security are threatened by combined complex impacts of climate change, land degradation, and water scarcity—a consequence of growing population and diet shifts. To mitigate and minimize these impacts, several CSA options were experimented and implemented by research and development consortia with support from international organizations, and are the high priority of the recently launched national agricultural strategy called "Green Generation 2020–30". The authors focus their review mainly on conservation agriculture (CA), organic farming (OF), and agroforestry (AF) as three pillars of regenerative agriculture and climate change mitigation and resilience. Resultantly, the Green Generation strategy is aiming to reach 100,000 ha and 1 million ha by 2030, respectively, for OF and CA. The authors review knowledge advances with relation to food sustainability and environment stewardship. They claim that ensuring farmers and producers of all scales and sizes adopt CSA systems and gain access to innovation, technology, training, and knowledge will help foster greater productivity, secure food and feed, lessen impacts on soil, water and air quality, and ensure an efficient use of external resources and scarce labor. According to the authors, efficient, sustainable and productive use of inputs, resources, and capital under CSA systems help farmers to control costs during volatile business cycles and increasing climate uncertainty. The authors analyze and discuss major challenges, drivers, potential, and perspectives from CSA mainstreaming while asserting that CSA brings, for all these stakeholders, changes in behavior and lifestyle as well as new insights and paradigms. The authors recall that the goal of achieving sustainability, heedlessly of the farming system, should be based on science and impartial, risk-based decision-making process.

In Chap. 6, Bhutto et al. investigate the ways that may help *Integrate Climate-Smart Agriculture and the Water, Energy, and Food Nexus Approach to Address Sustainability Challenges Under a Changing Climate.* The water, energy, and food (WEF) resources are essential for sustainable well-being and the WEF Nexus directly correlates to numerous Sustainable Development Goals (especially SDGs 2, 6, 7, 9, and 12), whereas CSA is an integrated approach to address the interlinked challenges of food security and climate change (especially SDG 13, *Take urgent action to combat climate change and its impacts*). Undeniably, climate change threatens people with food insecurity, water scarcity, flooding, infectious diseases, extreme heat, economic losses, and displacement. According to the authors, global challenges, though interconnected, have been traditionally addressed singly, at times reducing one problem while exacerbating the others. To reverse such a trend, the WEF nexus approach seeks to look at the three elements as an interrelated system through the identification of interactions and the assessment of achievements. It focuses on strengthening synergies and reducing trade-offs among these sectors. Similarly, SDGs require us to work in an integrated way and not in silos. Based on this concept, the authors discuss the adoption of CSA in the context of the progress on multiple SDGs covering the WEF nexus to mitigate adverse effects of climate change on agriculture. The authors also attempt to identify the knowledge gaps, the interlinked roles of institutions, infrastructure, and information in developing relevant policies, tools and resources essential for stakeholders in decision-making. Their work has implications for research, innovation, and policy design.

In Chap. 7, *Food and Water Security through Climate-Smart Agriculture in Asia and Africa,* Baig focuses on the case of Asian and African countries, which are mostly agricultural economies, dependent on specific climate variables for food production. Due to climate change and the high vulnerability of these countries to its impacts, their economies could decline in the future along with a decrease in water availability and food security. Increasing temperatures, weather variability, shifting agro-ecosystem boundaries, invasive crop diseases and pests, and more frequent extreme weather events will have a negative impact on crop yields, fisheries, livestock, water availability, and nutritional quality of food items. This may affect the capacity of food production systems to sustain the needs of a growing population. At the same time, agriculture contributes to climate change and is thus a major part of the climate problem. Climate adaptation and mitigation strategies are therefore required to ensure continued livelihoods, food security, and water availability. In this regard, the author presents CSA as an integrative approach that incorporates the management of landscapes, ecosystems, cropland, livestock, forests, and fisheries to raise agricultural productivity, enhance resilience to climate change, increase adaptation, and address the contribution of agriculture to GHG emissions.

In Chap. 8, *Energy-Economy-Environment Nexus: Moving Towards an Innovative Framework for Food-Secure Future,* Naqvi et al. focus on the 3-E framework (Energy-Economy-Environment) and develop an understanding of the world's energy state and the doctrines of exhaustible resource economics. This derives from the assumption that energy is the basic driving force of an economy, and there exists a robust linkage between economic development and energy consumption. More specifically, the authors provide a basic knowledge about judicious decision-making related to food value chain and energy markets, with particular reference to developing nations of the Asian region. The authors also discuss fundamental principles underlying the non-renewable energy system, their impacts on marginal populations, and possible solutions. The analysis helps the readers understand the prevailing food security scenarios, renewable and non-renewable energy, and their allied impacts on the environment.

Qureshi et al. address the case of the Cholistan Desert in Chap. 9, *Climate Change and Innovative Ecologic Interventions Through Climate-Smart Agriculture.* The Cholistan Desert is one of Pakistan's largest and most hyper-arid deserts, covering 2.6 million hectares and housing 0.155 and 2.0 million people and livestock, respectively. Livestock rearing is the primary land use in the area, and the livelihood of the local populations is mostly reliant on this activity. Nonetheless, the region is dealing with serious problems like desertification, acute water scarcity, drought, excessive grazing, deforestation, wind erosion, salinity, and encroachment for growing agriculture, which depletes natural resources including plant biodiversity. As a result, local communities are mostly vulnerable to the effects of climate change on crops and livestock. Based on this concept, the authors seek to implement climate-smart practices, particularly climate-resilient crops, as an effective means of mitigating the effects of climate change and managing food security challenges. According to the authors, climate-smart measures include reclaiming desert lands as productive ecosystems by maintaining existing natural resources such as lands, water, and biodiversity, as well as suitable alternatives to improve local communities' livelihoods.

Climate change, one of the main issues facing the sub-Saharan countries, is discussed by Wamwea and Culas in Chap. 10, *Enhancing Food Security through Climate-Smart Agriculture in Kenya.* In addition, the region's food security has also been threatened by numerous economic and climatic shocks. Kenya's economy, which is said to be developing at the highest rate among the African nations, is likewise seeing stable expansion. However, eighty percent of the country's surface area is made up of arid and semi-arid areas where the majority of the people reside. The main source of livelihood in Kenya is rain-fed subsistence agriculture, which makes the situation worse. Additionally, the effects of climate change are undermining food production systems. As a result, the authors promote CSA, which offers an opportunity to change these dynamics and raise food production in a sustainable manner. The conceptual framework for food and nutritional security, which describes the connections between the three pillars of food security, is adopted by the authors. In order to promote the SDGs for poverty, hunger, nutrition, and environmental sustainability, the Kenyan government has tried to develop a number of policy initiatives over the years, which are examined in this chapter. The importance of NGOs in Kenya in transforming and realigning agricultural systems to effectively implement climate-smart practices is also highlighted. It is claimed that Kenya is achieving sustainable food and nutrition security.

In Chap. 11, Thariq et al. examine the importance of *Ergonomics for Sustainable Agriculture and Food Security*. According to them, agricultural development is essential to minimize poverty, enhance broader prosperity, and supply food to the increased population, for which labor-intensive agriculture is common in low- and middle-income countries. The majority of farms in these countries are small and medium-sized businesses that are gradually moving towards technological applications. In addition, farmworkers are at a higher risk of developing musculoskeletal diseases (MSDs) due to the high degree of physical activity and manual labor involved in their job. Since the most precious resource in every production process is labor force, in order to increase output quality and productivity, working conditions need to be carefully considered. According to the authors, long-term detrimental effects of MSDs can include a decrease in labor productivity, a decrease in farm income, additional health problems, such as stress or depression, and a lower standard of living. These unfavorable outcomes have important impact on food security by reducing the efficiency and productivity of farming operations. Several ergonomics initiatives in the agricultural and animal production sector have resulted in improvements in performance and productivity while maintaining the health of employees. In addition, ergonomic measures contribute to creating a shared environment in which businesses and farmers work together. Employers will benefit from lower costs and increased production, while employees will benefit from better health, safety, comfort, work-life balance, as well as improved tools, equipment, and software use. A profitable and sustainable agricultural sector depends on a healthy workforce that maintains human well-being. Many studies have been conducted on agricultural ergonomics to improve farm workers' health, happiness, and productivity. Nevertheless, the authors recommend additional investigations in this area, especially the micro-ergonomics.

In Chap. 12, *Precision Agriculture as a Coping Strategy for Climatic Challenges*, Ahmad et al. consider agriculture, livestock, and health as vulnerable sectors to changing climate, and especially in the poor regions which are more at risk. According to the authors, Precision Agriculture (PA) is a leading management strategy to deal with unexpected climatic challenges and mitigating their implications. Incorporating precision agriculture concepts, improving technology, adopting new policies, and using indigenous knowledge is an effective way to achieve this. It is because PA strategies are based on technologies with emphasis on digitalization and the Internet of Things (IoT) in order to provide rapid and precise information that will improve productivity and farm sustainability. However, in many developing countries, these modern and state of the art technologies, which are available in developed countries, still lack. Without these sophisticated technologies, agricultural production could not be increased, and food security would not be guaranteed in the future. Thus, there is a need for initial investment in order to achieve the target through adoption of recommended practices and technologies.

In Chap. 13, *Contribution of Precision Agriculture to Drought and Food Security*, Mehmet Metin Ozguvens considers agriculture as one of the most important fields of activity where the essential and basic needs of humanity are produced. However, this activity is currently suffering from many obstacles and uncertainties that directly impact productivity of resources. Climate changes increase the incidence of diseases, pests, weeds, drought, economic crises, pandemic, hail, floods, tornado, decrease in agricultural land and natural resources as well as fragmentation of agricultural lands. Moreover, agriculture, which is often a set of difficult and laborious activities, is also associated with some risks such as the inability to obtain efficient production or sufficient income to meet the expenditures. Therefore, the author believes that the aim in agricultural production should be to increase productivity and product quality using minimum inputs, to ensure food safety, protect natural resource, and the environment, and to provide economic, sustainable, and efficient productive management. To achieve such objectives, it is necessary to use quality seeds, efficient irrigation techniques, and appropriate fertilizer application and plant protection practices. In this perspective, the use of precision agriculture technologies not only makes productivity more efficient but also prevents environmental pollution, protects natural resources, and improves food safety by preventing unnecessary input use. In this study, the author highlights the benefits of variable rates of fertilizer and pesticide applications that resulted in the development of precision agriculture technology, while the precision irrigation applications help in controlling drought, which, in turn, contributes to food security.

Udayanga and Najim have contributed their findings in Chap. 14, *The impact of climate change on the agricultural sector in Sri Lanka and the role of Precision Agricultural Technologies (PAT)*. Their findings are based on their studies on the impact of climate and environmental changes on agricultural productivity in Sri Lanka, as well as recent policy decisions taken to promote environment friendly agricultural strategies. According to them, it is still difficult to maintain higher productivity with minimum environmental impacts and to use optimal inputs. In this regard, the adoption of PAT can be seen as the most viable solution, since their different combinations could effectively support monitoring and respond to variability in soil, environment, and crops, thus ensuring optimal resource utilization. According to the authors, the combined use of these technologies would enable Sri Lankan farmers to detect and monitor soil and crop conditions, enabling customized automation of irrigation, fertilizer application, and pest and disease management. Appropriately calibrated crop modeling applications can also support decision-making and operational control in agriculture. However, PAT adoption in Sri Lanka remains relatively low due to numerous obstacles, such as lack of technological expertise and physical resources, poor socio-economic conditions of farmers, gaps in policies and implementation processes, poor coordination between the government and other entities, and lack of government support. Therefore, the government and other key agricultural entities must make efforts to promote PAT through enhancement of physical infrastructure facilities, dissemination of knowledge and technology, empowerment of farmers and Agricultural Instructors (AIs), and increasing farmers' technological readiness (TR). In addition, it is necessary to develop and implement long-term policies aimed at promoting PAT to ensure higher agricultural productivity in Sri Lanka with minimum environmental impacts.

In Chap. 15, a study by Adnan et al. titled "*Climate Change Challenges and its Preparedness towards Agriculture using Climate-Smart Agriculture in Potohar Plateau of Pakistan*" examines the challenges posed by climate change to agricultural practices in Pakistan's Potohar Plateau. The authors contend that even minor temperature fluctuations induced by climate change could have significant ramifications, potentially causing disruptive shifts within the ecosystem and society over the coming decades. The research emphasizes the inherent difficulties associated with rainfed agriculture in the Potohar Plateau, suggesting that these challenges may become even more pronounced under future climate change scenarios. While the study focuses on the challenges, it also acknowledges the potential of Climate-Smart Agriculture (CSA) techniques and practices as a way to manage these challenges. This option could benefit climate change mitigation, reduced soil erosion, drought buffering, stable yields, and reduced field preparation costs.

In Chap. 16, Umar and Haruna discuss the *Dynamics of Agricultural Extension for Nigeria's Rural Areas under the Current Climate Change scenario*. Since the colonial era, agricultural extension dynamics has shown a shift from advisory to facilitation or empowerment in the extension paradigms. Therefore, the vulnerability of agriculture and rural life in Nigeria is a compelling reason to think beyond technology transfer and perform better under new pressures or changes. In this context, a paradigm shift from "technology" to "knowledge", from "research centric" to "farmer centric", and from "top-down" to "bottom-up" solutions must be designed. To revitalize agriculture and manage many agrarian challenges in the developing world, it is time to develop the mechanism of extension pluralism and foster knowledge empowerment of key extension system players, especially in the context of climate change. In the near future, potential synergies between adaptation to climate change and mitigation must be reconsidered as Nigeria's ultimate goal, which is eager to progress to higher stages of development and competitive power. According to their findings, recommendations to ensure effective and efficient performance of the agricultural extension system in Nigeria, especially in transferring knowledge to farmers, should include: (1) climate change scientific research should have a real impact on farm productivity and livelihoods; (2) new methodologies for dissemination of information have to be developed or adapted; (3) reforms in agricultural extension should be oriented towards learning rather than teaching paradigm; (4) roles and responsibilities of researchers, extension workers and community should be re-defined and shared; (5) socio-economic analysis for each methodology should be performed before adapting any method, approach or strategy; and (6) an integrated approach to be implemented in diverse socio-cultural, economic, and political situations in order to achieve desired goals.

In Chap. 17, Ashraf et al. report the status of *Climate-Smart Water Management Technologies in Pakistan*, which is one of the most climate-vulnerable countries. The impact of climate change on ecosystems, in particular water resources, leads to frequent droughts and floods, which have devastating effects on infrastructure, agriculture, livestock, and humans. In addition to the increase in crop water requirements, climate change has a significant impact on crops through changes in sowing and harvesting times. It has not only varying impacts in the upper, middle, and lower riparian areas but also a greater impact on drylands compared to irrigated areas. The authors propose simple and climate-smart water management techniques that have the potential to trigger adaptation at larger scale, to reverse such trends and cope with climate change for the upper Indus basin, the plain irrigated area, the lower Indus basin, and the rainfed areas.

In a similar perspective, Amarasingam et al. focus in Chap. 18 on the *Applications of Smart Agriculture in Irrigation Water Management in Developing Countries.* According to them, climate change affects agriculture, which is the main source of global food production, in different ways such as changes in typical temperatures, precipitation, and atmospheric conditions. Fortunately, advanced automation of agronomic activities for land preparation, irrigation water management, application of fertilizers and pesticides, and harvesting has substantially improved productivity. In many agricultural areas, however, the problem of determining the exact amount of water needed for many crops remains unresolved, especially in the context of a changing climate. According to the authors, the use of precision farming technologies to increase agricultural productivity, reduce pollution, and lower production costs while overcoming climate change issues, is one of the best solutions. Smart agriculture plays an important role in maintaining the sustainability of agricultural systems, addressing water scarcity due to climate change, and improving the country's economic growth. Smart farming promotes agronomic practices such as assessing crop areas, monitoring crop growth, estimating soil moisture, evaluating soil fertility, detecting crop stress, detecting diseases and pests, and forecasting weather. Therefore, the authors address the importance of smart farming applications in irrigation water management to increase farm productivity by overcoming climate change impact, using less human energy, increasing efficiency, and improving management techniques through precision farming technologies.

In Chap. 19, Jemziya et al. discuss *Trending Smart Agricultural Techniques in Postharvest Handling to Mitigate Climate Change Impacts* with focus on *Sri Lanka.* According to them, postharvest losses are very common and are mainly due to poor infrastructure and logistic arrangements, lack of technology, and insufficient knowledge, skills and management capacity of supply chain stakeholders. Additional factors leading to high levels of losses include the lack of awareness and concern about proper post-harvest practices and incorrect storage, harvesting methods, and packaging techniques. Most agricultural systems are particularly sensitive to the impact of climate change which increases the cost of agricultural management and affects not only crops but also their post-harvest life. These dynamics lead to food insecurity and seriously jeopardize farmers with small holdings. Therefore, food losses in the post-harvest supply chain need to be reduced using advanced and innovative technologies that include climate-smart processing technologies, solar energy technologies, thermal and non-thermal technologies, cutting-edge technologies, ZigBee and sensor technologies, zero-energy cooling chambers, novel and intelligent packaging technologies, Internet of Things (IoT), advanced imaging technologies, automation, innovative plant breeding technologies, and advanced food quality management strategies. In addition, increasing the awareness and education of farmers through extension services and the modern use of post-harvest operations will reduce overall food losses and ensure food security.

In Chap. 20, Keskin and Sekerli address the issue of *Mitigation of the Effects of Climate Change on Agriculture through the Adoption of Precision Agriculture Technologies*. According to them agriculture is one of the main sources of greenhouse gas (GHG) emissions and has a significant impact on climate change, especially due to water scarcity, which reduces crop yields. The new technologies offer improved precision management from tillage to harvesting. Precision agriculture (PA) aims to manage crops in a more efficient manner, and PA technologies have the potential to mitigate the impacts of climate change by reducing the use of agricultural inputs, fuel savings, lower GHG emissions, less environmental pollution, lower costs, higher yield, increased profitability, and sustained natural resources. The use of PA technologies has been increasing, not only in the developed countries with large farms but also in the developing countries. However, there are concerns about threats to PA technologies, such as data protection, data theft, signal disruption, cyber-attacks on agricultural systems, etc. The authors attempt to present some basic information about PA technologies to highlight their benefits, adoption rate, and future trends in terms of climate change mitigation and adaptation, sustainability, and food security.

In Chap. 21, Fang et al. have tried to analyze the *Effects of Climate Change and Food Security on Sustainable Agricultural Development and Potential Solutions* in the context of China. They identified climate change and food insecurity as the two important global issues affecting national economies and people's livelihood, particularly in developing countries. Considering these adverse effects, the authors investigated the impacts of climate and food security on China's production efficiency, productivity growth, and sustainable agricultural development. Possible solutions which may significantly impact efficiency promotion and technological progress, have been elaborated. These include poverty alleviation, awareness creation, intelligent agricultural extension, and biotechnology generalization. In this regard, the Slack-based approach has been introduced with the aim to quantify the inefficiency for the overall agricultural production. Meanwhile, the authors have analyzed the variable-specific decomposition. According to the findings, China's agricultural inefficiency is 0.61, mainly attributed to food security constraints (0.29) in the changing climate context.

In Chap. 22, Diallo and Ndiaye attempt to answer the question "*Can Seasonal Climate Predictions Improve Household Health*?" with reference to the case of *Rural Senegal*. The authors highlight the uncertainty as to what extent climate predictions contribute to improving the rural well-being despite their potential to manage climate risks and build resilience. The authors, therefore, investigated the use of seasonal climate forecasts in production decisions and their association with better quality of life. Their empirical study assesses whether improvements in seasonal climate forecasts are associated with, or even lead to, improvements in rural household welfare using a propensity score adjustment method. According to their findings, there is no evidence that the use of seasonal climate forecasts in agricultural activities contributes to improving household welfare in rural Senegal. Not only did the findings hold regardless of the well-being indicator chosen but they were further confirmed when different matching algorithms were used. Research suggests that co-production of climate services through partnerships between producers (scientists) and users (farmers) can be an effective way to improve the welfare impacts of climate predictions.

In Chap. 23, Irfeey et al. assess the *Contribution of Renewable Energy Resources in Food Production and Food Security* by focusing on the case of Sri Lanka. Growing demand for food and fluctuating prices of fossil fuels have prompted the country to explore and invest in sustainable energy sources. The authors consider the integration of renewable energy into agricultural production processes to be a promising solution, especially since the country has significant potential to produce this type of energy. However, the implementation of this solution is limited due to economic, institutional, technological, social, and cultural factors. Therefore, efforts to promote all renewable energy sources among farmers and food production companies will reduce the burden of national energy needs. Implementing policies to promote the use of renewable energy with better energy production technologies is essential to ensure future food production in Sri Lanka with the lowest possible impact on the economy and the environment.

The chapters in "The Climate-Smart and Resilient Food Systems and Security" emphasize the urgent need for a multi-pronged approach to global food systems. This includes ensuring food security for all, adapting to climate change, and significantly reducing greenhouse gas emissions. CSA emerges as a key solution across the volume, offering a holistic approach to tackle food insecurity, promote sustainable development, and address climate challenges. Notably, the volume's focus aligns perfectly with the recently adopted UAE Declaration at UNFCCC COP 28 on "Sustainable Agriculture, Resilient Food Systems, and Climate Action". This declaration aims to bolster adaptation, resilience, and responses that lessen the vulnerability of food producers facing climate change. It emphasizes financial and technical support, infrastructure consolidation, and innovation—including early warning systems—to promote long-term food security, sustainable production, and healthy diets, all while safeguarding the environment (UNFCCC, 2023).

References

Qureshi and Barrett-Lennard (1998). *Saline* Agriculture for Irrigated land in Pakistan: A Handbook, Australian Centre for International Agriculture Research, Canberra, Australia.

UNFCCC (2023). COP28 UAE Declaration on Sustainable Agriculture, Resilient Food Systems, and Climate Action, https://www.cop28.com/en/food-and-agriculture.

World Bank (2023). Climate-Smart Agriculture, https://www.worldbank.org/en/topic/climate-smart-agriculture.

Transforming Agri-Food Systems Through Climate-Smart Practices: Ending Hidden Costs, Maximizing Benefits

Mohamed Behnassi, Aisha Baig, and Mirza Barjees Baig

Abstract The agri-food system encompasses the interconnected activities of production, processing, distribution, and consumption of food and other agricultural products. Currently, this system contributes more than 10% of global GDP, with a market value of over $10 trillion annually. However, inefficiencies in the system generate hidden costs of more than $12.7 trillion. While most of these costs are difficult to quantify, a large portion are attributed to labour and environmental productivity losses. The agri-food system is driving many environmental externalities such greenhouse gas emissions, land-use conversion, pollution, biodiversity loss, and excessive freshwater consumption. Therefore, the system is highly contributing to climate change, environmental degradation, and the depletion of natural resources. The agri-food system is also linked to many social externalities such as high levels of diet-induced diseases, undernourishment, and poverty. Richer nations are primarily responsible for these hidden costs, with poorer nations often bearing the brunt of the consequences. This mismatch reinforces global inequalities, heightens social and economic disparities, and complicates food, environmental, and climate crises. This chapter emphasizes the critical need to address hidden costs and trade-offs in the current agri-food system, unlocking the multidimensional benefits of sustainable and climate-smart practices.

Keywords Agri-food system · Climate change · Consumption · Market value · Global GDP · Food insecurity · Social inequalities · Biodiversity

M. Behnassi (✉)
Economic Social and Environmental Council (CESE), Rabat, Morocco
e-mail: behnassi@gmail.com

Center for Environment, Human Security and Governance (CERES), Agadir, Morocco

A. Baig
Master of Arts Degree in International Affairs, Carleton University, Ottawa, ON, Canada
e-mail: aishabg31@gmail.com

M. B. Baig
Prince Sultan Institute for Environmental, Water and Desert Research, King Saud University, Riyadh, Saudi Arabia
e-mail: mbbaig@ksu.edu

M. Behnassi et al. (eds.), *Climate-Smart and Resilient Food Systems and Security*,
https://doi.org/10.1007/978-3-031-65968-3_2

1 Overview

The agri-food system encompasses the interconnected activities of production, processing, distribution, and consumption of food and other agricultural products. The agri-food industry boasts a market value of nearly $10 trillion, representing more than 10% of the global GDP (FOLU, 2019:38). This system is invaluable, particularly in feeding and employing much of the world's population. However, the current agri-food system is not without drawbacks. While the capacity and productivity of this system have increased, its inefficient resource use and various environmental and social impacts are increasingly impacting the transition to global human security and inclusive, sustainable and resilience development. Climate change, food and water insecurity, and growing social and environmental inequalities, among others, demonstrate the flaws of the current system.

Therefore, there is a pressing need to decipher the links between conventional agri-food system and the negative costs imposed on social-ecological systems across the world. Most of these hidden costs are externalities, described by the FAO (2023:10) as "the effects of transactions on third parties", which are not accounted for in current agri-food markets, meaning the prices paid by consumers do not always reflect the actual cost of food. These externalities are created across the supply chain, extending from primary food production to food loss and waste (Fig. 1). The repercussions echo across ecosystems, economies, and societies, shaping the dynamics of our interconnected world. A complete understanding of these costs is necessary to maximize the potential of food production and to meet Sustainable Development Goals (SDGs) and the targets of the Paris Agreement.

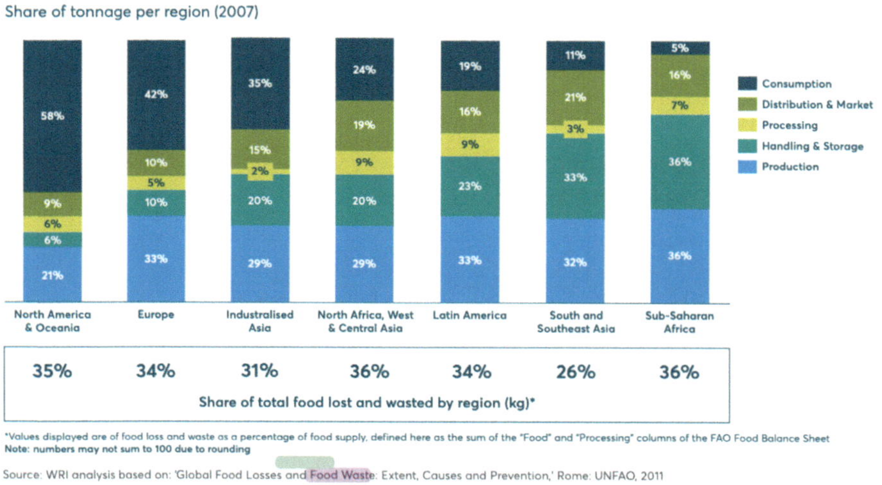

Fig. 1 Distribution of total global food loss and waste by region and stage across the supply chain. *Source* FOLU (2019)

In response to these considerations, the Food and Agriculture Organization (FAO) recently published *The State of Food and Agriculture 2023: Revealing the True Cost of Food to Transform Agrifood Systems.* Using the true-cost accounting approach, the FAO calculated that the quantifiable hidden costs of the agri-food system reach an overwhelming amount of $12.7 trillion (2020 PPP) (FAO, 2023:32). Due to the challenges associated with quantification, the actual costs are likely much higher. The bulk of this chapter builds on the FAO report by providing a deeper analysis of the hidden costs of the agri-food system and providing insight on their environmental, climatic, health, and socio-economic dimensions. Additionally, it concludes with a discussion on how climate-smart solutions can foster more sustainable practices and policy recommendations to address hidden costs within the agri-food system.

2 The Hidden Costs of the Agri-Food System

2.1 Environmental and Climate Impacts

The hidden costs of agri-food systems related to the environment are multifaceted and have far-reaching consequences extending beyond the agriculture sector. While some farmers act as stewards, supporting carbon capture and the sustainability of local ecosystems, most conventional farming is accelerating global warming, resource depletion, and environmental degradation. According to the FAO's estimations, environmental damage accounted for 20% of the total quantifiable hidden costs of agri-food systems (FAO, 2023:35). These costs are often externalized to the broader public, harming human health, and eroding the overall resilience of food systems.

2.1.1 Climate Change and Emissions

Agri-food systems produce one-third of global greenhouse gases (GHGs), along with other harmful emissions (FOLU, 2019:12). GHGs and emissions such as nitrogen are released from land-use conversion (discussed in the next section) and through agriculture production activities, including processes such as livestock rearing, rice cultivation, fertilizer production and usage, and energy consumption.

These emissions exacerbate global warming and increase the frequency and severity of extreme weather events, including wildfires, droughts, and floods, which create major losses in the agriculture sector. Extreme weather leads to crop failure, livestock destruction, and damaged agricultural infrastructure, resulting in food insecurity and economic losses for farmers and rural communities. Agricultural losses are especially concerning for subsistence farming and rural communities, where extreme weather events can amplify food, water, health insecurity, and economic instability. Moreover, the disruption of food supply chains can lead to food shortages and volatile prices, impacting consumers globally.

The increased frequency of extreme weather conditions is already having disastrous effects in middle- and low-income countries. For instance, in 2011, cyclones across the global South decimated crop yields in both Asia and Africa. In Sri Lanka, this extreme weather destroyed almost one-third of the country's rice paddies. Similarly, in early 2017, the country experienced a 40% decline in rice production due to drought conditions. In the later part of the year, heavy rainfall and floods caused mass displacement on the island (Gunaratne et al., 2021). These types of varying and extreme weather events not only undermine the food security and livelihoods of vulnerable populations but also have broader implications for the economies and overall development of poorer nations. Furthermore, the large-scale displacement and damaged infrastructure of extreme weather events undermine the ability of these countries to cope with future climate shocks.

2.1.2 Soil

Modern farming practices, characterized by excessive use of external inputs and intensive mechanization, impose significant costs on soil health, with far-reaching consequences for livelihoods, biodiversity, and food security. While these practices are intended to enhance agricultural productivity, they unintentionally disrupt the functions of soil microorganisms, which are crucial for promoting plant growth and nutrient cycling (WWF, 2018). These activities also deplete organic matter and diminish the soil's long-term water retention capacity. Over time, this leads to soil erosion, reduced soil productivity, increased need for external inputs, and even desertification.

The consequences of conventional farming practices are already evident, with improper farming practices having already compromised half of the world's topsoil (FOLU, 2019:42). In Africa, more than 60% of agriculture soil is experiencing some forms of degradation due to overgrazing and other agriculture activities. Reduced soil capacity poses a considerable threat to the millions of people on the continent who are dependent on agriculture as a form of livelihood. Moreover, credible links exist between the decline in soil quality and the prevalence of hunger and malnutrition across the region (Tilahun et al., 2015). Reduced agricultural productivity ultimately leads to a decreased food supply, which in turn raises prices and compromises the food security and well-being of already vulnerable populations.

In the sub-region of Near East and North Africa, natural resource degradation, especially where agriculture is practiced, is a real threat and remains a major limitation to the reliable supply of food in the region (FAO & ITPS, 2015). In most countries, salinization, water and wind erosion, loss of vegetation cover, soil physical degradation (including compaction and surface crusting) are the main threats to the soil's capacity to provide ecosystem services. The expansion of agriculture into marginal lands has greatly aggravated water erosion and consequently soil degradation. In almost all countries in the region, extreme climatic conditions, overgrazing, unsuitable cropping patterns and accumulation of salts have rendered large areas of land unproductive (Abahussain et al., 2002).

Soil degradation's consequences extend far beyond reduced agricultural productivity. It can trigger a cascade of environmental problems, including: Soil runoff (degraded soil loses its ability to absorb water, leading to increased surface runoff and potential flooding); biodiversity loss (healthy soil teems with life, crucial for nutrient cycling and plant growth. degradation disrupts this delicate ecosystem, leading to biodiversity loss); erosion-related air pollution (as wind and water erode soil, fine particles are released into the atmosphere, contributing to air pollution). Furthermore, degraded land becomes more vulnerable to natural disasters: i.e. floods can cause more damage on eroded land with poor water retention. The economic costs are significant and while The FAO Report (2023) doesn't provide a specific figure, earlier estimates by Eswaran et al. (2001) suggest land degradation could cost us around $400 billion annually. These challenges underscore the urgent need for sustainable land management practices. By protecting soil health, we safeguard our food systems and ensure a healthy environment for future generations.

2.1.3 Land Conversion

Agricultural activities are the leading driver of land-use conversion, harming critical natural habitats including tropical forests, drylands, and wetlands. Much of the transformed land is used for livestock production, including the conversion of land for grazing. Currently, 32% of the world's arable land is used for agriculture, 62% of that land is used for livestock production, while livestock provides less than 20% of global calories (FOLU, 2019:40). As global income rises, so will the demand for animal protein, fuelling the pressure to convert rainforests into agricultural land. This conversion will have devastating effects on global biodiversity and elevate the already substantial level of GHG emissions present in the atmosphere.

Globally, agriculture is responsible for an estimated 80% of deforestation, with a large portion of this deforestation occurring in tropical regions (Kissinger et al., 2012:11). Nearly 70% of the Amazon rainforest is now used for pasture (WHO & CBD, 2015:80). Remarkably, approximately 16% of the total carbon footprint of European diets is due to the deforestation occurring in developing countries, demonstrating the complexity of global food production and environmental issues (Pendrill et al., 2019). Tropical rainforests serve a vital role in supporting the Earth's ecosystem, providing crucial services such as water and climate cycle regulation. Deforestation disrupts these natural processes and compromises their role as carbon sinks, reducing the Earth's capacity to absorb carbon dioxide. Moreover, improper clearing techniques, such as burning, worsen this disruption by releasing previously trapped carbon into the atmosphere.

As tropical rainforests are hotspots for biodiversity, deforestation is driving the ongoing "sixth extinction", a phenomenon in which species are disappearing at an alarming rate (Kolbert, 2014). This loss of biodiversity not only affects the animals and plants that call these forests home but also has implications for human health and well-being. Plants in these ecosystems are important food and traditional medicine sources, and this extinction means a tremendous loss of knowledge and potential scientific discoveries for humanity. Surprisingly, the main cause of disease transmission from wildlife to humans is also land-use conversion (WHO & CBD, 2015).

2.1.4 Water

The world is facing a water crisis, with more than 2 billion people under water stress. A significant contributor to this crisis is the agricultural sector, which annually consumes 70% of global freshwater resources (FOLU, 2019:43). This percentage is expected to increase due to shifts in diets, inefficient farming practices, and climate change. Most agriculture-related water usage is for irrigation, which has surged in recent decades. Land requiring irrigation increased from 39 million hectares in 1961 to 320 million hectares in 2012 (FAO & IWMI, 2018:5). Unsustainable practices are also expanding the amount of land requiring irrigation annually. This depletion of groundwater reserves, particularly in arid regions, poses a threat to long-term agricultural productivity and food and water security.

The hidden costs of blue water usage in current agri-food systems, particularly in the context of future water scarcity, are estimated to be 105 billion 2020 PPP dollars, or about 4% of the total hidden costs of agri-food systems (FAO, 2023:35). This calculation is based on the decline in future labour and agricultural productivity. Additional hidden costs not captured in this figure include the pollution of water sources from fertilizers and pesticides and the adverse impacts on biodiversity. Agriculture runoff, containing agrochemicals, sediments, waste, and organic matter, is the main polluter of freshwater sources and a major polluter of saltwater bodies. This pollution is leading to the rise of dead zones, areas with low levels of oxygen that are unable to support marine life (FOLU, 2019). Notably, these dead zones are concentrated along the coastlines of areas with the most intensive agriculture, including Europe, Japan, and America, including a particularly massive dead zone in the Gulf of Mexico (WHO & SCBD, 2015). This pollution is not only intensifying global water scarcity but also poses a major cost to aquatic ecosystems, human health, and overall productive capacities.

2.1.5 Biodiversity

Food production activities are driving biodiversity loss, responsible for 70% of on-land and 80% of aquatic biodiversity depletion (UNCCD, 2022). As the previous sections have mentioned, the use of pesticides in agriculture and habitat loss are key contributions to this decline. Building upon these earlier sections, this section delves into a specific example to illustrate how these chemical inputs can adversely harm important pollinators. It will also touch upon the impact on crop diversity.

The hidden costs of chemical inputs on biodiversity became evident by the locust upsurge in the Horn of Africa from 2019 to 2021. In response to the spike in locust breeding, a by-product of climate variability, the FAO provided malathion and chlorpyrifos pesticides to combat the locusts in the region. Honeybees, exposed to these chemicals, either through direct contact with sprays or residues on plants, abandoned their hives (FAO, 2023:52). In Ethiopia, the largest producer of honey in Africa, this resulted in a 78% reduction in honey production in 2020 and a US$500 million loss in revenue. This loss posed a significant threat to the domestic honey industry and the livelihood of beekeepers. Moreover, given the important role that bees play in crop production and the fact that more than 70% of Ethiopia's population is employed in agriculture, the decline in bee productivity likely imposed even higher costs on food security and the overall economy of the country. The loss of pollinators like honeybees can lead to reduced crop yields and increased food prices, exacerbating the challenges faced by farmers and consumers. It's worth nothing that in Somalia, the use of bio-insecticides to combat the locusts did not harm the apiculture production in the country (Lazutkaite et al., 2023). This case demonstrates the ripple effect of hidden costs within agri-food systems and their implications for community well-being and the economy.

Furthermore, crop diversity plays a critical role in ensuring food security and adapting to climate change. However, the pursuit of short-term gains through monoculture farming has resulted in a significant loss of agrobiodiversity, which is the variety of plant and animal life used for food production. This represents another significant hidden cost of conventional agri-food systems. The mass adoption of monoculture farming, intended to maximize short-term yields through the cultivation of vast fields with a single crop variety, while it may appear efficient initially, has dramatically reduced the variety of crops cultivated by farmers. A sobering fact highlights this issue: presently, just four crops—rice, wheat, maize, and soybean—are responsible for a staggering 60% of global caloric intake (FOLU, 2019:42). This overdependence on a limited number of crops creates a precarious situation with far-reaching consequences for the resilience of our food systems, environment, public health, and human well-being. In this perspective, FAO (2019:102) and Brookfield (2014) had emphasized that the lack of crop diversity makes the global food supply extremely vulnerable to threats like pest and disease outbreaks. If a new disease emerges that targets a specific crop variety, entire harvests can be wiped out, leading to food shortages and price spikes. Monoculture farming practices, in combination with climate change, can exacerbate heat stress on crops and reduce agricultural productivity. Tilman et al. (2001) assert that monoculture farming also depletes soil nutrients and requires heavy reliance on pesticides and fertilizers, creating a cycle of environmental degradation and potential health risks.

FAO (undated) and Fremmpong et al. (2023) made a positive correlation between farm diversity and dietary diversity. Other research explored the importance of diverse crops for providing essential nutrients and showed a positive correlation between agrobiodiversity and dietary diversity. However, the decline of agrobiodiversity has contributed to nutritional deficiencies and related health challenges. Many traditional crops, particularly those rich in essential vitamins and minerals, have been abandoned in favour of high-yielding monoculture staples. This shift has resulted in a decline in the overall nutritional quality of our diets, contributing to problems like micronutrient deficiencies, particularly in developing countries.

2.2 Poor Diets and Malnutrition

The globalization of food systems has led to profound shifts in dietary patterns. Over the past two decades, there has been a dramatic increase in the global consumption of heavily processed, high calorie, low nutrient-density imported foods. The availability of inexpensive edible oils and artificial sweeteners has accelerated this trend, along with the dependence on a limited number of staple crops (Popkin et al., 2012). Consequently, most communities are restricted in their access to diverse, nutrient-rich food choices. Moreover, the aggressive marketing by transnational corporations to promote preprepared and fast-food products, along with the prevalence of modern supermarkets, further encourages unhealthy eating habits. As a result, many countries are experiencing rising rates of obesity, malnutrition, and other diet-related health issues.

While it can be generalized that in low-income countries, populations are experiencing energy and micronutrient deficiencies, and in middle- and high-income countries, obesity is prevalent, many countries are experiencing a simultaneous public health crisis of non-communicable diseases. This crisis is becoming more common in low- and middle-income countries, where people are increasingly adopting a "Westernized diet", displacing traditional, healthier diets rich in fruits, vegetables, and whole grains. For example, in urban areas in the Global South, such as India, increased consumption of processed foods has led to accelerating levels of diabetes and heart disease (Popkin et al., 2012).

Additionally, while high-income countries like the United States may have easy access to fresh produce and other nutrient-dense foods, obesity rates are still considerable due to the abundance of cheap fast-food options high in sodium and unhealthy fats. Type 2 diabetes is especially a concern for affluent populations in North America GBD 2017 Diet Collaborators (2019). These conditions not only impact individual mental and physical health but also lead to increased rates of workplace absences, reduced work performance, and greater healthcare costs, which, excluding undernourishment, result in an estimated annual loss of $9,310 billion (2020 PPP).

2.3 Food Insecurity

One of the major failures of current agri-food systems is the pronounced distributional inequalities of calories, leading to both undernutrition and obesity. While 680 million adults are experiencing obesity, upwards of 820 million are experiencing food insecurity (FOLU, 2019:36). Both diet-related obesity and undernourishment are linked to a lack of access to nutritious foods, with undernourishment primarily being driven by socioeconomic factors.

The root causes of food insecurity are poverty, social inequalities, and inappropriate governance settings, but they are also heightened by issues such as inadequate agriculture infrastructure and climate change. Women and rural populations, particularly in the Global South, experience the highest rates of food insecurity (Gillespie & Mara, 2017). Prolonged undernourishment in adults increases the risk of chronic diseases, susceptibility to infections, and diminished cognitive ability, which threatens their earning capacity. Poor diets can thus trap vulnerable people in a cycle where limited financial resources further undermine their ability to purchase more nutrient-dense food (see Fig. 2) (Perez-Escamilla, 2018).

This cycle not only affects adults but also has implications for the health and development of children in these communities. Globally, food insecurity and undernourishment have resulted in the stunting of an estimated 159 million children under the age of five, with malnutrition responsible for 45% of all under five mortality rates (Gillespie & Mara, 2017). An improper diet not only has physical consequences but also undermines cognitive development and educational achievement. This carries lifelong consequences, where children struggle to meet their full potential, perpetuating poverty into adulthood and leading to inequality across generations.

On a national level, a significant portion of the population being undernourished translates into a less productive workforce and decreased economic output. FAO (2023:23) estimates annual productivity losses of $51 billion (2020 PPP) due to undernourishment. Reduced productivity limits a country's ability to generate income

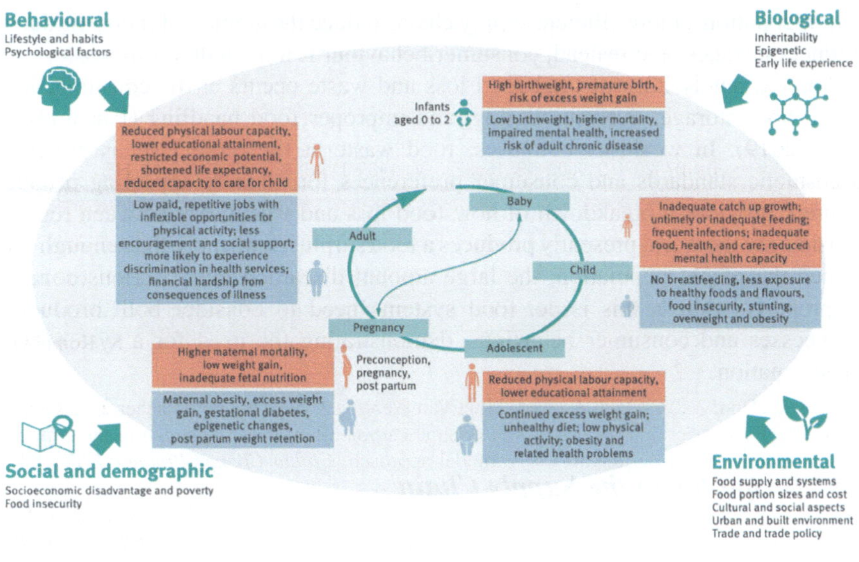

Fig. 2 The double burden of malnutrition through the life cycle and across generations and shared driver. *Source* Perez-Escamilla (2018)

and invest in critical sectors such as education, infrastructure, and healthcare. Since most food-insecure people are in Asia and Sub-Saharan Africa, this has had a substantial negative economic impact on those regions. Without transformation of agri-food systems, around 50% of the world's population will experience malnutrition-related health challenges by 2030. This would further exasperate the economic burden on vulnerable regions and have profound ramifications for the global economy.

2.4 *Food Loss and Waste*

Approximately one-third of all global food production is lost or wasted, which amounts to a staggering 1.3 billion tons of food each year (FAO, 2023). When food is discarded, all the resources required to produce it are essentially wasted, leading to hidden economic and environmental costs. The discarded food accounts for 24% of freshwater resources, 23% of arable crop area, and 23% of applied fertilizer (FAO & IWMI, 2018:18). Additionally, food loss and waste also release greenhouse gas emissions into the atmosphere, exacerbating climate change. Beyond environmental concerns, food waste intensifies issues of food insecurity and hunger around the world.

Food loss and waste occur along supply chains but are influenced by specific factors in each country. In low-income countries, particularly in the Global South, food loss is most prevalent during the processing and post-harvest phases. Contributing factors include poor practices, inadequate handling, and technological constraints, such as a lack of proper storage infrastructure. In contrast, in wealthier nations, more efficient supply chains reduce the amount of food loss during production stages, and instead, consumer behaviour is a main driver of waste.

Approximately 22% of total food loss and waste occurs at the consumer level due to poor storage, over-purchasing, and improper food handling (Ishangulyyev et al., 2019). In wealthier countries, food waste at the retail level is attributed to cosmetic standards and consumer preferences for visually appealing produce. Figure 1 provides a breakdown of how food loss and waste vary between regions. Considering the world presently produces a food surplus, with more than enough food to feed the global population, the large amount discarded raises serious concern. To properly address this issue, food systems need to consider both production weaknesses and consumer behaviour, demonstrating the need for a system-wide transformation.

2.5 *Inequality in the Supply Chain*

Agri-food systems are increasingly concentrated on vertical and horizontal lines, undermining the overall resilience of the food system. In the upstream agri-food system, "The Big 6" agrochemical firms that have dominated the agriculture market over the last few decades have consolidated into the Big 4 through mergers and acquisitions. These corporations control upwards of 60% of the world's seed sales. A small number of companies also control financial and technological service providers. Through such a power concentration, these firms influence prices, determine access to inputs and advanced technology, and shape farmer choices. As a result, farmers have limited bargaining power and the ability to adopt sustainable practices.

Similarly, downstream consolidation of exporters and retailers allows these firms to dictate market access and international food trade. Retailers determine pricing and product selection that limit farmers' influence over their income (IPES-FOOD, 2015). This environment inherently favours larger producers that can invest in the necessary infrastructure and technology, leading to the exclusion of smallholders from value chains. Consequently, smallholders are largely excluded from the benefits of participating in the agri-food system, including better prices for produce, long-term contracts, and agriculture services to improve their productivity (FOLU, 2019:46).

Both upstream and downstream agriculture firms use their financial resources to push for favourable economic policies. These firms have successfully influenced public decision-making through campaign donations, outreach initiatives, lobbying, etc. Agribusinesses also heavily invest in research and development, which can stifle public innovation and discourage major reform in the agri-food systems (IPES-FOOD, 2015).

2.6 Poverty Challenges for Farmers

The consolidation in both upstream and downstream value chains has led to poverty and income challenges for farmers. Most of the world's extreme poor are concentrated in rural areas, particularly in the Global South, where agriculture continues to be the main source of livelihood security. While food commerce is a lucrative business for suppliers, processors, and retail stores, smallholders are largely excluded from these profits. A significant portion of the 2 billion smallholder farmers live on less than $2 a day (World Bank Group, 2016). Many farmers do not make a fair wage and work long hours in unsafe conditions.

Several challenges further increase the vulnerability of these farmers. Smallholders are disproportionally impacted by climate change, unfavourable public policies, informal contracts, a lack of networks, and, overall, an unfavourable business climate. Often, Smallholders in the Global South do not hold formal land tenure, which makes them vulnerable to landgrabing by agribusinesses (Diao et al., 2023). This has alienated many poor farmers and communities from their livelihoods while adding little benefit to the local economy (FOLU, 2019).

Moreover, smallholders that manage to participate in international value chains often still experience poverty and vulnerability. Markets for international trade undervalue the raw materials produced by small farmers. For example, cocoa producers in Western Africa produce the majority of the world's coca, collectively creating $50 billion in annual profit for processers and producers despite earning between $0.50

and $0.84 a day (FOLU, 2019:158). This disparity exposes the systemic issues in the supply chain, including unfair pricing, and exploitative middle-man cuts, that prevent farmers from benefiting relative to their contribution. By extension, most of these farmers cannot afford healthy diets and, along with poverty, they are also experiencing food insecurity and undernourishment.

2.7 Erosion of Indigenous and Local Knowledge

The erosion of Indigenous and Local Knowledge (ILK) within agri-food systems poses a myriad of hidden consequences for the world. While ILK and culture are unique, there is a prevailing worldview of coexisting with local ecosystems, minimizing waste, giving back what is taken, and sharing benefits among community members (WHO & CBD, 2015). Aligned with this view, indigenous and local communities have often refined farming methods to grow crops in challenging environments, from high-altitude zones to flood-prone areas. While this knowledge could be a crucial resource to build resilience in the face of climate change, the industrialization of food production is endangering both this information source and the ecological balance that indigenous and local communities have preserved for generations (IPES-FOOD, 2015).

The threat to ILK of modern agri-food systems has environmental, cultural, and socioeconomic ramifications. Government regulations and corporate interests are limiting the access of indigenous and local communities to ancestral land and resources (such as water and biodiversity) that are fundamental to their way of life. Moreover, modern agricultural policies are preventing age-old practices such as seed-sharing—a vital component of ILK systems (Melash et al., 2023). Seed-saving and sharing allow for the preservation and exchange of traditional crop varieties adapted to local environments and are a crucial part of traditional diets. Restricting these practices not only undermines biodiversity but also erodes the cultural heritage and resilience of indigenous and local communities.

The loss of ILK is compounded by the fact that young people are increasingly moving away from agriculture due to the challenges of small-scale farming in the modern landscape. ILK is shared by an entire community through proverbs, storytelling, and other oral traditions passed down through generations. Without the transmission of ILK from one generation to the next, there is a risk of valuable agriculture practices and insights being lost forever.

2.8 Cost Distribution

The final thing worth noting that has been touched on throughout this section is that the hidden costs of agri-food systems are not equally distributed among income groups and regions. Agri-food systems in high-income countries, characterized by higher levels of industrialization and input use, impose average hidden costs of $3,800 (2020 PPP) per person. Despite a significantly lower contribution to the quantified hidden costs, countries, primarily in the Global South, bear a disproportionate burden of the costs. On average, the lowest income nations generate around $75 (2020 PPP). This has tremendous implications for the development of these nations and the well-being of their populations.

Moreover, within these regions, individuals from the lowest-income groups are experiencing the most adverse effects. Hidden costs tend to increase as income rises, which inevitably exasperates the inequality between regions and income groups. This also perpetuates a cycle for disadvantaged communities that have the least resilience to respond to these costs.

The previous discussion on hidden costs, along with this discussion on distribution, emphasizes the need for a complete transformation of agri-food systems towards one rooted in inclusivity and sustainability. Smart food systems are rooted in these values and have the potential to mitigate these costs and ensure a more equitable allocation of both the costs and benefits of agri-food systems.

3 The Path Forward: Transforming Agri-Food Systems

Climate-Smart Agriculture (CSA) provides a transformative approach to reinvasion agriculture in a way that prioritizes both innovation and sustainability across the numerous dimensions of agriculture.

3.1 Positive Environmental and Climate Impact

The environment, climate, and agri-food systems are interdependent, and only through proper environmental stewardship can agri-food systems reach their potential to feed the global population. There are various innovative ways CSA can mitigate the environmental and climate damage of agriculture. In addition to the sustainable use of resources and the elimination of pollution, one method is to invest in carbon sequestration techniques, which not only reduce carbon present in the atmosphere but also improve the productive capacity of the land.

33,000 farmers in Vietnam have been able to increase rice productivity, lower costs, and lessen the burden on the environment thanks to a climate-smart initiative in the Mekong Delta. The farmers have transitioned from flood irrigation to interspersing wetting and drying techniques. This approach has increased rice yields, reduced the amount of water needed for irrigation, and cut methane emissions (World Bank Group, 2015:12). Extension services and farmer field schools have also played a crucial role in encouraging the adoption of the modified practices.

Additionally, CSA promotes practices such as precision agriculture and incorporates agroecological principles that can reduce the need for external inputs and instead capitalize on the natural synergies found in the environment. The CSA approach is built on the idea of complementary positive outcomes that encourage sustainability throughout supply chains (Klytchnikova et al., 2015).

Social Equity and Inclusivity

Smallholder farmers will be crucial to the transformation of the farming system. In the present agri-food system, farmers carry most of the risk while receiving the lowest returns. CSA strategies can support this transformation by creating an enabling environment that distributes risks across the value chain and reduces the burden on farmers. One approach has been providing farmers with accessible credit and finance, such as impact investing. This financial tool provides several sources of funding to encourage socially responsible agriculture practices. Another initiative involves using shared services, such as Hello Tractor, which allows farmers to collectively purchase tools and equipment, reducing costs for any individual farmer (FOLU, 2019:201).

Another strategy to improve the inclusivity of value chains is by using ICTs to increase farmers' access to information through technology adoption. The World Business Council for Sustainable Development and Output Insights (WBCSD) is currently piloting a cloud-based initiative tool that improves access for small-scale cocoa farmers. The service will provide information on weather, allow farmers to connect with one another, and access extension services. Along with providing valuable information, these innovations can enable farmers to have more consistent yields and income and better compete with larger firms (FOLU, 2019).

Moreover, smart technology can also empower small-scale farmers by providing them with additional authority and resources. In Kenya, TMG Research gGmbH (TMG) has launched a digital tool that connects women farmers with organizations that can support them through land conflicts (TMG Research, 2023). As mentioned, without secure land tenure, farmers are often subject to landgrabing by large companies (FAO, 2023:73). This digital tool not only helps women farmers resolve land conflicts but also strengthens their position in negotiations with firms. This initiative and other similar ones are helping to create a more equitable and empowering environment for small-scale farmers globally. Additional efforts to protect the livelihood of small-scale farmers should focus on policy advocacy for environmental conservation, strengthening of local labour laws, and the adoption of progressive seed laws.

Finally, a new agri-food model must abandon the regressive views of ILK and instead recognize its potential to strengthen the agriculture landscape. Treating and viewing local people as equal partners increase the sustainability of agriculture initiatives. CSA as a framework has the potential to successfully incorporate this knowledge to build climate resilience and improve wellbeing among farming communities.

Economic

The current state of agri-food systems is plagued by inefficiencies in resource use, placing a heavy economic burden on people around the world. As the discussion on hidden costs has revealed, tackling these inefficiencies can yield substantial benefits for both society and the environment. Smart platforms are tools that can empower farmers to optimize their resource use and minimize waste. This not only translates to cost savings but also fosters a more sustainable agricultural sector. Furthermore, smart food systems can act as a shield, protecting agriculture from the substantial economic costs linked to climate change and extreme weather events.

One such initiative was launched in Bangladesh, as the country has experienced numerous floods and significant crop damage over the past 70 years. The 2007 floods and cyclones created an estimated loss of upwards of $3 billion (IMF, 2017). The Emergency Cyclone Recovery and Restoration Project was launched in response to this crisis, with the dual purpose of improving agriculture infrastructure and disaster preparedness. The initiative involved rebuilding municipal and rural infrastructure, including cyclone shelters, and renovating more than 300 km of embankments, along with providing agriculture knowledge and resources to local populations (ADA, 2012). Consequently, it has both improved community resilience to future disasters and enhanced the agricultural productivity of the region.

The ripple effects of smart food technologies extend beyond the farm. These initiatives not only empower the agricultural sector but also generate a wave of job opportunities across other industries. For example, developing climate-smart public infrastructure creates jobs in technology development and research. Furthermore, both the construction and ongoing maintenance of this infrastructure inject new jobs into the infrastructure sector itself. In essence, these smart food technologies have the potential to trigger job growth across a broad spectrum of industries, strengthening economies in both the short and long term.

Resilience

The concentration of power and resources within the agri-food system creates a double threat: it narrows the variety of food sources and leaves the entire system exposed to external shocks like climate change, diseases, and economic turmoil. To build resilience, we need to diversify agriculture. This means supporting local and indigenous varieties, alongside sustainable farming practices that improve soil health and bolster the natural services ecosystems provide. These steps can lessen the impact of climate change and secure a more stable, sustainable food supply for future generations.

Building resilience starts with strong roots. CSA practices are proving their worth worldwide. They move away from large-scale monocultures, a risky approach, and promote diverse farming methods. Take Argentina, for example. Farmers there are intercropping wheat with legumes, a strategy that tackles yield gaps caused by heat. This practice has yielded multiple benefits: increased carbon storage, improved water and land use efficiency, and reduced vulnerability to climate risks (World Bank Group, 2015:27).

In Niger, smallholder farmers are using agroforestry to breathe new life into degraded land and boost crop yields. These farmers intercrop millet with trees, a technique with a triple win. The trees provide nitrogen to the soil, enhance carbon capture, and offer shade and improved water retention for the crops. Remarkably, this approach has led to the rehabilitation of 5 million hectares of degraded land and a staggering 300% increase in millet harvests (World Bank Group, 2015:7). These initiatives not only improve food security and livelihoods for farmers but also contribute significantly to the overall resilience of the agricultural sector.

The path forward is clear: empower local communities. Governments must adopt policies that give local communities a say in decision-making and resource allocation. By fostering such an enabling environment, governments empower communities to tailor their food systems to their specific needs, ultimately strengthening the overall resilience of the agri-food system. At the same time, continued investment in research and development of climate-smart technologies remains crucial. By diversifying agricultural practices and fostering continuous innovation, we can build robust and adaptable agri-food systems ready to face future challenges.

4 Policy and Additional Recommendations

4.1 Local Networks and Loops

While CSA has the potential to create entirely new food systems that prioritize sustainability and community well-being, it is also important to strengthen existing local networks and loops within communities. Building upon these networks can help enhance food security, reduce food miles, and stimulate the local economy. Governments and communities need to support farmer markets, strengthen cooperative networks, and implement pro-smallholder policies to increase the local availability of food. Governments can also provide incentives for farmers to sell their produce locally, invest in infrastructure, and promote education and awareness about the benefits of supporting local producers. The recommendations discussed below further facilitate the development of local food systems.

4.2 Addressing Policy, Market, and Institutional Weaknesses

To unlock the full potential of CSA, governments must clear the path for smallholder farmers. Current policies often create obstacles instead of opportunities. For instance, some subsidies meant to boost production have unintended consequences. They distort markets, encourage unsustainable practices, and worsen climate change. Market failures also hold back smallholder farmers. Limited access to information and unfair competition from monopolies drive up costs and make it difficult for these farmers to compete effectively in value chains. Additionally, weak institutions fail to enforce regulations, provide essential services, and protect farmers' rights.

Two critical areas for government action are seed and land rights. By addressing these issues, governments can empower smallholder farmers and create a more equitable and sustainable agri-food system.

The seed industry is plagued by both policy failures and unfair market practices. A handful of multinational corporations control seed production, limiting farmer choice and driving the loss of traditional, climate-resilient varieties. These monopolies charge exorbitant prices for seeds and often require farmers to use specific, expensive herbicides with them. This traps smallholder farmers in a cycle of debt and marginalization. Therefore, policy changes are urgently needed to support farmers. Initiatives like seed sharing programs and well-stocked seed banks can provide farmers with a wider range of affordable seed options, better suited to their local needs and unpredictable weather patterns. Additionally, standardizing seed regulations across regions can facilitate seed exchange between countries, offering further security for smallholder farmers facing local shortages.

Finally, increased transparency and stricter regulations in the input industry can help create a fairer playing field for all.

Land tenure is another critical policy area for empowering smallholder farmers. Unclear or informal land rights leave farmers vulnerable to land grabbing and hinder long-term planning. By addressing these challenges, governments can provide the security farmers need.

Land rights are another critical area for governments to champion smallholder farmers. Unclear or informal land ownership creates a double jeopardy for farmers: vulnerability to land grabs and difficulty planning for the future. By tackling these land tenure challenges, governments can provide the security farmers need. Formal land ownership offers several benefits:

- Protection: Secure land titles safeguard farmers from land grabs.
- Long-term planning: Secure ownership allows farmers to confidently invest in long-term practices that improve soil health and productivity.
- Sustainable practices: Farmers with a vested interest in their land's long-term health are more likely to adopt sustainable practices.
- Access to credit: Formal land titles empower farmers to access credit for sustainable farming practices or income diversification.

However, the success of these initiatives hinges on strong institutions with the capacity to effectively administer land titles, validate ownership claims, and prevent illegal land grabs.

4.3 Capacity Building

Capitalizing on CSA initiatives requires capacity-building on all levels, across stakeholders interested in improving agri-food systems. For smallholder farmers, capacity-building initiatives, including training, extension services, or demonstration farms, provide essential knowledge and skills to successfully implement climate-smart production. Participatory learning approaches and farmer-to-farmer exchanges are especially effective in facilitating the transfer of knowledge and encouraging the adoption of practices and technology among farmers. These services reduce the risks associated with shifting to a new production model and provide farmers with the confidence to make the transition. Local networks and partnerships between farmers, researchers, development workers, and extension agents can also facilitate technology transfer to ensure that climate-smart technologies are not only accessible but also tailored to the specific needs and challenges of communities.

Strengthening government support for farmers is key to empower change and bridging the knowledge gap is crucial for widespread adoption of new practices. However, many government institutions lack the capacity to provide the necessary resources and support to farmers. Investing in public extension services is, therefore, essential. This means training and equipping extension agents to act as effective bridges between researchers and farmers, ensuring innovations and best practices reach the hands of those who need them most. A focus on equity is also vital. Strengthening extension services should prioritize reaching rural women. Overrepresented in poverty, food insecurity, and smallholder farming, they often face challenges accessing vital support. By equipping extension agents to better serve this critical vulnerable group, we can unlock its full potential and drive positive change in the agricultural sector.

4.4 Infrastructure Development

Smallholder farmers, often located in remote areas, face significant challenges in accessing resources and markets. To achieve social and economic inclusion for these farmers, infrastructure development in rural areas is crucial.

Transportation infrastructure, like roads and ports, is essential. Improved access allows farmers to obtain necessary inputs and services, and efficiently deliver their goods to markets when prices are favourable. Additionally, proper storage facilities and cold chains minimize food waste by preserving perishable items. For regions facing water scarcity, irrigation systems are vital to adapt to climate change and maximize water use efficiency.

Furthermore, information and communication technology (ICT) infrastructure is essential. Access to weather updates and market data empowers farmers to make informed decisions. This allows them to capitalize on innovative initiatives from organizations like WBCSD and TMG, fostering a more sustainable and efficient agricultural sector.

Developing this infrastructure will be a significant undertaking requiring collaboration between various stakeholders. Public–private partnerships (PPPs) offer a promising solution for financing and implementing these projects. By working together, governments, NGOs, and private companies can leverage their expertise and resources. Additionally, collaboration with organizations focused on hidden costs within food systems can further this goal. Through these combined efforts, we can create a more inclusive and sustainable agricultural future for all.

5 International Cooperation and Trade Agreements

While Climate-Smart Agriculture (CSA) offers a promising path forward for global food production, its benefits risk being unevenly distributed unless we address existing disparities in international agriculture trade. Current trade agreements often distort markets. A prime example is the North American Free Trade Agreement, where subsidized American corn flooded the Mexican market, displacing millions of smallholder farmers.

To achieve a more equitable system, we need to prioritize the needs of small-scale farmers through global institutional reform. This is a complex task that demands a concerted effort from all relevant stakeholders. In addition, stakeholders across the food system must advocate for policy changes that promote fair trade practices grounded in human rights and security. This could include: eliminating subsidies that favour large, industrialized nations; enforcing and punishing violations by transnational corporations that disregard environmental or economic regulations; and negotiating trade agreements that safeguard the rights and livelihoods of small-scale farmers. By working together—governments, international organizations, farmer groups, and consumers—we can push for reforms that address these inequalities. This collective action can level the playing field for all farmers and ensure that the benefits of trade are more equitably distributed.

Collective action and advocacy are the initial steps towards dismantling systemic issues that perpetuate inequality and exploitation within the global food system. Ultimately, a coordinated effort from all parties involved is necessary to bring about meaningful change. Only then can we create a food system that is truly fair and sustainable for all.

References

Abahussain, A. A., Abdu, A. S., Al-Zubari, W. K., El-Deen, N. A., & Abdul Raheem, M. (2002). Desertification in the Arab region: analysis of current status and trends. *Journal of Arid Environments, 51*, 521–545. https://www.fao.org/3/bc602e/bc602e.pdf

ADA. (2012). *Emergency disaster damage rehabilitation project.* Asian Development Bank.

Brookfield, H. (2014). Overcoming systemic barriers to sustainable agricultural intensification. *Proceedings of the National Academy of Sciences, 111*(34), 12739–12744. https://doi.org/10.3389/fenvs.2022.861225

Diao, X., Reardon, T., Kennedy, A., DeFries, R. S., Koo, J., Minten, B., Takeshima, H., & Thornton, P. (2023). The future of small farms: Innovations for inclusive transformation. In *Springer eBooks* (pp. 191–205). https://doi.org/10.1007/978-3-031-15703-5_10

Eswaran, H., Lal, R., & Reich, P. (2001). Response to land degradation: An overview. In E. Bridges, I. D. Hannam, L. R. Oldeman, F. W. Penning de Vries, S. J. Scherr, & S. Sombatpanit (Eds.), *Response to land degradation* (pp. 21–35). CRC Press.
FAO (undated). How the world's food security depends on biodiversity. https://www.fao.org/3/cb0416en/CB0416EN.pdf
FAO. (2023). *The state of food and agriculture 2023: Revealing the true cost of food to transform agrifood systems*. Food and Agriculture Organization of the United Nations.
FAO and ITPS. (2015). *Status of the World's Soil Resources (SWSR)—main report*. Food and Agriculture Organization of the United Nations and Intergovernmental Technical Panel on Soils.
FAO and IWMI. (2018). *More people, more food, worse water? A global review of water pollution from agriculture*. FAO and IWMI.
FAO. (2019). *The state of the world's biodiversity for food and agriculture* (p. 102). https://www.fao.org/documents/card/en?details=ca3129en
FOLU. (2019). *Growing better: Ten critical transitions to transform food and land use*. The Food and Land Use Coalition.
Fremmpong, R. B., Gross, E., & Owusu, V. (2023). Crop diversity, sustainable food and nutritional security among smallholder farmers in Ghana. *British Food Journal, 125*(12), 4372–4395. https://doi.org/10.1108/BFJ-12-2022-1060
GBD 2017 Diet Collaborators. (2019). Health effects of dietary risks in 195 countries, 1990–2017: A systematic analysis for the global burden of disease study 2017. *Lancet*, 1958–1972.
Gillespie, S., & Mara, V. (2017). Agriculture, food systems, and nutrition: Meeting the challenge. *Global Challenge*, 1–12.
Gunaratne, M., Radin Firdaus, R., & Rathnasooriya, S. (2021). Climate change and food security in Sri Lanka: Towards food sovereignty. *Humanities and Social Sciences Communications*.
IMF. (2017). *Seeking sustainable growth: Short-term recovery, long-term challenges*. International Monetary Fund.
IPES-FOOD. (2015). *The new science of sustainable food systems*. International Panel of Experts on Sustainable Food Systems.
Ishangulyyev, R., Kim, S., & Lee, S. H. (2019). Understanding food loss and waste—why are we losing and wasting food? *Food*.
Kissinger, G., Herold, M., & De Sy, V. (2012). *Drivers of deforestation and forest degradation: A synthesis report for REDD+ policymakers*. Lexeme Consulting.
Klytchnikova, I. I., Sadler, M. P., Townsend, R., Edmeades, S., Choudhary, V., Hussain, S., & Pietrowski, M. S. (2015). *Future of food: Shaping a climate-smart global food system*. World Bank Group.
Kolbert, E. (2014). *The sixth extinction: An unnatural history*. Henry Holt and Company.
Lazutkaite, E., Mullié, W. C., Müller, A., & Prakash, A. (2023). Insecticide use against desert locust in the horn of Africa 2019–2021 reveals a pressing need for change. *Agronomy*, 1–23.
Melash, A. A., Bogale, A. A., Migbaru, A. T., Chakilu, G. G., Percze, A., Ábrahám, É. B., & Mengistu, D. K. (2023). Indigenous agricultural knowledge: A neglected human based resource for sustainable crop protection and production. *Heliyon*.
Pendrill, F. U., Persson, M., Godar, J., Wood, R., Schmidt, S., Moran, D., & Kastner, T. (2019). Agricultural and forestry trade drives large share of tropical deforestation emissions. *Global Environmental Change*, 1–10.
Perez-Escamilla, R. (2018). Nutrition disparities and the global burden of malnutrition. *BMJ*.
Popkin, B. M., Adair, L. S., & Ng, S. W. (2012). Global nutrition transition and the pandemic of obesity in developing countrie. *Nutrition Reviews*, 3–21.
Tilahun, M., Singh, A., Apindi, E., Barr, J., Zommers, Z., & Lund, G. (2015). *The economics of land degradation in Africa: Benefits of action outweigh the cost*. ELD Initiative & UNEP.
Tilman, D., Cassman, K. G., Matson, P. A., Naylor, R., & Polasky, S. (2001). Agricultural sustainability and intensive production practices. *Science, 292*(5518), 279–282. https://pubmed.ncbi.nlm.nih.gov/12167873/

TMG Research. (2023). *TMG's work in Kenya—Towards inclusive and climate resilient food systems*. TMG Research.

TMG Research gGmbH. (2023). *Insecticide use against desert locust in the Horn of Africa 2019–2021 reveals a pressing need for change* . Retrieved from Prevention Web: https://www.preventionweb.net/news/insecticide-use-against-desert-locust-horn-africa-2019-2021-reveals-pressing-need-change

UNCCD. (2022). *Land restoration for recovery and resilience.* United Nations Convention to Combat Desertification.

WHO & CBD. (2015). *Connecting global priorities: Biodiversity and human health.* World Health Organization and Secretariat of the Convention on Biological Diversity.

World Bank Group. (2015). *Future of food: Shaping A climate-smart global food system.* World Bank Group.

World Bank Group. (2016). *A Year in the lives of smallholder farmers.* Retrieved from World Bank: https://www.worldbank.org/en/news/feature/2016/02/25/a-year-in-the-lives-of-smallholder-farming-families

World Wildlife Fund (WWF). (2018). *Soil erosion and degradation.* Retrieved from World Wildlife Fund: https://www.worldwildlife.org/threats/soil-erosion-and-degradation

Climate-Smart Agriculture (CSA) Practices for Sustainable Intensification in Major Agri-Food Systems of South Asia

Hanuman Sahay Jat, Parbodh Chander Sharma, Madhu Choudhary, Tanuja Poonia, Jagdish Chandra Dagar, and Chiter Mal Parihar

Abstract Indo-Gangetic Plains (IGP) of South Asia is known for cereal (rice, wheat, and maize)-based major agri-food systems, which meet food requirement of 50% population in the region. Continuous cultivation of monotonous rice-wheat (RW) cropping system with traditional practices since the last six decades is showing threats to the systems' sustainability. Future increase in food grain production might not be expected with business-as-usual practices. As per the Intergovernmental Panel on Climate Change (IPCC), the crop production in South Asia is expected to decrease by 30% particularly in cereal crops (rice, wheat, maize) by end of twenty-first century. South Asian farmers are mostly smallholders, their adaptive capacities are limited, and hence they are most vulnerable to climate change. Therefore, climate-smart agriculture (CSA) practices, comprised of three pillars (food security, adaptive capacity, and mitigation potential), are essential to design the new generation cereal systems to reduce farmers' vulnerability to climate change. Sustainable intensification through the integration of short-duration legumes and by replacement of input exhaustive crops are required to maintain the systems' sustainability in the region. Based on the on-station and off-station trials in South Asia, CSA-based modules had a potential to increase the system productivity and profitability by 5–15% and 20–25% in major cereal (rice, maize, and wheat) based agri-food systems *vis-a-vis* reduced the global warming potential by 15–35%. CSA-based production systems are found to be more resilient by improving soil quality (SOC—50–100%; NPK—30–50%). Smart application of input (through subsurface drip irrigation) saved irrigation water (~45%) and

H. S. Jat (✉)
ICAR-Indian Institute of Maize Research (IIMR), Ludhiana, Punjab, India
e-mail: hsjat_agron@yahoo.com

P. C. Sharma · M. Choudhary · T. Poonia
ICAR-Central Soil Salinity Research Institute (CSSRI), Karnal, India

J. C. Dagar
Natural Resource Management, Indian Council of Agricultural Research, Krishi Anusandhan Bhawan-II, Pusa New Delhi, India

C. M. Parihar
ICAR-Indian Agricultural Research Institute (IARI), New Delhi, India

M. Behnassi et al. (eds.), *Climate-Smart and Resilient Food Systems and Security*,
https://doi.org/10.1007/978-3-031-65968-3_3

N (~20%) with slightly higher yields compared to conventional tillage (CT)-based systems. CSA layered with nutrient management tools like Nutrient Expert (NE) and GreenSeeker (GS) reduced the N input by 15–20%, increased crop yield by 4–8% and reduced the global warming potential by about ~3% in rice and ~15% in wheat. Results showed wider practical applicability and scalability, which may pave a way paradigm for scaling up cereal systems in South Asia in near future.

Keywords Climate smart agriculture · Food security · Water conservation · Adaptive capacity · Mitigation potential · Sustainable intensification · South Asia · Indo-Gangetic plains

1 Introduction

South Asian agriculture always remains central to the nations' economy and made a spectacular progress in food self-sufficiency and economic growth through remarkable increase in crop yields over the past few decades. Agriculture provides employment about ~60% of the labor force, livelihood to ~70% of the people, and contributes 22% in the regional GDP (gross domestic product) even though is a global 'hot spot' and is more vulnerable to climate change (Wang et al., 2017). South Asia is a habitat for 1.7 billion people and by 2050, that number is nearly expected to rise by 2.4 billion (FAO, 2017). The increasing population growth will require 43% more cereals between 2010 and 2050. To meet the global demand for food by 2050, food grain production will need to be increased by 70% by 2050 as per the current scenario of food consumption pattern and population growth. The expected climate change effects in near future will reduce systems productivity and lead to greater instability in agri-food systems of the different agricultural sectors viz., crop, animal production, marine system and agroforestry.

Currently, in the majority of the agro-ecological systems, high levels of food insecurity and environmental degradation prevail with the limited options for coping the adverse weather abnormalities (Jat et al., 2018c). Agriculture accounts for ~30% of the total global greenhouse gas (GHG) emissions with current agricultural practices (Sapkota et al., 2019). Therefore, business-as-usual requires dramatic changes in crop management practices to achieve sustainable development goals (SDGs) by 2030 without delimiting the input use efficiencies and natural resources degradation (Jat et al., 2019d). By promoting best management practices, integrated crop management, zero-tillage (ZT), crop residue management, intercropping, cash cropping, improved seeds, water, and fertilizer management practices are required for climate change solutions in a pragmatic and focused manner (Jat et al., 2021b; Kakraliya et al., 2018a). Crops and cropping system resilience will be key, but 'climate smart' is broader and underscores the need for improving the adaptability through innovative and proactive changes in the current way farming is done to mitigate GHGs and increase productivity sustainably (Jat et al., 2018c; Kakraliya et al., 2018a, 2018b).

In agricultural sector, best management practices help in capturing the synergies that exist among different production activities to develop more resilient and productive food systems. Therefore, effective and efficient management of natural resource may become an important part of the solution to cope up with the climate change effects. Mostly the conventional tillage (intensive tillage-based) practices are being followed in the South Asian region to achieve the high productivity over the years but all at the cost of over-exploitation of natural resources (soil, groundwater and energy) and high use of production inputs (irrigation, fertilizer, and pesticides) (Choudhary et al., 2018a; Jat et al., 2019d, 2020d). The sustainability of the RW system in the South Asia is doubtful due to the rapid degradation of soil, decline in groundwater resources, and environmental quality.

Furthermore, continued cultivation of monotonous cropping system (RW system) over the last six decades allowed the problem of higher infestation of weeds, insects, and pests (Sharma et al., 2015). Sustainable agriculture is rooted in conservation agriculture (CA) and social development objectives and goals which, if reached, would contribute towards achieving SDGs of reducing hunger and improved environmental resilience. Sustainable utilization of natural resources requires ecosystem approaches based on the multi-stakeholder and multi-sectoral coordination and cooperation for effective crop management and governance practices. Therefore, transformation of conventional agriculture/business-as-usual to climate-smart agriculture (CSA) is a crucial element for making agriculture more productive but sustainably. More productive, profitable and eco-friendly agriculture is built on the principles of sound management of available natural resources, including land, soil, water, energy, and biodiversity (Jat et al., 2020d; Ladha et al., 2009).

The production intensification in India during the 'Green Revolution' in South Asia has been guided by: (a) improvement in genetic potentials of crops; (b) greater application of external inputs for plant nutrition; and (c) increased network of irrigation and mechanization. During the 'Green Revolution' era, the implicit assumption or approach was that more input is required to produce more output in agriculture production systems. This approach is now economically as well as environmentally unsustainable and also known to be ecologically intrusive. Nowadays, it leads to the degradation of soil and environment and results into sub-optimal partial factor productivities and crop yield levels that are difficult and expensive to maintain over time (Jat et al., 2011). In 1990s, Conservation agriculture (CA) was practiced to fostering systems sustainability and natural ecological processes to improve crop yields by minimizing soil disturbance, maintaining permanent soil cover, and diversifying crop rotations. The approach is to manage agro-ecosystems for sustained productivity, higher profits and food security while conserving and maintaining the natural resource base and environment. This encompasses available resource management at the farm level, village level, and landscape level to increase synergies between food processes and ecosystem preservation. In twentieth century, the convergence of CA technologies with other technologies called CSA was realized to cope with climate change effects and to increase the adaptive capacity of crops, cropping/farming systems while ensuring the food security.

In twenty-first century, it was realized that a process of 'sustainable intensification of agriculture' is required for sustaining the production at lesser costs, without deteriorating soil and environmental quality while maintaining the flow of agro-ecosystem services. The scope for further expansion of cultivable area is limited in the majority of the South Asian countries, therefore, productivity gains can be accomplished through agriculture intensification by pursuing one or other approaches including: (i) increase in per unit area crop yields; (ii) increase in cropping intensity; and (iii) alternative land uses like agroforestry. The process of agricultural intensification increases crop production by utilizing fixed natural resources (land, light, temperature, etc.) in the defined agro-ecosystem with optimum use of external inputs for attaining sustainable yield levels. Increasingly, sustainable intensification in any production system is being considered as an important component to ensure food security, poverty alleviation, health for all, social development, etc. by accelerating crop productivity, improving environmental quality, and preserving the natural resources.

Currently, the South Asian agriculture is facing triple challenges of resource fatigue, decelerating productivity, and declining farm profitability. Also, there exist large yield gaps between potential yields and actual yields because of 'management yield gaps' that range from 18 to 70%, 14 to 47%, and 36 to 77% in rice, wheat, and maize, respectively (Jat et al., 2011). Cropping systems such as rice-wheat, still the backbone of food security to feed the millions, now face the serious challenges of rising production and environmental costs. Thus, the sustainable agricultural systems/resilient systems by definition should be less vulnerable to shocks and stresses and have to be more profitable to practitioners. Sustainable intensification opportunities are of utmost importance for national food security and in addressing the current challenges of agriculture related to water shortages, emerging climatic variability, and natural resource degradation.

In 1896, the Nobel Laureate Svante Arrhenius predicted that doubling of CO_2 concentrations (then about 260 ppm) in the atmosphere would raise the temperatures of the Earth by about 2–6 °C. At present, the atmospheric CO_2 concentration is over 400 ppm. All these support the estimate by the Intergovernmental Panel on Climate Change (IPCC) that the Earth will warm by 1.4–5.8 °C during the current century. It is estimated that about 75% of the total emissions from agriculture and land use originate in developing countries. Agriculture accounts for 13.5% of global GHG emissions or about 1.8 Gt carbon equivalent/year or 6.6 Gt of CO_2 equivalent/year, mainly in the form of methane (CH_4), and nitrous oxide (N_2O) from fertilized soils, enteric fermentation, biomass burning, rice production as well as manure and fertilizer production. Agriculture and forestry capture a great deal of carbon absorption and fixation through photosynthesis and offer a solution to climate change. The FAO profile on Climate Change-2009 notes that agriculture has the technical potential to mitigate between 1.5 and 1.6 Gt C equivalent/year (5.5–6.0 Gt of CO_2 equivalent/year), mainly through soil carbon sequestration in developing countries. Studies estimated that there would be at least 10% increase in irrigation water demand, in arid and semi-arid region of Asia, with a 1 °C rise in temperature. South Asia is one of the most densely populated regions in the world and is among the most vulnerable to the climate change and climate variability (Aryal et al., 2020). In the absence of adaptation and mitigation, climate change can have major consequences on food security and development in the decades to come.

2 Climate-Smart Agriculture (CSA)

The CSA concept was first launched by FAO in the 2010 Hague Conference on Agriculture, Food Security and Climate Change in the context of national food security and SDGs, to tackle three main objectives: (i) sustainably increase food security by enhancing agricultural productivity and farm incomes; (ii) build system resilience and adaptability to climate change effects; and iii) reduce and/or remove GHGs where possible. Sometimes, the SMART also elaborated as 'sustainable management of agricultural resources and techniques' (Dagar et al., 2016a, 2016b). Thus, SMART, if CA is implemented at right time with desired resources, techniques and knowledge in a specified typological domain, will lead to food security while improving adaptive capacity and mitigating potential for sustainable agriculture production.

CSA may be defined as an approach of agriculture that deals with available agricultural resources management with the latest techniques of crop production under a particular set of soil and ecological conditions. CSA seeks to increase sustainable farm productivity, strengthen crops and cropping system' resilience, reduce GHG emissions, increase carbon (C) sequestration, and deliver environmental benefits while strengthening food security. CSA includes the proven practical techniques viz., conservation agriculture, crop rotation, intercropping, mulching, crop-livestock integration, agroforestry, range lands, and improved nutrient and water management, and innovative techniques for timely weather forecasting, more resilient food crops and risk insurance (Jat et al., 2021b; Kakraliya et al., 2018b). Conservation agriculture (CA) works on the three principles of zero-tillage/minimum tillage, rational surface cover, and efficient crop rotations and addresses the major challenges of CSA except the reduction of GHG emissions from the agri-food systems.

Once these principles layered with the smart management practices, it targets all the sphere of crop production system (Fig. 1). CSA is a pragmatic approach to ensure food security in a particular changing climate condition. Adaptation strategies involves the sustainable management of natural resources through promotion of conservation agriculture practices and climate-smart crops to counter the adverse impacts of climate change. CSA is a holistic approach and unites numerous issues of global development objectives and agricultural development related to the natural resources like soil, water, energy, and social and economic issues. Achieving all the dimensions of food security like availability, accessibility, stability, and nutrition security needs to be the ultimate goal of agriculture production and public distribution systems to the poor people. Food security is governed by multiple components of food systems and climate change and requires a diversity of approaches, skills, management, and resources (Mukhopadhyay et al., 2020).

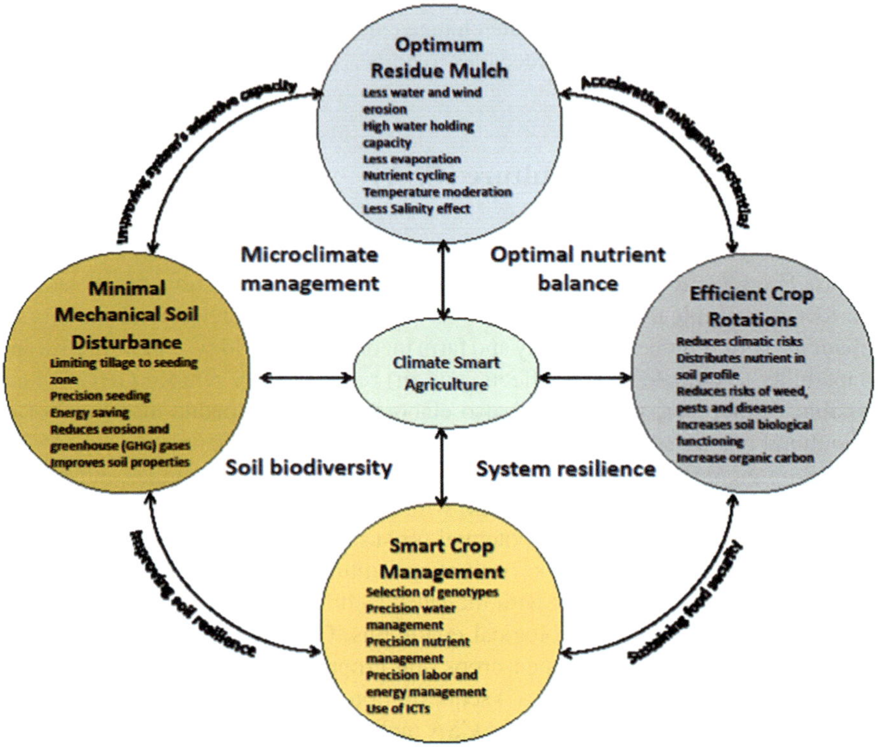

Fig. 1 Principles of CSA and management practices to achieve the same. *Source* Mukhopadhyay et al. (2020)

CSA is basically comprised of three pillars, i.e., food security, adaptive capacity, and mitigation potential (Sharma et al., 2015; Fig. 2). These pillars work in convergence to the food security and developmental goals while improving the quality of environment. CSA is location-specific and tailored to fit in the different agro-ecological and socioeconomic conditions by altering some modifications in the management practices as per the location. Sometimes, interventions that work in one ecology may not be applicable to another ecology because of specificity and available social and bio-physical conditions.

2.1 Food Security

CSA is a farming system that seeks to increase food security, eradicate poverty, conserve biodiversity, and protect environmental services. The negative effects of climate change are likely to be greatest in the currently food-insecure regions and may also be seen in those regions that have made significant gains in reducing food

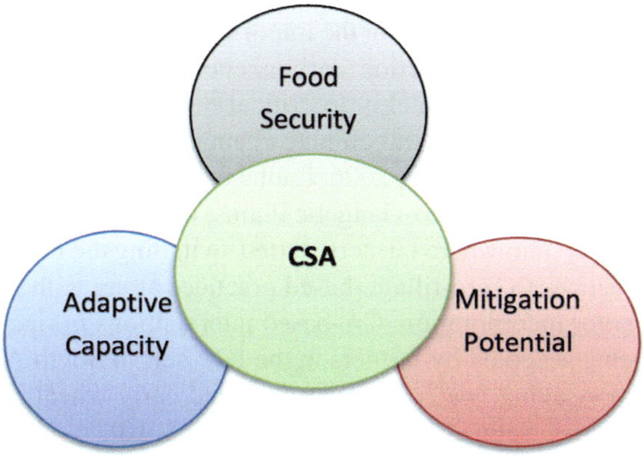

Fig. 2 Pillars of CSA. *Source* Jat et al. (2020a, 2020b, 2020c, 2020d, 2020e)

insecurity over the past 100 years. This significant achievement of the 'Green Revolution' in South Asia is largely due to the direct and sustained increase in food production due to the use of external inputs such as highly productive varieties, commercial fertilizers, and expansion of irrigation facilities. More recently, however (early in the twenty-first century), the problem of food insecurity with the added challenges of natural resource degradation has resurfaced and intensified through the indiscriminate use of resources, huge increase in input costs, diversification of agricultural inputs, and farm size reductions. In South Asia, natural resources are pressed by 3 to 5 times more due to increasing population growth and challenges of economic and political pressures compared to the rest of the world (Jat et al., 2011, 2016b, 2016c). In the region, the misuse and mismanagement of productive resources, especially land, water, energy, and agricultural chemicals, have had a significant impact on the health of natural resources and have contributed to global warmings that have led to climate change. Studies showed that at every 1 °C increase in temperature would increase the demand for irrigation water by at least 10% in the semiarid and arid regions of Asia (Sivakumar & Stefanski, 2011). Furthermore, higher temperature aggravates the resource instability and scarcity in the region (Aryal et al., 2020).

2.2 *Adaptation Capacity*

Climate change poses a major challenge to food security and agriculture development. CSA practices can contribute to making farming systems more adaptable and resilient to expected climate change effects. Adaptation to the agricultural sector is a major priority in this endeavor due to the natural sensitivity of food production to climate change and the strong links that exist between weather, agriculture, economic growth, and social development. The purpose is to understand and summarize the potential impacts of agricultural climate change in regions assessing risk factors, provide information on where investment is needed to improve climate-resilient agriculture, and explain the importance of current efforts to achieve sustainable agriculture. Water, nutrition, carbon, energy, knowledge, and good climate practices are able to respond quickly to critical needs that address the grievances (e.g., adaptation, productivity, and climate change) facing South Asian agriculture (Aryal et al., 2016, 2020; Barclay, 2006; Ladha et al., 2009; Saharawat et al., 2012; Sharma et al., 2002). In IGP (Indo-Gangetic Plains) of South Asia, rice-wheat is the dominant cropping system where farmers started switching the intensive tillage-based management practices to zero-tillage-based practices along with partial and full residue retention and/or incorporation. CA-based interventions in rice-wheat system are increasingly being accepted by farmers in the IGP belt of South Asia due to the many benefits of labor-saving, water conservation, and early wheat planting (Gupta & Sayre, 2007; Gupta & Seth, 2007; Saharawat et al., 2010).

2.3 *Mitigation Potential*

The problem of food security and the additional challenges of degrading natural resources, the huge increase in import costs, the diversification of human capital in agriculture, and the decline in farm size are the concerns of agriculture sustainability. In addition, while maintaining sustainable development, the region will also have to reduce its agricultural footprint. Agriculture has to address three intertwined challenges of ensuring food security through increased productivity and income, adapting to climate change, and contributing to climate change mitigation. Agricultural activities and land-use changes contribute to about one third of the total GHGs emissions and are the largest source of N_2O. However, agriculture also plays an important role as a carbon sink through its capacity to sequester and store carbon in soils and in plants and trees. Reducing agricultural emissions and increasing carbon sequestration in the soil and biomass have the potential to reduce agriculture's contribution to climate change. Therefore, it is necessary to identify management practices that provide opportunities to reduce GHG emission yet providing greater resilience in production systems, food security, and rural welfare. Reducing agricultural emissions and increasing carbon sequestration in the soil and biomass have the potential to reduce agriculture's contribution to climate change by 5.5–6.0 Gigatons (Gt) of carbon dioxide equivalent (CO_2 eq) per year (Olander et al., 2013). Depending upon management practices, the agricultural sector can be significant emitter to GHGs neutral.

There are a variety of CSA-based interventions that have the potential to increase the flexibility of the production system, reduce emissions, or improve carbon storage while increase the food grain production. In CSA, in addition to adaptability, we need to look at mitigation strategies that include agricultural practices. Improved agricultural management improves resource efficiency, often reducing GHG emissions (Sapkota et al., 2017, 2019, 2020, 2021). The effectiveness of these practices depends on factors such as climate, cropping system, soil type, input sources, and socio-economic condition of farmer. Approximately, 90% of total mitigation results from improved sink (C sequestration) and approximately 10% from reduction of emissions (Ortiz-Monasterio et al., 2010). Puddled transplanted rice with continuous flooding is a common practice in South Asian IGP that not only increases the CH_4 emission from rice fields but also adds the emission from fuel used in the pumping of groundwater. The water management practices like alternate wetting and drying instead of continuous flooding, ridge and furrow planting, and direct seeded rice are the option to mitigate GHG emissions to some levels. Reducing CH_4 emissions in agriculture can be largely achieved by following drying and wetting in rice, bed planting of rice, and cultivation of direct rice (DSR) (Sapkota et al., 2017, 2020; Tirol-Padre et al., 2016). This not only reduces the amount of irrigation water inputs but also lowers the CH_4 emissions (Hobbs & Govaert, 2010). Nitrogen oxide (N_2O) emissions can also be reduced by using best nitrogen (N) fertilizer management methods (33% at planting and post-planting stay) compared to farmers' practice. Best N management practices are able to reduce the combined (NOx and N_2O) emissions by more than 50% and NO_3- leaching by >60% (Matson et al., 1998). Optimizing N-fertilizer application rates and synchronizing them to crop growth and development will improve yields while reducing costs and N_2O emissions (Sapkota et al., 2014; Verhulst et al., 2011). In the rice-wheat system, the substitution of rice with corn, pigeon pea, and soybeans significantly reduce GHG in the northwest of the South Asian IGP (Jat et al., 2020a, 2020b, 2020c, 2020d, 2020e).

3 CSA and Their Portfolio of Practices

In agriculture, risks are exacerbated by a variety of factors, ranging from weather variability to climate change, uncertainties in crop yields and prices, poor resources and infrastructure, imperfect markets, and lack of financial services. These factors not only endanger the farmer's livelihood security but also undermine the viability of the agriculture sector and its potential to become a part of the solution to the problem of endemic poverty of the farmers and the agricultural labor. The poor penetration and development of various risk management tools open the opportunities for the emerging agricultural insurance and commodity markets. Risk management solutions help harness the technological advances in climate science, remote sensing technologies, climate-smart management practices and use of ICTs in agriculture to increase the effectiveness for pooling, sharing and transfer of risks, enhancing the coping capabilities of the farmers. Climate risk management can be overcome by targeting the smart management practices for adapting to climate extremes and to develop the resilience in crops and cropping systems. Mitigation, which means avoiding and reducing GHG emissions from the fields, is possible by targeting the right practices at right time under specific climatic conditions. Both adaptation and mitigation practices of crop production are required to climate risk management.

CSA welcomes a portfolio of smart/strategic interventions that include all domestic farm household activities. These include smart practices related to carbon, water, energy, nutrients, climate and knowledge. These portfolios of interventions work together to boost community resilience to climate stress while ensuring livelihood security. In cereal-based agri-food systems, the success of efficient and climate-resilient farming systems depends on activities performed by farm families and available resources, and their interactions. The choice of CSA techniques is identified by village farmers' committees, researchers, and planners in consultation with the local farming community with the objective to prioritize and implement important climate-smart interventions that are appropriate for the community. The potential benefits of specific interventions in climate change adaptation and mitigation are presented in Table 1.

In the IGP of South Asia, despite the apparent success of all agricultural sectors, it is increasingly being realized that rice-wheat cropping system is the best planting pattern for IGP and the technologies adopted in the past for rapid agricultural growth will need to be looked to address the problems of the second generation triggered by the Green Revolution. These problems are exacerbated by the misuse of production resources, especially water, energy, and agricultural chemicals. The sharp rise in energy costs, the shift in labor towards non-farm sectors, food price instability, and climate change risks pose significant challenges to farmers and our society. Overall, the challenge of achieving sustainable productivity with less available resources can only be achieved by adopting CSA technologies, which include agronomic management for increasing resource (water, nutrient, energy) use efficiency, laser land leveling, direct seeded rice (DSR), use of micro-irrigation, zero or strip tillage, bed planting, crop diversification, sustainable intensification and developing crop cultivars to adapt under a particular set of environment to tolerate biotic and abiotic stresses. Climate change in the arid and semi-arid regions of Asia can lead to an increase in demand for irrigation water (a 10% increase in constant temperature of 1 °C), and further exacerbating the shortage of resources (Sivakumar & Stefanski, 2011). This will also increase the instability in the region through completion of irrigation water and makes the smallholder agriculture riskier. Wheat is currently close to its high-temperature range, with high temperatures during production and grain filling, which represent a critical factor in reducing wheat yields for most IGP (Jat et al., 2018c). Increasing temperatures can have a profound effect. For example, Ortiz et al. (2008) estimated that by 2050 about half of IGP's wheat-producing areas could be reclassified as heat-stress short-term production area. Rice production is also expected to be affected by at least 10% in reduction for every 1 °C rise in night temperature (Peng et al., 2004).

Given the many challenges listed above, CA-based agricultural technologies of sustainable resilience and adaptability to emerging climate change while reducing GHG emissions (climate agricultural practices) are priorities for the R4D in RW system. To support food security and increase farmers' income, agricultural production systems are under pressure to increase productivity and strengthen agricultural resilience. Advanced agricultural systems can also emit low levels of greenhouse gases. The existing strategies for carbon sequestration and GHG reduction in the agricultural sector are not sufficient. Many of these CSA-based available strategies are able to improve food security, promote rural development, and help communities adapt to climate change (Table 1).

Table 1 The key smart practices in terms of food security (FS), climate risk management (CR), adaptation (A), and mitigation (M) potential to conventional practices

Climate-smart practices	Potential benefits relative to conventional practices	FS	CR	A	M
Laser land levelling (LLL)	Increase water and crop productivity, reduce GHG emissions, and increase area for cultivation	xx	xx	xx	xxx
Zero tillage (ZT)/ minimum tillage	Reduce irrigation water use, carbon sequestration, similar or higher crop yields, increase profit, reduce energy consumption and GHG emission, improve crop resilience	xx	xx	xxx	xx
Precise crop establishment (direct seeded of rice; DSR)	Irrigation water saving (20–30%), saving in time, reduce methane (CH_4) emission, reduce labor requirement, more tolerance to abiotic stress (water and heat)	x	xx	xxx	xx
Alternate wetting and drying in rice	Reduces methane (CH_4) emission by 40–50% compared to continuous ponding, lowered the irrigation requirement by 15–20%, improve soil physical conditions	xx	x	xxx	xxx
Sustainable intensification	Increase cropping intensity, efficient resource (water, soil, and energy) use, increase income and nutrition, improve soil fertility, reduce risk	xx	xxx	x	xx
Raised bed planting	Irrigation water saving (25–35%), more adaptable to water stress, less crop lodging, improve input use efficiencies	xx	xx	xxx	xx
Leaf color chart (LCC) and Green Seeker	Reduces N-fertilizer requirement, N-losses and environmental pollution, reduces nitrous oxide emission and nitrate leaching	xxx	–	x	xxx
Nitrification and urease inhibitors	Increase nitrogen (N) use efficiency, reduce N losses and environmental pollution	xx	x	–	xxx
Nutrient Expert-decision support tool	Optimize nutrient requirement, reduce nutrient losses and environmental pollution, reduce GHG emission	xxx	–	x	xxx
Crop residue management/soil mulching	Reduces soil erosion and evaporation losses, conserves soil moisture, buffers soil moisture and temperature, improves soil quality, increases carbon sequestration, avoids burning thereby reducing air pollution, reduces weed infestation	xx	xx	xx	xxx
Precise water management	Increase water and nutrient use efficiency, reduce GHG emissions, increase crop and water productivity	xx	x	xxx	xxx
Agroforestry	Sequester carbon into the soil, reduce soil erosion, enhance the biodiversity, improve the ecosystem	xx	xx	xx	xxx

(continued)

Table 1 (continued)

Climate-smart practices	Potential benefits relative to conventional practices	FS	CR	A	M
Information and communication tools (ICTs)	Provide timely information on climate change, weather forecasts, new seed varieties and improved farming practices	x	xxx	–	x
Knowledge services	Improve behavioral change towards climate change adaptation and in adoption of new technologies at community level	–	xx	xx	–

Where, xxx-high potential; xx-medium potential; x-reasonable potential. *Source* Wassmann et al., (2009a, 2009b), Jat et al., (2016b, 2021b)

4 Climate-Smart Agricultural Practices (CSAPs)

South Asia, especially India has made rapid strides to improve food production and the country, has not only adequately maintained food production but is now exporting large quantities to many other countries and receiving foreign exchange. In fact, most countries are already experiencing the effects of climate change on food grain production. Rising temperatures, incessant rainfall, abnormal weather conditions (such as floods and prolonged droughts), changes in soil fertility, and new insect-pest infections are major contributors to unsustainable agricultural growth. Climate-smart agricultural practices (CSAPs) include farming technologies/practices that continuously increase productivity, improve resilience, reduce global warming (GHG), and to attain national food security. CSAPs mainly include conservation agriculture (zero tillage, residue retention/incorporation and efficient crop rotation), site-specific nutrient and water management, and the use of improved stress-tolerant crop varieties. CSA options include water, energy, nutrients, carbon, climate, and knowledge-appropriate practices suitable for a variety of crops and planting systems in the region (Kakraliya et al., 2018a, 2018b; Table 2). These practices can also bring combined benefits in terms of reducing GHG emissions and improved soil carbon stocks and crop yields (Datta et al., 2021; Jat et al., 2019a, 2019b). Studies conducted on rice-wheat system have shown the benefits of CASPs in terms of yields, input use efficiencies, carbon sequestration, and mitigation potential.

Globally, research results suggest that the adoption of CA/CSA-based management techniques under different crop production systems and ecologies can address the foremost challenges of natural resource degradation (soil, energy, water), instability in crop yields, high production costs, labor crisis, low input-use efficiency, and adverse climate change effects (Choudhary et al., 2018a; Jat et al., 2021b; Kumar et al., 2018).

Table 2 CSA technologies to address the adaptation and mitigation potential in agriculture

Technologies	Management practices	Adaptation/mitigation potential
Carbon smart	Legume integration, agroforestry, zero tillage, residue management, land use system, livestock management, etc	Practices to increase carbon sequestration, reduce GHG emissions, reduce nutrient losses and less chemical fertilizer uses, improve soil physico-chemical and biological properties
Water smart	Rainwater harvesting, micro-irrigation, laser land leveling, furrow irrigation, cover/intercrops, mulch, crop diversification, alternate wetting and drying in rice, direct seeded rice (DSR), etc	Practices to improve water use efficiency, e.g., minimization of rainwater runoff and water loss from agricultural fields, effective control of water for irrigation and drainage
Nutrient smart	Site-specific nutrient management (SSNM) through Nutrient Expert decision support tools, Green Seeker and Leaf color chart (LCC), green manuring, intercropping/integration of legumes, fertigation, etc	Practices to improve nutrient use efficiency, e.g., optimum fertilizers application to match crop demand, improvement in soil health and balanced nutrient supply
Energy smart	Zero/minimum tillage, direct seeded rice, crop residue management, permanent beds, short duration legumes, DSR precision water management etc	Practices to improve energy use efficiency, e.g., reduced petroleum product use and commercial fertilizers
Weather smart	ICT-based weather forecasts, crop advisory, index-based insurance, climate analogs, stress-tolerant crops and varieties, etc	Practices to provide weather advice to farmers and services related to income security, crop-specific insurance to cover weather vagaries
Knowledge smart	Improved crop varieties, community seed banks and cooperatives, farmer-farmer learning, capacity development, etc	Use of indigenous technical knowledge backed with science-based evidences to cope up climatic risks and weather abnormalities viz., drought, flood, heat/cold stresses

5 CSA and Potential Benefits

5.1 CSA and System Productivity and Profitability

Sustainable intensification of intensive tillage-based RW system through CSA-based management practices provides excellent opportunities to achieve a 'fourfold victory' for the efficient use of crop residue energy otherwise lost during burnings, ensuring food security (~13% high plant productivity) and farm profits (~35%), and to stop groundwater depletion through reduced pumping (Jat et al., 2019d, 2020d; Kumar et al., 2018). CSA-based practices generally help improve the crop productivity and profitability in low and variable rainfall under both irrigated and rainfed ecosystem. Layering of CSAPs viz., carbon, nutrient, water, energy, weather, and knowledge-intensive techniques are more productive compared to the use in isolation. In RW system, Kakraliya et al. (2018b) reported that 8% increase in production, 23% in profit, 31% in total water productivity, and 53% in energy productivity with the implementation of CSAPs while saving 24% irrigation water and reducing GWP by 40% compared to farmers' practice. CSA-based practices in rice-wheat-mungbean (RWMb) systems improved the productivity and profitability by 11 and 24%, respectively, compared to conventional rice-wheat (RW) system/farmers practice (12.3 Mg ha^{-1} and INR 85,800 ha^{-1}, respectively) (Choudhary et al., 2016b; Kumar et al., 2018). However, maize-wheat-mungbean (MWMb) systems on flats (sowing crops on the flat surface) have the potential to increase the productivity by 16% and profitability by 50% compared to conventional RW system/farmers practice (12.3 Mg ha^{-1} and INR 85,800 ha^{-1}), and hence provide scalable technology package for much needed diversification in RW system of NW India. Jat et al. (2018d) recorded the 38% higher system productivity and 84% system profitability over CT-based RW system (11.78 Mg ha^{-1} and INR 1,00,400 ha^{-1}, respectively) in MWMb system on permanent beds. However, CSA-based MWMb system enhanced the system productivity by 28–30% and net returns by 37–41% on flat beds compared to their CT practices (Choudhary et al., 2018a). CA layered with SDI (subsurface drip irrigation) increased the RWMb system productivity and farm profitability by ~11 and 29% and for MWMb system 20 and 49%, respectively, compared to farmers' practice/conventional flood irrigated RW system (Jat et al., 2019d).

Zero-till direct seeded rice (ZT-DSR) has also shown to be an economically viable alternative to transplanted puddled rice (PTR) as it helped in reductions of production cost by 11–17% with 25–30% less irrigation water at similar yield levels (Kumar et al., 2018; Sharma et al., 2019). Compared to prevailing farmers' practice (FP), results suggest that CSA-based rice management improved the profitability by 13% while reducing irrigation requirement by 19% (Jat et al., 2021b). By substituting CSA-based maize for rice, similar profitability was realized (16%) while saving in irrigation (84%) compared to FP (Gathala et al., 2013; Jat et al., 2019d; Kumar et al., 2018). Inclusion of mungbean (green gram) in the rotation (i.e. maize-wheat-mungbean) for sustainable intensification increased the system productivity by 10–15% and system productivity by 15–23% in major agri-food systems (Choudhary et al., 2018a; Jat et al. 2018d, 2019d, 2020d). CSA-based sustainable intensification helped in mitigating the negative effects of terminal heat stress on wheat and increased the wheat yields by 0.7–0.8 Mg ha^{-1} than conventional- till (CT) wheat (5.0 Mg ha^{-1}) (Sharma et al., 2015). Inclusion of a short-duration leguminous crop (e.g., mungbean, urdbean, sesbania, cowpea) in RW system is found beneficial to improve the soil quality and breaks the monotony of cereal systems (Gathala et al., 2013; Jat et al., 2016c).

5.2 CSA and Resource (Water, Energy, and Nutrient) Use Efficiency

Adoption of CSA-based practices has shown a remarkable potential to grab the problems of declining groundwater, nutrient, water, and energy use efficiencies as well as other associated input management issues in RW/MW system. Results showed that CSA-based sustainable intensification saved irrigation water and energy and improved water productivity and weed control efficiency (Jat et al., 2019d, 2020a, 2020c, 2020d, 2021c). Zero-tillage direct-seeded rice (ZT-DSR) has been found as an economically viable and environmentally sustainable alternative to PTR as it saved 20–35% of irrigation water and 25–40% energy inputs. Kumar et al. (2019) reported that maize with CSA-based management appeared as a suitable diversification alternative to transplanted rice because it reduces the demand for irrigation water and energy use by 89 and 66%, respectively, compared with conventional PTR (2650 mm ha^{-1} and 79.2 GJ ha^{-1}, respectively). CSA-based rice-wheat-mungbean (RWMb) systems saved ~25% irrigation water and energy consumption while increasing the water productivity by 58% (0.74 kg grain m^{-3}) over CT-based RW system. CA-based maize-wheat-mungbean (MWMb) systems on flats improved the irrigation water productivity by 312% (1.94 kg grain m^{-3}) by using less irrigation water (71%) and energy input (47%) (42.3 GJ ha^{-1}) over conventional RW system (Choudhary et al., 2018a). CA-based MWMb on permanent beds (PBs) improved the irrigation water productivity by 245% (1.52 kg grain m^{-3} water) over conventional RW system (2005 mm ha^{-1}) and by 24.3% (1.13 kg grain m^{-3} water) over MW system (Choudhary et al., 2018d). Maize planted on PBs with different residue mulches improved irrigation water productivity (WP_I) by 16–25% and saved 11–29% of irrigation water over conventional system (Jat et al., 2015, 2018b).

On a system mean basis, subsurface drip irrigation (SDI) saved 47 (93.6 cm yr^{-1}) and 45% (29 cm yr^{-1}) irrigation water under rice-wheat (RW) and maize-wheat (MW) system integrated with mungbean, respectively, compared to flood irrigation system under the same management level in silty loam soils of Haryana (Jat et al., 2018d). However, CSA-based practices layered with SDI in sandy loam soils of Punjab saved the irrigation water by 48–53% in rice and 42–53% in wheat compared to flood irrigation system (Sidhu et al., 2019). The on-farm irrigation efficiency of a properly designed and managed drip irrigation system can be as high as 90%, compared with 35–40% efficiency in surface method of irrigation, and in semi-arid areas, saving of water usages with micro-irrigation has been reported to be 63.15% as compared to conventional methods of flood irrigation (Grewal et al., 2021). CSA practices in RW system increased the total water and energy productivity by 31 and 53%, respectively, and saved 24% irrigation water (Kakraliya et al., 2018b, 2019). CSA-based RW systems enhanced the crop productivity by 10%, water productivity by 56%, and profitability by 34%, while saving in irrigation water by 38%, compared with their respective CT-based systems. However, CSA-based MW systems enhanced the crop productivity by 16%, water productivity by 33%, and profitability by 36%, while saving in irrigation water by 32%, compared with their respective CT-based systems (Choudhary et al., 2018a; Jat et al., 2019c, 2019d, 2020).

Site-specific nutrient management (SSNM) is an approach to 'feed' the crops with the nutrients when they are actually needed (Jat et al., 2016a; Parihar et al., 2020). The SSNM method is not intended to reduce or increase the use of fertilizers. Instead, it aims to recommend nutritional supplements the right source in right amount at the right times and at the right place to achieve maximum benefits for farmers. Adoption of better practices is required to prevent the leakage of excess nutrients to the soil and environment for achieving high efficiency of nutrient in both spatial and temporal scale. The basic steps to be followed in SSNM are: (a) set an attainable target yield, (b) estimate the indigenous supply of nutrients, (c) estimate crop response to nutrients, and (d) estimate the nutrient application rate (Peng et al., 2010). Peng et al. (2010) also reported that SSNM reduced N-fertilizer by 32% and increased rice grain yield by 5% as compared to farmers' N management practices. Similarly, Bijay-Singh et al. (2015) conducted 7 field experiments of SSNM in IGP and obtained a similar yield of rice as compared to the blanket fertilizer application, but with reduced N application. The SSNM enhances the N recovery efficiency and agronomic efficiency. Khurana et al. (2008) reported higher grain yield of wheat and rice with lower application of N in the SSNM system as compared to farmers' fertilization practice. Parihar et al. (2020) also reported that SSNM in the ZT-system of maize-wheat-mungbean rotation resulted in better crop productivity, soil quality, and C-sequestration than unfertilized (control) and conventional farmers' practices, and recommended fertilizer doses (RDF). Jat et al. (2018a) reported 30 and 50% savings in N and K, respectively, in wheat crop under CA-based system as compared to conventional system (Table 3). Further, CA-based systems had higher total soil nitrogen and available nitrogen than conventional practice.

The application of the right source of N at the right time and rate in the right place is best accomplished with right tools. Several tools and techniques are available to help in making decisions regarding N management (Table 4). These tools increase the ability to fine-tune the N management decision and to develop SSNM plan for every field. Various tools of N management and their principal results in CA-based systems are given in the following sections.

CSA-based nutrient management technologies (e.g., fertilizer sources, fertigation) and CA technologies are needed to adapt to climate change and to increase the efficiency of nutrients. Balanced fertilization with conjoint use of nutrient expert® (NE) as decision support tool accommodating the SSNM strategy, newer concept of fertilizer application, and variable rate application are important for improving nutrient-use efficiency. The application of nitrogen through SDI reduced the fertilizer N requirement by 20% (30 kg N ha^{-1}) under each crop of rice, wheat, and maize (Jat et al., 2019d). On a system basis, RW and MW systems recorded 45 and 50% higher PFP_N with SDI compared to CT flood (Jat et al., 2019d; Sidhu et al., 2019; Kakraliya et al., 2021). Overall, NE-based recommendations reduced nitrogen dose by 15–35%, enhanced crop yield by 4–8%, and reduced global warming potential by 2–20% in wheat crop (Sapkota et al., 2021). NE-based fertilizer management reduced N application in rice and wheat by 5–25%. NE also increased rice and wheat yield by

Table 3 Effect of conservation agriculture-based practices in rice-wheat and maize-wheat based system on grain yield of wheat, total N, and available N content in soil

Treatment	Grain yield of wheat (Mg ha^{-1})				Total N (%)	Available N (kg ha^{-1})
	160 kg N ha^{-1}	112 kg N ha^{-1}	60 kg K ha^{-1}	30 kg K ha^{-1}		
Conventional rice-wheat system	5.33^{A}	4.63^{B}	5.00^{A}	4.52^{B}	0.14 ± 0.02^{B}	117 ± 1.52^{D}
Partial CA-based rice-wheat-mungbean system	5.23^{B}	5.37^{AB}	5.25^{A}	5.10^{B}	0.18 ± 0.01^{A}	132 ± 0.96^{C}
CA-based rice-wheat-mungbean system	4.99^{BC}	5.32^{AB}	5.01^{A}	5.06^{A}	0.19 ± 0.02^{A}	156 ± 2.29^{B}
CA-based maize-wheat-mungbean system	5.30^{AB}	5.16^{B}	5.35^{A}	5.40^{A}	0.19 ± 0.02^{A}	197 ± 2.40^{A}

Mungbean is synonym for green gram. The same letters superscripted in a column show that the data are statistically not significantly different from each other.
Source Jat et al. (2018a)

Table 4 Tools and techniques for precision N management in context of CA system

Serial number	Tools/techniques	Principle and functioning	Applications in N management
1	Chlorophyll/soil plant analysis development (SPAD) meter	Used for non-destructive determination of leaf chlorophyll content	Leaf greenness measurement by SPAD meter can signal N deficiency early enough without causing any yield loss (Jat et al., 2016a)
2	Leaf color chart (LCC)	LCC is a tool for rapid assessment of leaf N status and used for the determination of the need of N fertilizer	LCC reduced N requirement of rice-wheat system by 12–25% (Bijay-Singh et al., 2002)
3	Green seeker (GS)	It is a tool for measurement of crop vigor based on NDVI as the basis for N prescription rates	Green Seeker sensor-based N application enhanced the NUE and wheat yield with lower N application N use efficiency (Bijay-Singh et al., 2015; Sapkota et al., 2019)
4	Nutrient expert (NE)	It is a computer-based decision support tool that can quickly provide nutrient recommendation for an individual field both in absence and presence of soil testing data	SSNM using NE in maize-wheat system improved crop productivity and net returns by 13.4% and 15.3%, respectively, as compared to farmer's fertilizer practices (Jat et al., 2016a) NE-based N management resulted in higher maize yield (Sapkota et al., 2021; Satyanarayana et al., 2012), better NUE and crop yield in CA-based maize-wheat-mungbean system (Parihar et al., 2020) and better partial factor productivity of N (Sapkota et al., 2014)
5	Mechanization for precision N application	Turbo-happy seeder a mechanical tool for placement of N at right place	Proper placement of urea with seed drill resulted in 10–15 higher NUE in RW system as compared to broadcasting of urea (Jat et al., 2016a; Parihar et al., 2020). It also resulted in higher crop yield and partial factor productivity of nutrients (Sapkota et al., 2019, 2020)
6	Fertigation with Mini-sprinkler irrigation system	Application of water-soluble N fertilizers through mini-sprinkler	Fertigation with mini-sprinklers in rice and wheat crop saved 26 and 50% N as compared to RDF (Singh et al., 2020)
7	Fertilizer with subsurface drip irrigation (SDI)	Application of urea solution through SDI	Fertigation saved 20% N and improved the PFP_N by 40–50% in rice/maize-based systems (Jat et al., 2019d; Kakraliya et al., 2021; Sidhu et al., 2019)

2–2.5% in western IGP and reduced the global warming potential (GWP) by about 2.5 and 15% in rice and wheat, respectively. GreenSeeker increased the RW system productivity and profitability to the tune of 5–8% (Parihar et al., 2020).

In CA systems, plant residues have contributed significantly (~76%) to total energy input (167,995 MJ ha^{-1}); however, fertilizers (unavailable energy source) contributed significantly (43%) to total energy expenditure (47,760 MJ ha^{-1}) in CT-based systems. CA-based cropping system (rice/maize based) recorded higher energy and energy intensiveness (EI) levels by 251 and 300%, respectively, compared to CT-based rice-wheat system (Choudhary et al., 2018a; Jat et al., 2020a, 2020c, 2020d). The herbicide consumption in wheat is negligible after 3 years of continuous cultivation with CA-based management system in RWMb system (Sharma et al., 2015). A residue load of 6 Mg ha^{-1} with early sowing (last week of October) was optimized to reduce the weed *Phalaris minor* population by 75%. The most abundant weed species CT-based wheat systems were in order of *Phalaris minor* > *Anagallis arvensis* > *Melilotus indica* > *Chenopodium album.* Species like *Cronopus didyma, Spergula arvensis, Cyperus rotundus, Chenopodium murale, Rumex dentatus, Fumaria indica, Oxalis arvensis, Vernonia anagallia-aquatica,* etc. are also common weeds of wheat fields, particularly along water channels. Weeds in CT-based systems are suppressed by CSAPs in the long term, as evidenced from the results reported by Jat et al. (2021c). The lowest total weed density was recorded under the full CSA-based maize systems, where seven weed species disappeared, and the density of *P. minor* decreased to 0.33 instead of 25.93 plant m^{-2} after 8 years of continued planting. A complete CSA-based maize system could be another promising alternative to a well-selected wheat rice system in weed control in northwestern India (Jat et al., 2021c).

5.3 CSA and Soil Quality

Soil organic carbon is an important determinant of soil quality, playing a crucial role in food production, mitigation and adaptation to climate change as well as performs many ecosystem functions. Intensive tillage coupled with crop residue burning in rice-wheat (RW) system is a serious issue in IGP that causes soil degradation and environmental pollution. In post-Green Revolution era, intensive cropping has resulted in soil degradation mainly through the decline in soil organic carbon from 0.75 to 0.40%. Deteriorating soil health coupled with multi-nutrient deficiencies is a major problem in NW-IGP. Converging of improved climate-smart management practices helped improve the soil organic carbon, soil physico-chemical properties, and overall soil fertility in the region (Jat et al., 2019a, 2019b). CSA utilizing the principles of minimum soil disturbance, crop residue management, and suitable crop rotation along with mitigation co-benefits is getting popular among farmers. Zero-tillage has been widely recommended as an alternative for improving carbon sequestration in soil under different ecologies. Various studies across IGP showed the benefits of smart-management practices on soil quality parameters, especially on soil organic carbon (SOC). Soil organic carbon was doubled in rice/maize-based cropping systems after

5 years on continuous cultivation in IGP (Jat et al., 2019a, 2019b; Sharma et al., 2015). Results showed that CA-based management improved soil properties and availability of essential nutrients (N, P, K, Ca, Mg, S, Zn, Fe, and Mn) to crop plants which helped cut down the nutrient doses (N and K). CSA-based management practices improved the soil physical conditions, which resulted in higher nutrient availability to crop plants.

In CSA-based systems, crop residue management in RW system improved overall soil health (Choudhary et al., 2018b, 2018c; Jat et al., 2018a). It improved soil physical (BD, infiltration rate, aggregate stability), chemical (OC, all nutrients), and biological properties (microbial population and enzymatic activity). The bulk density of soil was improved by ~5% with residue incorporation compared to residue retention (Jat et al., 2018a). The infiltration rate was doubled with residue retention as compared to rice residue incorporation (0.15 cm hr^{-1}) in rice-wheat system (Patra et al., 2019; Roy et al., 2021). At 0–10 cm soil depth, the highest value of soil penetration resistance (SPR) was recorded with conventional tillage (1.35 MPa) at 5–10 cm, and the lowest with residue incorporation (0.46 MPa) at 0–5 cm depth (Abdallah et al., 2021). Beyond 20–25 cm lower SPR was recorded with residue retention as compared to residue incorporation (Jat et al., 2019a, 2019b). Higher near-saturated hydraulic conductivity was observed with residue retention due to better formation and continuity of macropores with larger size and numbers as compared with residue incorporation, which showed a transitional effect with respect to the different soil hydraulic characteristics (Patra et al., 2019). Highest aggregate associated C (8.94 g kg^{-1}) was retained in the 0.5–1 mm size class among all the size classes of aggregates (Jat et al., 2019a, 2019b). The higher infiltration rate in direct seeded rice (DSR) favors the groundwater recharge due to faster percolation rate water which reduces the possibility of runoff (Kumar et al., 2018). Crop diversification of rice-wheat system with other remunerative system and inclusion of leguminous crop (mungbean, berseem) improved the soil physico-chemical properties significantly (Srivastava et al., 2016).

In a laboratory experiment conducted to study the C mineralization with rice residue retention and incorporation, it was found that soils with residue incorporation observed the higher (~12%) cumulative C mineralization than placed on the soil surface (Datta et al., 2019). Soil organic carbon was increased by >75% after 6 years of cultivation under both residue retention and incorporation conditions from the initial values (0.45). However, the available nitrogen, phosphorus, and potassium availability were also increased by 40–50% with both residue retention and incorporation (Jat et al., 2019a, 2019b). With residue incorporation, SOC and available NPK were found higher in sub-surface layer (15–30 cm). Overall improvement in nutrient availability is more or less the same across the soil profile under both the residue placement (Choudhary et al., 2020; Jat et al., 2018a). Under the CSA-based management system with both rice residue incorporation and retention, 30 and 50% of N and K fertilizers can be saved after 4–5 years of continuous cultivation (Jat et al., 2018a). With residue retention, 10–15% higher earthworm population was recorded compared to residue incorporation during rice season (Choudhary et al., 2018c). In CA-based RWMb systems, available N, P, and K was increased by 33, 38, and 18%, respectively, however in CA-based MWMb systems, it was increased by 68, 25, and 73%, respectively, compared to conventional RW system (117:15.7:183.4 kg NPK ha^{-1}) after 4 years (Jat et al., 2018a).

The microbial population varies with the different management practices, soil conditions, soil moisture, soil temperature, etc. In general, higher microbial population was recorded under CA-based management practices with either residue incorporated or retained. With residue retention, higher (15–20%) fungi population was recorded and there was no significant effect on bacteria and actinomycetes population (Choudhary et al., 2018b, 2018c; Jat et al., 2019b, 2020b, 2021a). Efficient lignocellulosic crop residue degrading fungi belonging to genera *Aspergillus* and *Alternaria* that were able to degrade crop residue by 26–31% (Choudhary et al., 2016a). CA-based MW system registered 208, 263, 210, and 48% improvement in soil microbial biomass C (MBC) and N (MBN), dehydrogenase activity (DHA) and alkaline phosphatase activity (APA), whereas CA-based RW system registered 83, 81, 44, and 13%, respectively, compared with farmers' practice of RW system (MBC-646 μgg^{-1} dry soil; MBN-201μgg^{-1} dry soil; DHA-180 μg TPF g^{-1}soil 24 h^{-1}; APA-144 μg p-NP g^{-1} h^{-1}) after 3 years (Choudhary et al., 2018d, 2018e, 2018f). At phylum level, relative abundance of *Ascomycota* fungi was ranged from 55 to 74%, with the highest dominance found in MWMb (74%), followed by RWMb (71%), and the lowest with conventional RW system (55%). However, at class level, *Sordariomycetes* had the highest abundance followed by *Dothideomycetes* and *Eurotiomycetes* (Choudhary et al., 2018d, 2018f). Under bacterial diversity, *Proteobacteria, Acidobacteria, Actinobacteria,* and *Bacteroidetes* accounted for more than 70% of the identified phyla at phylum level. Rice and maize-based systems were dominated by phylum *Proteobacteria* and *Acidobacteria,* respectively. At class level, *Proteobacteria* was dominated by *Alphaproteo* bacteria, and it was closely followed by *Deltaproteobacteria, Betaproteobacteria,* and *Gammaproteobacteria* (Choudhary et al., 2018d, 2018e). Microbial population was increased by 28%, 68%, and 98%, respectively, of bacteria, fungi, and actinomycetes under CA-based MW management system than RW system (Choudhary et al., 2018b, 2018c). CA-based MW system recorded the highest (1.45) SQI and closely followed by RW system (0.58), and the lowest score (0.29) with conventional RW system after 3 years of cultivation. Mungbean integration in MW system recorded higher SQI (0.76) than other RW systems (Choudhary et al., 2018b, 2018c). The CSA practices influenced enzyme activities significantly and helped in more mobilization of nutrient for higher plant growth and development (Choudhary et al., 2020; Datta et al., 2021; Jat et al., 2020b, 2021a). Improved CSA-based management practices also help improve the availability of major nutrients like nitrogen and phosphorus by ~20% (Haokip et al., 2020; Tigga et al., 2020).

Global warming and the resulting climate change are threatening almost all ecosystems on the planet. The main cause of global warming is the continuous increase in the atmospheric concentration of GHGs due to various human activities. Therefore, reducing the increased concentration of GHGs is necessary to slow global warming and climate change. Among the various greenhouse mitigation options, soil biochar application is gaining popularity due to several advantages over other options (Gupta et al., 2020). Biochar is a highly stable form of carbon derived from the pyrolysis of biomass at relatively low temperatures. Applying biochar in soil has been reported to provide multiple benefits such as increased crop yields, efficient use of nutrients and water, and several environmental benefits. The recalcitrant nature, relatively higher carbon content, and readily available raw materials make biochar a highly sustainable and rapid option for carbon sequestration in soil. The application of biochar to the soil not only contributes to carbon sequestration but also offers a better option for the management of agricultural residues. It has also been reported that the application of biochar reduces a significant amount of methane and nitrous oxide emissions from agricultural fields due to its priming effect on the soil. Biochar yield, physical properties, and carbon content vary depending on the type of raw material and the pyrolysis conditions. The application of biochar ultimately leads to a positive contribution to climate change mitigation. It typically leaves 30–35% recalcitrant carbon and lignin which can remain in the soil for hundreds of years and sequester carbon. Most of the nutritional values of biochar are low, but it improves the physicochemical properties of the soil, increases CEC and moisture retention which ensure efficient use of nutrients and microbes (Alotaibi & Shoenau, 2019; Bertola et al., 2019). However, inefficient roasting can also increase the GHG load and global warming potential. According to the latest roasting technologies, lignin has become an important source of many industrial chemicals and, if exploited, can reduce stubborn residues by up to 510% compared to 40% for digestate or manure. The use of lignin will significantly reduce the potential sequestration.

5.4 CSA and GHG Emissions

Conventional management practices of RW system in NW India involve intensive tillage and favor open-field burning of crop residues resulting in loss of lives, soil nutrient depletion, respiratory diseases in human beings and environmental pollution through GHG emissions (Lohan et al., 2018). More than 30 million tons of cereal crop residues are burnt in NW India annually. Labor scarcity and escalating fuel costs are multiplying the overall cost of cultivation, making conventional cereal-based agriculture a less profitable enterprise compared to multi-enterprise systems (Yadav et al., 2021). Cultivation of high-yielding varieties of rice and wheat results in production of large quantities of crop residues in this region. Combined harvesting of rice and wheat predominates in this region, with large amounts of loose and anchored crop residues left in the fields after harvest (Groot et al., 2019). The majority of farmers currently choose to burn rice straw, plow fields, and sow wheat using conventional seeders. Rice residue burning in NW India contributes to air pollution, health hazards, disruption of transportation, school closures, and soil degradation. Significant amount (25 million metric tons) of rice residues is being burnt leading to a huge loss of C, N, P, and S and the destruction of beneficial microflora in RW cropping system (~4.1 million ha) to clear the field for wheat sowing in the available narrow window (~10 to 20 days) between rice harvesting and wheat sowing.

Business as usual in RW system not only requires intensive use of resource (labor, water, and energy) but also emits significant amounts of GHGs (CO_2, CH_4, and N_2O) depending on the conditions (anaerobic/aerobic) of the soil (Shirsath et al., 2020). CA-based systems have the potential to reduce environmental footprint by using less irrigation water, and energy and help reduce the total global warming potential while increasing the productivity and profitability (Sapkota et al., 2019, 2020, 2021). The GHG emission with residue incorporation and retention was 1063 and 1045 kg CO_2 eq ha^{-1}, respectively (Shyamsundar et al., 2019). When conventional practices along with residue burning were compared to ZT wheat with residue retention, it reduced GHG emissions by ~80% compared to those of farmers' common practice of burn+disc (4757 CO_2 eq kg ha^{-1} $year^{-1}$). Shifting from CT-based tillage and crop establishment in rice to conservation practice (CT-PTR to ZT-DSR) reduced the CH_4 emissions by ~56% (Tirol Padre et al., 2016). Kumar et al. (2018) also showed that zero-till DSR reduced the global warming potential (GWP) by 32% compared to conventional PTR (4713 kg CO_2 eq ha^{-1}). CA-based RW systems have the potential to reduce GWP by 15–30% (1.3–2.0 Tg CO_2 eq yr^{-1}) compared with business as usual/farmers' practice (Sapkota et al., 2017). CSA lowered GWP and greenhouse gas intensity by 40 and 44% respectively, compared to the farmers' practice (7653 kg CO_2 eq ha^{-1} yr^{-1} and 0.64 kg kg^{-1} CO_2 eq ha^{-1} yr^{-1}) (Kakraliya et al., 2018a, 2018b).

6 CSA and Improved Resilient Cropping System Under Location-Specific Climate Change Scenario

As South-East Asia is particularly vulnerable to climate change, the identification of location-specific improved resilient cropping system is very important for the sustainability of the production. Considering the specific needs of various regions under a changing climatic situation and socio-economic status, AICRP on Integrated Farming system and AICRP for dryland agriculture is engaged in identifying this type of cropping system. Various criteria like replacing existing more susceptible crop by climate-resilient crops suitable for that region, short duration cultivars, heat stress resisting varieties, photoperiod insensitive cultivars, cultivars with high tillering ability and rooting ability, more efficient photosynthesis ability with less photo respiration, various abiotic and biotic stress resistance, etc. can be considered to decide the cropping system considering location specific needs and socio-economic status. For example, CRIDA identified, potential rainfed cropping system for varying range of rainfalls, soil, agricultural drought situation based on length of growing period (LGP).

LGP	Crops and cropping system suggested
<75 days	Monocropping of short duration crops like pulses and perennial vegetation

(continued)

(continued)

LGP	Crops and cropping system suggested
75–140 days	Monocropping of short duration crops like pulses, pearl millet, castor, sorghum, sesame or monocropping during post rainy season with safflower, chickpea, etc
140–180 days	Intercropping
>180 days	Double cropping with winter sorghum, chickpea, safflower, barley, lentil, mustard

Source Velayutham (1999)

Similarly, in irrigated agro-ecosystem also, some efficient alternative cropping systems for different regions have been identified by Gangwar and Singh (2011).

Agro-climatic zone/region	Existing cropping system	Alternative cropping system
North-western plains of Uttarakhand	Rice-wheat	Rice-potato-green gram Rice-wheat-green gram
Eastern Himalayan region (Asom)	Rice-rice	Rice-toria-cowpea Rice-oat-green gram
Agroclimatic zone of Bihar	Rice-wheat	Rice-wheat-green gram Rice-potato-oat
Eastern plain zone of Rajasthan	Pearl millet-wheat	Cluster bean-onion
Northern Telengana	Rice-rice	Soybean-sunflower

Source Compiled from Gangwar and Singh (2011)

Thus, adaptation of climate resilient and efficient cropping systems can be an option for tackling the impact of climate change and ensure the livelihood security of farmers.

7 Harnessing Genotype × Environment × Input Management Interactions

In South Asia, the adoption of suitable management practices for attaining maximum yield potential of newly developed varieties under diverse ecological condition has massive potential for future food and nutritional security of the country by breaking the yield barrier. In our decadal studies, we found that maize hybrids respond differently to nutrient management practices and had a significant effect on grain yield, which varied across the agro-ecologies (Jat et al., 2017, 2018e). With SSNM-based nutrient management, maize hybrid PMH 3 produced higher grain yield which was statistically equivalent to S 6217, but significantly higher compared to other treatments. Hybrid S 6217 gave significantly higher grain yield over all other hybrids and absolute control (no nutrient application). Further, at 100% RDF, hybrid CMH 08–292 produced much higher grain yield as compared to all other hybrids (Kumar et al., 2015). The field experiments were conducted at 11 locations across the country to evaluate the response of five maize hybrids to site-specific nutrient management (SSNM) in contrast with blanket recommendations (RDF) and farmers' fertilization practices (FFP).

The SSNM doses for hybrid maize were calculated based on Nutrient Expert by considering the previous year's potential yield of the respective hybrids of that particular location. Grain yield enhancement to the tune of 8% (1.4–22.1%) was found with SSNM compared to RDF at 10 locations. The yield responses from FFP were highly erratic, while the SSNM resulted in yield enhancement of 29% (6.6 to 84%) at 10 locations. The highest percent yield enhancement by SSNM over RDF and FFP was recorded in Northern Hill Zone (13.1 and 69) while the lowest in Peninsular Zone (3.1 and 9.2), respectively. The nutrient prescription doses showed that change in SSNM over RDF and FFP was 11.7 and 26.3 kg N ha^{-1}, −0.8 and 18.6 kg P_2O_5 ha^{-1}, and 49.9 and 72.5 kg K_2O ha^{-1}, respectively. It shows that more K needs to be recommended in all zones with SSNM over RDF and FFP. Significant genotypes × nutrient management interactions were observed and PMH 3 and DHM 117 maize hybrids outperformed at most of the locations with SSNM. The net returns over fertilizer cost with SSNM were 6.9 and 21.9% higher over RDF and FFP, respectively. The SSNM-based nutrient application gave 6,263 and 19,792 rupees more net returns per hectare over fertilizer costs in maize cultivation compared to RDF and FFP, respectively.

CA is considered a promising solution for long-term sustainability of cereal-based cropping systems of South Asia. Traditionally, wheat genotypes are tailored under conventional tillage management system. However, genotypes bred under CT system may differ in their performance under upcoming CA system. Among different agronomic practices, the availability of precise genotypes of wheat and other crops adapted to CA is important for realizing the full potential. Directed crossing and continuous selection under CA environment in wheat led to the development of CA adapted genotypes, namely HDCSW 18 and HD 3117, with yield gains of 9.7 and 9.4% over the respective best checks have been released for early sown and late sown condition respectively, in India (Yadav et al., 2018). Twenty-one different genotypes were tested over three years (2012–2014) under different days of sowing and cultivation in three agro-ecological locations (Bihar, Madhya Pradesh, and Punjab) of India. The results showed that the effect of sowing dates on yield stability was spread unevenly across all management practices (Jat et al., 2017, 2018e). The genotypes 'CSW 18' (G03), 'DPW 621–50' (G05), 'BAZ' (G01) were the best and most stable cultivars in the early, normal and later sowing days, respectively across the locations. Studies suggest that genotypes that slow down during the early days of planting tend to be a higher producer during the late sowing days. The genotypes 'Munal' and 'HD-2967' were the best producers and most stable with CT and ZT, respectively across the locations. The genotypes 'HD-2824' and 'DPW-621-50', and 'Munal' can serve as a variety for diverse parents to improve high quality, climate-smart, locally adapted genotypes suited for Bihar under both CT and ZT practices. Genotypes 'HD-2932', 'BAZ' and 'JW-3288', and 'GW-322' and 'HD-2967' are suitable genotypes for developing locally adapted stress-tolerant under both CT and

ZT. These studies suggest the need to integrate genotype × sowing dates × tillage management system.

8 Agroforestry Interventions

Agroforestry is now considered an effective land-use system, which contributes to food, nutritional, and environmental security, and has a great potential to restore degraded landscapes (Dagar et al., 2020). To meet the socio-economic needs of the ever-increasing human population as well as mitigate and adapt to the prevailing climate change, there is a need to restore or rehabilitate all kinds of degraded lands. Agricultural systems need fertile land for optimum production, but agroforestry systems have unique characteristics that provide nutritional and livelihood security besides providing food, fodder, fuel, timber, and several other products from the degraded lands. Agroforestry, as subject of scientific investigation, assumes wider recognition in view of the need to maximize biological productivity based on sustainable land management.

During the past four decades, agroforestry has come of age and begun to attract the attention of the international scientific community, primarily as a means for sustaining the productivity of degraded lands besides conserving biodiversity and enhancing environmental services. Research efforts have shown that most of the degraded areas prone to soil erosion, salinization, waterlogging and mining, and coastal mangrove areas can be made productive by adopting suitable agroforestry practices and techniques (Dagar & Minhas, 2016; Dagar & Singh, 2018; Dagar & Tewari, 2016, 2017; Dagar et al., 2014, 2020). The possibilities are developing plantation-based cropping systems, using fertilizer trees in dry lands and integrating high-value medicinal and aromatic plants, livestock, poultry, forest and fruit trees, and vegetables in agroforestry. The faulty methods of canal irrigation have led to second-generation problems of waterlogging and salinity. The application of bio-drainage technology has proved that farmers can get two to three times yield of RW system besides getting additional income from wood products and sequestering carbon both above and below-ground biomass and other environmental benefits (Dagar et al., 2016a, 2016b). Thus, agroforestry practices and technologies and their service functions can have a focus on biomass production, sustainability indicators, carbon sequestration, soil fertility improvement, biodiversity conservation, ecosystem services, and climate change mitigation. To implement the agroforestry as a practicable tool for restoring degraded landscapes and CSA, we need appropriate policies, strategies, plans, programs, and viable long-term projects.

9 Decision Support System (DSS)-Based Modeling on Climate Change and Future Crop Production

DSS-based crop modeling is an important tool for predicting climate change and related anomalies and plan accordingly so that the sustainability of production system can be maintained. For example, by using InfoCrop models, the regional impact of climate change and its vulnerability on irrigated wheat were assessed. It was projected that climate change will reduce late sown and very late sown irrigated wheat production by 23% by 2050 and 25% by 2080 scenario, if no adaptation measures are followed. However, by adopting some climate-smart technologies like sowing improved variety and efficient input management practices, this negative impact can be minimized along with marginal yield improvement by 2050, while in 2080, the wheat yield will decline by 6% in spite of the above adaptation strategy. Similarly, in climate change scenario of MIRO and PRECIS, the irrigated rice yields are projected to reduce by 7% by 2050 and 10% by 2080. In case of rainfed rice, this decrease seems to be marginal (<2.5%). The predicted yield loss will be more by 2050 in north-west India (15–17%), while some parts of central India (Maharastra and Madhya Pradesh) also projected to face >5% of yield loss.

By adopting climate-smart technologies, and additional 25% nitrogen application can not only offset the climate change impacts but also can improve the production by 6–17% in irrigated condition and 20–35% in rainfed condition. DSS-based simulation study is also used to predict the capacity of systems technology resiliency. Various simulation models like CESAR (Carbon Emission and Sequestration by Agricultural land use (Vleeshouwers & Verhagen, 2002), CAMF (Carbon Accounting Model for Forest) (Brack & Richards, 2002), etc. can be used to predict the carbon sequestration potential of soil system and helps in carbon budgeting. Simulation modeling-based study found that carbon sequestration potential of agro-forestry practices in middle Gangetic plains ranged between 28 and 38 Mg C ha^{-1} in the next 24 years (Naresh Kumar et al., 2012). Thus, DSS-based modeling can be used in assessing the system's resiliency as well.

10 Conclusions

Climate change may threaten food security and the health of millions of people around the world. Rising temperatures, increasing climate variability, and extreme weather events could have a significant impact on food grain production for decades to come. Associated warming habits along with frequent and intense climatic events have been observed in the developing world in recent decades. Climate events such as a cold wave, hot summers, endless heat, droughts, unusual rains, and floods have shown significant weather forces to influence the production of food crops. Therefore, there is a need to use modern science combined with the traditional wisdom of farmers to increase the climate resilience of modern agriculture. Overcoming the dreaded problems of the cereal system, a sustainable CSA-based approach has emerged as another important means of achieving the goals of productivity growth, job creation, and the sustainability of natural resources in South Asia. CSA-based agro-technology package, robust planting system, and holistic farming not only save natural resources but also can help produce more at a lower cost, improve soil health, promote timely planting, ensure crop diversification, and reduce environmental pollution and the adverse effects of climate change. With limited scope of continued agricultural expansion, agricultural productivity benefits can be achieved through CSA-based agricultural intensification through one or more strategies that include: (i) increasing yields per unit area; (ii) increasing cropping intensity per unit of land; and (iii) changing land use.

Rice-wheat cropping system, which is still the backbone of food security, is now facing major challenges due to rising production costs and environmental losses. With the development of efficient and diverse farming systems, CSA management can help improve the flexibility of the cropping system and reduce the negative impact of climate change. Technology savvy practices and alternative land uses and other carbon capture biological systems can also help in both adaptation and mitigation while maintaining food security. Accurate and reliable weather forecasting of environmental change will be used extensively with supportive policies for the dissemination of such information that are needed to assist farmers in a significant way. There is a need to develop a comprehensive adaptation strategy to reduce the negative impact of climate change. Policy decisions to promote a smart agricultural portfolio of interventions, capacity building for climate forecasting, risk management processes, and familiarity with climate-resilient technologies must be made at the local and regional level. CSA is seen as a means of managing resources, increasing farm productivity and profitability, and improving soil health.

References

Abdallah, A. M., Jat, H. S., Choudhary, M., Abdelaty, E. F., Sharma, P. C., & Jat, M. L. (2021). Conservation agriculture effects on soil water holding capacity and water-saving varied with management practices and agroecological conditions: A review. *Agronomy, 11*(9), 1681.

Alotaibi, K. D., & Shoenau, J. J. (2019). Addition of bio char to a sandy desert soil: Effect on crop growth, water retention and selected properties. *Agronomy, 9*, 327.

Aryal, J. P., Jat, M. L., Sapkota, T. B., Rai, M., Jat, H. S., Sharma, P. C., & Stirling, C. (2020). Learning adaptation to climate change from past climate extremes: Evidence from recent climate extremes in Haryana, India. *International Journal of Climate Change Stratergies Management, 12*(1), 128–146.

Aryal, J. P., Sapkota, T. B., Stirling, C. M., Jat, M. L., Jat, H. S., Rai, M., Mittal, S., & Sutaliya, J. M. (2016). Conservation agriculture-based wheat production better copes with extreme climate events than conventional tillage-based systems: A case of untimely excess rainfall in Haryana, India. *Agriculture, Ecosystems & Environment, 233*, 325–335.

Barclay, A. (2006). The direct approach: A return to the ways of their forefathers has seen Indian and Bangladeshi rice farmers reduce their cost for water and address the growing problem of labor shortages. *Rice Today, 5*(2), 12–18.

Bertola, M., Mattarozzi, M., Sanangelantoni, A. M., Careri, M., & Visioli, G. (2019). PGPB colonizing three-year biochar-amended soil: Towards biochar-mediated biofertilization. *Journal of Soil Science and Plant Nutrition, 19*(4), 841–850.

Bijay-Singh, V.-S., Purba, J., Sharma, R. K., Jat, M. L., Yadvinder-Singh, T. H. S., Gupta, R. K., Chaudhary, O. P., Chandna, P., Khurana, H. S., Kumar, A., Jagmohan-Singh, U. H. S., Uppal, R. K., Vashistha, M., & Gupta, R. (2015). Site-specific fertilizer nitrogen management in irrigated transplanted rice (*Oryza sativa*) using an optical sensor. *Precision Agriculture, 1*, 1–21.

Bijay-Singh, Y.-S., Ladha, J. K., Bronson, K. F., Balasubramanian, V., Singh, J., & Khind, C. S. (2002). Chlorophyll meter- and leaf color chart-based nitrogen management for rice and wheat in north-western India. *Agronomy Journal, 94*, 821–829.

Brack, C. L., & Richards, G. P. (2002). Soil organic matter quality interpreted thermodynamically. *Soil Biology and Biochemistry, 31*, 1889–1891.

Choudhary, M., Sharma, P. C., Jat, H. S., Nehra, V., McDonald, A. J., & Garg, N. (2016a). Crop residue degradation by fungi isolated from conservation agriculture fields under rice–wheat system of North-West India. *International Journal of Recycling Organic Waste in Agriculture, 5*(4), 349–360.

Choudhary, R., Singh, P., Sidhu, H. S., Nandal, D. P., Jat, H. S., & Jat, M. L. (2016b). Evaluation of tillage and crop establishment methods integrated with relay seeding of wheat and mungbean for sustainable intensification of cotton-wheat system in South Asia. *Field Crops Research, 199*, 31–41.

Choudhary, K. M., Jat, H. S., Nandal, D. P., Bishnoi, D. K., Sutaliya, J. M., Choudhary, M., Sharma, P. C., & Jat, M. L. (2018a). Evaluating alternatives to rice-wheat system in western Indo-Gangetic Plains: Crop yields, water productivity and economic profitability. *Field Crops Research, 218*, 1–10.

Choudhary, M., Datta, A., Jat, H. S., Yadav, A. K., Gathala, M. K., Sapkota, T. B., Das, A. K., Sharma, P. C., Jat, M. L., Singh, R., & Ladha, J. K. (2018b). Changes in soil biology under conservation agriculture based sustainable intensification of cereal systems in Indo-Gangetic Plains. *Geoderma, 313*, 193–204.

Choudhary, M., Jat, H. S., Datta, A., Yadav, A. K., Sapkota, T. B., Mondal, S., Meena, R. P., Sharma, P. C., & Jat, M. L. (2018c). Sustainable intensification influences soil quality, biota, and productivity in cereal-based agroecosystems. *Applied Soil Ecology, 126*, 189–198.

Choudhary, M., Jat, H. S., Garg, N., & Sharma, P. C. (2018d). Microbial diversity under conservation agriculture (CA)-based management scenarios in reclaimed salt-affected soils: Metagenomic approach. *Journal of Soil Salinity and Water Quality, 10*(1), 87–94.

Choudhary, M., Sharma, P. C., Jat, H. S., Dash, A., Rajashekar, B., McDonald, A. J., Jat, M. L. (2018e). Soil bacterial diversity under conservation agriculture-based cereal systems in Indo-Gangetic Plains. *3 Biotech, 8*(7), 1–11.

Choudhary, M., Sharma, P. C., Jat, H. S., McDonald, A., Jat, M. L., Choudhary, S., & Garg, N. (2018f). Soil biological properties and fungal diversity under conservation agriculture in Indo-Gangetic Plains of India. *Journal of Soil Science and Plant Nutrition, 18*(4), 1142–1156.

Choudhary, M., Jat, H. S., Datta, A., Sharma, P. C., Rajashekar, B., & Jat, M. L. (2020). Topsoil bacterial community changes and nutrient dynamics under cereal based climate-smart agri-food systems. *Frontiers in Microbiology, 11*, 1812.

Dagar, J. C., Singh, A. K., & Arunachalam, A. (Eds.) (2014). Agroforestry systems in India: Livelihood security and environmental services. In *Advances in agroforestry* (Vol. 10, p. 400). Springer.

Dagar, J. C., & Minhas, P. S. (Eds.). (2016). Agroforestry for the management of waterlogged saline soils and poor-quality waters. In *Advances in Agroforestry* (vol. 13, pp. 10–23). The Springer, New Delhi.

Dagar, J. C., & Tewari, J. C. (Eds.). (2016). *Agroforestry research developments* (p. 578). Nova Science Publishers Inc.

Dagar, J. C., Khajanchi-Lal, M.-K., Jeet-Ram, C. S. K., Yadav, R. K., Singh, G., Ahamad, S., & Kaur, A. (2016a). Impact of eucalyptus geometry on biomass production, water table drawdown,

carbon sequestration and inter-crop yield on waterlogged saline soils of North-West India. *Agriculture, Ecosystem & Environment, 233*, 33–42.
Dagar, J. C., Sharma, P. C., Chaudhari, S. K., Jat, H. S., & Ahamad, S. (2016b). Climate change vis-à-vis saline agriculture: Impact and adaptation strategies. In P. C. Sharma, D. K. Sharma, & A. K. Singh (Eds.), *Dagar JC* (pp. 5–53). Innovative saline agriculture.
Dagar, J. C., & Tewari, V. P. (2017). Agroforestry: Anecdotal to modern science (p. 879). Springer Nature Singapore Pte Ltd.
Dagar, J. C., & Singh, A. K. (Eds.) (2018). *Ravine lands: Greening for livelihood and environmental security* (p. 636). Springer Nature Singapore Pte Ltd.
Dagar, J. C., Gupta, S. R., & Teketay, D. (Eds.) (2020). *Agroforestry for degraded landscapes: Recent advances and emerging challenges* (vol. 1 and 2, pp. 1–554 and 1–475) Springer Nature Singapore, Pte Ltd.
Datta, A., Choudhary, M., Jat, H. S., Sharma, P. C., Kakraliya, S. K., Dixit, B., & Jat, M. L. (2021). Climate smart agriculture influences soil enzymes activity under cereal-based systems of North-West India. *Journal of the Indian Society of Soil Science, 69*(1), 86–95.
Datta, A., Jat, H. S., Yadav, A. K., Choudhary, M., Sharma, P. C., Rai, M., Singh, L. K., Majumder, S. P., Choudhary, V., & Jat, M. L. (2019). Carbon mineralization in soil as influenced by crop residue type and placement in an Alfisols of Northwest India. *Carbon Manage, 10*(1), 37–50.
FAO. (2017). The Future of Food and Agriculture: Trends and Challenges.
Gangwar, B., & Singh, A. K. (2011). Efficient alternative cropping system (p. 339). Project Directorate for Farming System Research, Modipuram, Meerut, Uttar Pradesh, India.
Gathala, M. K., Kumar, V., Sharma, P. C., Saharawat, Y. S., Jat, H. S., Singh, M., Kumar, A., Jat, M. L., Humphreys, E., Sharma, D. K., & Sharma, S. (2013). Optimizing intensive cereal-based cropping systems addressing current and future drivers of agricultural change in the northwestern Indo-Gangetic Plains of India. *Agriculture, Ecosystems & Environment, 177*, 85–97.
Grewal, S. S., Lohan, H. S., & Dagar, J. C. (2021). Micro-irrigation in drought and salinity prone areas of Haryana: Socio-economic impacts. *Journal of Soil Salinity and Water Quality, 13*(1), 94–108.
Groot, A. E., Bolt, J. S., Jat, H. S., Jat, M. L., Kumar, M., Agarwal, T., & Blok, V. (2019). Business models of SMEs as a mechanism for scaling climate smart technologies: The case of Punjab, India. *Journal of Cleaner Production, 210*, 1109–1119.
Gupta, D. K., Gupta, C. K., Dubey, R., Fagodiya, R. K., Sharma, G., Mohamed, N., Dev, R., & Shukla, A. K. (2020). Role of biochar in carbon sequestration and greenhouse gas mitigation. *Biochar applications in agriculture and environment management* (pp. 141–165). Springer.
Gupta, R. K., & Sayre, K. (2007). Conservation agriculture in South Asia. *Journal of Agricultural Science Cambridge, 145*, 207–214.
Gupta, R. K., & Seth, A. (2007). A review of resource conserving technologies for sustainable management of the rice–wheat cropping systems of the Indo-Gangetic Plains. *Crop Protect, 26*, 436–447.
Haokip, I. C., Dwivedi, B. S., Meena, M. C., Datta, S. P., Jat, H. S., Dey, A., & Tigga, P. (2020). Effect of conservation agriculture and nutrient management options on soil phosphorus fractions under maize-wheat cropping system. *Journal of the Indian Society of Soil Science, 68*(1), 45–53.
Hobbs, P. R., & Govaerts, B. (2010). How conservation agriculture can contribute to buffering climate change. In M. P. Reynolds (Ed.), *Climate change and crop production* (pp. 177–199). CAB International.
Jat, M. L., Saharawat, Y. S., & Gupta, R. (2011). Conservation agriculture in cereal systems of south Asia: Nutrient management perspectives. *Karnataka Journal of Agricultural Sciences, 24*(1), 100–105.
Jat, H. S., Singh, G., Singh, R., Choudhary, M., Jat, M. L., Gathala, M. K., & Sharma, D. K. (2015). Management influence on maize–wheat system performance, water productivity and soil biology. *Soil Use and Management, 31*(4), 534–543.

Jat, H. S., Jat, R. K., Singh, Y., Parihar, C. M., Jat, S. L., Tetarwal, J. P., Sidhu, H. S., & Jat, M. L. (2016a). Nitrogen management under conservation agriculture in cereal-based systems. *Indian Journal of Fertilisers, 12*(4), 76–91.

Jat, M. L., Dagar, J. C., Sapkota, T. B., Yadvinder-Singh, G. B., Ridaura, S. L., Saharawat, Y. S., Sharma, R. K., Tetarwal, J. P., Hobbs, H., & Stirling, C. (2016b). Climate change and agriculture: Adaptation strategies and mitigation opportunities for food security in South Asia and Latin America. *Advances in Agronomy, 137*, 1–115.

Jat, M. L., Jat, H. S., Jat, R. K., Tetarwal, J. P., Jat, S. L., Parihar, C. M., & Sidhu, H. S. (2016c). Conservation agriculture-based sustainable intensification of cereal systems for enhancing pulse production and attaining higher resource-efficiency in India. *Indian Journal of Argonomy, 61*, 182–198.

Jat, M. L., Jat, R. K., Singh, P., Jat, S. L., Sidhu, H. S., Jat, H. S., Bijarniya, D., Parihar, C. M., & Gupta, R. (2017). Predicting yield and stability analysis of wheat under different crop management systems across agro-ecosystems in India. *American Journal of Plant Sciences, 8*(8), 1977–2012.

Jat, H. S., Datta, A., Sharma, P. C., Kumar, V., Yadav, A. K., Choudhary, M., Choudhary, V., Gathala, M. K., Sharma, D. K., Jat, M. L., & Yaduvanshi, N. P. (2018a). Assessing soil properties and nutrient availability under conservation agriculture practices in a reclaimed sodic soil in cereal-based systems of North-West India. *Archives of Agronomy and Soil Science, 64*(4), 531–545.

Jat, H. S., Singh, G., Singh, R., & Choudhary, M. (2018b). Effect of planting and *ex-situ* mulching on growth, productivity, profitability and nutrient uptake in maize-wheat cropping sequence. *Journal of Soil Salinity and Water Quality, 10*(1), 95–103.

Jat, M. L., Stirling, C. M., Jat, H. S., Tetarwal, J. P., Jat, R. K., Singh, R., Lopez-Ridaura, S., & Shirsath, P. B. (2018c). Soil processes and wheat cropping under emerging climate change scenarios in South Asia. *Advances in Agronomy, 148*, 111–171.

Jat, R. D., Jat, H. S., Nanwal, R. K., Yadav, A. K., Bana, A., Choudhary, K. M., Kakraliya, S. K., Sutaliya, J. M., Sapkota, T. B., & Jat, M. L. (2018d). Conservation agriculture and precision nutrient management practices in maize-wheat system: Effects on crop and water productivity and economic profitability. *Field Crops Research, 222*, 111–120.

Jat, R. K., Singh, P., Jat, M. L., Dia, M., Sidhu, H. S., Jat, S. L., Bijarniya, D., Jat, H. S., Parihar, C. M., Kumar, U., & Ridaura, S. L. (2018e). Heat stress and yield stability of wheat genotypes under different sowing dates across agro-ecosystems in India. *Field Crops Research, 218*, 33–50.

Jat, H. S., Datta, A., Choudhary, M., Sharma, P. C., Yadav, A. K., Choudhary, V., Gathala, M. K., Jat, M. L., & McDonald, A. (2019a). Climate Smart Agriculture practices improve soil organic carbon pools, biological properties and crop productivity in cereal-based systems of North-West India. *CATENA, 181*, 104059.

Jat, H. S., Datta, A., Choudhary, M., Yadav, A. K., Choudhary, V., Sharma, P. C., Gathala, M. K., Jat, M. L., & McDonald, A. (2019b). Effects of tillage, crop establishment and diversification on soil organic carbon, aggregation, aggregate associated carbon and productivity in cereal systems of semi-arid Northwest India. *Soil and Tillage Research, 190*, 128–138.

Jat, H. S., Kumar, P., Sutaliya, J. M., Kumar, S., Choudhary, M., Singh, Y., & Jat, M. L. (2019c). Conservation agriculture based sustainable intensification of basmati rice-wheat system in North-West India. *Archives of Agronomy and Soil Science, 65*(10), 1370–1386.

Jat, H. S., Sharma, P. C., Datta, A., Choudhary, M., Kakraliya, S. K., Sidhu, H. S., Gerard, B., & Jat, M. L. (2019d). Re-designing irrigated intensive cereal systems through bundling precision agronomic innovations for transitioning towards agricultural sustainability in North-West India. *Science and Reports, 9*(1), 1–14.

Jat, H. S., Choudhary, K. M., Nandal, D. P., Yadav, A. K., Poonia, T., Singh, Y., Sharma, P. C., & Jat, M. L. (2020a). Conservation agriculture-based sustainable intensification of cereal systems leads to energy conservation, higher productivity and farm profitability. *Environmental Management, 65*(6), 774–786.

Jat, H. S., Choudhary, M., Datta, A., Yadav, A. K., Meena, M. D., Devi, R., Gathala, M. K., Jat, M. L., McDonald, A., & Sharma, P. C. (2020b). Temporal changes in soil microbial properties and nutrient dynamics under climate smart agriculture practices. *Soil and Tillage Research, 199*, 104595.

Jat, H. S., Jat, R. D., Nanwal, R. K., Lohan, S. K., Yadav, A. K., Poonia, T., Sharma, P. C., & Jat, M. L. (2020c). Energy use efficiency of crop residue management for sustainable energy and agriculture conservation in NW India. *Renewable Energy, 155*, 1372–1382.

Jat, H. S., Kumar, V., Datta, A., Choudhary, M., Kakraliya, S. K., Poonia, T., McDonald, A. J., Jat, M. L., & Sharma, P. C. (2020d). Designing profitable, resource use efficient and environmentally sound cereal based systems for the Western Indo-Gangetic plains. *Science and Reports, 10*(1), 1–16.

Jat, M. L., Jat, H. S., Agarwal, T., Bijarniya, D., Kakraliya, S. K., Choudhary, K. M., Kalvaniya, K. C., Gupta, N., Kumar, M., Singh, L. K., & Kumar, Y. (2020e). *A compendium of key climate smart agriculture practices in intensive cereal-based systems of South Asia* (p. 42). International Maize and Wheat Improvement Center (CIMMYT), New Delhi, India.

Jat, H. S., Datta, A., Choudhary, M., Sharma, P. C., Dixit, B., & Jat, M. L. (2021a). Soil enzymes activity: Effect of climate smart agriculture on rhizosphere and bulk soil under cereal based systems of north-west India. *European Journal of Soil Biology, 103*, 103292.

Jat, H. S., Datta, A., Choudhary, M., Sharma, P. C., & Jat, M. L. (2021b). Conservation Agriculture: Factors and drivers of adoption and scalable innovative practices in Indo-Gangetic plains of India–a review. *International Journal of Agricultural Sustainability, 19*(1), 40–55.

Jat, H. S., Kumar, V., Kakraliya, S. K., Abdallah, A. M., Datta, A., Choudhary, M., Gathala, M. K., McDonald, A. J., Jat, M. L., & Sharma, P. C. (2021c). Climate-smart agriculture practices influence weed density and diversity in cereal-based agri-food systems of western Indo-Gangetic plains. *Science and Reports, 11*(1), 1–15.

Kakraliya, S. K., Ishwar, S., Dadarwal, R. S., Singh, L. K., Jat, R. D., Jat, H. S., & Jat, M. L. (2018a). Impact of Climate-Smart agricultural practices on growth and crop yields of rice (*Oryza sativa*)-wheat (*Triticum aestivum*) system in north-western Indo-Gangetic plains. *The Indian Journal of Agricultural Sciences, 88*(10), 1543–1551.

Kakraliya, S. K., Jat, H. S., Singh, I., Sapkota, T. B., Singh, L. K., Sutaliya, J. M., Sharma, P. C., Jat, R. D., Choudhary, M., Lopez-Ridaura, S., & Jat, M. L. (2018b). Performance of portfolios of climate smart agriculture practices in a rice-wheat system of western Indo-Gangetic plains. *Agricultural Water Management, 202*, 122–133.

Kakraliya, S. K., Jat, H. S., Singh, I., & Jat, M. L. (2019). Effect of climate smart agriculture practices on crop yields and factor productivity of rice-wheat cropping system in Indo-Gangetic plains of India. *Indian Journal of Fertilisers, 15*(8), 852–858.

Kakraliya, S. K., Jat, H. S., Kakraliya, M., Gora, M. K., Choudhary, K. K., Parihar, C. M., Bijarniya, D., Jat, R. D., Sidhu, H. S., Sharma, P. C., & Jat, M. L. (2021). Subsurface drip fertigation: A milestone technology in sustainable crop production. *Indian Journal of Fertilisers, 17*(6), 582–589.

Khurana, H. S., Bijay-Singh, A. D., Phillips, S. B., & Sidhu, A. S. (2008). Site-specific nutrient management performance in a rice-wheat cropping system. *Better Crops, 92*(4), 26–28.

Kumar, S. N., Singh, A. N., Aggarwaal, P. K., Rao, V. U.M., & Venateswarlu, B. (2012). *Climate change and Indian agriculture: Impact, adaptation and vulnerability* (p. 32). Salient Achievements from ICAR Network Project, IARI Publication.

Kumar, S., Narjary, B., Kumar, K., Jat, H. S., Kamra, S. K., & Yadav, R. K. (2019). Developing soil matric potential based irrigation strategies of direct seeded rice for improving yield and water productivity. *Agricultural Water Management, 215*, 8–15.

Kumar, V., Jat, H. S., Sharma, P. C., Gathala, M. K., Malik, R. K., Kamboj, B. R., Yadav, A. K., Ladha, J. K., Raman, A., Sharma, D. K., & McDonald, A. (2018). Can productivity and profitability be enhanced in intensively managed cereal systems while reducing the environmental footprint of production? Assessing sustainable intensification options in the breadbasket of India. *Agriculture, Ecosystems & Environment, 252*, 132–147.

Kumar, V., Singh, A. K., Jat, S. L., Parihar, C. M., Pooniya, V., Singh, B., & Sharma, S. (2015). Precision nutrient and conservation agriculture practices for enhancing productivity, profitability, nutrient-use efficiencies and soil nutrient status of maize (*Zea mays*) hybrids. *The Indian Journal of Agricultural Sciences, 85*(7), 926–930.

Ladha, J. K., Kumar, V., Alam, M. M., Sharma, S., Gathala, M., Chandna, P., Saharawat, Y. S., Balasubramanian, V. (2009). Integrating crop and resource management technologies for enhanced productivity, profitability, and sustainability of the rice-wheat system in South Asia. In J. K. Ladha, Y. Singh, O. Erenstein, & B. Hardy (Eds.), *Integrated crop and resource management in the rice–wheat system of South Asia* (pp. 69–108). International Rice Research Institute, Los Banos, Philippines.

Lohan, S. K., Jat, H. S., Yadav, A. K., Sidhu, H. S., Jat, M. L., Choudhary, M., Peter, J. K., & Sharma, P. C. (2018). Burning issues of paddy residue management in north-west states of India. *Renewable and Sustainable Energy Reviews, 81*, 693–706.

Matson, P. A., Naylor, R., & Ortiz-Monasterio, I. (1998). Integration of environmental, agronomic and economic aspects of fertilizer management. *Science, 280*, 112–115.

Mukhopadhyay, R., Sarkar, B., Jat, H. S., Sharma, P. C., & Bolan, N. S. (2020). Soil salinity under climate change: Challenges for sustainable agriculture and food security. *Journal of Environmental Management*, 111736.

Naresh Kumar, S., Singh, A. K., Aggarwal, P. K., Rao, V. U., & Venkateswarlu, B. (2012). *Climate change and Indian agriculture: Impact, adaptation and vulnerability*. Indian Agricultural Research Institute, New Delhi, p. 32.

Olander, L., Wollenberg, E., Tubiello, F., & Herold, M. (2013). Advancing agricultural greenhouse gas quantification. *Environmental Research Letters, 8*, 011002.

Ortiz, R., Sayre, K. D., Govaerts, B., Gupta, R., Subbarao, G. V., Ban, T., Hodson, D., Dixon, J. M., Ortiz-Monasterio, J. I., & Reynolds, M. (2008). Climate change: Can wheat beat the heat? *Agriculture, Ecosystems & Environment, 126*, 46–55.

Ortiz-Monasterio, I., Wassmann, R., Govaerts, B., Hosen, Y., Katayanagi, N., & Verhulst, N. (2010). Greenhouse gas mitigation in the main cereal systems: Rice, wheat and maize. In M. P. Reynolds (Ed.), *Climate change and crop production* (pp. 151–176). CABI.

Parihar, C. M., Jat, H. S., Yadvinder-Singh, J. S. L., Kakraliya, S. K., & Nayak, H. S. (2020). Precision nutrient management for higher nutrient use efficiency and farm profitability in irrigated cereal-based cropping systems. *Indian Journal of Fertilisers, 16*(10), 1000–1014.

Patra, S., Julich, S., Feger, K. H., Jat, M. L., Jat, H., Sharma, P. C., & Schwärzel, K. (2019). Soil hydraulic response to conservation agriculture under irrigated intensive cereal-based cropping systems in a semiarid climate. *Soil and Tillage Research, 192*, 151–163.

Peng, S., Buresh, R. J., Huang, J., Zhong, X., Zou, Y., Yang, J., Wang, G., Liu, Y., Hu, R., Tang, Q., & Cui, K. (2010). Improving nitrogen fertilization in rice by sitespecific N management. A review. *Agronomy for Sustainable Development, 30*(3), 649–656.

Peng, S. B., Huang, J. L., Sheehy, J. E., Laza, R. C., Visperas, R. M., Zhong, X. H., Centeno, G. S., Khush, G. S., & Cassman, K. G. (2004). Rice yields decline with higher night temperature from global warming. *Proceedings of the National Academy of Sciences, 101*, 9971–9975.

Roy, D., Datta, A., Jat, H. S., Choudhary, M., Sharma, P. C., Singh, P. K., & Jat, M. L. (2021). Impact of long-term conservation agriculture on soil quality under cereal based systems of North West India. *Geoderma, 405*, 115391.

Saharawat, Y. S., Ladha, J. K., Pathak, H., Gathala, M., Chaudhary, N., & Jat, M. L. (2012). Simulation of resource-conserving technologies on productivity, income and greenhouse gas emission in rice-wheat system. *Journal of Soil Science and Environmental Management, 3*(1), 9–22.

Saharawat, Y. S., Singh, B., Malik, R. K., Ladha, J. K., Gathala, M., Jat, M. L., & Kumar, V. (2010). Evaluation of alternative tillage and crop establishment methods in a rice-wheat rotation in North Western IGP. *Field Crops Research, 116*, 260–267.

Sapkota, T. B., Jat, M. L., Rana, D. S., Khatri-Chhetri, A., Jat, H. S., Bijarniya, D., Sutaliya, J. M., Kumar, M., Singh, L. K., Jat, R. K., & Kalvaniya, K. (2021). Crop nutrient management using Nutrient Expert improves yield, increases farmers' income and reduces greenhouse gas emissions. *Science and Reports, 11*(1), 1–11.

Sapkota, T. B., Majumdar, K., Jat, M. L., Kumar, A., Bishnoi, D. K., McDonald, A. J., & Pampolino, M. (2014). Precision nutrient management in conservation agriculture-based wheat production of Northwest India: Profitability, nutrient use efficiency and environmental footprint. *Field Crops Research, 155*, 233–244.

Sapkota, T. B., Shankar, V., Rai, M., Jat, M. L., Stirling, C. M., Singh, L. K., Jat, H. S., & Grewal, M. S. (2017). Reducing global warming potential through sustainable intensification of basmati rice-wheat systems in India. *Sustainability, 9*(6), 1044.

Sapkota, T. B., Singh, L. K., Yadav, A. K., Khatri-Chhetri, A., Jat, H. S., Sharma, P. C., Jat, M. L., & Stirling, C. M. (2020). Identifying optimum rates of fertilizer nitrogen application to maximize economic return and minimize nitrous oxide emission from rice–wheat systems in the Indo-Gangetic Plains of India. *Archives of Agronomy and Soil Science, 66*(14), 2039–2054.

Sapkota, T. B., Vetter, S. H., Jat, M. L., Sirohi, S., Shirsath, P. B., Singh, R., Jat, H. S., Smith, P., Hillier, J., & Stirling, C. M. (2019). Cost-effective opportunities for climate change mitigation in Indian agriculture. *Science of the Total Environment, 655*, 1342–1354.

Satyanarayana, T., Majumdar, K., Pampolino, M., Johnston, A. M., Jat, M. L., Kuchanur, P., Sreelatha, D., Sekhar, J. C., Kumar, Y., & Maheswaran, R. (2012). Nutrient expert: A tool to optimise nutrient use and improve productivity of maize. *Better Crops—South Asia, 6*, 18–21.

Sharma, P. K., Bhushan, L., Ladha, J. K., Naresh, R. K., Gupta, R. K., Balasubramanian, B. V., & Bouman, B. A. (2002). Crop-water relations in rice-wheat cropping under different tillage systems and water-management practices in a marginally sodic, medium-textured soil. In *Water-wise rice production* (Vol. 8, pp. 223–235). International Rice Research Institute, Los Baños, Philippines.

Sharma, P. C., Jat, H. S., Kumar, V., Gathala, M. K., Datta, A., Yaduvanshi, N. P. S., Choudhary, M., Sharma, S., Singh, L. K., Saharawat, Y., Yadav, A. K., Parwal, A., Sharma, D. K., Singh, G., Jat, M. L., Ladha, J. K., & McDonald, A. (2015). Sustainable intensification opportunities under current and future cereal systems of North-West India. Technical bulletin (Central Soil Salinity Research Institute (Karnāl, India), 2015/4, pp. 1–46.

Sharma, P. C., Datta, A., Yadav, A. K., Choudhary, M., Jat, H. S., & McDonald, A. (2019). Effect of crop management practices on crop growth, productivity and profitability of Rice-Wheat System in Western Indo-Gangetic Plains. *Proceedings of the National Academy of Sciences, India Section B: Biological Sciences, 89*(2), 715–727.

Shirsath, P. B., Jat, M. L., McDonald, A. J., Srivastava, A. K., Craufurd, P., Rana, D. S., Singh, A. K., Chaudhari, S. K., Sharma, P. C., Singh, R., & Jat, H. S. (2020). Agricultural labor, COVID-19, and potential implications for food security and air quality in the breadbasket of India. *Agricultural Systems, 185*, 102954.

Shyamsundar, P., Springer, N. P., Tallis, H., Polasky, S., Jat, M. L., Sidhu, H. S., Krishnapriya, P. P., Skiba, N., Ginn, W., Ahuja, V., Cummins, J., Datta, I., Dholakia, H. H., Dixon, J., Gerard, B., Gupta, R., Hellmann, J., Jadhav, A., Jat, H. S., Keil, A., Ladha, J. K., Lopez-Ridaura, S., Nandrajog, S. P., Paul, S., Ritter, A., Sharma, P. C., Singh, R., Singh, D., & Somanathan, R. (2019). Fields on fire: Alternatives to crop residue burning in India. *Science*, 365(6453), 536–538.

Sidhu, H. S., Jat, M. L., Singh, Y., Sidhu, R. K., Gupta, N., Singh, P., Singh, P., Jat, H. S., & Gerard, B. (2019). Sub-surface drip fertigation with conservation agriculture in a rice-wheat system: A breakthrough for addressing water and nitrogen use efficiency. *Agricultural Water Management, 216*, 273–283.

Singh, R., Singh, A., Kumar, S., Rai, A. K., Rani, S., Sharma, D. K., Joshi, P. K., Chaudhari, S. K., Dey, P., Thimmappa, K., & Tripathi, R. S. (2020). Feasibility of mini-sprinkler irrigation system in direct seeded rice (Oryza sativa) in Indo-Gangetic plains of India. *The Indian Journal of Agricultural Sciences, 90*(10), 1946–1951.

Sivakumar, M. V. K., & Stefanski, R. (2011). Climate change in South Asia. In R. Lal, M. V. K. Sivakumar, S. M. A. Faiz, A. H. M. M. Rahman, & K. R. Islam (Eds.), *Climate change and food security in South Asia* (pp. 13–30). Springer.

Srivastava, K., Jat, H. S., Meena, M. D., Choudhary, M., Mishra, A. K., & Chaudhari, S. K. (2016). Long term impact of different cropping systems on soil quality under silty loam soils of Indo-Gangetic plains of India. *Jouranl of Applied and Natural Science, 8*(2), 584–587.

Tigga, P., Meena, M. C., Dey, A. B., Dwivedi, B. S., Datta, S. P., Jat, H. S., & Jat, M. L. (2020). Effect of conservation agriculture on soil organic carbon dynamics and mineral nitrogen under different fertilizer management practices in maize (*Zea mays*)-wheat (*Triticum aestivum*) cropping system. *The Indian Journal of Agricultural Sciences, 90*(8), 1568–1574.

Tirol-Padre, A., Rai, M., Kumar, V., Gathala, M., Sharma, P. C., Sharma, S., Nagar, R. K., Deshwal, S., Singh, L. K., Jat, H. S., & Sharma, D. K. (2016). Quantifying changes to the global warming potential of rice wheat systems with the adoption of conservation agriculture in North-Western India. *Agriculture, Ecosystems & Environment, 219*, 125–137.

Velayutham, M. (1999). Crop and land use planning in dryland agriculture. In H. P. Singh, Y. S. Ramakrishna, K. L. Sharma, & B. Venkateswarlu (Eds.), *Fifty years of dryland agricultural research in India* (pp. 73–80). ICAR-CRIDA, Hydrabad, India.

Verhulst, N., Nelissen, V., Jespers, N., Haven, H., Sayre, K. D., Raes, D., Deckers, J., & Govaerts, B. (2011). Soil water content, rainfall use efficiency and maize yield and its stability as affected by tillage practice and crop residue management in the rainfed semi-arid highlands of Mexico. *Plant and Soil*. https://doi.org/10.1007/sII104-011-o728-8

Vleeshouwers, L. M., & Verhagen, A. (2002). Carbon emission and sequestration by agricultural land use: A model study for Europe. *Global Change Biology, 8*, 519–530.

Wang, S. W., Lee, W. K., & Son, Y. (2017). An assessment of climate change impacts and adaptation in South Asian agriculture. *International Journal of Climate Change Stratergies and Management, 9*, 517–534.

Wassmann, R., Jagadish, S. V. K., Heuer, S., Ismail, A., Redoña, E., Serraj, R., Singh, R. K., Howell, G., Pathak, H., & Sumfleth, K. (2009a). Climate change affecting rice production: The physiological and agronomic basis for possible adaptation strategies. *Advances in Agronomy, 101*, 59–122.

Wassmann, R., Jagadish, S. V. K., Sumfleth, K., Pathak, H., Howell, G., Ismail, A., Serraj, R., Redoña, E., Singh, R. K., & Heuer, S. (2009b). Regional vulnerability of rice production in Asia to climate change impacts and scope for adaptation. *Advances in Agronomy, 102*, 91–133.

Yadav, G., Jat, H. S., Raju, R., Yadav, R. K., Singh, S. K., Chaudhari, S. K., Sharma, D. K., Singh, G., & Sharma, P. C. (2021). Enterprise mix diversification: An option for ecologically sustainable food and nutritional security of small holders in Indo-Gangetic plains. *International Journal of Agricultural Sustainability*, 1–11.

Yadav, R., Gaikwad, K., Bhattacharyya, R., Bainsla, N. K., Kumar, M., & Yadav, S. S. (2018). Breeding new generation genotypes for conservation agriculture in maize-wheat cropping systems under climate change. *Food Security and Climate Change, 19*, 189–228.

Climate-Smart Agriculture: Potential Role in Carbon Sequestration and to Address Climate Change Under Irrigated Agro-Ecosystems

Nayab Komal, Qamar uz Zaman, Anastasia Pantera, Ghulam Yasin, Kamran Ashraf, Saba Nazir, and Mirza Barjees Baig

Abstract Ecosystems and their associated living organisms are struggling for their survival and existence due to climate change. The atmospheric concentration of greenhouse gasses (GHG) increases considerably, which has now crossed the digit of 397.60 ppm^3 and set to rise further. The climate scenario in Pakistan has posing negative impacts due to persistent change in the climate. In the coming years, i.e., 2025 if the present rate of increases continues, the atmospheric concentration of carbon dioxide would be increasing the globe's temperature up to 2 °C–4 °C. Minimizing the concentration of atmospheric carbon and sequestering it into the agricultural landscape (soil, trees, crops) is the most effective, cheaper eco-friendly approach. Agroforestry (AF) is an integral part of climate-smart agriculture that involves the growing

N. Komal · Q. Zaman (✉) · S. Nazir
Department of Environmental Sciences, The University of Lahore, Lahore, Pakistan
e-mail: qamar.zaman1@envs.uol.edu.pk

N. Komal
e-mail: nayabkomal5@gmail.com

S. Nazir
e-mail: sabawarraich989@gmail.com

A. Pantera
Agricultural University of Athens, Athina, Greece
e-mail: panteranatassa@gmail.com

G. Yasin
Department of Forestry, Islamia University of Bahawalpur, Bahawalpur, Pakistan
e-mail: ghulam.yasin@iub.edu.pk

K. Ashraf
State Key Laboratory of Bioreactor Engineering, East China University of Science and Technology, Shanghai 200237, China
e-mail: kamran2417@gmail.com

M. B. Baig
Prince Sultan Institute for Environmental, Water and Desert Research, King Saud University, Riyadh, Saudi Arabia
e-mail: mbbaig@ksu.edu.sa

M. Behnassi et al. (eds.), *Climate-Smart and Resilient Food Systems and Security*,
https://doi.org/10.1007/978-3-031-65968-3_4

of forest trees and raising agricultural crops and livestock on a particular piece of land. Agriculture happens to be the major contributor to the atmospheric concentration of carbon dioxide; however, soil carbon sequestration is the most efficient tool to enrich soil organic carbon pools, which depend on improved agricultural practices. Trees when grown in combination with agricultural crops store more carbon range 34.61 t C ha^{-1} than the monocrop system with 18.74 t C ha^{-1}. This chapter provides analyzes of the current knowledge about soil organic carbon (SOC) sequestration processes and their management in agroforestry systems. A detailed discussion on the useful C sequestration tree species (E. camaldulensis, S. cumin, P. ciliata, A. Acacia, Z. manritiana, C. sinensis, A. indica, D. sisso, Bambusoideae, M. azedarach, and M. alba) in the agroforestry systems and their role in climate change mitigation, enhancing soil organic carbon sequestration in agroforestry soils and sequestration rates, and the processes contributing to the stabilization of SOC in agroforestry soils has been presented. The analysis led to revealing that climate-smart agriculture has great potential to lockup more carbon and helps in the reduction of CO_2 emissions to the atmosphere.

Keywords Climate change · Agroforestry system · Carbon sequestration · Soil organic carbon · Trees and crop biomass carbon

1 Introduction

Carbon dioxide is just one of the major greenhouse gases (GHG) among all gases that is causing severe environmental and climatic effects. Global climate change and the increase in the trend of carbon dioxide gas emissions are a growing concern today (Mondal et al., 2017; Qurieshi & Rasool, 2021). Climate change is an important environmental issue that has captured the world's attention during the recent past. Due to several activities like combustion of fossil fuels and deforestation during the last 100 years, chemical makeup of this flimsy layer of the atmosphere has been extremely altered (Pold et al., 2017; Zinedine & El Akhdari, 2021). Pakistan's contribution to global greenhouse gas (GHG) emissions represents only 0.8%, estimated at 355 megatons of CO_2 equivalent annually, including emissions from the Land Use Change and Forestry (LUCF) sectors. Consequently, the country is ranked low at 148th in the world on the list of global emitters, based on a per capita GHG emission of 1.96 tons of CO_2 equivalent (Chandio et al., 2021).

Pakistan as a developing country is facing many problems like high population growth and their subsequent requirements such as the need for shelter, food, and energy (Chandio et al., 2021; Kamal et al., 2014). The mechanism of pollution monitoring, especially for development projects in the country, is not very strict, which has adversely affected the ambient air quality in many areas. This has had a significant impact on plant morphology and physiology (Bales, 2012; Krayem et al., 2021). One of the most common sources of gaseous air pollution in Pakistan is the traditionally small-scale and unorganized brick kiln industries, which are spread in the rural and peri-urban areas of the country (Jahan et al., 2016).

Agroforestry systems can attain higher yields than monoculture production systems by increasing the capture of solar radiation, water and nutrients by all crops in the same plot. Trees can also provide additional marketable food products such as fruits and nuts. Agroforestry protects croplands from hot and desiccating winds as land use system (Abbas et al., 2017; Arévalo-Gardini et al., 2021). It provides food to humans and feed to livestock, pollen for honeybees, wood for fuel, and timber for shelters and construction work. Similarly, Pakistani farmers adopting sustainable production techniques through appropriate agroforestry systems have also been able to enhance their incomes (Ahmad et al., 2017; Essa et al., 2011; Magsi et al., 2014). The sustainable mitigation model is represented in Fig. 1.

Pakistan is predicted to be among the 10 countries most affected by climate change, according to the 2019 Global Climate Risk Index (Eckstein et al., 2017). The average temperatures are predicted to rise faster than most other locations, increasing 4ºC by the year 2100 (Gillespie et al., 2012). The country also faces an increased incidence of severe droughts and floods threatening health, agriculture, and water supplies (Salam, 2018). Agroforestry can facilitate adaptation to climate change (Vignola et al., 2015) by protecting understory crops from the increasing temperatures

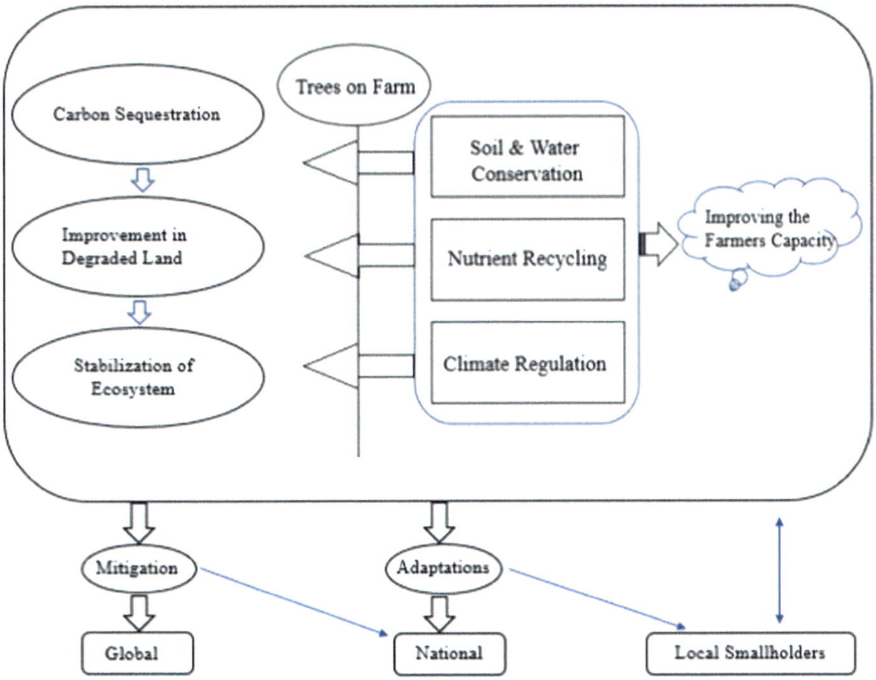

Fig. 1 Climate change adaptation and mitigation potential of agroforestry systems. *Source* Lasco et al., (2014)

while reducing crop transpiration, soil evaporation, and wind speeds. Trees increase aboveground carbon sequestration and the soil carbon on cultivated cropland (Abbas et al., 2017). Agroforestry can offset or reduce emissions caused by forest harvest or degradation through increased production of on-farm timber and fuelwood (Abbas et al., 2017; Minang et al., 2011). Agroforestry, as a sustainable land-use system, has a substantial potential to sequester atmospheric carbon into the soil and the above- and below-ground biomass (Abbas et al., 2017; Lipper et al., 2014).

Climate-smart agriculture (CSA) proposes a framework that supports decision-making in the agriculture sector by considering three foundational outcomes while fully accounting for the trade-offs and synergies among them. It is comprised of agricultural systems that contribute to sustainable and equitable increases in agricultural productivity and incomes. This chapter provides discussion on the useful C sequestration tree species in the agroforestry systems and their role in climate change mitigation, enhancing soil organic carbon sequestration in agroforestry soils and sequestration rates, and the processes contributing to the stabilization of soil organic carbon (SOC) in agroforestry soils has been presented. The analysis has led to revealing that CSA has great potential to lockup more carbon and helps reduce CO_2 emissions to the atmosphere.

2 Climate Change

2.1 Climate Change Sources

Humanity is witnessing, in the past few years, unparalleled changes in the climate. If we continue on our present course, life on the Earth will be modified to such an extent that it will be very difficult to reverse (IPCC, 2000; Mahowald et al., 2017). Human activities, such as fossil fuel combustion and deforestation during the last 100 years, have altered the chemical composition of the atmospheric layers (Pold et al., 2017). Such modifications in the chemical composition in turn resulted in a wide range of significant harmful effects on the planet's long-term weather conditions, the ecological systems supported by the climate of the Earth, and the welfare of human beings and economy (CDIAC, 2000; Harsch et al., 2017).

Global warming is among the greatest horrors of the modern times (Pascale et al., 2017). It is believed that carbon is among the most significant casual factors of global warming (Jones et al., 2017). Air pollution, derived from increasing industrialization and urbanization during the last few decades, has become a serious environmental stress to crop plants (Patil et al., 2021; Rajput & Agrawal, 2004). GHG emissions (CO_2, CH_4, N_2O, and H_2O) are the primary source of changes in the climate since GHGs can influence the climate significantly. The gradual increase in negative atmospheric events caused by climate change across the globe has led to the underestimation of climate change. Aside from human activities, natural phenomena like volcanic explosions, seismic activities, and solar cycles play a key part in environmental degradation (Hussain et al., 2020).

In 2010, there was an unprecedented increase of 5.8% in GHG emissions worldwide (Hussain et al., 2019), and Pakistan's emissions rose at a similar rate. Due to the extreme impacts of climate change, Pakistan is facing potential endangerment to the economy and social and environmental development (Khan et al., 2016). Rapid urban development accompanied by transportation and, the subsequent, energy usage as well as waste increases, are predicted to worsen this global ranking by 2030 (Hussain et al., 2018). In the past, natural sinks have kept GHG emissions below hazardous levels; however, through continual human interventions, i.e., vehicular usage, agricultural activities, forestry (Hussain et al., 2020), industrial development, and combusting fossil fuel (Ahmed et al., 2020), the global GHG emissions are increasing at an accelerating pace with the exception of H_2O vaporization (US EPA, 2016). The two main reasons for the rise in anthropogenic CO_2 emissions in the atmosphere are fossil fuel combustion through human activities and land-use changes, which are contributing to global warming since the mid of twentieth century (Yousaf et al., 2017) (Fig. 2).

Climate change has now become probably the most highlighted topic, and this may pose the largest threat that human civilization will face in the near future (Dalby, 2021). The impacts of climate change can be categorized into extreme and non-extreme types (Davenport & Diffenbaugh, 2021). The extreme impacts are those which may be very significant, resilient implications on the biotic landscape (i.e., natural ecosystems), natural environment, and human societies caused due to a single

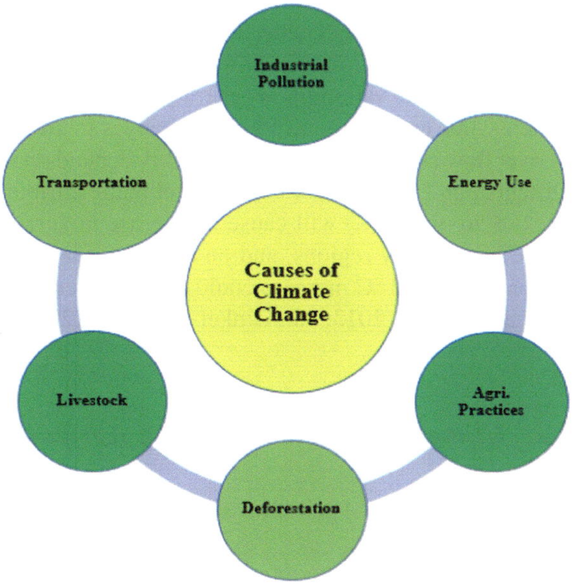

Fig. 2 Causes of climate change. *Source* (Smith et al., 2020)

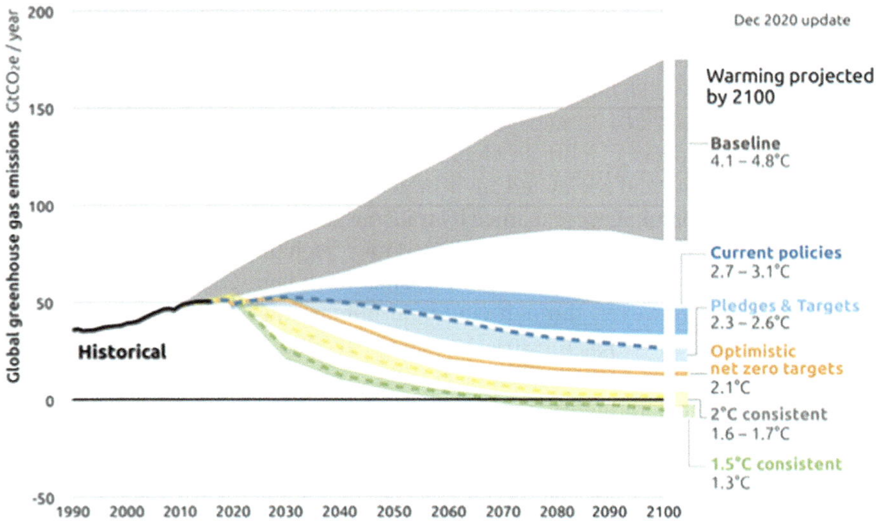

Fig. 3 Emissions and expected warming by 2100. *Source* Climate action tracker (Temperature) (2020)

extreme event or by consecutive extreme or non-extreme events (Dalby, 2021; Field et al., 2012).

A higher atmospheric concentration of carbon dioxide enhances plant growth and increases water-use efficiency (CO_2 fertilization) and so affects water availability. Climate variability, especially changes in rainfall patterns, is particularly important for rainfed agriculture (Dalby, 2021). Decreased rainfall and the resultant deficiency in soil moisture will reduce crop productivity in case of rainfed farming systems. Although rainfall variability's generated risks are significant for the irrigated farming systems, this can also be exposed to changes in the spatial and temporal distribution of river flow as irrigation systems entirely depend on run-off regimes (Davenport and Diffenbaugh 2021). Minor change in climatic attributes (a global mean temperature rises of around 2 °C) will cause a decrease in agricultural production, particularly in the hot and dry regions, and an increase in areas within middle to high latitude regions. Even a 2 °C increase could culminate enough risks in the crop productivity (Calzadilla et al., 2013; van Ginkel & Biradar, 2021) (Fig. 3).

2.2 Climate Change Impacts on Agriculture

2.2.1 Impacts on Agricultural Productivity

Agriculture is a key economic sector that contributes by 21% to the gross domestic product (GDP), employs 45% of the total workforce, and contributes about 60% to

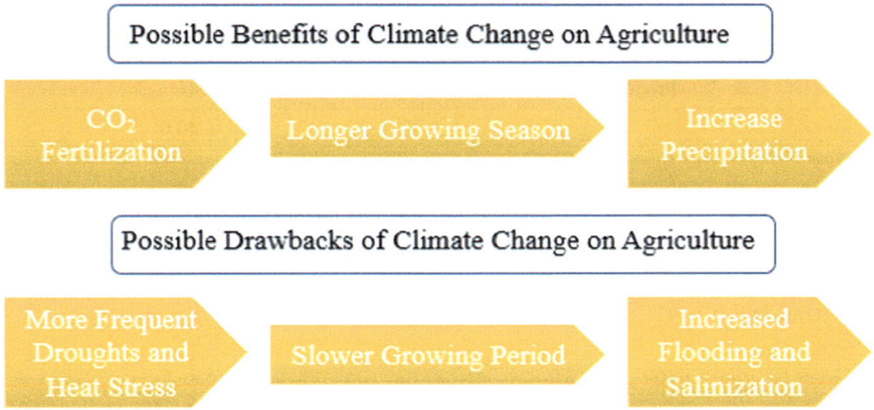

Fig. 4 Possible effects of climate change on agriculture. *Source* (Smith et al., 2020)

exports (Ullah et al., 2021). The total cropped area is 23.4 million hectares (Mha), representing 29% of the total reported area of which irrigated areas covers 18.63 Mha (24% of the total irrigated area), with the percentage by province of 77% in Punjab, 14% in Sindh, 5% in Khyber Pakhtunkhwa, and 4% in Balochistan (Agricultural Statistics of Pakistan 2011; Ullah et al., 2021). Currently, 3.8 Mha are under Sailaba/Rod-Kohi riverine and Barani farming systems, commonly called as the spate irrigation farming system. The potential area under spate irrigation is estimated to be around 6.935 Mha, divided as follows: 4.68 Mha in Balochistan, 0.862 Mha in Khyber Pakhtunkhwa, 0.571 Mha in Punjab, and 0.551 Mha in Sindh).

There are two crop categories, the winter and the summer crops. Winter crops (wheat is the major one) are sown in autumn (October–December) and harvested in spring (March–April). Crops sown in summer are called summer crops. The summer crop season is generally longer in Pakistan, starting with sugarcane in February, cotton in March–May, rice in June–July, and maize in July–August (Fig. 4).

The major cropping patterns are: (i) rice–wheat, (ii) maize-wheat, (iii) cotton-wheat, (iv) sugarcane-wheat, and (v) coarse grain-wheat. Other minor cropping patterns exist as well. Crops grown in both irrigated areas and those under spate farming systems are highly sensitive to the amount of available water and temperature variability. It is estimated that a temperature rises by + 0.50 °C–20 °C would decrease agricultural productivity by around 8%–10% by 2040. Different simulation studies, using the crop–growth simulation model, estimate a decrease in yield of major crops, specifically for wheat and rice, and the length of growing season in four agroclimatic zones of Pakistan (Table 1). The model predicted the largest decrease of around 14 days for 10 °C rise in temperature in the growing season's length of wheat in the northern mountainous region compared to southern Pakistan (Dalby, 2021; Dehlavi et al., 2015).

Table 1 Impact of climate change on wheat production by 2080

Climatic Regions	% Share in National Production	Current Yield (kilogram/hectare)	% Change in Yield in 2080
Northern mountainous	2	2658	+ 40.0
Northern submountainous	9	3933	–11.0
Southern semiarid plains	42	4306	–80.0
Southern arid plains	47	4490	–6.0
Pakistan	100	4326	–6.4

Source Iqbal et al., (2010)

Climate change is considered to be a global phenomenon; however, its impacts are more likely to severely affect developing countries due to their greater vulnerabilities and reduced adaptation capacities. Since most developing countries, including Pakistan, are agriculture-based economies, their agricultural sector is highly affected by natural phenomena (Kurukulasuriya et al., 2006; Lee, 2021). Therefore, the major impact of climate change is on agricultural production due to changes in precipitation pattern, temperature, floods, droughts, and subsequent negative effects on water and land resources (Ahmed et al., 2016).

Extreme weather events do not have a single cause, rather they result from various possible contributing factors including human-induced climate change such as global warming. The changing climatic variables, particularly temperature and rainfall, will introduce several challenges to agriculture in future. Climate variabilities are projected to bring changes in the agricultural sectors through changes in land and water regimes. Changes in the frequency and intensity of droughts, flooding, and storm damage are anticipated to occur (Ullah et al., 2021). The yield changes, particularly in wheat production, are alarming for Pakistan, and likely require significant adaptation interventions (Table 2).

Table 2 Projected percent changes in major crops yield (2020–2080) (site-based, regional and global scales)

Crops	2020	2050	2080
Wheat	–3.2%	–11.0%	–27.0%
Rice	0%	–0.8%	–1.9%
Maize	–2.4%	–3.3%	–4.3.0%

Source Chaudhry (2017)

2.2.2 Impacts on Food Security

The two main important factors in food supply are natural resources and agriculture industrialization. In the case of natural resources, the degradation of water, soil, forest, and fisheries resulting in declined production capacity directly affects food security (Fanzo, 2015; Maxwell et al., 2010). Similarly, improvements in fertilizer, pesticide, technology, and agriculture research will boost the production capacity of agriculture industrialization to benefit food security. Deforestation, forest conversion, overgrazing, productivity losses, unsuitable agricultural practices, overharvesting, restrained access, and disputes over fish stocks are the reasons for challenged natural resources (Fanzo, 2015). Climate-related events, especially droughts, floods, and tsunami can damage food resources, resulting in risks to food security.

Increased food production must come from land already under cultivation employing economical and environmentally feasible options (Scarlat & Fahl, 2021). In the case of fisheries improvement, proper management, maintaining the harvest level and legitimate access to improved sustainable aquaculture production will meet the food security vision (van Ginkel & Biradar, 2021). Climate change moves millions of people into a cycle of food insecurity and poverty. However, we can fight food insecurity and climate change by adopting climate-improved agriculture practices (FAO, 2018; Ferranti, 2016). Depletion of soil nutrients, pests' induced crop losses, fertilizer effects on water quality, environmental concerns regarding fertilizers, pesticide effects on human health and existing technology, all impact food production (Ferranti, 2016). Fertilizer input needs to be efficient for plant nutrient uptake, available pest control needs to be improved with no negative environmental effects, and biotechnology research is necessary to support the technology gap in sustainable agriculture (Schuenemann et al., 2018).

2.3 Climate Change in Pakistan

Pakistan is also going to face the brunt of climate change due to its hydrological reserve's shrinkage, rapid glacier melting floods, and droughts (Chaudhry et al., 2009; Siddique, 2021). Pakistan's temperature is going to rise from 0.9 °C to 1.5 °C by 2020 until 2050. Pakistan has faced the worst droughts in its history in 1998 and 2004 (Hussain & Mumtaz, 2014), and its largest province Baluchistan was badly affected. In particular, 84% of its population was directly affected and 76% of its livestock was killed. Similar to these effects, the whole country suffered greatly by massive floods displacing a large number of people in its northern as well as central parts (Carignan et al., 2017; Ullah et al., 2018a, 2018b). Sizzling temperature, severe droughts, pest diseases, health-related problems, and lifestyles switches are likely to remain in the coming years as well (Hussain et al., 2018).

Pakistan ranks relatively low among countries on a per capita GHG emissions basis, due to its relatively low level of development in infrastructure and high population (Siddique, 2021). According to a preliminary projection, GHG emissions levels for Pakistan are expected to increase manyfold in the coming decades. This is based on the assumption that GHG emissions intensity for the above-indicated five main sectors will remain essentially the same as during 1994–2008. Thus, the projected total GHG emissions of Pakistan, in line with the government's economic growth strategy, will be more than double by 2020 (compared to the emissions in 2008), and increase by around 14 times by 2050 (The Global Economy, 2015) (Table 3).

In the previous five decades, there was a rise of 0.5 °C in the annual mean temperature of Pakistan. In the last century, there was approximately 10 cm increase in the sea level along the coastal areas of Karachi. Global emissions can cause an increase of 3 °C–5 °C in the annual mean temperature of Pakistan by the end of this century (Mustafa et al., 2021). Melting glaciers in the mountainous region of Himalayas are affecting the volume of water flowing in the rivers at an alarming rate and these impacts are threatening the lives of millions. As a consequence of constant climate change, Pakistan's climate has become gradually more unpredictable, and this trend is anticipated to persist (Mustafa et al., 2021).

Fuels are burned to produce energy, which releases GHGs such as methane, nitrous oxide, and carbon dioxide into the atmosphere. Presently, the quantities of these gases in our atmosphere are increasing on a daily basis (Preethika et al., 2021). Heat waves, arising from global warming and unpredictable weather conditions, continue to expand to areas of favorable temperatures for human existence. In 2015, the State of Food Insecurity report by the International Fund for Agricultural Development (IFAD) and World Food Program (WFP) indicated that more than 795 million of the global population lack access to adequate and quality food. Changing precipitation patterns, water availability periods, droughts, intensity and frequency of heat waves and weather-induced natural disasters all are contributing to deter our weak economy (Kiran, 2017; Preethika et al., 2021). Table 4 shows a decrease in production levels recorded in Pakistan and a projection on the likely scenario in the future, should the mitigation measures against climate change fail.

Pakistan is one of the countries with a lower altitude that suffers from output losses as a result of rising temperatures. From Table 4, it can be seen that only the mountainous regions in the north will experience crop yield increase by 2085. All

Table 3 Projected greenhouse gas emissions by sector in 2020 and 2050 under business-as-usual scenarios ($MtCO_2e$ = million ton of carbon dioxide equivalent)

Sector	1994	2008	2012	2020	2050
Energy	86	157	169	358	2685
Agriculture	72	120	165	245	1395
Industrial processes	13	18	14	26	67
Land use change and forestry	7	9	10	14	38
Wastes	4	6	10	7	15
Total national emissions	182	309	369	650	4200

Source Ijaz et al., 2021

Table 4 Decrease and Projection in Production Levels

Region	Share in national production (2016) (%)	Baseline yield (kg/h)	Projected change in yield in 2085 (%)
Northern mountains	2	2658	+ 50
Northern submountainous	9	3933	−11
Southern semi-arid plains	42	4306	−8
Southern arid plains	47	4490	−5

Source Iqbal and Arif (2010)

the remaining Pakistani areas, 98% of agricultural land, will experience a reduction in production levels (Iqbal & Arif, 2010; Shahzad & Yasin, 2021).

2.4 Climatic Zones of Pakistan

Pakistan is divided into six major geographical zones based on physiographic and climatic factors, as shown in Table 5.

Table 5 Climatic zones of Pakistan

Regions	Meteorological stations in the regions
Zone I(a): Greater Himalayas (Winter dominated)	Astor, Bunji, Chilas, Chitral, Dir, Drosh, Gilgit, Gupis, Skardu
Zone II: Western Highlands	Cherat, D.I. Khan, Kohat, Parachinar, Peshawar, Risalpur
Zone III: Central & Southern Punjab	Bahawalnagar, Bahawalpur, Faisalabad, Khanpur, Mianwali, Multan, Raffique, Sargodha
Zone IV: Lower Indus Plains	Chhor, Hyderabad, Jacobabad, Nawabshah, Padidan, Rohri
Zone V(a): Balochistan Plateau (Northern) (Sulaiman & Kirthar Ranges)	Barkhan, Kalat, Khuzdar, Lasbela, Quetta, Sibbi
Zone V(b): Balochistan Plateau (Western)	Dalbandin, Nokkundi, Panjgur
Zone VI: Coastal Belt	Badin, Jiwani, Karachi, Pasni

Source Iqbal and Arif (2010)

3 Climate-Smart Agriculture (Agroforestry)

Climate change drives many direct and indirect stressors of agricultural production as well as the ability of ecosystems to provide goods and services (Maia et al., 2021). As temperature increase and extreme climate events become more severe, global crop yields may start to decrease from the 2030s; ecosystems and biodiversity may also experience potential losses (IPCC, 2014; Preethika et al., 2021).

Agroforestry systems, as part of CSA, are agricultural systems that use trees and shrubs in land management, crop and/or animal production practiced by an estimated 30% of the global rural population (Zomer et al., 2009; Awazi et al., 2021). At global and national scale, agroforestry can be considered as a mitigation strategy through carbon sequestration in biomass and soil. Deforestation and land use conversion from forest to agriculture have been significant sources of GHG emissions. Abating these processes by incorporating more trees in farms will likewise result in benefits for local climate, water resources, and biodiversity and community livelihood, thereby increasing resilience to the impacts of climate change (Awazi et al., 2021; Lipper et al., 2014) (Fig. 5).

3.1 *Agroforestry in Pakistan*

Natural disasters, economic instability, and militancy have challenged food security in Pakistan over the past years, despite significant increases in staple crop yields.

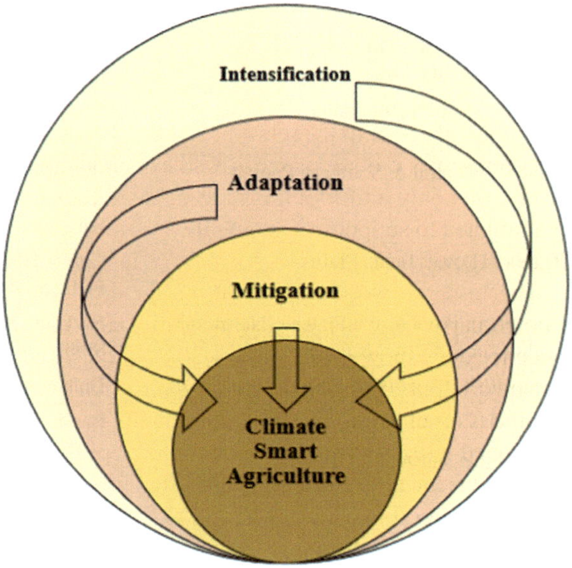

Fig. 5 Brief Introduction of Climate-Smart Agriculture. *Source* (Smith et al., 2020)

Pakistan ranks 78th out of 113 countries in the Global Food Security Index (EIU, 2015; Muzamil et al., 2021), with more than half of the population (60%) experiencing food insecurity. The average food supply in the country is estimated at 2,440 kcal/person/day, yet this is rated insufficient to meet demand (the country boasts an index score of 39.7, where 100 indicates sufficiency of supply) given high geographical disparities in food production and supply. From a nutrition perspective, over one-fifth (22%) of the total population is undernourished, one-third (31%) of children are underweight for their age (15% are affected by wasting) and diets generally lack diversity. Low nutritional quality is associated with economic hardship, low literacy levels, poor sanitary conditions, and inadequate access to health services (FAO 2017; Muzamil et al., 2021).

In Pakistan, agroforestry, a traditional land management approach, provides an opportunity to achieve these objectives. Planting trees on private farmland provides clear ownership of any timber resource while a well-managed system has great potential both to boost agriculture and reduce wood shortages (De Leijster et al., 2021). About 64% of Pakistan's population live in rural areas and over 50% are engaged in agriculture. Agriculture accounts for 18.9% of the GDP and 42.3% of total employment (Baig et al., 2021). Well-managed agroforestry systems can maximize production in a given space and time, whilst providing benefits such as soil stabilization and wildlife habitat. These practices are not new to Pakistan whose people traditionally grow trees in their home courtyards and on their land for multiple purposes (De Leijster et al., 2021). Historically, farmers often retained trees on their farms to sustain agricultural production, reduce soil erosion, retain water, provide shade, and generate income. Most commonly trees are found on the farmlands in the Province of Punjab (Awazi et al., 2021; Farooq et al., 2018). Sindh Province sustains the forests on huge areas; however, farmers raise trees in clumps to sale the poles and reclaim the degraded lands. In lower Sindh, farmers plant *Acacia nilotica* (Babul) as woodlots, locally called '*hurries*'. The Punjab Province is known as the 'food and fruit basket' for the whole country, sustaining various irrigated, dry, rainfed and desert areas and producing different types of crops and trees. It has a very high potential of supporting various forms of agroforestry systems (Preethika et al., 2021).

3.2 Agroforestry and Its Benefits

Agroforestry is the management of trees, crops and livestock on the same piece of land, thereby diversifying the agricultural productivity (Karki, 2019). Agroforestry has the potential to offer both economic and environmental reimbursement (Gupta et al., 2017). Some of the major benefits of maintaining agroforestry systems are discussed below.

3.2.1 Enhanced Soil Fertility

Soil fertility is a vital component in the food security, falling poverty, preserving environment and for sustainability (Pandey, 2007). Agroforestry land-use systems like silvoarable (agricultural or horticultural crops), silvopastoral (combine tree growing with the production of livestock), agrosilvopastoral system (a collective name for land-use systems), etc., are proficient ways of restoring soil organic matter (De Leijster et al., 2021; Pandey, 2007). Jama et al., (2006) reported that yield is higher with improved crop rotation than with continuous cropping. The leaf litter from agroforestry practices forms humus after decomposition and improves various soil properties (ICAR, 2006). Agroforestry systems can also reduce water and nutrient losses, as well as soil material and organic matter loss, by controlling runoff and soil erosion (Verma et al., 2021).

3.2.2 Improved Income

The varied component of agroforestry provides multiple harvests at different times of the year (Hidayat & Sianipar, 2021). It increases food production, improves supply of fodder for fish and livestock, increases supply of fuel wood, improves soil fertility and water supply, habitats, etc. Therefore, it reduces the risk of crop failure and ensures alternate income for the farmers (Pandey, 2007; Verma et al., 2021).

3.2.3 Augmented Carbon Stock

Agroforestry has a huge potential as a climate mitigation strategy because of its ability to sequester carbon in its several plant species and soil (ICAR, 2006). The normal carbon sequestered by this practice (agroforestry) has been estimated to be 9, 21, 50, and 63 Mg C ha^{-1} in semiarid, sub-humid, humid, and temperate regions. In the tropics, for small agroforestry systems, it has been found to be ranging from 1.5 to 3.5 Mg C ha^{-1} yr^{-1}, and as a consequence, it can be a viable strategy for carbon storage (Hidayat & Sianipar, 2021; Montagnini & Nair, 2004; Roshetko et al., 2002).

3.2.4 Decreased Vulnerability

Agroforestry improves the flexibility of farming systems besides defending against different risks, both biophysically (hydraulic lift, soil fertility) and economically by diversification and decreasing income risk (Verchot et al., 2007).

3.3 *Constraints to Agroforestry in Pakistan*

The major constraints to agroforestry in Pakistan include concerns about: (i) agricultural production, (ii) a lack of markets and supplies, (iii) land designation, and (iv) technical assistance. In some locations, water shortage is a major constraint to the adoption of agroforestry.

3.3.1 Concerns About Agricultural Production

Shading by trees can reduce the level of heat stress on understory crops as well as the harmful desiccating effects of hot winds. Trees can also increase the organic matter content of cultivated soils whereas nitrogen-fixing trees can enhance soil nitrogen status (Baig et al.,identified a lack of capital 2020). Despite these benefits, some farmers believe that trees depress crop production due to shade, nutrient competition, and crop damage from insect pests, diseases, and birds (Hidayat & Sianipar, 2021; Jamilu et al., 2014). In windy locations, increased yield of crops distant from the tree line can offset yield depression close to the trees. Also, most trees in Pakistan are deciduous and thus leafless during winter when the sunlight is most limited. The use of tree species with narrow upright canopies, e.g., Kikar (*A. nilotica*) and black poplar (*P. nigra*), can further minimize light competition.

3.3.2 Lack of Markets and Supplies

Current market structure and institutional policies limit agroforestry opportunities. Farooq et al., (2017) identified a lack of capital and extra cash to buy saplings, limited transportation facilities, and unreliable root stock as important constraints. Many farm foresters and craftsmen are not trained in the skilled use of wood carving, which limits opportunities for value addition and marketing. Although farmers can access incentives like free seedlings and saplings, they typically want more incentives (Hasan, 2014) like low-cost irrigation water on their farms and no tax on agricultural crops. Farmers also want some form of security or assurance from the government that markets will be available to buy their trees and logs with appropriate and reasonable prices and will assist them in finding suitable buyers (Baig et al., 2021).

3.3.3 Land Designation

Some farmers are reluctant to plant trees as they fear that foresters will take away or get control of their farmlands once the trees are established (Baig et al., 2021). This concern may be warranted, as most of the farmers in the uplands are not allowed to cut trees without permission from the forest department, even when planted on their own land. Similarly, in the plains, where social forestry programs are well-established, the transportation of wood from land to the sale depot or market is not allowed without a permit. Farmers view this as a major barrier and disincentive. Such policies will require immediate and suitable measures to overcome the problem and to improve the situation (Baig et al., 2020).

3.3.4 Poor Communication and Links Between Research and Extension

A lack of technical assistance can be a major constraint to agroforestry adoption (Farooq et al., 2017). The constraints include the absence of technology transfer mechanisms, unsupportive attitudes of foresters and administrators, limited or non-availability of trained staff, negative client responses and human behaviors, and the lack of follow-up visits by the extensionist and proper communication, which prevent better technologies from reaching end-users (Baig et al., 2020).

3.4 Agroforestry System in Irrigated Ecosystem

Irrigation has many benefits for nature as well as for the living things that live in nature, including us humans, launching the flow of water, soil fertilization with essential nutrients, quality of water and enabling us to cultivate certain plants or raise animals in areas that would have been impossible otherwise (Saco et al., 2021). Irrigation systems represent a human effort to modify the distribution of water contained in natural channels by using constructions and artificial channels, for all or part of the water, to support and/or enhance the production of agricultural crops (Umum, 2012). The irrigated plains agricultural area (14.6 million ha) lies in the valley of the river Indus and its tributaries and is a large geographical subdivision of Pakistan comprising the provinces of Punjab, Sindh, and some parts of Khyber Pakhtunkhwa. However, the agricultural sector is shrinking (−0.19% rate of growth) compared with the rest of the economy (2.53% rate of growth) due to extreme variations in climatic conditions, which loom food security (Ali et al., 2017).

In the Pothwar Plateau of the Punjab, *Accacia nilotica* and *Zizyphus muritiana* are sometimes raised in the fields for fuelwood, timber, shade, and fodder. Similarly, in irrigated areas of the Punjab province, *Dalbergia sissoo* is planted for the same purposes. Such trees typically are planted on the borders of agricultural fields as shelterbelts or windbreaks. Agroforestry has been an integral part of the farming systems in all the provinces of the country. Currently, eucalyptus species are gaining popularity given their rapid growth and ability to thrive in areas receiving less than 300 mm of average annual rainfall (Baig et al., 2020). Agroforestry systems are recognized as an integrated approach for sustainable land use aside from their contribution to climate change adaptation and mitigation (Schoeneberger et al., 2012). Globally, an estimated 700, 100, 300, 450, and 50 M ha of land are used for tree intercropping, multistate systems, protective systems, and tree woodlots, respectively (Nair and Garrity, 2012).

3.5 Successes and Challenges for Agroforestry Extension in Pakistan

Agriculture in some locations was only possible if trees were used to fix the sand dunes and mitigate the effects of the desert's hot, sand-laden winds. A successful afforestation extension campaign in the Thal desert was launched in the 1950s (Baig et al., 2021). Dry desert was gradually transformed into a green landscape as millions of trees were planted. The same model was replicated in Sindh Province when water from the Indus River was made available for managing the province's agricultural lands (Podger et al., 2021). Based on these successes, most tree extension programs in Pakistan make cuttings/plants and seedlings available to the farmers free or at nominal cost. Numerous similar afforestation/reforestation initiatives were undertaken in the Himalayan foothills (high altitude areas in Northern Pakistan) and the upper plains in the Bhurban, Gullies, and Daur of Murree Hills. Another success was the expansion in the use of poplar trees in Peshawar and Mardan (both cities in the north-western Pakistan, in the KPK Province) that began in 1971 where private forestry was supported with good technical support and extension services. Wattoo et al., (2010) stressed the need for strong and meaning interactions among the stakeholders to the forestry extension programs successful in the KPK province. Extension by forest officials has been less successful. Farmers in Pakistan often have negative views about foresters and many foresters do not like to carry out extension activities despite incentives. The current relationship between foresters and farmers is often challenging and may be antagonistic, as many foresters have a police/control role to keep people away from state forests. Moreover, most foresters have no training in extension systems and methods (Baig et al., 2020; Saco et al., 2021).

In rural areas of Pakistan, the lack of awareness surrounding a problem may be related to low education levels and literacy rates. In addition, a perceived problem such as deforestation may be less important than the simple day-to-day struggle of survival and the inherent requirement for firewood. In a similar way, people may be closely linked to specific geographic areas and unwilling or unable to travel from their villages to see reforestation programs in their regions. Extension needs to successfully identify such constraints and progress can be further enhanced by producing a creative vision for positive impact (Baig et al., 2020, 2021; Saco et al., 2021).

3.6 Carbon Sequestration Potential Through Agroforestry

The planting of trees together with crops has many advantages, which enable to enhance soil fertility, limit soil erosion, lower water logging, decrease fermentation and eutrophication of streams and rivers, enhance local biodiversity, and reduce pressure on common forests for fuel, etc. Preservation of existing tree cover and advancement of agroforestry are the land-use management measures that will not only have positive impacts on biodiversity but also advance the use of biomass fuels, changing the fossil fuels, hence contributing to net reduction in CO_2 emissions (Baig et al., 2020, 2021; Murthy et al., 2013; Saco et al., 2021) (Table 6).

Gnanavelrajah et al., (2008) carried out a study to estimate and map carbon stock of different agricultural land uses in a sub-watershed of Thailand and to assess the land-use sustainability with respect to carbon management. They estimated a total carbon stock of agricultural lands to be 20.5 Tg out of which 41.49% was biomass carbon and 58.51% was soil carbon. Among the land uses, rubber showed the highest average biomass C (136.34 Mg C ha^{-1}) while minimum was in paddy (7.08 Mg C ha^{-1}). They concluded that such information on carbon stock could be valuable to develop viable land-use options for agricultural sustainability and carbon sequestration.

Global agricultural production is challenged by climate change (Godfray et al., 2010). Climate change affects different crops and regions differently, but it is generally expected that agricultural productivity will decline (Lobell & Field, 2007). Several researchers have determined that planting trees with crops results in higher carbon sequestration compared with croplands depending upon the environmental and socioeconomic conditions of the area (Nair et al., 2009; Sanchez, 2000). Around 45.3 PgC of the world's terrestrial carbon is currently stored in agricultural lands, with trees contributing 33.9 PgC (Zomer et al., 2016), and the potential could increase up to another 586 Tg $year^{-1}$ by 2040 if 630 million ha of unproductive croplands were converted into agroforestry globally (Smith, 2004).

Table 6 Carbon sequestration potential by 2040

Sources	Mt C $year^{-1}$
Water land restoration	20
Restoration of degraded land	50
Agroforestry	600
Forest management	250
Grazing management	375
Rice management	20
Cropland management	150

Source Abbas et al., (2017)

3.6.1 Accumulation of Carbon in Block and Boundary Plantations

Soil and above-ground carbon stock in block and boundary plantations in agroforestry systems may influence soil fertility and carbon build-up. For example, in India, there are several studies on carbon sequestration in tree biomass. In a study by Kumar (2010), results reported from 9-year-old block and boundary plantations of four different agroforestry systems (*Populus deltoides* block plantation + wheat, *Eucalyptus* hybrid boundary plantation + wheat, *Populus deltoides* boundary plantation + wheat and *Populus deltoides* block plantation + lemon grass). The author found that total carbon sequestration in trees was 70.59, 21.38, 116.29, and 18.53 t ha^{-1} in system *Populus deltoides* + wheat followed by 68.53, 20.63, 113.03, and 17.60 t ha^{-1} in system *Populus deltoides* + Lemon grass, respectively. It was noticed that agroforestry systems, integrating tree production with crop and animal production systems, are believed to have a higher potential to sequester C than pastures or field crops, based on the notion that tree incorporation in croplands and pastures would outcome in greater net aboveground as well as belowground C sequestration. These are indications for the characterization of agroforestry as a promising practice to sequester C while providing numerous environmental, economic, and social benefits. In addition, agroforestry protects existing C stocks if improved fallows and similar agroforestry practices could provide food and fuel wood, thereby reducing the rate of deforestation (Table 7).

4 Impact of Carbon Sequestration of Agroforestry System

Afforestation plays an important part in locking the carbon dioxide produced from human activities into plant biomass. In any case, by changing the biomass inputs into soil, afforestation can have huge impacts on soil organic carbon (SOC) pools, a few of which have exceptionally longer turnover times than plant biomass (Berthrong 2012; Baig et al., 2021). Agroforestry promotes the proper use of land to produce crops and livestock on a viable basis and can also help avoid land degradation (Baig et al., 2021).

Table 7 Carbon sequestration rates for selected agroforestry practices

Agroforestry practice (tones C eq ha^{-1} $yr^{-1)}$	Carbon sequestration rate	Region
Complex agroforestry	11	Central Africa
Complex agroforestry	8–21	Costa Rica
Boundary planting Hedgerow	35–50	Central Africa
Intercropping	1–37	Indonesia and India
Home gardens	145	Panama

Source Hutchinson et al., (2017)

Carbon sequestration in agroforestry system (AFS) is composed by the estimates of carbon stocks in its components: how much C is, or potentially could be, accumulated and stored in above- and below-ground compartments of AFS under different ecological and management conditions. It is estimated that it ranges from 0.29 to 15.21 Mg ha^{-1} $year^{-1}$ aboveground and 30–300 Mg C ha^{-1} up to 1-m depth in the soil (Nair et al., 2010). The United Nations Framework Convention on Climate Change (UNFCCC) defines carbon sequestration as "the process of removing C from the atmosphere and depositing it in a reservoir", and it entails the transfer of atmospheric CO_2, and its secure storage in long-lived pools (UNFCCC, 2007). From an agroforestry perspective, carbon sequestration primarily involves the uptake of atmospheric CO_2 during photosynthesis and the transfer of fixed C into the vegetation, detritus, and soil pools for "secure" (i.e., long-term) storage (Baig et al., 2021; Lorenz & Lal., 2014) (Fig. 6).

There are two major segments of the AFS: aboveground and belowground. Each can be partitioned into sub-segments: the former into specific plant parts (stem,

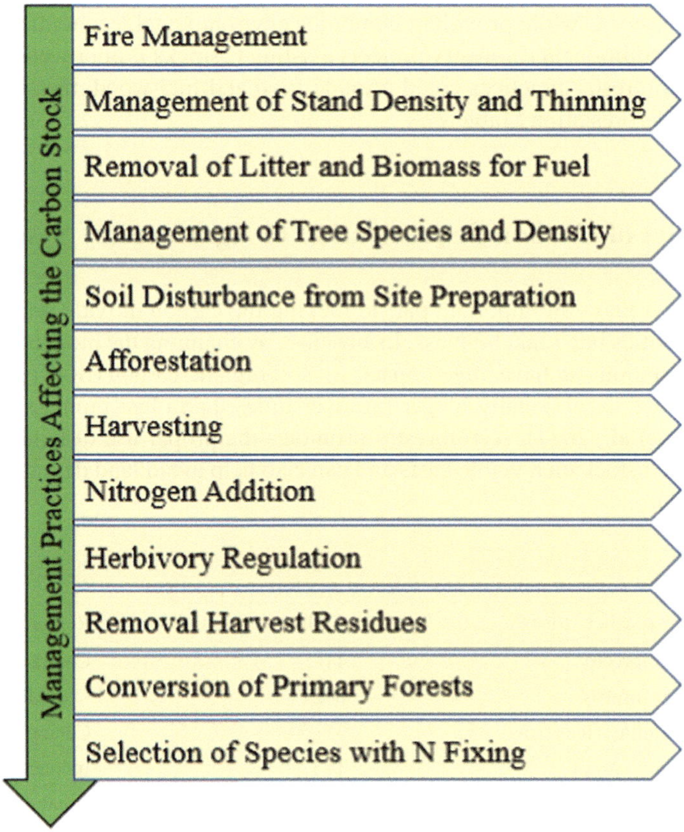

Fig. 6 Practices affecting the carbon stock

leaves, etc., of trees and herbaceous components), and the latter into living biomass such as roots and other belowground plant parts, soil organisms, and C stored in various soil horizons. On average, the aboveground parts and the soil (including roots and other living biomass) are estimated to hold roughly one-thirds and two-thirds, respectively, of the total C stored in tree-based land use systems (Lal, 2011). Based on the notion that tree incorporation in croplands and pastures would result in greater net C storage above- and below-ground (Haile et al., 2008; Palm et al., 2004), AFS are believed to have a higher potential to sequester C than pastures or field crops growing under similar ecological conditions (Kirby & Potvin, 2007; Roshetko et al., 2002).

4.1 Carbon Sequestration Potential of Soil

Enriching soil organic carbon (SOC) pools in agriculture by encouraging soil C sequestration is an efficient way towards diminishing atmospheric carbon dioxide (CO_2) level and inducing soil health (Bronick & Lal, 2005; Lal, 2011). By declaring 2015 as the "International Year of Soils", the United Nations General Assembly (UNGA) has recognized the major importance of the soil resource for human life and the need to maintain it for future generations (Baig et al., 2021; Stockmann et al., 2015). Soil, a natural four-dimensional body at the atmosphere-lithosphere interface, is organic carbon-mediated realm that generates numerous ecosystem goods and services. Soil carbon pool consists of two distinct pools: soil organic carbon (SOC) and soil inorganic carbon (SIC), whereas SOC is the main component of soil organic matter, and it strongly impacts soil quality, functionality and health (Baig et al., 2021; Lal, 2016).

The global magnitude (Pg) of soil organic carbon is 1.505–1.0 m depth (Baig et al., 2021; Lal, 2018), thus, even a small change in soil C stock can strongly impact the atmospheric stock (1 Pg of soil C stock = 0.47 ppm of atmospheric concentration of CO_2) (Trenberth and Guillemot, 1994). For instance, the decade of 2007–2016, the average anthropogenic emission of carbon has been estimated 1.3–0.7 Pg C/year from land use conversion. The continuous increase in the world's food-insecure population to 1.02 billion (Baig et al., 2021; Lal, 2011) is seriously threatening global food security and degradation of soil and water resources (Lal, 2009) only because soils are severely depleted of their soil organic carbon and nutrient reserves and are strongly degraded. Carbon dioxide (CO_2) and other greenhouse gases have been recognized as a serious environmental issue of the twenty-first century (Baig et al., 2021; Kumar and Nair 2011).

Although agricultural soils can be a good sink of atmospheric CO_2 (Al-Kaisi et al., 2005; Baig et al., 2021; Chen et al., 2009), land misuse and mismanagement can cause depletion of soil organic carbon stocks with a subsequent emission of CO_2 into the atmosphere (Chen et al., 2009; Lal, 2004). There is a strong link between food insecurity, soil degradation, and climate change, because of a relation among the relevant processes. Yet, the combined crisis of climate change and food insecurity, which may characterize the future (Baig et al., 2021; Lal, 2011) can be addressed through the restoration of the SOC pool and improvement in soil quality. Among several solutions being debated to mitigate climate change (Jacobson, 2009), an important option is sequestration of carbon in agro-ecosystems, especially in agricultural soils since soil C sequestration can be considered as a cost-effective strategy for mitigating climate change (Chen et al., 2009).

Soil organic carbon (SOC) depleted soils yield less but also have low-use efficiency of added inputs/nutrients and sequester less atmospheric CO_2. Thus, the process of sequestering atmospheric CO_2 will itself enhance SOC pool and off-set anthropogenic emissions while benefitting both agricultural productivity and mitigating temperature rise. Soils have a huge potential for either sequestering or releasing carbon into the atmosphere (Joschko et al., 2012).

Wang et al., (2018) conducted a study to predict and map the spatial distribution of soil organic carbon and total nitrogen in the north-eastern coastal areas of China. They generated a C/N ratio map from the predicted maps of SOC and TN. The same authors collected 149 soil samples from 0 to 20 cm depth and twelve environmental variables, including climatic variables, were assessed. The Boosted Regression Tree (BRT) model explained approximately 64% and 56% of the total SOC and TN variability. Topographic variables showed a maximum influence in the prediction of SOC and TN, followed by vegetation and climate. The spatial distribution of C/N ratio was closely related to land use types in a decreasing order: woodland > orchard > cultivated land > grassland. They found that topographic variables are the main environmental indicators of SOC and TN distribution and recommended to include them in future SOC and TN mapping studies.

4.1.1 Soil Organic Carbon Stabilization

SOC sequestration primarily depends on soil C input and stabilization processes (Mitchell et al., 2021). Plant root and rhizosphere inputs, in particular, make a large contribution to SOC (Schmidt et al., 2011). Accumulation of SOC is mainly the result of partial degradation, microbial products, and fire residues rather than humic substances. Physical disconnection, e.g., from enzymes, decomposers, electron acceptors, and sorption/desorption, i.e., organo-mineral associations, and freezing/thawing, govern SOC cycling and these processes are shaped by environmental conditions (Mitchell et al., 2021; Schmidt et al., 2011). Some surface residue C may be incorporated into the mineral soil by physical mixing and solubilization, transport, and subsequent adsorption (Lorenz & Lal, 2005). Plant roots, i.e., litter and rhizodeposition, are the primary vector for most C entering the SOC pool (Rasse et al., 2005).

The relative importance of root litter and rhizodeposition versus other incorporation processes for profile SOC distribution and dynamics depend on climate, soil, and vegetation types (Rumpel & Kögel-Knabner, 2011). The SOC stability represents an ecosystem property since it depends on the biotic and abiotic environment

(Schmidt et al., 2011). In contrast, the molecular structure of plant inputs and OM plays only a secondary role in determining SOC residence times over decades to millennia. However, microbially derived materials may play a crucial role in SOC stabilization (Kleber et al., 2011). The SOC turnover appears to be a function of microbial ecology and the resource availability within a given physical soil environment. Thus, processes that slow down mineralization are major centennial-scale stabilization mechanisms for SOC (Mitchell et al., 2021; Sanderman et al., 2009).

The two important groups of processes for long-term stabilization of SOC are: (a) processes that lead to physical protection, rendering OM spatially inaccessible to decomposers or their water-soluble degradative enzymes; and (b) organo-mineral complexes and organo-metal interactions, i.e., interactions of OM with minerals, metal ions, and other organic substances (Lützow et al., 2006). Physical protection may retard decomposition for decades to centuries, whereas organo-mineral complexes or organo-metal interactions may be responsible for stabilization of most of the non-charred SOC for centuries to millennia (Kögel-Knabner et al., 2008). Thus, stabilization of SOC in agroforestry systems is a combination of these short- and long-term processes (Mitchell et al., 2021; Six et al., 2000).

4.1.2 Reducing Soil Organic Carbon Loss

The biomass C input to the soil in agroforestry systems can be increased and its decomposition rate decreased by adoptions of conservation-effective measures that reduce losses of nutrients and water, increase biomass production, and protect SOC against losses through enhancing biological, chemical, and physical stabilization mechanisms (Heger et al., 2021). Thus, a decrease in cultivation intensity may result in an increase in SOC levels in agroforestry systems (Nair et al., 2010). Similar to other agricultural systems, adding amendments such as fertilizers and composts, supplying water through irrigation, and incorporating organic residues into soil may enhance SOC sequestration in agroforestry systems. However, fertilization studies on most tropical tree species are rare, thus preventing any conclusions on fertilization effects (Nair et al., 2010).

The management of agroforestry systems for SOC sequestration includes the selection of tree species and their silvicultural management such as stand density and rotation length (Nair et al., 2009). Functionally important tree species, i.e., those having deep and extensive root systems to enhance C input into the soil, may have a high potential to enhance SOC sequestration in agroforestry systems (Kell, 2012). Broadleaf trees, in particular, have a larger and more deeply anchored root system, i.e., higher root biomass/aboveground biomass ratios than coniferous tree species, and may, therefore, generate higher SOC inputs from roots at soil depth (Laganière et al., 2010). Mixed plantings including N-fixing trees produce more biomass, and this may result in increased SOC sequestration (Nair et al., 2009). However, positive, accretion is reported. Thus, appropriate tree species must be selected to enhance SOC sequestration (Heger et al., 2021; Oelbermann et al., 2006).

4.2 Carbon Sequestration Potential of Trees

Agroforestry system (AF) consists of mixture of trees, agricultural crops, and livestock to exploit the economic and ecological benefits of agro-ecosystem. It is a crucial leader of terrestrial C sequestration containing about 12% of the global terrestrial C (Baig et al., 2021). The store C in trees growing rapidly by exploring the benefits of favorable temperature at early growth stage (Rotenberg & Yakir, 2010). The roots of forest tress and perennial crops penetrate deeper subsurface horizons, thus placing SOC at deeper horizons far away from the range of tillage implements (Baig et al., 2021; Lorenz & Lal, 2014) (Table 8).

Nair et al., (2010) estimate the C sequestration potential of AFS under varied ecological and management environment ranged from 0.29 to 15.21 Mg ha^{-1} $year^{-1}$ in aboveground plant biomass and 30–300 Mg ha^{-1} $year^{-1}$ in belowground plant parts up to a depth of 1.0 m. Aboveground biomass is a direct measure of C sequestration, assuming that 50% of the biomass is made up by C. Carbon sequestration is a long-term storage of carbon and is an important way to reduce the concentration of carbon dioxide in the atmosphere. There are several possibilities for carbon sinks, such as soil, oceans, rocks, atmosphere, and biosphere (VijayaVenkataRaman et al., 2012). Photosynthesis is a process that contributes to the assimilation and retention of carbon in the plant. This carbon can be distributed in four pools (Bolinder et al., 2007).

Carbon was estimated as 50% of the biomass of the tree (Afzal & Aqeela, 2013; Nawaz et al., 2017a). Field trends are changing for planting farm trees due to dieback disease in shisham (*D. Sissoo*) (Nawaz et al., 2016). Shisham was a very popular tree among farmers and was intensively planted with crops due to its growing characteristics (nitrogen fixing timber tree species) and value in market. However, recently, *E. camaldulensis* (Sufaida) is replacing the Shisham only due to fast growing nature and better profitability. Other than sufaida, widely planted trees are Kikar (*Acacia nilotica*) and Poplar (*Populus deltoides*). Similar findings are reported in Nawaz et al., (2016, 2017b).

On irrigated farmlands, being fast-growing trees, both *D. sissoo* and *E. camaldulensis* show almost similar primary and secondary growth under semi-arid climatic conditions. However, contrary to forest or plantations, secondary growths are more linked with tree ages. Under favorable conditions, this tree can grow up to 30 m height. It belongs to the family *Fabaceae* (Papilionoideae) and commonly called as Indian rosewood, Shisham, *Sissoo* tree. *D. sissoo* is found in tropical to subtropical climates in natural and planted forests, and native to India, Nepal, and Pakistan. This tree has been widely used for afforestation in most parts of the country. There is remarkable variation in growth pattern and yield per unit area due to the wide adaptability of the tree in different ecological sites. *D. sissoo* has been used for reclamation of sodic soil (Tyagi et al., 2009) and mine spoil (Mukhopadhyay et al., 2013).

D. sissoo has been widely used for reclamation of different types of soil. Carbon stock was 29.5 t/ha for 9 years old stand on a sodic soil (Tyagi et al., 2009); 97.8 t/ha (16 years) in alluvium soil (Gulzar et al., 2014); 35.8 t/ha (10 years) in Tarai soil (Kanime et al., 2013); 51.5 t/ha (8 years) for afforested site in plain area. In

Table 8 Some common tree's names and characteristics

Scientific name	Common name	Characteristics
A. nilotica	Kikar	*A. nilotica* is a tall tree (6–8 m high) with compound leaves, tiny leaflets, yellow and fragrant round flowers, sharp thorns, ball-like crown, stem in form of legume which flourishes in spring and autumn. It grows in semi-arid, hot and wet regions (Dorostkar, 2015)
A. indica	Neem	Neem tree belongs to the family Meliaceae, which is found in abundance in tropical and semitropical regions like India, Bangladesh, Pakistan, and Nepal. It is a fast-growing tree with 20–23 m tall and trunk is straight and has a diameter around 4–5 ft. The leaves are compound, imparipinnate, with each comprising 5–15 leaflets (Alzohairy, 2016)
E. camaldulensis	Safeeda	Large evergreen tree 50 m high with stout trunk often short and crooked, to 2 m in diameter; crown open, widely spreading, irregular. Leaves grey-blue, alternate, drooping, 8–22 cm long, 1–2 cm wide (Orwa et al., 2009)
M. azedarach	Dahreek	*M. azedarach* can attain a height of 15 m. The leaves are up to 50 cm long, alternate, long-petioled, two- or three-times compound (odd-pinnate); the leaflets are dark green above and lighter green below, with serrate margins. *M. azedarach* has a high fruit and seed output, and the fruits are consumed by birds which then disperse the seeds (Rahman et al., 2014)
M. alba	Shehtoot	*Morus alba* is a fast-growing shrub or moderate-sized tree with a fairly cylindrical, straight bowl, up to 35 m high and 1.8 m in girth. The species is now extensively planted and widely naturalized throughout the warm temperate world and in subarctic regions as well. The fruits are also eaten, often dried or made into wine (Burgess et al., 2005)
P. deltoides	Popular	Populus deltoides is a deciduous Tree growing to 30 m (98ft) by 20 m (65ft) at a fast rate. *P. deltoides* tolerates frost, heavy soil, sand, slope, and waterlogging. It is one of the fastest growing trees, often planted where fast growth is the main requirement (Orwa 2009)
S. cumini	Jaman	*Syzygium cumini* is a multipurpose large, evergreen native tree species distributed in south and Southeast Asia from Pakistan to Australia, growing up to 30 m tall, Flowers are white, 7.5–13 mm across, in branched clusters at stem tips. Fruits are variable in size, up to 2.5 cm long, ellipsoid. It grows well in sandy soils, riverbeds and clayey loam soils, poorly drained soil too. This tree is a pollutant tolerant tree (Stephen, 2012)
Z. mauritiana	Beeri	*Z. mauritiana* species has a wide range of morphologies from shrubs to small or medium-sized trees which might be erect, semierect or spreading. Height can vary from 3–4 to 10–16 m or more although trees of 20 m are rare, the bark has deep longitudinal furrows and is greyish brown or reddish in color (Jin, 2018)

(continued)

Table 8 (continued)

Scientific name	Common name	Characteristics
D. sissoo	Shesham	Dalbergia sissoo is a medium to large tree of about 10–15 m meters high in dry areas, and up to 25 m in wet areas. It is deciduous, with a light crown trunk. Leaves are compound, with about five alternates leaflets (Naqvi et al., 2019). Shesham is a cash tree because it financially supports the farmer by selling wood which is used in making furniture, plywood industries, construction work and fuel purposes (Azaizeh et al., 2003)
C. sinensis L	Malta	*Citrus sinensis L.* could create a sustainable ecosystem with favorable C sequestration potential due to their year-round activity and tree longevity (>30 years), vigorous growth, large root system, high leaf area index (Azaizeh et al., 2003)
P. ciliata	Shehtoot	*Populus ciliata* is a deciduous tree growing to 20 m (65ft 7in) at a fast rate and an easily grown plant. It is harvested from the wild for local use as a medicine and source of materials. Used for timber for construction and making furniture. It is the source of fiber for various grades of papers namely, fine paper, packing paper, newsprint, etc. (Orwa 2009)
Bambusoideae	Bans Bamboo	Bamboos are the fastest growing plants in the world having growth up to 60 cm or more and it belongs to grass family (Bhonde, 2014). Bamboo resources have a considerable potential to socio-economic development of the local people; environmental protection such as adaptation to global climatic change through carbon sequestration; biodiversity conservation and wider ecological adaptation (Ceccon and Ruiz, 2019). Bamboo can stock more carbon than fast-growing tropical and subtropical trees (Thokchom & Yadava, 2015)

reclamation studies, the superior performance of *Dalbergia* in comparison with other trees has been reported (Shankar et al., 2014). The carbon stock calculated for the mining sites in the present study is lower than that of other sites (Kanime et al., 2013; Lodhiyal & Lodhiyal, 2003; Tyagi et al., 2009; Yadava, 2011).

Arora and Chaudhry (2017) explained that the soil respiration or CO_2 evolution rates from the soil of *A. nilotica* + *D. sissoo* plantation ranged from 1.16 to 5.04 g CO_2-C $m^{-2}d^{-1}$ with the highest rate of 5.04 CO_2-C $m^{-2}d^{-1}$ in the month of August than other months indicating that rainy season accounted for higher soil respiration as compared to the dry months during the study period. (Rojas et al., 2010) estimates both the carbon stored aboveground, and belowground in the Bamboo; the majority find that approximately two-thirds of the carbon sequestered is found in aboveground biomass, with the remainder below ground. The average aboveground carbon density reported by these studies ranges from roughly 15 to 65 tons ha^{-1}; below ground the range is roughly 10 to 35 t ha^{-1}. The culm density ranges from approximately 2,000–4,500 culms ha^{-1}.

Orange (*Citrus sinensis* L.) is one of the main fruit crops worldwide and its evergreen orchards may have a great potential for carbon (C) sequestration. Evergreen

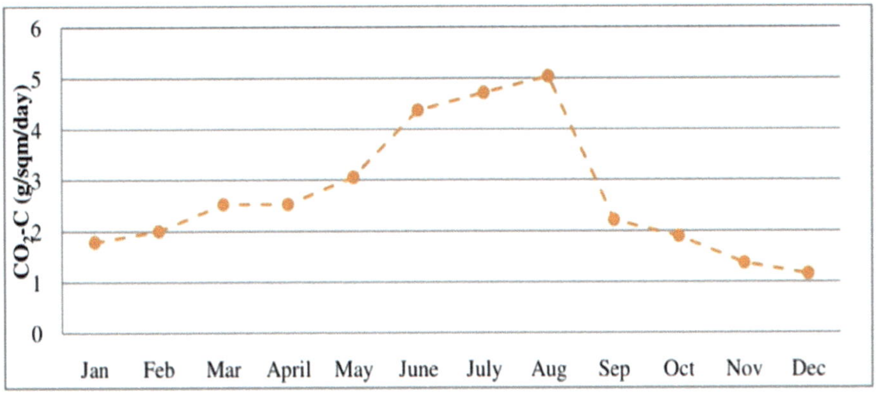

Fig. 7 Monthly variation in CO_2 evolution rates from the soil of *A. nilotica* + *D. sissoo* plantation. *Source* Arora and Chaudhry (2017)

species such as orange (*Citrus sinensis* L.)—3.9 million ha globally (FAO, 2007)—could create a sustainable ecosystem with favorable C sequestration potential, due to their year-round activity and tree longevity (>30 years), vigorous growth, large root system, high leaf area index and slow leaf turnover. Carbon fixation in the fruits and in the canopy of single trees was almost twice as much (10.7 kg C/tree) in the traditional than in the intensive system (5.5 kg C/tree); however, total NPP of the orchard did not change with planting density, being 5.3 t C/ha/year in the traditional system and 5.5 t C/ha/ year in the intensive one. Carbon fixation by the ground cover was higher in the traditional (1.1 t C/ha/year) than in the intensive system (0.5 t C/ ha/year) (Fig. 7).

4.2.1 Trees and Soil Organic Carbon

Trees have extensive root systems which can grow deep into the mineral soil. The root-derived C inputs are critical sources for the SOC pool in deeper soil horizons (Kell, 2012). Specifically, root-derived C is more likely to be stabilized in the soil by physicochemical interactions with soil particles than shoot-derived C (Lantini, 2021). Agroforestry systems store more C in deeper soil layers near trees than away from trees (Nair et al., 2010). Mixed plantings with nitrogen fixing trees may cause higher biomass production and, thus, SOC sequestration and pools, particularly in deeper soil horizons as N may promote humification rather than decay, but SOC and N interactions are not entirely understood (Gärdenäs et al., 2011; Samariks et al., 2021).

Also, changes in microbial decomposer community composition under N-fixing trees may result in greater retention of relatively stable SOC (Resh et al., 2002). Among the reasons for the positive effects of trees on SOC sequestration are that trees modify the quality and quantity of belowground litter C inputs and modify microclimatic conditions such as soil moisture and temperature regimes (Laganière et al., 2010). In addition to fixing N, fertilizer trees may recycle the soil's phosphorus, calcium, magnesium, and potassium (Ajayi et al., 2011; Lantini, 2021).

4.2.2 Afforestation Effects on Soil Organic Carbon

The introduction of trees on previously treeless cropland, pasture, or natural grassland, may provide some insight on the potential effects of agroforestry trees on SOC sequestration (An et al., 2021). In temperate regions, afforestation of former cropland caused a long-lasting SOC sink, but the majority of afforested grasslands lost SOC (Poeplau et al., 2011). The planting of N-fixing trees for afforestation can increase the SOC pool as indicated by the more than 30% increase in SOC pools when N fixers are present in forest stands (Resh et al., 2002). However, the long-term tree productivity in plantations and, thus, C inputs to soils may be reduced as trees take up considerable amounts of nutrients from the soil which may be partially removed by repeatedly harvesting tree biomass (An et al., 2021; Laganière et al., 2010).

4.3 Carbon Sequestration Potential of Crops

There are more than 50,000 edible plants but only three of them, namely, wheat, rice, and maize provide 60% of the world's food. Wheat was one of the first domesticated plants and it continues to be the main food grain crop consumed by humans in the past 8000 years. Presently, the world's population is more than eight billion, which is growing by 83 million people each year, and it is considered to reach nine billion by the year 2044. Thus, a projection of 786 million tons of wheat will be needed annually for human use in the year 2025. This scenario calls for a rapid and continuous increase in wheat production (Rao et al., 2009).

Pakistan's *per capita* wheat production has been static in the range of 120–140 kg *per annum* since 1980. Thus, if the population keeps on growing at its present trend by the year 2030, the country will need another 8 million tons of wheat *per annum* (Ali et al., 2013). Punjab is the major wheat producer in the country and here more than 60% of the area under wheat crop comes after the cultivation of cotton or rice crops. In the southern zone of the province where cotton is grown, especially Vehari and Khanewal districts, only 20% of wheat is sown at optimum sowing time, i.e., first fortnight of November, while the remaining sowing is done in late November (30%) to December (50%) (Khan et al., 2002).The country's agriculture is solely based on irrigation water but unfortunately water availability *per capita* is declining at an alarming rate from 5260 m^3 in 1951 to 1066 m^3 in 2010 and will be less than 870 m^3 *per capita* by 2025 (Shaukat, 2011).

Nawaz et al., (2017b) stated that farmers are mostly planting diverse trees species (not mono-species per plot) along with all crops. However, four types of tree-crop combinations were frequent as compared to all other types. Farmers preferred to plant trees intercropped with wheat (*Triticum aestivum*), sugarcane (*Saccharum officinarum),* berseem (*Trifolium alexandrinum*) and sarsoun or mastard (*Brassica rapa*). At the soil depth of 0–15 cm, it was estimated that maximum amount of soil carbon was present under the combination of trees plus berseem crop (0.45%). Trees grown in combination with wheat crop captures less soil carbon than tree plus sugarcane but more carbon than the sarsoun plus trees. At 15–30 cm soil depth, the minimum carbon (%) was observed for sarsoun plus trees with no significant difference for other three combinations. The overall carbon percentage in the soils was less than 0.5, signifying that studied soils were carbon deficient and had poor organic matter contents.

5 Conclusion

The evidence suggests that conversion of forest to agriculture land use results in land degradation with huge losses of soil organic carbon stocks. Cultivation of land releases about 20–70% of the stored carbon within two decades depending on the climatic conditions. Resultantly, CO_2 is building up in atmosphere. Agroforestry systems retain much higher quantities of carbon in above and belowground biomass in comparison to crop and grazing land-use systems. At the global scale, 630 million ha of unproductive croplands could be used for agroforestry as part of an ecological engineering practice to potentially sequester 586,000 MgC year^{-1} by 2040. Moreover, in current global and national carbon monitoring protocols, there is a need to incorporate agroforestry in carbon stocks to precisely estimate the contribution of this neglected pool. To simulate the potential of agroforestry systems in sequestering carbon, new models are needed that can precisely predict net uptake of atmospheric CO_2 compared to treeless systems especially under the IPCC scenarios of projected global climate change. Given appropriate incentives, Punjab's farmers could help Pakistan meet its commitments in Paris Agreement through reasonable changes in tree planting on existing agroforestry systems.

References

Abbas, F., Hammad, H. M., Fahad, S., Cerdà, A., Rizwan, M., Farhad, W., & Bakhat, H. F. (2017). Agroforestry: A sustainable environmental practice for carbon sequestration under the climate change scenarios a review. *Environmental Science and Pollution Research, 24*(12), 11177–11191.

Afzal, M., & Aqeela, M. A. (2013). Factors Affecting Carbon Sequestration in Trees. *Journal of Agricultural Research, 51*(1), 61–68.

Ahmad, A., Shahbaz, B., Shehzad, M., Khursheed, K., & Aftab, M. (2017). Analysis of performance of farm forestry and its role in household income in district Faisalabad, Pakistan. *Journal of Global Innovations in Agricultural and Social Sciences, 5*, 39–42.

Ahmed, R., Liu, G., Yousaf, B., Abbas, Q., Ullah, H., & Ali, M. U. (2020). Recent advances in carbon-based renewable adsorbent for selective carbon dioxide capture and separation-A review. *Journal of Cleaner Production, 242*, 118409.

Ahmed, T., Scholz, M., Al-Faraj, F., & Niaz, W. (2016). Water-related impacts of climate change on agriculture and subsequently on public health: A review for generalists with particular reference to Pakistan. *International Journal of Environmental Research and Public Health, 13*(11), 1051.

Ajayi, O. C., Place, F., Akinnifesi, F. K., & Sileshi, G. W. (2011). Agricultural success from Africa: The case of fertilizer tree systems in southern Africa (Malawi, Tanzania, Mozambique, Zambia and Zimbabwe). *International Journal of Agricultural Sustainability, 9*(1), 129–136.

Ali, H., Iqbal, N., Ahmad, S., Shahzad, A. N., & Sarwar, N. (2013). Performance of late sown wheat crop under different planting geometries and irrigation regimes in arid climate. *Soil and Tillage Research, 130*, 109–119.

Ali, S., Liu, Y., Ishaq, M., Shah, T., Ilyas, A., & Din, I. U. (2017). Climate change and its impact on the yield of major food crops: Evidence from Pakistan. *Foods, 6*(6), 39.

Al-Kaisi, M. M., Yin, X., & Licht, M. A. (2005). Soil carbon and nitrogen changes as influenced by tillage and cropping systems in some Iowa soils. *Agriculture, Ecosystems & Environment, 105*(4), 635–647.

Alzohairy, M. A. (2016). Therapeutics role of Azadirachta indica (Neem) and their active constituents in diseases prevention and treatment. *Evidence-Based Complementary and Alternative Medicine.*

An, Z., Bernard, G. M., Ma, Z., Plante, A. F., Michaelis, V. K., Bork, E. W., & Chang, S. X. (2021). Forest land-use increases soil organic carbon quality but not its structural or thermal stability in a hedgerow system. *Agriculture, Ecosystems & Environment, 321*, 107617.

Arévalo-Gardini, E., Farfán, A., Barraza, F., Arévalo-Hernández, C. O., Zúñiga-Cernades, L. B., Alegre, J., & Baligar, V. C. C. (2021). Growth, Physiological, Nutrient-Uptake-Efficiency and Shade-Tolerance Responses of Cacao Genotypes under Different Shades. *Agronomy, 11*(8), 1536.

Arora, P., & Chaudhry, S. (2017). Vegetation and soil carbon pools of mixed plantation of *Acacia nilotica* and *Dalbergia sissoo* under social forestry scheme in Kurukshetra, India. *Journal of Materials and Environmental Science, 8*, 4565–4572.

Awazi, N. P., Tchamba, M. N., Temgoua, L. F., & Tientcheu-Avana, M. L. (2021). Agroforestry as an adaptation option to climate change in Cameroon: Assessing farmers' preferences. *Agricultural Research*, 1–12.

Azaizeh, H., Fulder, S., Khalil, K., & Said, O. (2003). Ethnomedicinal knowledge of local Arab practitioners in the Middle East Region. *Fitoterapia, 74*, 98–108.

Baig, M. B., Burgess, P. J., & Fike, J. H. (2021). Agroforestry for healthy ecosystems: Constraints, improvement strategies and extension in Pakistan. *Agroforestry Systems, 95*(5), 995–1013.

Bales, K. (2012). *Disposable people.* University of California Press.

Berthrong, S. T., PIneiro, G., Jobbágy, E. G., & Jackson, R. B. (2012). Soil C and N changes with afforestation of grasslands across gradients of precipitation and plantation age. *Ecological Applications, 22*(1), 76–86.

Bhonde, D., Ngarnaik, P. B., Parbhat, K. D., Waghe, P. U. (2014). Physical and Mechanical Properties of Bamboo (*Dendrocalmus Strictus*). *International Journal of Scientific & Engineering Research 5*(1).

Bolinder, M. A., Janzen, H. H., Gregorich, E. G., Angers, D. A., & VandenBygaart, A. J. (2007). An approach for estimating net primary productivity and annual carbon inputs to soil for common agricultural crops in Canada. *Agriculture, Ecosystems & Environment, 118*(1–4), 29–42.

Bronick, C. J., & Lal, R. (2005). Soil structure and management: A review. *Geoderma, 124*(1–2), 3–22.

Burgess, K. S., Morgan, M., Deverno, L., & Husband, B. C. (2005). Asymmetrical introgression between two *Morus* species *(M. alba, M. rubra)* that differ in abundance. *Molecular Ecology., 14*, 3471–3483.

Calzadilla, A., Rehdanz, K., Betts, R., Falloon, P., Wiltshire, A., & Tol, R. S. (2013). Climate change impacts on global agriculture. *Climatic Change, 120*(1), 357–374.
Carignan, C. C., Punshon, T., Karagas, M. R., Cottingham, K. L., & Hanover, N. H. (2017). HHS Public Access. *Ann Glob Health.* Author manuscript; available in PMC.
CDIAC. (2000). Carbon Dioxide Information Analysis Center, US Department of Energy, Oak Ridge National Laboratory, http://cdiac.esd.ornl.gov. Accessed August 11, 2017.
Ceccon, E., & Gómez-Ruiz, P. A. (2019). Bamboos ecological functions on environmental services and productive ecosystems restoration. *Revista De Biología Tropical, 67*(4), 679–691.
Chandio, A. A., Jiang, Y., Akram, W., Adeel, S., Irfan, M., & Jan, I. (2021). Addressing the effect of climate change in the framework of financial and technological development on cereal production in Pakistan. *Journal of Cleaner Production, 288*, 125637.
Chaudhry, Q. U. Z. (2017). *Climate change profile of Pakistan.* Asian Development Bank.
Chaudhry, Q. U. Z., Mahmood, A., Rasul, G., & Afzaal, M. (2009). Climate change indicators of Pakistan. *Pakistan Meteorological Department.*
Chen, H., Marhan, S., Billen, N., & Stahr, K. (2009). Soil organic-carbon and total nitrogen stocks as affected by different land uses in Baden-Württemberg (southwest Germany). *Journal of Plant Nutrition and Soil Science, 172*(1), 32–42.
Dalby, S. (2021). *Global climate change and security threats.* Edward Elgar Publishing.
Davenport, F. V., & Diffenbaugh, N. S. (2021). Using machine learning to analyze physical causes of climate change: A case study of US Midwest extreme precipitation. *Geophysical Research Letters*, e2021GL093787.
De Leijster, V., Santos, M. J., Wassen, M. W., García, J. C., Fernandez, I. L., Verkuil, L., & Verweij, P. A. (2021). Ecosystem services trajectories in coffee agroforestry in Colombia over 40 years. *Ecosystem Services, 48*, 101246.
Dehlavi, A., Groom, B., & Gorst, A. (2015). Climate change adaptation in the indus ecoregion: a microeconometric study of the determinants, impacts, and cost effectiveness of adaptation strategies. *Islamabad: World Wide Fund for Nature (WWF) Pakistan.*
Dorostkar, A. (2015). Investigating the properties of "Acacia nilotica" as a species with capability of utilization in furniture industry. *International Journal of Innovative Science, Engineering & Technology. Eng Tech, 1*(10), 748–752.
Eckstein, D., Künzel, V., & Schäfer, L. (2017). *Global climate risk index 2018.* Germanwatch, Bonn.
EIU. (2015). Food Security Index. The Economist Intelligence Unit (EIU). Available at http://foo dsecurityindex.eiu.com
Essa, M., Nizami, S. M., Mirza, S. N., Khan, I. A., & Athar, M. (2011). Contribution of agroforestry in farmers' livelihood and its impact on natural forest in northern areas of Pakistan. *African Journal of Biotechnology, 10*(69), 15529–15537.
Fanzo, J. (2015). Ethical issues for human nutrition in the context of global food security and sustainable development. *Global Food Security, 7*, 15–23.
FAO (2007). Foastat. Available online at http://faostat.fao.org (verified 5 May 2009).
FAO (2018). *Food security and nutrition around the world: hunger is on the rise.* Rome: FAO. https://www.fao.org/state-of-food-security-nutrition/en. Accessed November 2019.
Farooq, T. H., Gautam, N. P., Rashid, M. H. U., Gilani, M. M., Nemin, W., Nawaz, M. F., & Wu, P. (2018). Contributions of agroforestry on socioeconomic conditions of farmers in central Punjab, Pakistan–a case study.
Farooq, T. H., Nawaz, M. F., Khan, M. W., Gilani, M. M., Buajan, S., Iftikhar, J., & Wu, P. (2017). Potentials of agroforestry and constraints faced by the farmers in its adoption in District Nankana Sahib, Pakistan. *International Journal of Sustainable Development, 6*, 586–593.
Ferranti, P. (2016). Food sustainability, security, and effects of global change. *Ref. Mod. Food Sci.*
Field, C. B., Barros, V., Stocker, T. F., & Dahe, Q. (Eds.). (2012). *Managing the risks of extreme events and disasters to advance climate change adaptation: special report of the intergovernmental panel on climate change.* Cambridge University Press.

Gärdenäs, A. I., Ågren, G. I., Bird, J. A., Clarholm, M., Hallin, S., Ineson, P., & Stendahl, J. (2011). Knowledge gaps in soil carbon and nitrogen interactions–From molecular to global scale. *Soil Biology and Biochemistry, 43*(4), 702–717.

Gillespie, T. W., Lipkin, B., Sullivan, L., Benowitz, D. R., Pau, S., & Keppel, G. (2012). The rarest and least protected forests in biodiversity hotspots. *Biodiversity and Conservation, 21*, 3597–3611.

Gnanavelrajah, N., Shrestha, R. P., Schmidt-Vogt, D., & Samarakoon, L. (2008). Carbon stock assessment and soil carbon management in agricultural land-uses in Thailand. *Land Degradation & Development, 19*(3), 242–256.

Godfray, H. C. J., Beddington, J. R., Crute, I. R., Haddad, L., Lawrence, D., Muir, J. F., & Toulmin, C. (2010). Food security: The challenge of feeding 9 billion people. *Science, 327*(5967), 812–818.

Gulzar, S., Nizami, S. M., Saleem, A., Mirza, S. N., & Mahmood, T. (2014). Allometric relationships for predicting the biomass in *Dalbergia sissoo roxb.* plantation in Punjab, Pakistan. *Pakistan Journal of Agricultural Sciences, 51*(3), 525–529.

Gupta, R. K., Kumar, V., Sharma, K. R., Buttar, T. S., Singh, G., & Mir, G. (2017). Carbon sequestration potential through agroforestry: A review. *International Journal of Current Microbiology and Applied Science, 6*(8), 211–220.

Haile, S. G., Nair, P. R., & Nair, V. D. (2008). Carbon storage of different soil-size fractions in Florida silvopastoral systems. *Journal of Environmental Quality, 37*(5), 1789–1797.

Harsch, M. A., Phillips, A., Zhou, Y., Leung, M. R., Rinnan, D. S., & Kot, M. (2017). Moving forward: Insights and applications of moving-habitat models for climate change ecology. *Journal of Ecology, 105*(5), 1169–1181.

Hasan, G. (2014). Present and future prospects of agroforestry as perceived by farmers in Punjab, Pakistan. *The International Asian Research Journal*, 11–17.

Heger, A., Becker, J. N., Navas, L. K. V., & Eschenbach, A. (2021). Factors controlling soil organic carbon stocks in hardwood floodplain forests of the lower middle Elbe River. *Geoderma, 404*, 115389.

Hidayat, N., & Sianipar, J. (2021). The Potential of Agroforestry in Supporting Food Security for Peatland Community–A Case Study in the Kalampangan Village Central Kalimantan. *Journal of Ecological Engineering, 22*(8), 123–130.

Huang, J. K. (2014). Climate change and agriculture: Impact and adaptation. *Journal of Integrative Agriculture, 13*(4), 657–659.

Hussain, M., & Mumtaz, S. (2014). Climate change and managing water crisis: Pakistan's perspective. *Reviews on Environmental Health, 29*(1–2), 71–77.

Hussain, M., Butt, A. R., Uzma, F., Ahmed, R., Irshad, S., Rehman, A., & Yousaf, B. (2020). A comprehensive review of climate change impacts, adaptation, and mitigation on environmental and natural calamities in Pakistan. *Environmental Monitoring and Assessment, 192*(1), 1–20.

Hussain, M., Butt, A. R., Uzma, F., Ahmed, R., Islam, T., & Yousaf, B. (2019). A comprehensive review of sectorial contribution towards greenhouse gas emissions and progress in carbon capture and storage in Pakistan. *Greenhouse Gases: Science and Technology, 9*(4), 617–636.

Hussain, M., Liu, G., Yousaf, B., Ahmed, R., Uzma, F., Ali, M. U., & Butt, A. R. (2018). Regional and sectoral assessment on climate-change in Pakistan: Social norms and indigenous perceptions on climate-change adaptation and mitigation in relation to global context. *Journal of Cleaner Production, 200*, 791–808.

Hutchinson, J. J., Campbell, C. A., & Desjardins, R. L. (2007). Some perspectives on carbon sequestration in agriculture. *Agricultural and Forest Meteorology, 142*(2–4), 288–302.

ICAR. (2006). Handbook of Agriculture (5thedn). Indian Council of Agricultural Research, New Delhi.

IPCC. (2000): Summary for Policymakers: The Science of Climate Change. Intergovernmental Panel on Climate Change. IPCC Working Group I, 26 Feb.

IPCC. (2014) Chapter 7: Food security and food production systems. Contribution of Working Group II to the Fifth Assessment Report of the Intergovernmental Panel on Climate Change

(IPCC). Available at: http://ipcc-wg2.gov/AR5/images/uploads/WGIIAR5-Chap7_FGDall.pdf (Accessed April 11, 2014a).

Iqbal, M. M., & Arif, M. (2010). Climate-change aspersions on food security of Pakistan. *A Journal of Science for Development, 15.*

Jahan, S., Falah, S., Ullah, H., Ullah, A., & Rauf, N. (2016). Antioxidant enzymes status and reproductive health of adult male workers exposed to brick kiln pollutants in Pakistan. *Environmental Science and Pollution Research, 23*(13), 12932–12940.

Jama, B., Kwesiga, F., & Niang, A. (2006). Agroforestry innovations for soil fertility management in sub-Saharan Africa: Prospects and challenges ahead. *World Agroforestry into the Future, 60.*

Jamilu, A., Ammar, H., & Gardish, D. M. (2014). Factors upsetting agroforestry system in Swat, Pakistan. *Int J Agrofor Silvic, 1*, 86–92.

Jacobson, M. Z. (2009). Review of solutions to global warming, air pollution, and energy security. *Energy & Environmental Science, 2*(2), 148–173.

Jin. X., (2018). Jujuba Ziziphus jujube. Exotic Fruits. Pages 263–269

Jones, S. K., Helfter, C., Anderson, M., Coyle, M., Campbell, C., Famulari, D., & Skiba, U. M. (2017). The nitrogen, carbon and greenhouse gas budget of a grazed, cut and fertilised temperate grassland. *Biogeosciences, 14*(8), 2069–2088.

Joschko, M., Barkusky, D., Rogasik, J., Fox, C. A., Rogasik, H., Gellert, R., & Gerlach, F. (2012). On-farm study of reduced tillage on sandy soil: Effects on soil organic carbon dynamic and earthworm abundance. *Archives of Agronomy and Soil Science, 58*(supl), S252–S260.

Kamal, A., Malik, R. N., Martellini, T., & Cincinelli, A. (2014). Cancer risk evaluation of brick kiln workers exposed to dust bound PAHs in Punjab province (Pakistan). *Science of the Total Environment, 493*, 562–570.

Kanime, N., Kaushal, R., Tewari, S. K., Raverkar, K. P., Chaturvedi, S., & Chaturvedi, O. P. (2013). Biomass production and carbon sequestration in different tree-based systems of Central Himalayan Tarai region. *Forests. Trees and Livelihoods, 22*(1), 38–50.

Karki, M. B., & Chowdhary, C. L. (2019). Non-timber Forest Products (NTFP) and Agro-forestry Subsectors: Potential for Growth and Contribution in Agriculture Development. *Agricultural Transformation in Nepal: Trends, Prospects, and Policy Options*, 385–419.

Kell, D. B. (2012). Large-scale sequestration of atmospheric carbon via plant roots in natural and agricultural ecosystems: why and how. *Philosophical Transactions of the Royal Society B: Biological Sciences, 367*(1595), 1589–1597.

Khan, M. A., Ahmad, N., Khan, M. A., & Sikandar, Z. (2002). Wheat planting pattern in Punjab and its economic implications. *Journal of Agricultural Research, 40*(3–4), 249–258.

Khan, M. A., Khan, J. A., Ali, Z., Ahmad, I., & Ahmad, M. N. (2016). The challenge of climate change and policy response in Pakistan. *Environmental Earth Sciences, 75*(5), 412.

Kiran, A. (2017). Climate Change: Implications for Pakistan and Way Forward. *ISSRA PAPERS, 9*(II).

Kirby, K. R., & Potvin, C. (2007). Variation in carbon storage among tree species: Implications for the management of a small-scale carbon sink project. *Forest Ecology and Management, 246*(2–3), 208–221.

Kleber, M., Nico, P. S., Plante, A., Filley, T., Kramer, M., Swanston, C., & Sollins, P. (2011). Old and stable soil organic matter is not necessarily chemically recalcitrant: Implications for modeling concepts and temperature sensitivity. *Global Change Biology, 17*(2), 1097–1107.

Kögel-Knabner, I., Guggenberger, G., Kleber, M., Kandeler, E., Kalbitz, K., Scheu, S., & Leinweber, P. (2008). Organo-mineral associations in temperate soils: Integrating biology, mineralogy, and organic matter chemistry. *Journal of Plant Nutrition and Soil Science, 171*(1), 61–82.

Krayem, M., El Khatib, S., Hassan, Y., Deluchat, V., & Labrousse, P. (2021). In search for potential biomarkers of copper stress in aquatic plants. *Aquatic Toxicology*, 105952.

Kumar, A. K. (2010). Carbon sequestration: Underexplored environmental benefits of Tarai agroforestry. *Indian Journal of Soil Conservation, 38*, 125–131.

Kumar, B. M., & Nair, P. R. (Eds.). (2011). Carbon sequestration potential of agroforestry systems: opportunities and challenges.

Kurukulasuriya, P., Mendelsohn, R., Hassan, R., Benhin, J., Deressa, T., Diop, M., & Dinar, A. (2006). Will African agriculture survive climate change. *The World Bank Economic Review, 20*(3), 367–388.799–822.

Laganière, J., Angers, D. A., & Pare, D. (2010). Carbon accumulation in agricultural soils after afforestation: A meta-analysis. *Global Change Biology, 16*(1), 439–453.

Lal, R. (2009). Soil carbon sequestration for climate change mitigation and food security.

Lal, R. (2004). Soil carbon sequestration to mitigate climate change. *Geoderma, 123*(1–2), 1–22.

Lal, R. (2011). Sequestering carbon in soils of agro-ecosystems. *Food Policy, 36*, S33–S39.

Lal, R. (2016). Soil health and carbon management. *Food Energy SecurIty, 5*(4), 212–222.

Lal, R. (2018). Digging deeper: A holistic perspective of factors affecting soil organic carbon sequestration in agroecosystems. *Global Change Biology., 24*(8), 3285–3301.

Lantini, L. (2021). *Health monitoring of trees and investigation of tree root systems using ground penetrating radar (GPR)* (Doctoral dissertation, University of West London).

Lasco, R. D., Delfino, R. J. P., & Espaldon, M. L. O. (2014). Agroforestry systems: Helping smallholders adapt to climate risks while mitigating climate change. *Wiley Interdisciplinary Reviews: Climate Change, 5*(6), 825–833.

Lee, C. (2021). Quantifying Effects of Spatiotemporal Changes of Urban and Green Areas on Regional Climate Change: South Korean Cities from the 1980s to the 2010s. *Urban Forestry & Urban Greening*, 127286.

Lipper, L., Thornton, P., Campbell, B. M., Baedeker, T., Braimoh, A., Bwalya, M., & Torquebiau, E. F. (2014). Climate-smart agriculture for food security. *Nature Climate Change, 4*(12), 1068–1072.

Lobell, D. B., & Field, C. B. (2007). Global scale climate–crop yield relationships and the impacts of recent warming. *Environmental Research Letters, 2*(1), 014002.

Lodhiyal, N., & Lodhiyal, L. S. (2003). Biomass and net primary productivity of Bhabar Shisham forests in central Himalaya India. *Forest Ecology and Management, 176*(1–3), 217–235.

Lorenz, K., & Lal, R. (2005). The depth distribution of soil organic carbon in relation to land use and management and the potential of carbon sequestration in subsoil horizons. *Advances in Agronomy, 88*, 35–66.

Lorenz, K., & Lal, R. (2014). Soil organic carbon sequestration in agroforestry systems A Review. *Agronomy for Sustainable Development, 34*(2), 443–454.

Lützow, M. V., Kögel-Knabner, I., Ekschmitt, K., Matzner, E., Guggenberger, G., Marschner, B., & Flessa, H. (2006). Stabilization of organic matter in temperate soils: Mechanisms and their relevance under different soil conditions–a review. *European Journal of Soil Science, 57*(4), 426–445.

Magsi, H. U., Lohano, H. D., & Mirani, Z. U. (2014). Marketing system and structure for agroforestry products in Sindh Pakistan. *European Academic Research, 2*(7), 9509–9522.

Maia, A. G., dos Santos Eusebio, G., Fasiaben, M. D. C. R., Moraes, A. S., Assad, E. D., & Pugliero, V. S. (2021). The economic impacts of the diffusion of agroforestry in Brazil. *Land Use Policy, 108*, 105489.

Maxwell, D., Webb, P., Coates, J., & Wirth, J. (2010). Fit for purpose? Rethinking food security responses in protracted humanitarian crises. *Food Policy, 35*(2), 91–97.

Mahowald, N. M., Scanza, R., Brahney, J., Goodale, C. L., Hess, P. G., Moore, J. K., & Neff, J. (2017). Aerosol deposition impacts on land and ocean carbon cycles. *Current Climate Change Reports, 3*, 16–31.

Minang, P. A., Bernard, F., Noordwijk, M. V., & Kahurani, E. (2011). Agroforestry in REDD+: Opportunities and Challenges. ASB Policy Brief 26.

Mitchell, E., Scheer, C., Rowlings, D., Cotrufo, M. F., Conant, R. T., & Grace, P. (2021). Important constraints on soil organic carbon formation efficiency in subtropical and tropical grasslands. *Global Change Biology, 9*(5), 95–103.

Mondal, M., Goswami, S., Ghosh, A., Oinam, G., Tiwari, O. N., Das, P., & Halder, G. N. (2017). Production of biodiesel from microalgae through biological carbon capture: a review. *3 Biotech, 7*(2), 1–21.

Montagnini, F., & Nair, P. K. R. (2004). Carbon sequestration: an underexploited environmental benefit of agroforestry systems. In *New vistas in agroforestry* (pp. 281–295). Springer, Dordrecht.
Mukhopadhyay, S., Maiti, S. K., & Masto, R. E. (2013). Use of Reclaimed Mine Soil Index (RMSI) for screening of tree species for reclamation of coal mine degraded land. *Ecological Engineering, 57*, 133–142.
Murthy, I. K., Gupta, M., Tomar, S., Munsi, M., Tiwari, R., Hegde, G. T., & Ravindranath, N. H. (2013). Carbon sequestration potential of agroforestry systems in India. *Journal of Earth Science & Climatic Change., 4*(1), 1–7.
Mustafa, U., Baig, M. B., & Straquadine, G. S. (2021). Impacts of Climate Change on Agricultural Sector of Pakistan: Status, Consequences, and Adoption Options. In *Emerging Challenges to Food Production and Security in Asia, Middle East, and Africa* (pp. 43–64). Springer, Cham.
Muzamil, M. R., Tschakert, P., Boruff, B., & Shahbaz, B. (2021). Institutional dimensions of post-flood recovery in war-affected Swat, Pakistan. *International Journal of Disaster Risk Reduction*, 102526.
Nair, P. K. R., & Garrity, D. (2012). Agroforestry research and development: the way forward. *Agroforestry—the future of global land use. Springer, Dordrecht*, 515–531.
Nair, P. R., Nair, V. D., Kumar, B. M., & Haile, S. G. (2009). Soil carbon sequestration in tropical agroforestry systems: A feasibility appraisal. *Environmental Science & Policy, 12*(8), 1099–1111.
Nair, P. R., Nair, V. D., Kumar, B. M., & Showalter, J. M. (2010). Carbon sequestration in agroforestry systems. *Advances in Agronomy, 108*, 237–307.
Naqvi, H. A. S., Mushtaq, S., Malik, T. M., & Umar., Rehman., Fareed, S., Zulfiqar. A.M. (2019). Factors Leading Towards *Dalbergia Sissoo* Decline (Syndrome) in Indian Sub-Continent: A Critical Review and Future Research Agenda. *Pakistan Journal of Agricultural Research., 32*(2), 302.
Nawaz, M. F., Shah, S. A. A., Gul, S., Afzal, S., Ahmad, I., & Ghaffar, A. (2017b): Carbon sequestration and production of Eucalyptus camaldulensis plantations on marginal sandy agricultural lands. *Pakistan Journal of Agricultural Sciences* 54:335–342.
Nawaz, M., Gul, S., Farooq, T., Siddiqui, M. T., Asif, M., Ahmad, I., & Niazi, N. K. (2016). Assessing the actual status and farmer's attitude towards agroforestry in Chiniot, Pakistan. *International Journal of Biology, Biomolecular, Agriculture, Food, Biotechnological Engineering, 10*, 465–469.
Nawaz, M.F., Mazhar, K., Gul, S., Ahmad, I., Yasin, G., Asif, M., Tanvir, M. (2017a): Comparing the early-stage carbon sequestration rates and effects on soil physico-chemical properties after two years of planting agroforestry trees. *Journal of Basic and Applied Sciences* 13: 527–533.
Oelbermann, M., Voroney, R. P., Thevathasan, N. V., Gordon, A. M., Kass, D. C., & Schlönvoigt, A. M. (2006). Soil carbon dynamics and residue stabilization in a Costa Rican and southern Canadian alley cropping system. *Agroforestry Systems, 68*(1), 27–36.
Orwa, C., Mutua, A., Kindt, R., Jamnadass, R., Anthony, S. (2009). Agroforestree Database: A tree reference and selection guide version 4.0.
Palm, C., Tomich, T., Van Noordwijk, M., Vosti, S., Gockowski, J., Alegre, J., & Verchot, L. (2004). Mitigating GHG emissions in the humid tropics: Case studies from the Alternatives to Slash-and-Burn Program (ASB). *Environment, Development and Sustainability, 6*(1), 145–162.
Pandey, D. N. (2007). Multifunctional agroforestry systems in India. *Current science*, 455–463.
Pascale, S., Boos, W. R., Bordoni, S., Delworth, T. L., Kapnick, S. B., Murakami, H., & Zhang, W. (2017). Weakening of the North American monsoon with global warming. *Nature Climate Change, 7*(11), 806–812.
Patil, J. R., Sharma, P. P., & Mhatre, K. J. (2021). Global warming Induced Stress and its Impact on Biodiversity. *Science and Technology, 6*(3), 34–45.
Podger, G. M., Ahmad, M. U. D., Yu, Y., Stewart, J. P., Shah, S. M. M. A., & Khero, Z. I. (2021). Development of the Indus River System Model to Evaluate Reservoir Sedimentation Impacts on Water Security in Pakistan. *Water, 13*(7), 895.

Poeplau, C., Don, A., Vesterdal, L., Leifeld, J., Van Wesemael, B. A. S., Schumacher, J., & Gensior, A. (2011). Temporal dynamics of soil organic carbon after land-use change in the temperate zone–carbon response functions as a model approach. *Global Change Biology, 17*(7), 2415–2427.

Pold, G., Grandy, A. S., Melillo, J. M., & DeAngelis, K. M. (2017). Changes in substrate availability drive carbon cycle response to chronic warming. *Soil Biology and Biochemistry, 110*, 68–78.

Preethika, D. D. P., Wadanambi, R. T., Wandana, L. S., Chathumini, K. K. G. L., Dassanayake, N. P., & Arachchige, U. S. (2021). Energy Management Contribution for Green House Gas Mitigation. *Journal of Research Technology and Engineering, 2*(2), 1–31.

Qurieshi, M. A., & Rasool, M. (2021). Climate Change and Population Health. In *Microbiomes and the Global Climate Change* (pp. 105–122). Springer, Singapore.

Rahman, M. K., Asaduzzaman, M., Rahman, M. M., Das, A. K., & Biswas, S. K. (2014). Physical and mechanical properties of ghora neem (*Melia azedarach*) plywood. *Bangladesh Journal of Scientific and Industrial Research, 49*(1), 47–52.

Rajput, M., & Agrawal, M. (2004). Physiological and yield responses of pea plants to ambient air pollution. *Indian Journal of Plant Physiology, 9*, 9–14.

Rao, S. P., Rao, S. S., Seetharama, N., Umakath, A. V., Reddy, P. S., Reddy, B. V. S., & Gowda, C. L. L. (2009). *Sweet sorghum for biofuel and strategies for its improvement*. International Crops Research Institute for the Semi-Arid Tropics.

Rasse, D. P., Rumpel, C., & Dignac, M. F. (2005). Is soil carbon mostly root carbon? Mechanisms for a specific stabilisation. *Plant and Soil, 269*(1), 341–356.

Resh, S. C., Binkley, D., & Parrotta, J. A. (2002). Greater soil carbon sequestration under nitrogen-fixing trees compared with Eucalyptus species. *Ecosystems, 5*(3), 217–231.

Rojas Quiroga, R. A., Li, T., Lora, G., & Andersen, L. E. (2013). *A measurement of the carbon sequestration potential of Guadua angustifolia in the Carrasco National Park, Bolivia* (No. 04/2013). Development Research Working Paper Series.

Roshetko, J. M., Delaney, M., Hairiah, K., & Purnomosidhi, P. (2002). Carbon stocks in Indonesian homegarden systems: Can smallholder systems be targeted for increased carbon storage? *American Journal of Alternative Agriculture, 17*(3), 138–148.

Rotenberg, E., & Yakir, D. (2010). Contribution of semi-arid forests to the climate system. *Science, 327*(5964), 451–454.

Rumpel, C., & Kögel-Knabner, I. (2011). Deep soil organic matter—a key but poorly understood component of terrestrial C cycle. *Plant and Soil, 338*(1), 143–158.

Saco, P. M., McDonough, K. R., Rodriguez, J. F., Rivera-Zayas, J., & Sandi, S. G. (2021). The role of soils in the regulation of hazards and extreme events. *Philosophical Transactions of the Royal Society B, 376*(1834), 20200178.

Salam, A. (2018). Pakistan is ground zero for global warming consequences. Article Published in "USA TODAY" on July 24, 2018. https://www.usatoday.com/story/news/world/2018/07/24/pakistan-one-worlds-leading-victimsglobal-warming/809509002/. Accessed 22 Feb 2019.

Samariks, V., Īstenais, N., Seipulis, A., Miezīte, O., Krišāns, O., & Jansons, Ā. (2021). Root-soil plate characteristics of silver birch on wet and dry mineral soils in Latvia. *Forests, 12*(1), 20.

Sanchez, P. A. (2000). Linking climate change research with food security and poverty reduction in the tropics. *Agriculture, Ecosystems & Environment, 82*(1–3), 371–383.

Sanderman, J., Farquharson, R., & Baldock, J. (2009). Soil carbon sequestration potential: a review for Australian agriculture. *Soil carbon sequestration potential: a review for Australian agriculture.*

Scarlat, N., & Fahl, F. (2021). Geospatial analysis approaches for assessing biomass potentials from agricultural landscapes. In *Exploring and Optimizing Agricultural Landscapes* (pp. 303–328). Springer, Cham.

Schmidt, M. W., Torn, M. S., Abiven, S., Dittmar, T., Guggenberger, G., Janssens, I. A., & Trumbore, S. E. (2011). Persistence of soil organic matter as an ecosystem property. *Nature, 478*(7367), 49–56.

Schoeneberger, M., Bentrup, G., De Gooijer, H., Soolanayakanahally, R., Sauer, T., Brandle, J., & Current, D. (2012). Branching out: Agroforestry as a climate change mitigation and adaptation tool for agriculture. *Journal of Soil and Water Conservation, 67*(5), 128A–136A.

Schuenemann, F., Msangi, S., & Zeller, M. (2018). Policies for a sustainable biomass energy sector in Malawi: Enhancing energy and food security simultaneously. *World Development, 103*, 14–26.

Shahzad, L. A. I. L. A., & Yasin, A. (2021). Analyzing Resilience and Food Insecurity of Drought Prone Communities of Tharparkar Desert, Pakistan. *Journal of Research Technology and Engineering, 2*(3), 21–31.

Shankar, R., Chandra Sekhar, C., Joseph, B., Sunitha Devi, K. B., & Aarif Khan, M. A. (2014). Carbon Sequestration in multipurpose agroforesty plantations by using monoculture agroforestry models. *International Journal of Science and Research, 3*(9), 355–359.

Siddique, A. (2021). Contemporary Ecological Paradigms and Emerging Challenges of Water Security Nexus for Pakistan. *Journal of Political Studies, 28*(1), 34–56.

Six, J. AET, Elliott, E. T., & Paustian, K. (2000). Soil macroaggregate turnover and microaggregate formation: A mechanism for C sequestration under no-tillage agriculture. *Soil Biology and Biochemistry, 32*(14), 2099–2103.

Smith, P. (2004). Carbon sequestration in croplands: The potential in Europe and the global context. *European Journal of Agronomy, 20*(3), 229–236.

Smith, G. S., Anjum, E., Francis, C., Deanes, L., & Acey, C. (2022). Climate change, environmental disasters, and health inequities: The underlying role of structural inequalities. *Current Environmental Health Reports, 9*(1), 80–89.

Stephen A., (2012). *Syzygium cumini* (L.) Skeels: A multipurpose tree, its phytotherapic and pharmacological uses. *Journal of Phytotherapy and Pharmacology*. 1(4), 22–32

Stockmann, U., Padarian, J., McBratney, A., Minasny, B., de Brogniez, D., Montanarella, L., & Field, D. J. (2015). Global soil organic carbon assessment. *Global Food Security, 6*, 9–16.

The Global Economy. (2015). Country Ranking using Data on Carbon Dioxide Emissions.

Thokchom, A., & Yadava, P. (2015). Bamboo and its role in climate change. *Current Science, 108*, 762–763.

Tyagi, K., Sharma, S. D., & Tyagi, P. (2009). Development of biomass and productivity in an age series of *Dalbergia sissoo* plantations in sodic lands of Uttar pradesh. *Annals of Forestry, 17*(2), 219–237.

U.S. Census Bureau. (2011). International Programs. International Data Base. World Population: 1950–2050.

Ullah, S., Ahmad, W., Majeed, M. T., & Sohail, S. (2021). Asymmetric effects of premature deagriculturalization on economic growth and CO2 emissions: fresh evidence from Pakistan. *Environmental Science and Pollution Research*, 1–15.

Ullah, W., Mohammad, N., & Zaman, R. (2018). Understanding climate change vulnerability, adaptation and risk perceptions at household level in Khyber Pakhtunkhwa. *Pakistan International Journal of Climate Change Strategies and Article information.*

Ullah, W., Nihei, T., Nafees, M., Zaman, R., & Ali, M. (2018). Understanding climate change vulnerability, adaptation and risk perceptions at household level in Khyber Pakhtunkhwa, Pakistan. *International Journal of Climate Change Strategies and Management.*

Umum, B. P. (2012). Operation and Maintenance of Participatory Irrigation in Chea. *Kementerian Pekerjaan Umum, Jakarta.*

UNEP Emissions Gap Report 2013. Available online: http://www.unenvironment.org/resources/emissions-gap-report-2013 (accessed on 22 December 2020).

UNFCCC (2007). Report of the conference of parties on its thirteenth session, Bali, Indonesia. In United Nations Framework Convention on Climate Change: Geneva.

US EPA. (2016). Climate change indicators in the United States.

van Ginkel, M., & Biradar, C. (2021). Drought Early Warning in Agri-Food Systems. *Climate, 9*(9), 134.

Verchot, L. V., Van Noordwijk, M., Kandji, S., Tomich, T., Ong, C., Albrecht, A., & Palm, C. (2007). Climate change: Linking adaptation and mitigation through agroforestry. *Mitigation and Adaptation Strategies for Global Change, 12*(5), 901–918.
Verma, A., Kumar, P., & Saresh, N. V. (2021). Secondary metabolites: Harvesting short term benefits from arid zone agroforestry systems in India. *Agroforestry Systems, 95*(3), 515–532.
Vignola, R., Harvey, C. A., Bautista-Solis, P., Avelino, J., Rapidel, B., Donatti, C., & Martinez, R. (2015). Ecosystem-based adaptation for smallholder farmers: Definitions, opportunities and constraints. *Agriculture, Ecosystems & Environment, 211*, 126–132.
VijayaVenkataRaman, S., Iniyan, S., & Goic, R. (2012). A review of climate change, mitigation and adaptation. *Renewable and Sustainable Energy Reviews, 16*(1), 878–897.
Wang, S., Adhikari, K., Wang, Q., Jin, X., & Li, H. (2018). Role of environmental variables in the spatial distribution of soil carbon (C), nitrogen (N), and C: N ratio from the northeastern coastal agroecosystems in China. *Ecological Indicators, 84*, 263–272.
Wattoo, M. A., Ali, T., Khan, M. A. J., & Shahbaz, B. (2010). Stakeholder's role and interactions in context of forestry extension interventions in the Northwest Pakistan. *Pakistan Journal of Agricultural Sciences, 47*(2), 173–177.
Yadava, A. K. (2011). Potential of agroforestry systems in carbon sequestration for mitigating climate changes in Tarai region of central Himalaya. *Nature and Science, 9*(6), 72–80.
Yousaf, B., Liu, G., Abbas, Q., Wang, R., Ali, M. U., Ullah, H., & Zhou, C. (2017). Systematic investigation on combustion characteristics and emission-reduction mechanism of potentially toxic elements in biomass-and biochar-coal co-combustion systems. *Applied Energy, 208*, 142–157.
Zinedine, A., & El Akhdari, S. (2021). Food safety and climate change: case of mycotoxins. In *Research Anthology on Food Waste Reduction and Alternative Diets for Food and Nutrition Security* (pp. 39–62). IGI Global.
Zomer, R. J., Neufeldt, H., Xu, J., Ahrends, A., Bossio, D., Trabucco, A., & Wang, M. (2016). Global Tree Cover and Biomass Carbon on Agricultural Land: The contribution of agroforestry to global and national carbon budgets. *Scientific Reports, 6*(1), 1–12.

Bundling Climate Smart Agriculture Systems in Morocco: Insights, Advances and Perspectives in a Changing Climate

Rachid Mrabet, Si Bennasseur Alaoui, and Khalid Daoui

Abstract In Morocco, food and nutrition security are threatened by the combined complex impacts of land degradation processes, climate change and water scarcity, ballooning demography and diet shifts. In order to mitigate and curb these impacts, several climate-smart agriculture (CSA) options were experimented and implemented by research and development consortia with support from international organizations and are at the high priority of the recently launched agricultural strategy termed Green Generation 2020–30. The review will focus mainly on conservation agriculture (CA), organic farming (OF) and agroforestry (AF) as pillars of regenerative agriculture and climate change mitigation and resilience. According to the Ministry of Agriculture, the area of cultivated organic crops in 2020 is estimated to 10,300 ha in addition to spontaneous picking of around 300,000 ha. While the area under conservation agriculture is approximatively 25,000 ha. Green Generation strategy is aiming to reach 100,000 ha and 1 million ha by 2030, respectively, for OF and CA. In this chapter, a review on knowledge advances is addressed with relation to food sustainability and environment stewardship. It is clear from the review that ensuring that farmers and producers of all scales and sizes adopt CSA systems and gain access to innovation, technology, training and knowledge will help foster greater productivity and secure food and feed and lessen impacts on soil, water and air quality, as well as effectively use external resources and scarce labor. In other terms, efficient, sustainable and productive use of inputs, resources and capital under CSA systems helps farmers control costs during volatile business cycles and increasing climate uncertainty. Major challenges, main drivers, potential and perspectives from CSA mainstreaming were also elaborated and discussed. For all these stakeholders, CSA brings changes in behavior and lifestyle as well as new insights and paradigms.

R. Mrabet (✉) · K. Daoui
National Institute of Agronomic Research (INRA), Rabat, Morocco
e-mail: rachid.mrabet@inra.ma

K. Daoui
e-mail: khalid.daoui@inra.ma

S. B. Alaoui
Hassan II Institute of Agronomy and Veterinary Medicine, Rabat, Morocco
e-mail: b.alaoui@iav.ac.ma

M. Behnassi et al. (eds.), *Climate-Smart and Resilient Food Systems and Security*,
https://doi.org/10.1007/978-3-031-65968-3_5

Still stepping up adoption of these options is no regret with little significant trade-offs. Overall, the goal of achieving sustainability, heedlessly of the farming system, should be based on science and impartial, risk-based decision-making process.

Keywords Climate change · Regenerative agriculture · SDGs · Soil security · Adoption · Scaling

1 Introduction: Opportunities in Sustainability

In recent decades, the scientific community generated much new knowledge of the fundamental mechanisms of the Earth's climate system as well as the implications and impacts of climate change on ecosystems and sectors (IPCC, 2022). According to recent publication by Ortiz-Bobea et al. (2021), human-caused climate change has slowed global agricultural productivity growth by 21 percent since 1961. That is the equivalent of losing the last seven years of global productivity gains. In addition, small-scale farmers are the most threatened due to limited access to knowledge, technology and resources to allow them to adapt and increase their resilience.

An agricultural system is considered sustainable when it achieves four cornerstone objectives (i) conserve natural resources, (ii) ensure food production and supply, (iii) guaranty health and nutrition and (iv) alleviate poverty harnessed to wellbeing (FAO et al., 2020).

Morocco occupies an important and strategic position as an Afro-Mediterranean Country but also as a hot spot for climate change, biodiversity and desertification. Climate change and social and economic factors have largely contributed and are contributing to unsustainable land use and management practices. Continuous dwindling soil fertility and associated declining soil health are great challenges facing landowners, researchers and developers for several decades. In other terms, both rainfed and irrigated farming systems are operating at or over the limit of sustainability. According to a GAP report (Steensland, 2021), the productivity growth of Morocco is still low reaching 0.35% in 2020.

Agriculture is the backbone of Morocco's economy, which accounts for an average of 14% of the country's Gross Domestic Product (GDP) and hires about half of the countries workforce. More than 80% of rural families depend on agriculture for their livelihood (Al Balghitti et al., 2020). For decades, economic and political forces and growing populations needing more land to cultivate, have driven land expansion and the extension of irrigation. However, at the moment no more additional productive land to cultivate. Table 1 gives details on agricultural land uses in addition to the argan ecosystem which is mostly considered as agroforestry system. As shown in this table, cereals, fruit trees (mainly olive tree) and fallowing are dominating agricultural landscapes (85%). It is also interesting to show that continuous cereal cropping is the dominant farming system followed by cereal-fallow.

The Moroccan farmers are at the frontline of most vulnerabilities. The country is projected to receive less rainfall, which will negatively affect long-term crop yield in

Table 1 Agricultural land use in Morocco

	Area (millions ha)	(%)
Cereals	4,524	52
Fruit trees including date palm	1,305	15
Fodder crops	0,435	5
Leguminous crops	0,348	4
Vegetables	0,261	3
Industrial crops	0,174	2
Fallow	1,566	18
Oil seed crops	0,174	2
Fallow land	1,566	18
Total agricultural land	8,787	100
Argan	0,713	-

Source ADA (2018)

the production basins. Hence, accelerating agricultural productivity and enhancing farmer's livelihoods must be at the core of a comprehensive strategy to sustainably feed the Moroccan population. The latest launched Generation Green strategy (2020–2030) recognized that land productivity has generally fallen due to technological and institutional constraints, low and variable rainfall and climate extremes, reduced soil fertility and quality and improper management of crops and livestock with respect to changes in rainfall patterns. This initiative is a policy recognition of the urgent need to reverse increasing land degradation and environmental footprints, or at least to prevent their relentless progression across millions of hectares of agricultural and pastoral lands.

Morocco faces the challenge of fulfilling the demands of agricultural commodities for its ever-growing population (9.49% to 2030, Steensland (2021) and changing diets. Hence, the government has made science, innovation and technology a centerpiece of its agricultural strategy, based on the transformation of the food system, giving priority to the human element, combined to the sustainability and consolidation of agricultural development, with the defy of strengthening the resilience and sustainability of agricultural development and the preservation of natural resources in the face of the problem of climate change.

The chapter aims to support the on-going debate and chaining arguments with scientific evidence on implementing and spreading regenerative agriculture or climate-smart agricultural systems in Morocco. In this chapter, the focus will be on no-tillage/conservation agriculture, agroforestry and organic farming systems and over-arching question to be addressed is whether these systems can be applied simultaneously while regenerating soils and farming systems under changing and varying climate. The specific objectives are to provide synthesis of research findings and to deliver the values of these environmental options. The benefits and drivers of climate-smart agriculture (CSA) adoption in Morocco are discussed at length to understand the current advances and identify the knowledge gaps to formulate management, institutional and policy decisions to further spread CSA systems. In other terms, this chapter complements earlier publications by documenting the rationales that underpinned the re-development efforts in implementing adopting CSA systems.

2 Concepts and Terminologies: Minimizing Ambiguity and Uncertainty

The 2021 United Nations Food Systems Summit called to renew national and global priorities and to advance the transformation of agrifood systems to be more efficient, inclusive, resilient and sustainable. The summit also showed that there is no "one size fits all" solution and that a "full package" of solutions and measures is available to enhance food production and tackle the main threats from land degradation, climate change, increasing water scarcity and declining water quality. The summit proposed to focus on integrated approaches (options and solutions) to address these challenges. Among such approaches but still subject to debate and discussions: sustainable land management (SLM), climate-smart agriculture (CSA), sustainable intensification (SI), regenerative agriculture, nature-based solutions (NbS), and more effective and documented ones are agroecology, conservation agriculture, agroforestry, organic farming that are also considered regenerative, sustainable and climate smart (Kassam & Kassam, 2020). All these measures help to sequester carbon, improve soil quality, reduce erosion and runoff, and generally offer a more sustainable way of growing food.

In spite of the relevance to farmers, researchers, extension specialists and policy makers that CSA systems are gaining over the years, several aspects are misunderstood, and gaps remain under-investigated. Among these aspects, CSA concepts and principles are counterintuitive and contradict the traditional farming systems, which have worked for generations and contributed to cultural values and rural traditions. The term and definition as well as the principle and practices of CSA should be made more inclusive and precisely defined for the general public and decision-makers and practitioners. In addition, overlaps among sustainable intensification, climate-smart agriculture, and agroecology should be minimized. Another barrier is the insufficient experiential knowledge about CSA systems if mainstreaming simultaneously within the same landscape.

3 Conservation/No-Tillage Agriculture: A Moroccan Insight

In its simplest form, conservation agriculture (CA) is a farming system that is based on three principles (i) minimum or elimination of soil disturbance, (ii) maintenance of a permanent soil covers with crop residue and living plants, and (iii) crop diversification through rotation, intercropping or succession; it is promoted as a sustainable alternative to systems involving conventional tillage (CT) (FAO, 2014; Hobbs, 2007; Fig. 1). It enhances biodiversity and natural biological processes above and below the ground surface, which contribute to increased water and nutrient use efficiency and to improved and sustained crop production. The three principles are intended to help to protect the environment and to reduce both the impacts of climate change on agricultural systems (adaptation) and the contribution of the agricultural practices to greenhouse gases (GHG) emissions (mitigation) through sustainable land management (IPCC, 2019).

In other terms, optimal crop/animal assemblages and integration, skillful crop residue maintenance, cover cropping and well-planned rotations are instrumental for successful CA design and implementation and allow the farming system—over time—to sponsor self-enrichment of soil nutrient pools, enhanced water/soil productivity, crop protection and yields.

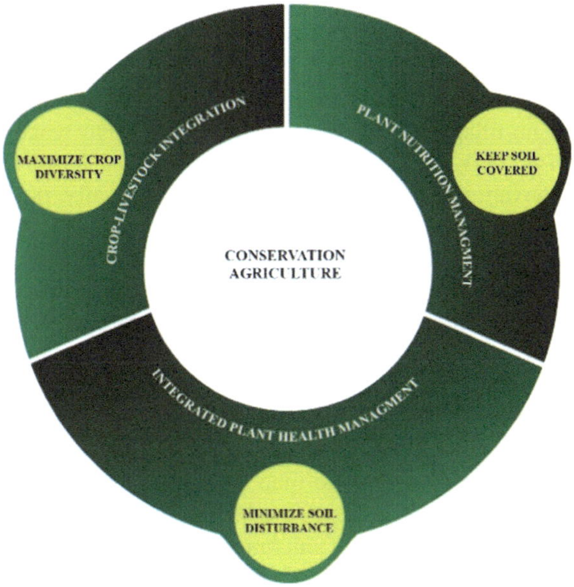

Fig. 1 Schematic presentation of fundamental principles of CA and complementary system's management

3.1 Historical Background and Adoption Pathways in Morocco

Since the droughts of 1980s, a new rhetoric leading to a profound revision of existing agricultural and rural development models was called for by researchers to reject the status quo that has sustained most policies without a clear positive impact. This revision emphasized the need to shift paradigms. The research consortium called for converting from conventional agriculture based on tillage and soil preparation to innovative and climate-smart systems centered around principles of conservation agriculture (Bouzza, 1990) in order to achieve and invest in resilient food and nutrient security. However, literature on the adoption and adaptation of CA are limited in Morocco. Despite some large development efforts, there is no comprehensive assessment of CA adoption in the country (Boulal et al., 2014).

Research on water conservation started early 1980s in semi-arid Morocco in order to propose to farmers the best options for managing soils and crops under erratic rainfall. The first adoption studies were initiated in 1998 with support from FAO, CIRAD and government's extension services and development agencies. In 2011, the area under no-tillage farming did not exceed 5,000 ha and concerned mainly large farms having imported direct seed drills.

According to El Gharras et al. (2017), no-tillage use by farmers spread over an area of 9,000 ha. The 2018 estimates of no-tillage area showed an increase of CA area to 12,826 ha, with 8,033 have being managed by 800 smallholder farmers (with less than 10 ha) and 4,793 ha by 120 large-farmers (with more than 10 ha). In 2019, OCP group and Mohammed VI Polytechnic University (Benguerir) have launched with the technical support from INRA, the El Moutmir initiative, a large-scale adoption of no-tillage systems in Morocco in order to reach 2,000 farmers for a land area of 10,000 ha. This program has mobilized 35 direct seeders and concerned 600 innovation platforms in 18 provinces. Actually, the no-tillage acreage in Morocco increased to 25,000 ha (Fig. 2; Mrabet et al., 2021). The diffusion of CA adoption decisions across Moroccan grain producing regions is shown in Table 2.

According to a study by The World Bank (2014a), in the Central Morocco region, the CA systems can be extended to 60,000 ha in 2025. The model by Bonzanigo et al. (2016) gave an estimate that 40% of farmers in the Central Morocco region would adopt no-tillage systems.

In 2022, the Ministry launched a program of 50,000 hectares of CA. In fact, based on research findings and accumulated experience, the department of agriculture proposed a road map to promote and implement NT systems in 1 million hectares during the next decade in cereal-based systems under the new agricultural strategy called Green Generation (2020–2030). This road map proposes CA development to ensure sustainability through an array of parallel and supporting interventions (i.e., investment in education and capacity development, rural community and farmer organizations empowerment and mobilization, awareness creation among

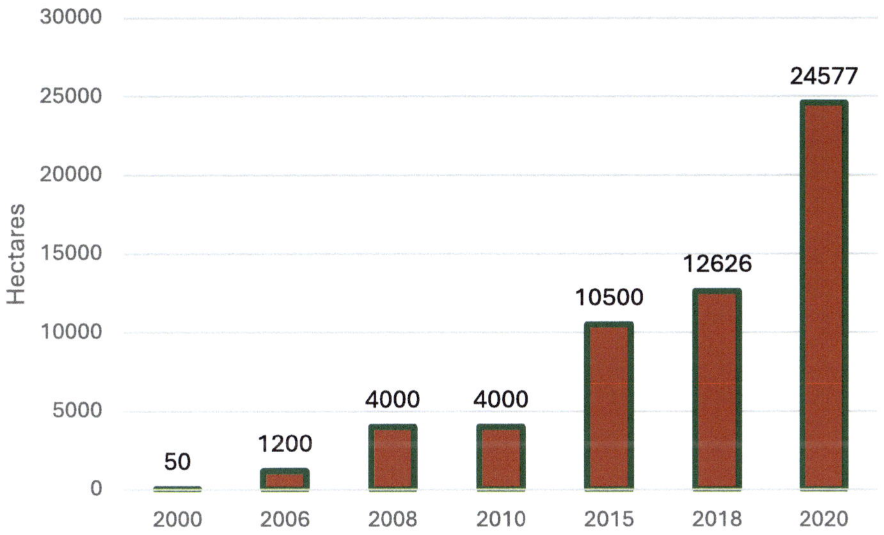

Fig. 2 Temporal trends in the adoption of no-till/CA farming in Morocco (hectare)

Table 2 Area under CA systems and number of direct seeders used in different regions of Morocco

Region	Area under CA		Seeders	
	Hectares	Percent	Number	Percent
Fes-Meknes	9000	36.6	36	34.3
Casablanca-Settat	7300	29.7	24	22.9
Rabat-Sale-Kenitra	4670	19	28	26.7
Beni Mellal-Khenifra	2100	8.5	6	5.7
Marakech-Safi	1144	4.7	7	6.7
Tanger-Tetouan-El Hoceima	315	1.3	3	2.9
Draa-Tafilelt	48	0.2	1	1
Total	24,577	100	105	100

Source AMAC (2020)

stakeholders, youth inclusion and entrepreneurship, legal and institutional piloting frameworks, research and extension, technology and outreach, communication and media development, service delivery, machinery and seeds, marketing, and enterprise diversification and public–private partnership strengthening).

The adoption and spreading of CA systems in Morocco is constrained by an array of barriers that are summarized in the following points:

- Short-term economic gains and profitability: Farmer's attitude to risk-taking and uncertainty
- Characteristics of farms: geophysical and socio-political indicators (farm size, crops and livestock, land tenure, income, type of ownership, soil fertility, slope, altitude, etc.)
- Cultural: Farmers' demographics, knowledge, and attitudes (Gender, poverty, land ownership, family size and structure, social class, farmer's profile)
- Social norms: Social influence and conformity, neighbor's acceptance of CA, Trust decisions, Generational renewal at the farm
- Communication gaps and lack of information and awareness on CA systems and their benefits
- Availability, access and costs of machinery (especially NT drills and sprayers) and inputs (seeds, fertilizers and pesticides)
- Changes in weed, disease and pest profiles, infestation and management, and emerging diseases and pests
- Livestock integration and scarcity of pastures placing pressure on crop residues and stubble
- Site-specific conditions, field-scale effectiveness, and difficulty for upscaling
- Insufficient endorsement and encouragement by private sector at national and local levels of CA systems
- Scarce and limited international cooperation for CA system adoption and upscaling

3.2 Advancing Food Security Indicators: Tracking Productivity

Options to expand cultivated land areas are limited so improved productivity should be aligned with the intensification of available land that is already exhausted and mined. Hence, crop yields are either stagnating or the yield growth rates are lower than required to meet the growing food demand and contributing to food and nutrient security. On the other side, the benefit–cost ratio under traditional farming systems is too low to encourage subsistence and smallholder farmers (<5 ha) to use improved seeds, fertilizers, and pesticides. The lack of adoption of modern innovation is also attributed to relatively high farm gate price, the low market price of food crops, and risk of low agronomic efficiency of applied input due to variable rainfall amount and distribution. Thus, there is a strong need to redesign such unsustainable farming systems to achieve high profitability and efficiency by implementing innovative incentives and policy measures. Although benefits are context specific, CA has been identified as an effective tool for sustainably increasing yields for various crops and under contrasting climate-soil conditions (Mrabet et al., 2021, 2022). Such improvements in the productive potential of soils under CA as compared to conventional tillage techniques are due to amelioration in soil health, functions and productivity, as is discussed in the section below. Research trials either on stations or in farmer's fields also confirmed that soil conditions under CA systems favor more vigorous and healthier plants that are resilient to a range of biotic or edaphic stresses.

Several authors reported that several crops can be successfully grown in sequence under CA systems in the contexts of the diverse climate of the country. Among such crops are oil seed (sunflower and sesame), food legumes (chickpea, bean, lentil, faba bean), forages (vetch) and most cereals (corn, bread and durum wheat, barley, and oat) (Mrabet, 2011). In addition, the author recommended that NT with partial residue cover could be adopted for the mixed crop-livestock systems in semi-arid areas for the purpose of guarantying the yield of both grains and feed.

A low adoption rate of improved varieties may reduce benefits from CA systems and keep the level of production insufficient to satisfy the food requirements of the growing population. Hence, it is of paramount importance that farmers satisfy the three fundamental principles while judiciously using complementary management practices (i.e., certified varieties, nutrient management and integrated plant health management). 4R Nutrient Stewardship, a science-based framework to utilize the Right nutrient source, at the Right rate, at the Right time, and in the Right place should be made accessible and deployed by CA farmers.

Overall, crop productivity under CA systems increased over time with a range of 2 to 146% compared to that under the CT systems (Mrabet et al., 2021). According to FERT (French Association for International Cooperation for Agricultural Development in Developing and Emerging Countries) (FERT, 2018), crop yield increases of 8 to 20% were observed by CA farmers in the Maghreb compared to conventional practices. Mrabet (2011) reported that CA treatments retained more soil moisture that helped mitigate drought impacts and helped to sustain yields of cereals. In other terms, compared to traditional farming systems, CA systems were particularly effective in reducing drought severity during wheat growth and development through higher water use efficiencies which is of high significance to secure food for a growing population (Mrabet et al., 2012).

Reduced yields under CA systems compared to that under CT systems were mostly explained by (i) partial adoption of CA principles and the prevalence of bare soil surface, (ii) by the necessity of a transition phase for soil recovery and yield improvement vis-à-vis rainfall variability, (iii) inappropriate use of direct seeders and (iv) insufficient weed control (Mrabet et al., 2021, 2022; Yigezu et al., 2021).

According to Friedrich (2005), increased dependence and overreliance on herbicides can be a limitation for small farmers to shift to CA systems and for the public at large as concerns for health and the environment are emerging. In addition, herbicide and weed mismanagement induce resistance and shift in weeds. From weed research studies, the focus should be on managing weed populations and flora under CA with integrated practices rather than eliminating weeds solely with herbicides (Hajjaj et al., 2016a, b; Tanji et al., 2017a, b). Changes in soil temperature and moisture regimes due to in-situ crop residue retention affect weed seed germination and seedling emergence, and hence, could either reduce, delay or favor weed growth and infestation. Overall, CA has been found to influence weed flora by favoring development over time of grass weeds and deep-rooted perennials if not appropriately controlled (Hajjaj et al., 2018). However, in their review work, Nichols et al. (2015) reported that the initially large number of weeds immediately following the transition to CA should not be discouraging, as it can be a transitory phenomenon. They also indicated that the principles of CA, particularly crop rotation and surface residue retention, are in themselves methods of weed control.

According to work by Hajjaj et al. (2018), the main reason for the decrease in weed infestation under CA over time is due to a reduction in weed seed banks in the soil. Adoption of CA systems changes weed dynamics and communities and therefore necessitates adjusting weed control methods and shifting to integrated weed management. Integrated weed management (IWM) under CA includes early seeding, combination or rotation of herbicides, diversified cropping systems, plant geometry, nutrient management, and improved crop varieties that are competitive with weeds (Mrabet, 2008; 2011; Tanji et al., 2017a, b). In these regards, developing crop genotypes with strong allelopathic potential against associated weeds is an important research issue. In addition, using IWM successful weed management can be achieved over time with low dependence on herbicide to improve the environmental footprint of CA systems (Sims et al., 2018).

CA systems accumulate seeds near the soil surface where they are more likely to germinate but are also exposed to greater mortality risks through drought, weather variability and predation. Assuming no seed input into the farming system, germinable seedbanks under CA decline more rapidly than under conventional tillage (Mrabet et al., 2022). This relaunches the opportunity of eliminating herbicides from CA systems over time and allows to ensure complementarity with organic farming.

3.3 Improving Soil Health and Sustaining the System Productivity

Soil health and functioning depend on the interaction between soil properties, water and nutrient availability, mineral matrixes, agricultural inputs, and management practices. Sequestration of soil organic carbon (SOC) is an important climate mitigation option with numerous co-benefits of enhancing agricultural production, improving water resources, and strengthening biodiversity. It is also one of a few strategies that could be applied on a large scale and at low cost (Mbow et al., 2019; Roe et al., 2021). Hence, implementation of conservation agriculture for soil improvement needs to take place across many aspects of the farming system over several years, and where the benefits may take decades to be fully quantifiable.

Mrabet et al. (2022) revealed that management practices integrating conservation agriculture can help to recover, keep, and build soil quality to meet food security issues. CA systems have the potential to provide many benefits to soil health in diverse cropping systems and climates.

Common soil chemical properties that have been investigated for their response to CA management include soil organic carbon or organic matter (both quality and quantity), nutrient concentrations (especially nitrogen, potassium and phosphorus), pH, cation exchange capacity, and electrical conductivity (Mrabet et al., 2012, 2022). Numerous studies have demonstrated the importance of proper CA management for optimized benefits to soil physical, chemical, and biological properties. CA systems increase soil carbon and nitrogen as well as other nutrients and improve their cycling within cropping systems as found by several authors (Laghrour et al., 2015, 2016, 2019; Moussadek et al., 2014; Mrabet et al., 2001a, b). Increases in SOC contents were also related to higher aggregation, more plant nutrient pools, reduced pH and enriched quality of soil organic matter. According to Belmekki et al. (2014), maintenance of high residue cover helps to increase sequestration rates due to reduced soil temperature amplitude, lower evaporation and accumulated water in the soil profile. Indeed, recent modeling work by Lembaid et al. (2022) shows that carbon sequestration and dynamics under CA are dependent both on climate and soil factors with temperature regime having more influence on carbon accumulation as well as the initial level of soil carbon and clay content of soils.

Mrabet et al. (2001b) expressed that the potential for CA systems including diversification of crops to maintain and build soil fertility may allow for reduced reliance on synthetic fertilizer inputs over time. This issue is of importance consideration in shifting to regenerative agriculture (combining organic farming and CA systems).

In CA, nutrient management can be implemented through changes in fertilizer rate, placement, type, or timing, and consequently, avoid losses of applied fertilizers that cause environmental pollution (i.e., encouraging 4R Nutrient Stewardship). The soil testing and fertilization recommendation technology can also prevent improper fertilization and low utilization ratio of synthetic fertilizers under CA systems. Development of direct seeding drills tried to take into consideration appropriate rates and banding of fertilizers to satisfy crop requirements in N, P and K. Split N application may increase efficiency, and precise banding to separate fertilizer from crop residues can off-set N immobilization. Much research is still needed to better understand the changes in nutrient cycling and efficiencies under CA cropping systems. The focus should be on developing nutrient management strategies that meet the CA principles while fitting the farmer's circumstances. In limited rainfed production basins, these strategies should stabilize crop productivity and enhance biomass production and quality (Mrabet et al., 2022).

On the other side, upon conversion to CA systems, changing from continuous mono-cropping to diversified crop rotations involving leguminous crops (lentils, vetch, faba bean, and chickpea) can increase the net SOC sequestration and improve nutrient budget. Crops with deep root systems (i.e., forage crops, safflower, rapeseed, etc.) can also sequester SOC in the deeper soil layers, which are not readily mineralized.

However, most of these changes in soil chemical properties with CA systems are limited to the upper soil surface (Mrabet, 2002). Additional research investigation may be needed to explore the influence of CA systems and principles on soil organic carbon stored in deeper soil depths and additional cropping systems and climatic conditions.

In addition to their ability to promote beneficial soil microbes for carbon and other nutrient management and accumulation, CA systems also affect microbial community structure and soil biological activity (Hamim et al., 2014). Crop residues maintained at the soil surface help to improve particulate organic matter content which is a signal of increased and diversified microbial activity (Bessam & Mrabet, 2003). However, more profound research is needed to understand the soil-microorganism-plant system and relationships when shifting to and implementing CA under dryland conditions in Morocco (Mrabet et al., 2022).

As CA system's effects on soil health are generally not realized exclusively in the short-term, additional experiments and modeling studies will be needed to determine and assess the long-term effects of CA principles on soil carbon management sequestration and fertility buildup and their potential benefits to the long-term sustainability in diverse cropping systems and climates (Lembaid et al., 2021). Using the Soil & Water Assessment Tool (SWAT) model, CA was found to be the best adaptation strategy against climate change for Moroccan systems (Brouziyne et al., 2018).

Climate change most likely will increase the threat from biotic and abiotic factors to farming systems in Morocco. With balanced crop nutrition practices—using the right mix of macro- and micro-nutrients to meet the needs of the crops and soils and appropriate plant health management, CA systems are capable to sustain and boost productivity and resilience of farming systems.

Livestock systems (based on grazing, mixed farming systems, or industrialized systems) are extremely important for farmers and strategic for the government but will be more and more negatively affected by climate change, especially global warming. Hence, it is clear that health and productivity of farm animals will be affected and therefore overall food security and resilience. In addition, the livestock sector requires a significant number of natural resources and is responsible for greenhouse gas emissions (methane and nitrous oxide) (Grossi et al., 2019). Available research calls for encouraging integrated CA systems where livestock is an essential component. Several practical options for durable integration of livestock in CA systems while guarantying crop productivity, soil health and farm sustainability were given by Mrabet et al. (2021). The options relate to both adaptation and mitigation measures which incorporate management, genetics and nutrition in order to improve productivity (milk and meat) and profitability of livestock systems while avoiding environmental trade-offs.

3.4 Environmental Stewardship

Productivity growth is a necessary priority to feed growing population but not sufficient to achieve sustainability and environmental stewardship (FAO et al., 2020). According to the last report on the world's state of natural resources by FAO (2021), land and water systems are at breaking point and future food and nutrient security will depend on safeguarding land, soil and water resources. At the same time, IPCC (2018) reported that mitigation pathways that limit warming to 2 °C or 1.5 °C should involve deep, rapid and sustained reduction of greenhouse gas (GHG) emission in all sectors and especially in agriculture, forestry and livestock.

On the other side, one of the most serious environmental consequences of the mechanical disturbance of soil by tillage agriculture has been the decomposition and mineralization of soil organic matter (SOM) and the accelerated emission of CO_2. Among mitigation portfolios, CA systems have been of interest for their ability to help mitigate the effects of climate change and global warming through their capability to sequester and fix the atmospheric C (Mrabet et al., 2021) and mitigate losses of CO_2 (Moussadek et al., 2011b).

From the research studies and investigations carried out in Morocco, no-tillage agriculture can meet the broad demands of sustainable intensification and provide multiple and diverse ecosystem services (Devkota et al., 2022). Reliance of CA on agrichemicals for augmenting crop nutrition and managing weed, insect, and soil-borne disease can be reduced over time if the three principles of CA are optimized (Ygizu et al., 2021). CA relates to the combination of yield increase with environmental stewardship, merging two apparently conflicting goals and ensuring that increased profitability contributes to local development and provides and sustains environmental benefits (Mrabet, 2008). Most authors in the country generally recognized that CA systems lead to significant benefits of an environmental nature (reduction of GHG emissions, wind and water erosion, increased soil water retention and storage capacity, mitigated soil water evaporation and enhanced infiltration and distribution of water) (Moussadek et al., 2011a, b; Mrabet et al., 2012). Especially, the importance of the herbicide glyphosate in the economics of no-till adoption has long been recognized but more research and investigation are underway to reduce such dependence and develop more integrated weed management or even limited use of herbicides.

The use of advanced and digital technology to improve the effectiveness and efficiency of CA farming practices has increased tremendously in several western countries (i.e., USA and Australia) and is one the important pillars of the Green Generation Strategy. This new strategy is giving great importance of precision agriculture and to emerging technologies such as geospatial technologies (GIS, GPS), Internet of Things (IoT), Big Data analysis, and artificial intelligence (AI) as management decision tools aiming to increase crop production and optimize use of external inputs (water, pesticides, fertilizers and seeds) while ensuring higher farm income or profitability and environmental protection.

3.5 Social and Economic Benefits: Considering Human-Centered Aspects

The primary requirement of the accelerated uptake and spreading of CA systems was to reassess current farming practices better while proposing more pertinent social and economic metrics (Findlater et al., 2019). Globally, the decision to remain approximately right rather than precisely accurate was mainly justified by cost-effectiveness concerns (Mrabet et al., 2012). In addition, both conventional agriculture and CA farming performances and their predictions remain highly uncertain, especially in dryland cropping conditions. Furthermore, an early limited or reduced CA adoption rate may not be as damaging for decision-making at long-term as perceived.

A team of researchers, including INRA, joined forces and adapted several no-till drill models that helped producers to maximize crop profit with limited resources and in water scarce and droughty regions (El Gharras et al., 2017). Consequently, huge efforts were made to design new seeding machines with a variety of end-users in mind, including farmers, service providers and development agents with minimal CA literacy and skills. To achieve this, the layouts were meant to be as versatile and intuitive as possible, with a focus on overall simplicity and costs. In fact, a mix of policy instruments, including research and development (R&D) and deployment incentives, has been effective in reducing technology costs, increasing adoption and addressing distributional effects related to CA systems.

Figure 3 gives more information on science-based benefits and impacts associated with CA adoption.

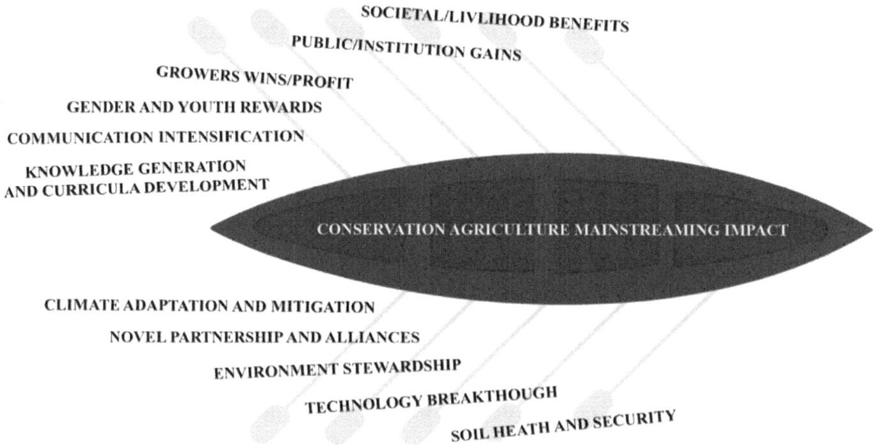

Fig. 3 Summary of CA science-based impacts

3.6 Trends and Drivers: Renouncing Added Complexity by Re-Asserting Objectives

Several projects and initiatives (Table) have been testing the effects of CA systems on farms, promoting their application for the achievement of food security, adaptation and mitigation objectives. However, such efforts were not designed for wide-spread and mainstreaming of CA systems.

Among the objectives set within the framework of the new country's agricultural strategy, the promotion of direct seeding to reach 1 million hectares of cereals in direct seeding by 2030; this technology is known to improve yields, especially under water-stressed conditions, improve carbon sequestration and maintain soil fertility and moisture. Hence, commitments are made to push the boundaries of research, innovation and technology transfer to meet the challenges of climate change and food security through the doubling of investments in climate-smart agriculture and innovation in sustainable food systems. This initiative aims to increase and accelerate innovation in agriculture and food systems through scientific collaboration on agriculture-related climate issues, capacity building, knowledge sharing and deployment of climate-smart solutions. It also aims to provide financial support for farmers and stockbreeders, the preservation of their production performances, make available support for investment in the agricultural value chains as well as a lessening of the resulting effects of climate change and price volatility on the creation of incomes, added value and employments. It takes up also the alleviation of the financial burdens of farmers and professionals in the sector through credits and subsidies. All these engagements and arrangements are solid and concrete drivers and enhancers for CA adoption and spreading in Morocco and within the diverse farming communities.

In dry areas of Morocco, at the frontline of the climate change and variability, crop rotations and hence system diversification are arguably the most effective way to pledge for sustainable production. However, producers adopting CA will face several managerial and technological changes that require additional skills and capacity development.

We argue that mainstreaming CSA relies on identifying and acting on a certain set of key drivers and forms of interventions—stepping stones—that can facilitate the embeddedness and maintenance of CA across farming systems and production basins (Table 3).

Ensuring that farmers of all sizes and scales get access to improved technology, better training and value-added knowledge will certainly assure large-scale adoption of CSA systems in Morocco and at the same time permit higher and diversified production and competitive business cycles without jeopardizing environment protection.

Long-term research and adoption platforms (over a decade) are needed to get consistent sustainability attributes of CA systems (i.e., slow-to-change attributes of soil quality and environment). In other terms, the main challenge with projects listed in Table 4 is their exit-strategy when farmers do not rely on resources from

Table 3 Summary of drivers influencing the adoption of CA by farmers

Drivers/ Enablers	– Polycentric policy approach – Insertion of CA principles in multi-level governance and policy: e.g., environmental and climate change programs, ecological and energy transition, youth and gender initiatives, political incentives and supports – Links and compliance with CA world science community, international organizations and financial agencies, e.g., South-South Cooperation mainly Brazilian CA experience – Creation of centers of excellence on CA research, development and extension – CA long-term capacity development initiatives and high standard educational platforms – Farmer to farmer learning–sharing experiences, leadership, and education, including synergetic and supportive process of communication – Farmer's and other stakeholder's willingness and compliance with CA principles – Local development of machinery manufacturing (drill, tractor, sprayer, etc.) including public–private partnerships – Developing institutional capacity: Strengthening and enforcement of extension and technical advisory services – Creation of employment and generation of wealth in rural communities – Financial investments, insurance, incentives, and subsidies
Facilitators/ Motivators	– Reinforcement and fine-tuning of research institutions and teams (innovative applied research approaches) – Gender mainstreaming in CA projects and initiatives – Adoption of CA can be facilitated by development of artificial intelligence, eResearch and high-tech – Payments for environment benefits and services associated with CA spread and adoption as environmentally friendly practices – Households engagement in supplementary or complementary off-farm activities to generate incomes to cushion the effects of CA risks and poverty – Social learning and training and capacity building – CA curricula development at different levels of the education system – Cost and benefit sharing among farmers using CA systems – Social reputation, commitment, sustainable attitudes, positive behavioral change of CA adopters – Ease of access to technology, technical assistance, and knowledge dissemination – Access to permanent and skilled labor (family or hired) for machinery operation – Market integration of CA productions – Encouragement of service and input providers (outsourcing) and crowdsourcing – Consumer demands for sustainable products and certification of CA systems – Business-science-community cooperation – Land stewardship and ethics

*Source*Mrabet et al. (2022)

Table 4 Major CA projects and initiatives in Morocco

Initiatives/Projects
– AAA initiative (Foundation): https://www.aaainitiative.org/organization
– CANA: Conservation agriculture in North Africa (2012–2015)
– FERT: Agriculture de Conservation au Maghreb
– KASSA Project, Knowledge Assessment and Sharing on Sustainable Agriculture, http://Kassa.cirad.fr
– 2004–2006
– CA2AFRICA: Conservation agriculture in Africa: Analyzing and foreseeing its impact – comprehending its adoption. (2009–2011) http://ca2africa.cirad.fr/
– Azehar Program (Comisión Interministerial de Ciencia y Tecnología, Plan Nacional I + D + I): Conservation agriculture in irrigated agricultural systems (2006–2008) (University of Cordoba, CSIC)
– CA2AFRICA: Conservation agriculture in Africa: Analyzing and foreseeing its impact – comprehending its adoption. (2009–2011) http://ca2africa.cirad.fr/
– PICCPMV: Integrating Climate Change in the implementation of The Plan Maroc Vert» (2011–2015): https://projects.worldbank.org/en/projects-operations/project-detail/P117081?lang=fr (World Bank, 2014a, b)
– ACCAGRIMAG: Adaptation to climate change of agriculture in the Maghreb (2013–2016) (https://www.ffem.fr/en/carte-des-projets/adaptation-climate-change-agriculture-maghreb-region-accagrimag
– CRP-Dryland Systems for North Africa and West Asia (2012–2015): Saiss, region – rainfed systems https://ccafs.cgiar.org/partners/crp-11-dryland-systems-0 & http://drylandsystems.cgiar.org/node/432
– Integrated Natural Resource Management: INRM project funded by INRA Morocco
– Arab Authority of Agricultural Investment and Development (AAAID)
– OCP & Mohammed VI Polytechnic University, INRA: Initiative El Moutmir: Climate change integration in Agricultural sector (2019—on-going)
– CAMA: Conservation Agriculture in Mediterranean Area. https://intranet.iamz.ciheam.org/Practicas/www.camamed.eu/index (PRIMA project)

Source Mrabet et al. (2022)

projects. Their limited duration restricts the continuous appropriation of CA systems by farmers, and hence, their long-term benefits and success of lasting adoption.

4 Status of Moroccan Organic Sector and Future Perspectives

Growth of organic agriculture has been accelerating worldwide, and over 71.5 Million Hectares of worldwide farmland are organic. The global organic market continues to grow worldwide and has passed the 100 Billion US Dollar Mark. Approximately, a quarter of all certified organic land is located in developing countries. Organic producers in developing countries constitute 82% of all organic producers worldwide. The share of the organic agricultural land in the total agricultural land is increasing in most of the world's countries, and there is a remarkable increase in the demand for organic products in recent years (IFOAM, 2020).

In Morocco, the sector of organic agriculture experienced a slow evolution at the beginning of the 1990s, then relatively faster; it has picked up a pace that has barely improved since the 2010s compared to similar countries. Its current rate of growth does not reflect the possibilities offered, given the assets available to the country. The number of operators who are active in organic remains very limited (maximum 400 all products combined) and the organic agricultural area is far from 1% of the total agricultural arable; moreover, the sectors of organic agriculture are not sufficiently well organized and are generally considered to be not very dynamic at the national level (Kabbaj & Alaoui, 2018).

The 2011–2020 program contract between the Ministry of Agriculture and the Federation of professionals FIMABIO, which was to lead to qualitative leaps in the organic farming sector and a significant improvement in the indicators, did not give all the hoped-for measures, likely to put Morocco on the map of active organic agriculture countries. According to Kabbaj and Alaoui (2019), the next few years are decisive to relaunch the dynamic and to achieve the necessary qualitative leap in terms of the mobilization and synergy of private actors, the structuring of all sectors, the public–private partnership, concrete and ambitious projects. Only an integrated and dynamic SMART vision can respond to the complex issues faced by the organic sectors (Fig. 4).

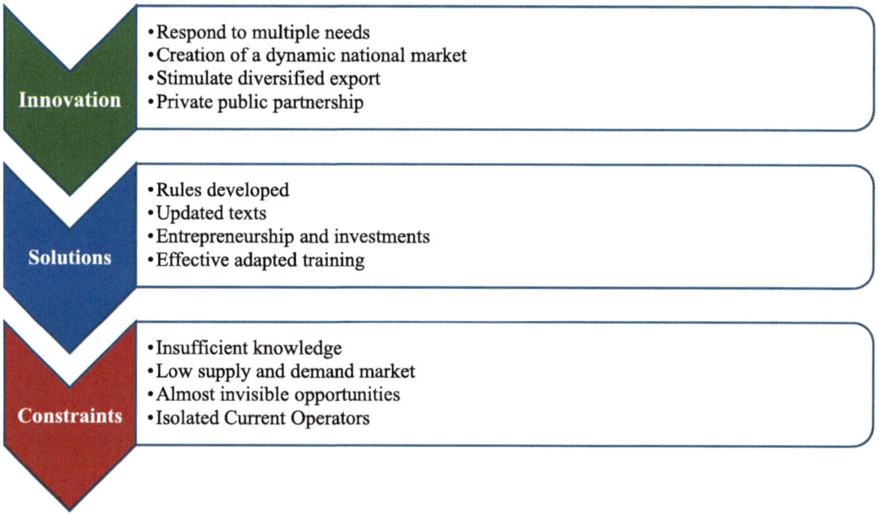

Fig. 4 Constraints and opportunities. *Source* Adapted from Kabbaj and Alaoui (2019)

4.1 Morocco's Current Organic Production

The diversity of soils and climate would make it possible to offer a wide range of products. Most of them are grown naturally without any treatment in various areas (mountains, oases, areas away from urban centers), which gives them a high potential to be easily certified as organic products (Azim & Ramli, 2017).

Nine main regions (Fig. 5) are involved in organic production, located in farms in the following regions: Marrakech-Safi, Rabat-Sale-Kenitra, Souss-Massa, Fez-Meknes, Tangier-Tetouan-Al Hoceima, Casablanca-Settat, Draa-Tafilalet, Eastern Morocco, and Beni Mellal-Khenifra. Medicinal and aromatic plants are found in almost all regions, with, however, specificities of quality for the region of Marrakech for verbena, that of Taroudant (Taliouine) for saffron and those of Fez and Safi for caper. The Souss valley stands out as the main market vegetables region, due to its subtropical climate favorable for off-season production, sometimes practiced in greenhouses. Some coastal regions (Azemmour and Rabat) are also qualified for market vegetables production. Fruit production, especially citrus fruits, comes mainly from two regions: Marrakech and Agadir (Kabbaj & Alaoui, 2018).

The distribution of the certified organic cultivated area by species in 2020 shows a predominance of olive trees, cultivated aromatic and medicinal plants, citrus fruits and almond trees (Fig. 6).

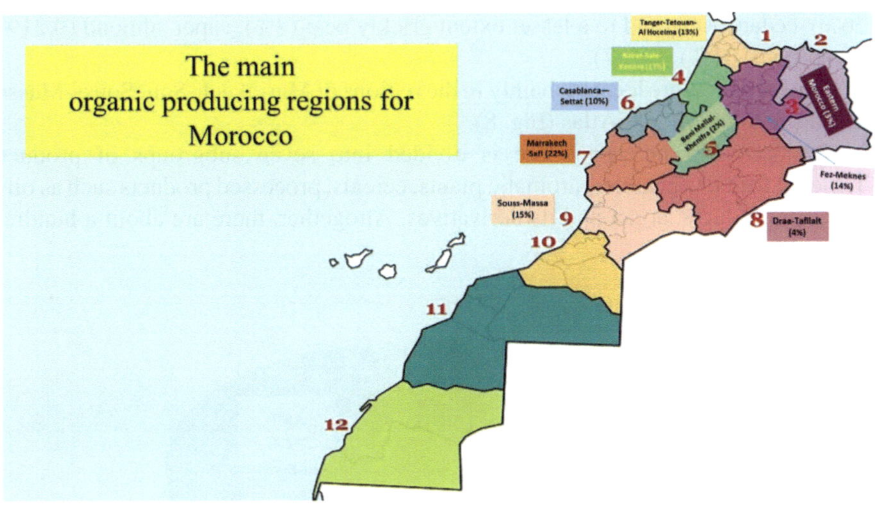

Fig. 5 The main organic-producing regions for Morocco

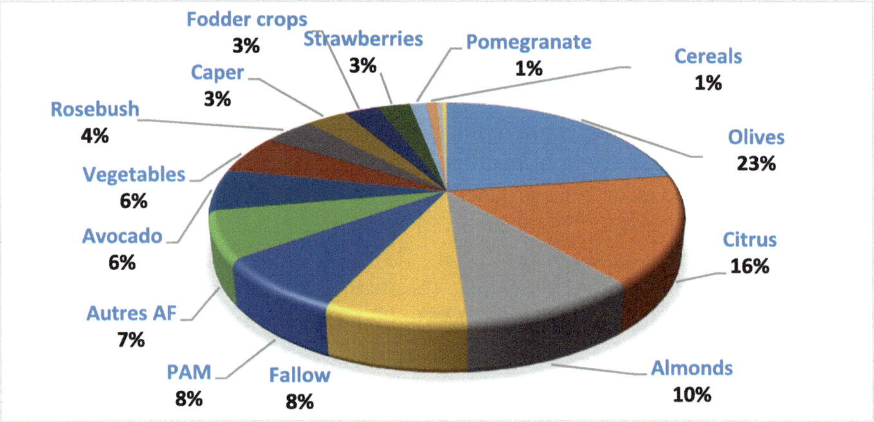

Fig. 6 Breakdown of organic cultivated areas by species. *Source* DDFP (2022b)

4.2 Breakdown of Spontaneous Plants Areas by Region and by Species

Concerning the distribution of the total area of spontaneous biological production by species, we notice the dominance of aromatic and medicinal plants (42%), argan (26%), cedar (30%) and to a lesser extent prickly pear (2%), caper, almond (0.21%) and carob (0.01%) (Fig. 7).

These products are located mainly in the regions of Marrakech-Safi, Souss-Massa, the East and the Middle Atlas (Fig. 8).

Moroccan organic agriculture is divided into seven subgroups of products (Table 5): fruits, vegetables, aromatic plants, cereals, processed products such as oils, livestock products, argan and its derivatives. Altogether, there are about a hundred

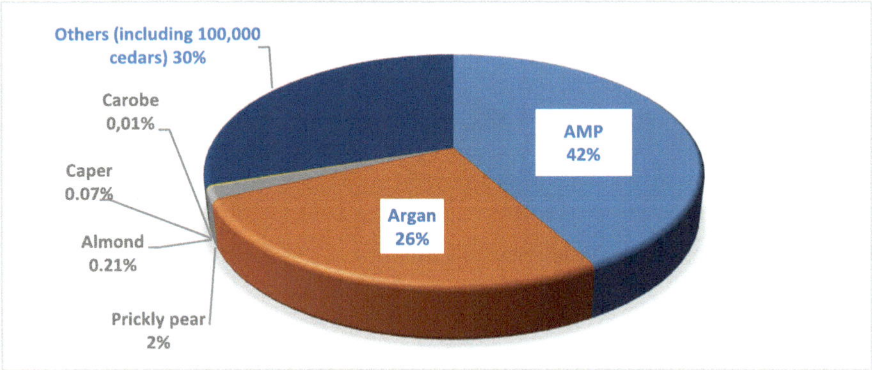

Fig. 7 The distribution of areas by spontaneous species. *Source* DDFP (2022b)

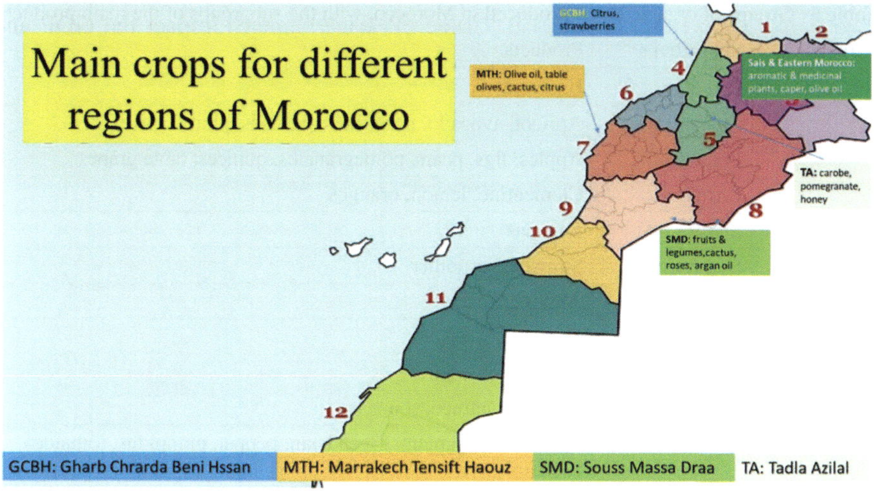

Fig. 8 Main crops grown organically by region in Morocco

products, the majority of which consist of fresh products (fruits and vegetables) or products with low added value (oils, dried fruits or with mixtures) (Kabbaj & Alaoui, 2019).

Figure 9 shows that the areas certified organic have tripled in nine years, going from 4,000 ha in 2010–12,000 ha in 2019. The growth is however relatively slow, compared to similar countries. At the start of the 2010s, the actual figures were close to the forecasts of the Ministry of Agriculture—FIMABIO program contract, then the gap widened, reaching less than 50% of the forecasts in 2017, and 39% in 2020 (DDFP, 2022b).

Given the many constraints facing this sector, the areas under organic farming at the national level have stagnated and even decreased between 2019 and 2020. According to the Moroccan Ministry of Agriculture, the area of cultivated organic crops in 2020 is estimated to 10,300 ha in addition to spontaneous picking of around 300,000 ha.

In terms of consumption, there are awareness-raising efforts to be made, based in particular on the pronounced tastes of Moroccans for "Beldi" products (produced traditionally without any chemical inputs), and defined as old local products, old breeds, or traditional seeds. The popular wisdom that gives "Beldi" and grandmother's recipes good health benefits represents a set of values to be converted into organic. We actually know that chemicals destroy what is beneficial in the Beldi, but that organic has the ability to adapt to local products for the best.

Table 5 Groups of organic crops produced in Morocco, with the subgroups of the main products

Groups	Sub-groups	Products
Fruits	Hazelnut	Hazelnut
	Stone fruits	Apricot, avocado, nectarines, peaches, plums, olives
	Pome fruits	Apples, figs, pears, pomegranates, quinces, table grapes
	Citrus	Clementine, lemon, oranges
	Starchy	Bananas
	Red fruits	Strawberries, other
	Dried fruits	Almonds, dates
	Pod fruits	Carobs
Vegetables	Bulbs	Garlic, leeks, onion
	Leafy	Cabbage, lettuce, spinach
	Fruits	Cucumber, eggplant, green bean, pepper, pumpkins, tomatoes, zucchini
	Grains	Beans, chickpeas, peas
	Roots	Carrots, radishes
	Tubers	Potatoes
Aromatic plants	Shrubs	Capers
	Spices, condiments and aromatic and medicinal plants	Anise, basil, bay leaf, chamomile, chives, coriander, dill, fennel, lavender, lemongrass, mint, nettle, oregano, parsley, roses, rosemary, saffron, sage, sorrel, tarragon, thyme, verbena,
	Other	Aloe vera
Cereals	Cereals	Durum wheat, oats, sweet corn
Processed	Oils	Olive oil, essential oils of aromatic plants
	Fruits	Apricots, olives
Livestock products	Meat	Chickens, cattle, sheep
	Honeybees	Honey
Argan	Oil seed	Argan nuts
	Oil	Argan oil (edible and cosmetic)

Source Kabbaj and Alaoui (2019)

4.3 Total Production of Certified Organic Products

The total production of certified organic products is estimated in 2020 at 130,000 tons (DDFP, 2022b) from both cultivated and spontaneous sectors (Fig. 10).

Figure 11 below describes the value chain of organic products in Morocco from production to consumption, either locally or in importing countries.

Table 6 gives the areas of spontaneous species, which have never undergone any chemical treatment and which could be easily certified.

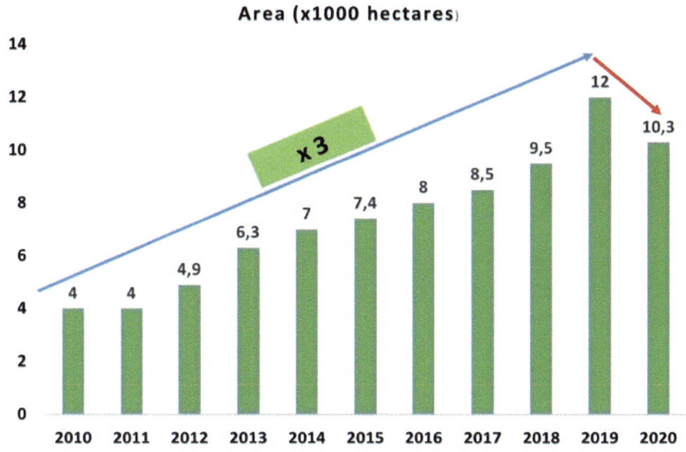

Fig. 9 Evolution of areas certified organic in Morocco for the last 10 years

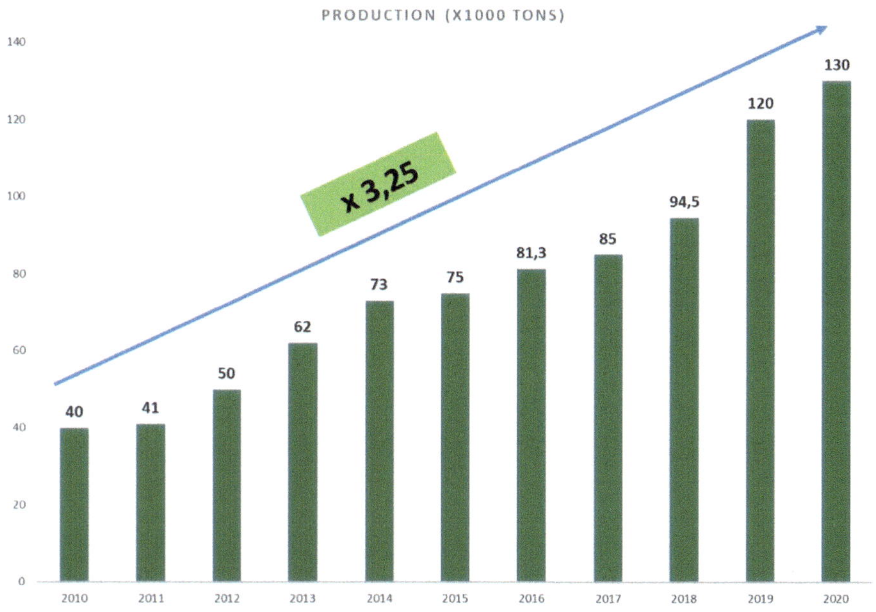

Fig. 10 Evolution of certified organic production (all products confounded) since 2010 (year of reference)

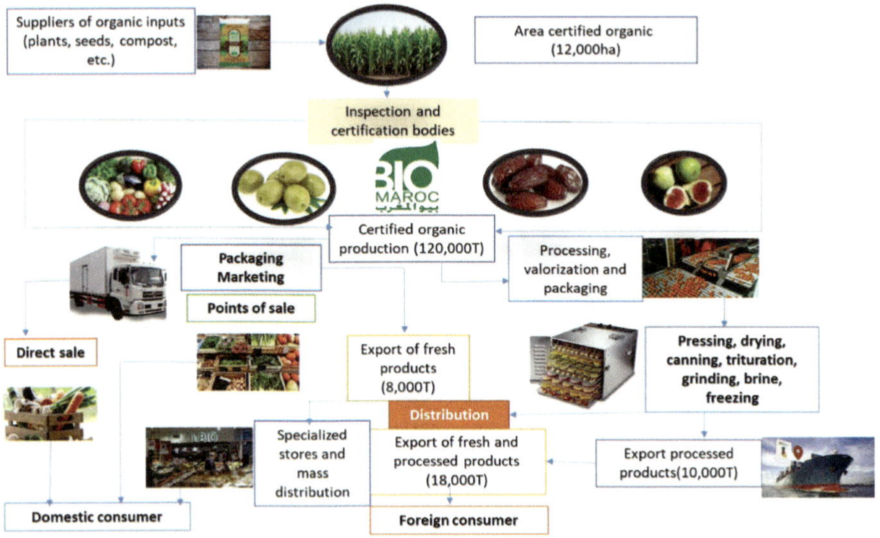

Fig. 11 Value chain of organic products for Morocco. *Source* Adapted from MAPMDREF (2022)

Table 6 Areas of spontaneous species certified organic

Species	Area (ha)
Argan	71.300,00
Prickly pear	41.000,00
Rosemary	56.000,00
Other aromatic & medicinal plants	1.700,00
Forest trees	35,00
Total of spontaneous species	170.035,00

4.4 Moroccan Legislation on Organic Farming

The organic sector also relies today on a regulatory legal framework. Moroccan organic production has in fact been governed since September 2018 by Law No. 39–12. This makes the kingdom the second country in Africa to adopt its own legislation on organic production.

On the regulatory level, the organic sector now has two approved certification and control bodies. Since the entry into force of the legal and regulatory framework of the organic sector, agricultural products are certified according to Moroccan regulations, and they are sold on the market under the Bio Maroc label (DDFP, 2022a).

4.5 *The Main Export Markets for Moroccan Organic Products*

In Morocco, the majority of organic agricultural products are intended both for local consumption and for export, and export volumes have grown remarkably, from 7,230 tons in 2007–18,000 tons in 2019, and 14,011 tons in 2020 (DDFP, 2022b).

They mainly concern fruits, fresh vegetables (early produced fruits and citrus fruits) and processed products, in particular frozen orange juice, food and cosmetic argan oil, canned green beans, aromatic and medicinal plant products, frozen strawberries and capers.

The main destination markets are France, Spain, Italy, Germany, the Netherlands, Scandinavia and Lithuania. Moroccan exporters also ship their organic products to the USA, Canada and some Asian countries such as Japan and South Korea.

4.6 *How to Export Moroccan Organic Products to European Countries?*

To export a product as coming from organic farming in Morocco, farmers and collection and processing companies must have their activity checked and certified each year by a control and certification body (OCC) accredited by the European Union (EU).

In order to operate in Morocco, OCCs must be approved by the Ministry of Agriculture and Maritime Fisheries, Rural Development and Water and Forests in accordance with the provisions of Order No. 270–15 relating to approval of organic production OCCs (AgriMaroc, 2021).

4.7 *The Assets of Morocco for the Development of Organic Agriculture*

As mentioned by Alaoui (2009), Morocco has all the ingredients necessary for the development of this type of agriculture, such as low use of chemical fertilizers and pesticides (less than 20% of cereal acreages are treated with herbicides). This means that a large acreage of this commodity is already free of all the banned chemicals in the certified organic agriculture, and that no conversion period is needed. All it will take is a marketing strategy. Much acreage of fruits and nuts have never been treated chemically, and the only fertilizers used have been animal manure. Other assets are mentioned in Fig. 12.

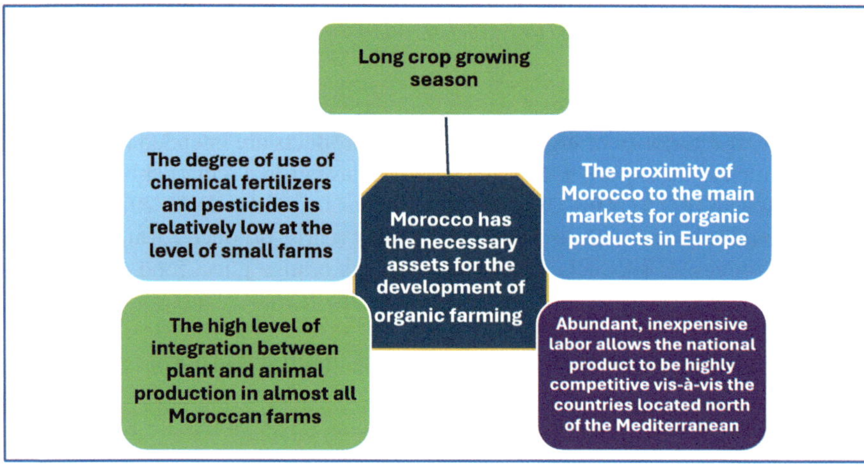

Fig. 12 The assets for the development of organic agriculture in Morocco. *Source* Adapted from Alaoui (2009)

4.8 Measures Put in Place to Encourage Organic Farming

To promote organic farming in Morocco, the government has developed strategies that aim both to encourage producers to move toward organic production, then to convert to organic, and to raise awareness among producers and the population in general on the importance of organic farming. The main measure is the granting of certification fees.

The forecasts were that subsidies for certification would be 70% with a threshold of 1,000 dh/ha for crops, possibly 700 dh/ton for animal products and 1,000 DH/ton for exported processed products. The subsidy should be granted only during the conversion period. The MAPMEFDR subsidizes agriculture in general with other financial aid for agricultural equipment (drip irrigation) and agricultural inputs, which are tax exempt when imported (Azim & Ramli, 2017).

For organic products of plant origin, subsidies are fixed according to the area of the production unit (MAPMEFDR, 2022). These vary between a rate of 90% for the unit whose area is equal to or greater than 0.5 ha and less than 5 ha with a ceiling of 10,000 dh for each unit per year, as well as a rate of 70% for that greater than 20 ha with an annual ceiling of 40,000 dh per unit (Table 7).

The incentive of the State, within the framework of the agricultural development fund (FDA), for certification/conversion to organic mode according to the following rates and ceilings (Table 7).

With regard to livestock and beekeeping products, the rate of the subsidy is 80% of the overall cost of certification of livestock and beekeeping products with a ceiling of 25,000 dh for the benefit of the production unit per year.

Kabbaj and Alaoui (2019) believed that these measures are still minimal to promote and support organic farming. Political ambition could lead to much more;

Table 7 The incentive of the State, within the framework of the FDA, for certification/conversion to organic mode of production

Acreage	Rate	Ceilings (dh[1])
Less than 5 ha	90%	10 000
Greater than 5 and less than 10 ha	70%	20 000
Greater than 10 and less than 20 ha		30 000
Greater than 20 ha		40 00

[1]One US dollar = 9.46 Moroccan dirham (Exchange rate on February 24, 2022)
Source DDFP (2022a)

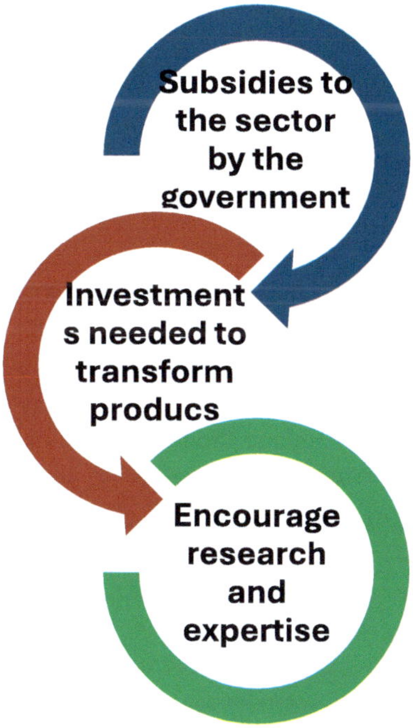

Fig. 13 Series of actions to be put in place for rapid development of the organic sector in Morocco

the next few months and years will be decisive from this point of view for the future of organic on a national scale. The authors consider that the following three series of actions should be put in place quickly in this regard and will condition future success (Fig. 13):

4.9 Obtaining Subsidies for Certification in Morocco

To promote organic farming, the Moroccan government and Fimabio have developed a strategy aimed both at encouraging producers to convert to organic farming and at raising awareness among producers and the population in general of the importance of organic farming. The main planned measure is the granting of subsidies. However, while no subsidy was provided for certification costs until the application decrees had been approved by the competent national authorities (Azim & Ramli, 2017), a government decree on the organic subsidy was issued and has already been adopted in March 2018 by Government Council and should create a certain momentum (Kabbaj & Alaoui, 2019).

To benefit from this subsidy on organic certification granted for a period of three years for each production unit, the application must be submitted to the competent services of the Provincial Directorate of Agriculture or the Regional Office for Agricultural Development, to which the production unit belongs.

This grant application is made on the basis of a document issued by the inspection and certification body. This must justify that the production unit is being converted to organic production. The request is made based on a certificate of conformity attesting that the production unit is certified according to the organic production method (AgriMaroc, 2021).

4.10 Organizations Representing the Organic Sector

On the organizational level, professionals in the organic sector have organized themselves into a national association created in 2010. This was the Moroccan Association of the Organic Production Sector (Amabio), which at the time brought together all the actors working in the organic sector, in particular:

- Producers and processors;
- Exporters of fresh and processed organic products;
- Control and certification bodies: suppliers of organic and compost inputs;
- Researchers and people from the knowledge society; associations and non-governmental organizations (NGOs) specializing in the promotion of the organic sector, environmental protection and sustainable development.

In order to comply with Law 03–12, in force since January 2015, relating to the organization of agricultural inter-professional organizations in Morocco and its implementing texts, the inter-professional organization of the organic sector (formerly AMABIO) created in June 2015 three associations representing the three main chain links in the sector, namely production, development and marketing (AgriMaroc, 2021).

The three associations created constituting the inter-professional sector are AnaproBio: National Association of Producers of the Organic Sector; Valbio Morocco: National Association for the Valorization of Biological Products; Anadexbio: National Association of Distributors and Exporters of Organic Products (AgriMaroc, 2021).

In June 2016, the three associations created the Moroccan Interprofessional Federation of the organic sector (Fimabio). It is in 2017 that the Ministry of Agriculture recognized this federation as the sole interlocutor vis-à-vis the public authorities for the treatment of all aspects relating to the development of organic agriculture in Morocco. However, it should be noted that in January 2021, the Ministry of Agriculture withdrew its recognition of Fimabio (AgriMaroc, 2021) for several reasons, mainly for the lack of progress in terms of indicators such as the improvement of acreages converted to organic and the number of farmers who converted their farms to organic.

4.11 Achievements and Impacts

In 2019, the organic area cultivated was multiplied by 3 to reach 12,000 ha against 4,000 ha in 2010. Organic production followed the same upward trend, rising from 40,000 tons in 2010 to around 130,000 tons in 2020 (DDFP, 2022b).

Likewise, exports of organic products experienced a net increase of about 17,000 tons exported in 2019 against 10,000 tons in 2010. In this context of increased production linked to organic farming, no less than 1.4 million working days were counted in 2019. The country has seen the emergence of specialized organic stores following the increase in national demand for organic products.[1]

4.12 The Challenges of Organic Farming in Morocco

Despite the growth of the organic sector in Morocco, many challenges and constraints plague the sector. Among them, Kabbaj and Alaoui (2019) cited the following:

- Unsatisfactory promotion of organic products: although the Moroccan organic sector is booming, the kingdom imports 90% of organic food products on the shelves;
- The increase in production costs caused by the high costs of certification, currently carried out by foreign regulators;
- The scarcity of authorized inputs, including biopesticides and composts on the domestic market as well as seeds and plants used in organic farming;
- Insufficient communication and promotion of organic products by government entities in domestic and foreign markets;
- Lack of a domestic market for organic products, although there is significant potential of the domestic demand. However, it should be noted that with the Covid-19 pandemic, there has been a growing increase in the consumption of organic products, with people moving toward healthy eating habits and diets.

[1] MAPMEFDR website [Accessed on February 24, 2022].

4.13 The Future of Organic Farming in Morocco

The Ministry of Agriculture aims to expand both acreage and organic production and at the same time increase exports (MAPMEFDR website, 2022).

Indeed, like the Green Morocco Plan, the Generation Green 2020–2030 (GG) agricultural strategy attaches great importance to the organic farming sector. According to an information note from the ministerial department of agriculture on this sector, two-thirds of the projected production are for export (AgriMaroc, 2021).

This aims to reach 100,000 ha of cultivated area and a production of 900.000 tons by 2030.

To reach these objectives, a development plan for the organic sector has been drawn up aiming at setting up strategic intervention levers for the development of this sector (DDFP, 2022b), in particular through (Fig. 14):

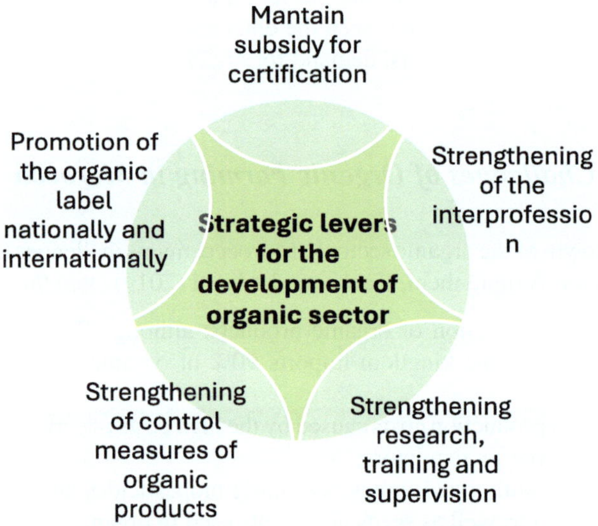

Fig. 14 Strategic levers for the development of the organic production sector. *Source* Adapted from DDFP (2022b)

5 Agroforestry Systems

Agroforestry is a set of diverse land management systems that integrate woody biomass (including trees and woody shrubs) with crops and/or livestock in space and/or time. AF systems are classified as agri-silviculture (crops, including shrubs/vines combined with trees), silvo-pastoral (pasture/animals and trees) and agro silvo-pastoral (crops, pasture/animals and trees). A wide varieties of AF practices exist across climatic regimes from arid to temperate and humid tropic conditions and offer a range of benefits, products and services for land users and their environments (Mbow et al., 2014). AF named by Garrity et al. (2010) as an Evergreen Agriculture and defined it as a holistic approach and cost-effective strategy to achieving food security and environmental resilience even in low-potential regions such as Africa.

Due to the negative impacts of intensification based on monoculture, agroforestry (association of perennial crops—annual crops and/or livestock) has attracted the interest of the international scientific community (Mbow et al., 2019; Daoui and Fatemi, 2014). It has been recognized as an integrated approach to sustainable land use because of its production and environmental benefits during the past three decades (Mbow et al., 2019; Ramachandran et al., 2009). This practice is adopted by farmers for various reasons related in particular to risk management and the enhancement of ecological niches. Indeed, associated species establish negative (competition) or positive (facilitation and niche complementarity) interactions to exploit the resources of the environment. According to Workman and Allen (2004) agroforestry is new way of thinking about an old way of farming. Ramachandran et al. (2009) estimate that the area currently under agroforestry worldwide is 1,023 million ha.

Agroforestry systems are presented in different forms and types: silvo-pastoral systems, alley cropping, windbreaks, riparian buffer strips, and agriculture in the forest (special forest products; mushrooms, etc....). These agroforestry systems are known to sequester large amounts of carbon in tree biomass. Several publications have also shown a positive impact of agroforestry systems on soil organic carbon stocks, especially after conversion of cultivated land. According to Montagnini and Nair (2004), the average carbon storage by agroforestry practices has been estimated at 9, 21, 50 and 63 Mg C ha-1 year-1 in semi-arid, sub-humid, humid and temperate regions. For smallholder agroforestry systems in the tropics, potential C sequestration rates range from 1.5 to 3.5 Mg C ha-1 yr-1. Agroforestry can also have an indirect effect on carbon sequestration, as it helps reduce pressure on natural forests, which are the largest terrestrial carbon sink. Another indirect means of carbon sequestration is through the use of agroforestry technologies for soil conservation, which could improve its storage in trees and soils (Montagnini & Nair, 2004). Indeed, Palma et al. (2007) estimated that this practice can reduce soil erosion by up to 65% and nitrogen leaching by up to 28% and increase carbon immobilization in trees (0, 1 to 3 t of carbon/ha/year) and up to 2.6 times of habitat index (indicator of biological diversity of the landscape).

Agroforestry systems in Morocco are characterized by their diversity according to geographical areas, climate, substrate, land morphology and history of landscape occupation by humans (Sabir et al., 2010). Silvo-pastoral systems essentially correspond to forests or matorrals of holm oaks, cork oaks, cedars, cedars and argan trees which are grazed extensively by the herds of users of these spaces. These systems occupy the forest areas of the Rif, Middle Atlas, High Atlas and Anti Atlas. In the western Rif, the trees most used on the cultivated plots are olive, fig, plum, pear and apricot trees.

In some valleys, on some plots of holm oak on limestone substrate and cork oak on sandstone substrate are found as well as ash and poplar in the wet bottoms of the valleys. In the drier Central and Eastern Rif, we first find the almond tree, then the olive tree and the fig tree. The vine is used on certain plots as the understory of the almond or fig tree.

In the Middle Atlas, rosaceous are dominant at altitude in intensive plantations (cherry, plum, apple, cognacs, etc.). The carob tree is beginning to become a widely used species, especially in the region between Beni Mellal and Khenifra. In the High Atlas, there is a dominance of the almond tree on the semi-arid and arid slopes and of the olive tree on the sub-humid slopes (Sabir et al., 2010). However, at the national level, few scientific studies have been undertaken to analyze these systems as a whole (Daoui & Fatemi, 2014). From this rough description of trees and in reference of agricultural land uses in the country (Table 1), It is possible to support that most traditional farming systems that are based on trees or perennial crops with annual crops and/or livestock can be assessed as agroforestry system. Figure 15 from different agro climatic regions of Morocco shows different agroforestry systems initiated by farmers.

Fig. 15 From the north to the south of Morocco different agroforestry systems exist depending on pedoclimatic conditions and associated crops: **a** mountainous region with different associations. **b** oasis system. **c** agricultural system and livestock (goats). **d** carob trees and annual crops. **e** almonds trees and peas. **f** olive trees and annual crops

At national level some scientific studies have been conducted concerning the characterization of agroforestry systems basis on olive tree (Daoui & Fatemi, 2014; Ammassaghrou et al., 2021). Those studies demonstrated the importance of the association of olive tree and annual corps comparatively to olive tree as monocrop. Cereals, legume or fodder crops are conducted as intercrops in olive orchard alley in rainfed conditions while when irrigation water is available vegetables, alfalfa or corn are more dominant. Daoui et al. (2012) estimated that agroforestry based on olive trees concerns approximately 75% of olive orchards while the remaining 25% concerns olive orchards as monocrop planted at higher slopes or very aged but where grazing of spontaneous herbs in the alley is common. Other studies (Temani et al., 2020; Ammassaghrou et al., 2021) demonstrated that under rainfed conditions with an annual rainfall of more than 450 mm association of olive tree and annual crops (cereal or legumes) is more profitable than sole crops. Calculated Land Equivalent Ratio under such condition exceed 1 and was more important when olive trees are combined with legume crops (Ammassaghrou et al., 2021).

Under semi-arid conditions, Chryaa and El Mzouri (2004) and Chebli et al., (2012; 2018) demonstrated the benefits of introducing forage shrubs. The adopted and tested system is based on the association of Atriplex (*Atriplex nummularia*) and barley (*Hordeum vulgare*). This combination allows an increase in yield per unit area and a diversification of species; a decrease in the cost of feed in addition to the increase in the economic efficiency of low-potential land; improved animal performance and finally rehabilitation of marginal land, fauna and flora (Chryaa & El Mzouri, 2004).

Agroforestry systems exit and are diverse under Moroccan conditions, nevertheless, government agricultural strategy does not yet take into account the systemic approach as it could tackle such system in practice by farmers or as a perspective in developing new designs for production systems. One simple step to overcome such constraints and reach development goals under the Generation Green strategy is rather than the development of fruit trees as monocrop is the development of agroforestry basis on those fruit trees. The questions to be raised are:

- Where can we adopt agroforestry systems combining fruit trees and annual corp and/or livestock according to pedoclimatic conditions?
- Which combination of annual and perennial crops to adopt?
- How can we manage such a system?

To answer those questions, investigation on the characterization of farmer's practices in the open field laboratories and scientific research at experimental stations may help to develop approaches and technical options for sustainable management of agroforestry systems under Moroccan conditions from design to exploitation.

6 Policy and Institutional Channels and Measures: Pathways for a Vibrant Future

The year 2015 has seen the advent of two important global policy frameworks the 2030 Agenda for Sustainable Development and the Paris Agreement on climate change. These frameworks have introduced the Sustainable Development Goals (SDGs), nationally determined contributions (NDCs) and land degradation neutrality (LDN). Many national, regional and international institutions (AFD, World Bank, FAO, EU, etc.) are encouraging important investment in sustainable farming methods, climate-smart agriculture and sustainable land management as an engine for economic and social growth in Morocco.

A number of interconnected barriers to wider adoption of CSA have been identified in the farming context of Morocco and described along the chapter. Under a rapidly changing climate and burgeoning social constraints, drivers of change from policymakers to the farming community are essential to overcome the complexity associated with CSA adoption. A polycentric policy approach should be favored and effected in order to permit increased cooperation, communication, interaction among local, national and international actors and instigate learning-by-doing to improve initiatives and policies at different levels of decision-making and overtime.

Improving agricultural sustainability requires multi-faceted solutions built on science-based public policies (Lal, 2021). In addition, any policy pertaining to water-climate-food nexus should put soils as central (Lal et al., 2017).

Morocco's sustainable natural resources pathway should include CSA strategies and actions that are known to efficiently and sustainably manage, use and maintain natural resources and the enabling factors and institutional enhancers that sustain their implementation and development across scales. The pathways include technical improvement and social commendation but also most importantly the social, political and economic barriers that need to be addressed and released. A better and vibrant future for Moroccan agriculture requires that the Green Generation initiative gathers CA, OF and AF practices in food production and natural resource valuation and develop an integrative and inclusive road Map while attracting innovative communication mechanism, blended and green funding, renewed partnership and fresh alliances, and novel governance and entrepreneurship systems.

Although CSA potential is large from a technical, economical, biophysical and ecological perspective, its feasibility can be hampered by a lack of durable institutional support, uncertainty over long-term benefits and trade-offs, weak governance, fragmented land ownership, and uncertain permanence of funds and incentives. In the next decade, a novel focus on sustainable growth, capacity building and development and innovative approaches should be carried by socially minded entrepreneurs and private businesses that can support in solving and relieving barriers for CA adoption mainly by smallholder farmers. Moroccan African farmers are embedded in multiple layers of institutional environment which affect their commitment, capacity, and decision to adopt new technologies and to change their paradigms. Securing resources for survival and growth are of first importance for farmers. Hence, the willingness of "innovative" CSA farmers and stakeholders to adopt and promote these systems is a prerequisite for large-scale dissemination and hence ensuring the creation of social movement for CSA acceptance at all levels of the society and government's hierarchy.

While trade-offs are inevitable, public and private insurance programs, such as crop insurance or weather index insurance, help preserve farmers and producers' incomes and enable them to expand risk management capacity while keeping their assets productive. These programs should guarantee the confidence of farmers in CSA systems mainly during the transition period and long-lasting impacts on the economic stability and prospects of farmers. Under situations of extreme drought or high risks due to shocks or stresses, social protection programs such as cash grants should be devoted to farmers' communities and value chain agents. However, such programs should facilitate CSA farmers to move from protection to productivity and stability in incomes.

Private sector should be incentivized to form durable partnerships for disseminating CSA systems and encouraging capacity development and improved nutrition to fully generate sustainability. Science-based regulatory systems and practices are of great importance and should be developed in order to ensure trust and shared responsibilities in shifting to and adopting CSA systems (Zeigler & Steensland, 2016).

Changes in farming systems will also need to be met with changes in market dynamics. Reliable admittance to stable, well-managed markets and financial systems incentivize growers (mainly smallholder farmers) to make sustained and sustainable investments in CSA systems and hence adopt them. Overall, enabling policy and instruments can ensure that CSA farmers stay productive under seasons of risks and increase their adaptability and resilience compared to traditional farmers.

For CSA options to be effectively and widely implemented in rural communities, the accompanying policy should promote rural prosperity and secure education and remunerative employment mainly among women and youth. In this sense, CSA systems should be connected to markets, value chains, entrepreneurship, off-farm opportunities in connection with urban neighbors and supporting services to further create jobs in rural communities.

Long-term implementation of CSA research and information and modeling platforms should avoid spurious and misleading trends while revealing consistent dynamics in benefits and outcomes. However, such interfaces require closer collaboration among scientists, stakeholders and farmers, stronger private–public partnerships, effective policies, and enhanced public awareness and engagement. On the other hand, replicating successful CSA projects requires institutional drivers and enabling conditions that promote social learning, strong leadership, transformational coalition and durable mobilization of resources and funds.

Last but not least, transition to CSA systems is a knowledge-intensive process. Hence, to be able to continue deep adoption and up/out scale of these systems, the science-policy gap should be narrowed, and new arrangements and approaches should be investigated while consolidating the science-policy interface. In order to build science-policy interfaces, raise awareness, communicate, advocate, leverage and promote CA in Morocco, it is necessary to create centers of excellence. Other functions of these centers include providing advice to national institutions, international cooperation, technical assistance, policy analysis, formulation, and monitoring, and project management (Mrabet et al., 2022).

7 Conclusions and Future Scope for Enhancing CSA Adoption Effectiveness

From the latest publication by Rattan Lal 2020 World Food Prize Laureate, Green Revolution (GR) of the twenty-first century must be: (i) soil-centric, based on soil health and resilience, (ii) ecosystem-centric, based on eco-efficiency of inputs, (iii) knowledge or innovation-centric, based on scientific principles, and (iv) nature-centric, based on nature-positive solutions which restore and enhance nature (Lal, 2021). The new GR should also recognize the "One Health" concept, which states that the "health of soil, plants, animals, people, ecosystems, and the planetary processes is one and indivisible" (Lal et al., 2017). Smith (2018) clearly showed the importance of and reliance on sustainable management of soils in delivering most SDGs. All these new paradigms and concepts should be rooted in the Green Generation strategy of Morocco. This interconnectivity is also the basis of the "Adapting African Agriculture" (AAA) initiative adopted by COP22 in Morocco (Badraoui et al., 2018).

It is clear from the remarkable achievements and undeniable progress from CSA systems that long-term country's economic growth is driven by innovation. An agricultural research consortium cannot provide the answers to all the societal concerns related to CSA uptake and mainstreaming (Gaffney et al., 2019). In here, it is of paramount importance to create strategic partnerships and platforms for deep-scaling CSA systems while encouraging shared learning and knowledge co-production and communication.

The new agricultural strategy can be seeing as an inclusive and responsible agricultural development narrative that can shift the focus to the real potential of the country to develop and thrive to reverse present-day and incipient negative trends due to climate change, population growth, limited resources, water scarcity, etc. The necessary steps to realize and unlock the potentialities require a shift to CSA systems and especially conservation agriculture, organic farming and agroforestry, and merge them in comprehensive and integrated deployment. Hence, deep and up-scaling agro-ecological approaches are of great importance in enhancing climate resilience of millions of Moroccan farmers across generations. According to Smith (2018), soil organic matter enhancement and sustainable intensification are among '*no regret*' options with little risk of significant trade-offs.

Public R&D provided foundational results and knowledge on CSA systems or ecologically-intensive systems (i.e., conservation agriculture, organic farming and agroforestry) even though in separate projects and research consortium. Through the chapter and minds assembled to prepare it, we have attempted to bring an updated body of insights and knowledge able to provide farmers, producers, development agents and decision-makers the necessary indicators or metrics to successfully prioritize CSA systems in their agendas. In this way, it is intended to deliver a system thinking that allows capturing benefits and mitigating trade-offs and a behavioral change that increases adoption rates. Such transformational change necessitates multi-faceted collaborative approaches involving all stakeholders (i.e., farmers, NGOs, agribusinesses, retailers, service providers, government and public agents, opinion and decision-makers, etc.).

For balanced plant nutrition, optimum fertilization is more critical with CA and AF, and soil analysis is necessary before applying fertilizers. However, with time the soils could develop their nutrient pools allowing unnecessary use of synthetic fertilizers as in OF. In the case of weed control which seems obligatory under CA systems, the judicious use and diversification of crop rotations and herbicide molecules could over time reduce weed seed banks and hence the reliance on such agrochemicals. CSA systems allow for minimizing the ecological footprint of food production. In other terms, these options permit to close yield and efficiency gaps simultaneously (van Noordwijk & Brussaard, 2014). It is worth saying, that Australia, a Mediterranean country like Morocco, is a world leader in both CA and OF with respectively 23.6 and 36 million hectares (Kassam et al., 2022; Willer et al., 2020).

Further research is needed to assess and quantify CSA resilience metrics at the watershed level through long-term experiments, living labs and advanced modeling and artificial intelligence tools. Such studies will help an integrated understanding of CSA impacts at different time horizons and across farming systems for their implementation at a large scale.

In many situations, crop rotation is considered to be incomplete if legumes are not included in a cropping system however in the country these crops are still limited in terms of acreage (Al Balghitti et al., 2020) (Table 1). Inclusion of food legumes in cropping systems has gained global attention because of their multiple benefits. Among these benefits: crop diversification, nutrient management and cycling, disease and pest control, soil health restoration, reduced dependence on agrochemicals and higher incomes to farmers (Praharaj & Maitra, 2020). In Morocco, leguminous crops merit wider interest by farmers, producers, extension agents and policymakers when shifting to organic farming, CA and intercropping systems and promoting use of biofertilizers and scheduled input use and management.

According to Lal (2016), sustainable intensification of agroecosystems (which includes CSA systems) can produce enough food grains to feed one person for a year on 0.045 ha of arable land. Further, the use of fertilizers, pesticides, and chemicals, along with the irrigation water must be reduced to the minimum required.

The knowledge, tools and experience with the adoption and spread of conservation agriculture, and especially the integration of farm trees with CA, has expanded substantially since 2000 in UN Development Organizations, development and conservation INGOs and the CGIAR. Hence, there is a great movement to merge conservation agriculture, AF and organic farming to reach at least a third of the cultivated area by 2030 (that is 500 million hectares) (WCCA, 2021; Evergreen Alliance, 2022; https://www.evergreening.org/home-2/greenuptocooldown/). In other words, when applying and implementing CSA systems at such a large scale, business as usual is no longer an option, and in-depth transformations to farming systems are needed. CSA systems must fit into emerging farming environments, as well as multi-government policy and both farmers and decision-makers should be motivated to transform current unsustainable farming systems into more durable forms. A common language is needed among all stakeholders to mainstream CSA options and upscale their benefits. In other words, high institutional strength is needed to foster adoption pathways and channel social, societal and economic concerns of farmers through the decision process. CSA feasibility should be intrinsically linked to substantial motivation and incentive mechanisms mainly through subsidies, micro-finance, carbon trading, access to markets, payments for ecosystem services and co-benefits, and compensatory costs for farmers.

In conclusion, the shift to climate-smart farming represents one of the most substantial landscape changes in Moroccan agriculture. In this regard, giving adequate attention to the range of factors that can determine whether CAS systems offer a real or perceived relative advantage to farmers compared to the status quo can allow a high proportion of adoption decisions to be explained. The following are the key factors (or requirements) to drive and promote CSA adoption in the country as summarized by Mrabet et al. (2022):

- Carrying out in-depth social, economic and risk analysis at farmer, community, and country scales (including livelihood resilience to shocks and stresses).
- Ensuring an integral ex-ante assessment (monitoring) of success and adoption (multi-criteria, multi-stakeholder evaluation including co-benefits and trade-offs) for tailoring CSA to a sustainable intensification pathway, and hence, ensuring long-lasting and large-scale commitments by CSA users.
- Government's initiatives to create a conducive environment for the adoption and development of CSA systems including tackling major barriers while creating inclusive and equitable learning, financial and institutional mechanisms in addition to policy support.
- Development of instruments in co-addressing societal challenges and financial incentives in CSA diffusion.
- Guaranteeing sharing costs and risks in addressing CSA uptake and up-scaling.
- Encouraging and supporting CSA-based farmer organizations (associations and cooperatives).
- Use of innovative tools for knowledge management and information sharing will be needed to further stimulate the demand for the technology from a wider range of clients (e.g., small and large farms, public and private, NGOs and other environmental activists).
- Transforming CSA research and science-based benefits to skills, curricula, awareness and positive changes aligned with the anticipated needs of users.
- Develop efficient and quality research and innovation systems and provide a system thinking forum to further embed CSA in development contexts and constraining climate and global changes.

- Creation and enforcement of centers of excellence, technology consortiums or competence centers.
- Develop strong and durable human capital and scientific capacity.
- Harnessing and influencing partners and resource mobilization to achieve the best resource allocation to farmers and their organizations to improve production capacity and environment stewardship over time and ensure focused effectiveness and tangible impacts of funds.
- Mounting innovative public–private partnership for converging visions and priorities.
- For confirming impacts, farmers should be mobilized in CSA research, promotion and development pathways and foresight processes.
- Inclusion of CSA systems in multi-government policies, framework and programs.
- Development of nature-positive legislation and regulatory systems.
- Farmers must be incentivized to adopt recommended CSA practices through payments for ecosystem services.
- Organic farming advocates healthy products free from components that may harm humans and nature. There is a great research need to close the gaps among OF and other CSA systems.

References

ADA. (2018). *Guide de l'investisseur dans le secteur agricole au Maroc* (p. 160). Edition MAPM.

AgriMaroc. (2021). *L'agriculture biologique au Maroc*. https://www.agrimaroc.ma/agriculture-biologique-maroc-2021/ [Accessed on February 22, 2022].

Alaoui, S. B. (2009). Organic farming in the world, and case study of Morocco: Achievements, drawbacks and future perspectives. In: Bouaziz et al. (Eds.), *Actes du Symposium International AGRUMED, Durabilité des Systèmes de culture en zone méditerranéenne: Gestion des ressources en eau et en sol* (p. 400).

Al Balghitti, A., Dahan, R., & Kradi, C. (2020). *Analyze de la chaîne de valeur des légumineuses alimentaires* (p. 172). INRA—Editions. ISBN: 978–9920–35–593–3.

AMAC. (2021). *'Rapport sur la dissémination du semis direct au Maroc', Atelier pour la présentation de la feuille de route systèmes de semis direct sur 1,5 Mha (2020–2030)*. INRA-Rabat, 28 Janvier, 2021.

Amassaghrou, A., Bouaziz, A., Daoui, K., Belhouchette, H., Ezzahouani, A., Barkaoui, K. (2021). Productivité et efficience des systèmes agroforestiers à base d'oliviers au Maroc: Cas de Moulay Driss Zerhoun. *Cahiers Agricultures, 2021*(30), 2.

Azim, K., & Ramli, H. (2017). *Country report: Organic agriculture development in Morocco*. ISOFAR. http://isofar.org/isofar/index.php/2-uncategorised/284-isofar-and-inra-morocco-organize-the-1st-eu-north-africanconference-on-organic-agriculture-eu-nacoa

Badraoui, M., Balaghi, R., Mouchfi, N. (2018). *Rapport du comité scientifique de l'Initiative sur l'adaptation de l'agriculture africaine au changement climatique (Initiative AAA) : Etat des lieux et propositions de solutions*. Publication INRA.

Belmekki, M., Mrabet, R., El Gharous, M., Moussadek, R., Iben Halima, O., Boughlala, M., & Bencharki, B. (2014). Effets du non-labour et de la gestion des résidus sur la conservation de l'eau et la qualité d'un sol calcimagnésique dans les zones semi-arides Marocaines. *International Journal of Innovation and Applied Studies, 6*(4),792–800.

Bessam, F., & Mrabet, R. (2003). Long-term changes in soil organic matter under conventional and no-tillage systems in semiarid Morocco. *Soil Use & Management, 19*(2), 139–143.

Bonzanigo, L., Giupponi, C., & Moussadek, R. (2016). Conditions for the adoption of conservation agriculture in Central Morocco: an approach based on Bayesian network modelling. *Italian Journal of Agronomy, 11*(1), 24–34. https://doi.org/10.4081/ija.2016.665

Boulal, H., El Mourid, M., Ketata, H., & Nefzaoui, A. (2014). Conservation agriculture in North Africa: Past and the future. In R. A. Jat, K. L. Sahrawat, & A. H. Kassam (Eds.), *Conservation agriculture: Global prospects and challenges* (pp. 293–310). CABI.

Bouzza, A. (1990). *Water conservation in wheat rotations under several management and tillage systems in semi-arid areas.* PhD dissertation, University of Nebraska, Lincoln, Nebraska.

Brouziyne, Y., Abouabdillah, A., Hirich, A., Bouabid, R., Zaaboul, R., & Benaabidate, L. (2018). Modeling sustainable adaptation strategies toward a climate-smart agriculture in a Mediterranean watershed under projected climate change scenarios. *Agricultural Systems, 162*, 154–163.

Chebli, Y., Mrabet, R., & Chentouf, M. (2012). Alley cropping as a durable alternative for pasture land development in the drought prone region of Eastern Morocco. *Options Méditerranéennes, 102*, 449–452.

Chriyaa, A. A., & El-Mzouri, H. (2004). Introduction des arbustes fourragers dans les systèmes de production en zones à faible pluviométrie. In A. Ferchichi (comp.), *Réhabilitation des pâturages et des parcours en milieux méditerranéens = Rangeland and pasture rehabilitation in Mediterranean areas* (pp. 203–206). Zaragoza (Spain): CIHEAM-IAMZ, 2004: (En). (Cahiers Options Méditerranéennes ; v. 62). 11. Réunion du Sous-Réseau Ressources Fourragères Méditerranéennes du Réseau Coopératif Interrégional FAO-CIHEAM de Recherche et Développement sur les Pâturages et les Cultures Fourragères, 2002/10/29–2002/11/01, Djerba (Tunisia).

Daoui, K., Fatemi, Z., Bendidi, A. R. Razouk, R., Chergaoui, A., & Ramdani, A. (2012). Olive tree and annual crops association's productivities under Moroccan conditions. In *Abstracts of the 12th congress of the European society of agronomy*, Helsinki, Finland, 20–24 August 2012 (pp. 586–587).

Daoui, K., & Fatemi, Z. (2014). Agroforestry systems in Morocco: The case of olive tree and annual crops association in Saïs region. In *Science, policy and politics of modern agricultural system* (pp. 281–289). Springer.

DDFP. (2022a). *Fiche sur la filière biologique au Maroc. Division de la Labellisation, Direction de Développement des Filières de Production.* Ministère de l'Agriculture, des Eaux et Forests, et du Développement Rural.

DDFP. (2022b). *Fiche relative aux Principaux indicateurs de la filière biologique: année 2020.* Direction des Filières de Production. Division des labels.

Devkota, M., Singh, Y., Yigezu Y. A., Bashour, I., Mussadek, R., & Mrabet, R. (2021). Conservation agriculture in the drylands of the Middle East and North Africa (MENA) region: Past trend, current opportunities, challenges, and future outlook. *Advances in Agronomy.* https://doi.org/10.1016/bs.agron.2021.11.001. https://hdl.handle.net/20.500.11766/66826 Netherlands: Elsevier Ltd.

El Gharras, O., El Mourid, M. & Boulal, H. (2017). Conservation Agriculture in North Africa: experiences, achievements and challenges. In A. Kassam, S. Mkomwa, & Friedrich, T. (Eds.), *Conservation agriculture for Africa: Building resilient farming systems in a changing climate* (pp. 127–138). Wallingford, UK: CAB International.

Evergreen Alliance. (2022). Annual report 2022/23 (29 p.). https://www.evergreening.org/annual-reports/

FAO. (2014). *What is conservation agriculture?* FAO CA website (http://www.fao.org/ag/ca/1a.html).

FAO, IFAD, UNICEF, WFP & WHO. (2020). *The state of food security and nutrition in the world 2020. Transforming food systems for affordable healthy diets.* Rome: FAO. https://doi.org/10.4060/ca9692en

FAO. (2021). *The state of the world's land and water resources for food and agriculture—Systems at breaking point*. Synthesis report 2021, Rome. https://doi.org/10.4060/cb7654en

FERT. (2018). *L'agriculture de conservation au Maghreb: Les agriculteurs font évoluer leurs pratiques*. https://www.agricord.org/sites/default/files/2018_med_capi_ac_maghreb_fr.pdf.

Findlater, K. M., Kandlikar, M., & Satterfield, T. (2019). Misunderstanding conservation agriculture: Challenges in promoting, monitoring and evaluating sustainable farming. *Environmental Science and Policy, 100*, 47–54.

Friedrich, T. (2005). Herbicides and no-till farming: Does no-till farming require more herbicides? *Outlooks on Pest Management, 16*(4), 188–191.

Gaffney, J., Bing, J., Byrne, P. F., Cassman, K. G., Ciampitti, I., Delmer, D., Habben, J., Lafitte, H. R., Lidstrom, U. E., Porter, D. O., Sawyer, J. E., Schussler, J., Setter, T., Sharp, R. E., Vyn, T. J., & Warner, D. (2019). Science-based intensive agriculture: Sustainability, food security, and the role of technology. *Global Food Security, 23*, 236–244. https://doi.org/10.1016/j.gfs.2019.08.003

Garrity, D. P., Akinnifesi, F. K., Ajayi, O. C., et al. (2010). Evergreen Agriculture: A robust approach to sustainable food security in Africa. *Food Security, 2*, 197–214. https://doi.org/10.1007/s12571-010-0070-7

Grossi, G., Goglio, P., Vitali, A., & Williams, A. (2019). Livestock and climate change: Impact of livestock on climate and mitigation strategies. *Animal Frontiers, 9*(1), 69–76.

Hajjaj, B., Bouhache, M., Mrabet, R., & Taleb, A. (2018). Stock semencier des mauvaises herbes dans le sol et évolution de la flore de surface sous système de semis direct. *Revue Marocaine De Protection Des Plantes, 12*, 1–17.

Hajjaj, B., Bouhache, M., Mrabet, R., Taleb, A., & Douaik, A. (2016a). Efficacité de quelques herbicides contre les mauvaises herbes du pois chiche et de la féverole conduits en semis direct. *Revue Marocaine des Sciences Agronomiques et Vétérinaires, 4*(3), 37–47.

Hajjaj, B., Bouhache, M., Mrabet, R., Taleb, A., & Douaik, A. (2016b). Efficacité de quelques herbicides des céréales dans une culture de blé tendre conduit en semis direct. *Revue Marocaine des Sciences Agronomiques et Vétérinaires, 4*(3), 48–56.

Hamim, A., Essahat, A., El Othmani S., et al. (2012). Etude de l'effet de la rotation et du travail du sol sur la densité spécifique des champignons mycorhiziens arbusculaires. In *International congress: Microbial biotechnology for development (MICROBIOD 2)*, Marrakech, Morocco.

Hobbs, P. R. (2007). Conservation agriculture: What is it and why is it important for future sustainable food production? *Journal of Agricultural Science, 145*, 127–137. https://doi.org/10.1017/S0021859607006892

IFOAM. (2020). *Global organic area continues to grow.* https://www.ifoam.bio/global-organic-area-continues-grow [Accessed in February 20, 2022].

IPCC. (2018). Summary for policymakers. In V. Masson-Delmotte, P. Zhai, H.-O. Pörtner, et al. (Eds.), *Global warming of 1.5°C. An IPCC Special Report on the impacts of global warming of 1.5°C above pre-industrial levels and related global greenhouse gas emission pathways, in the context of strengthening the global response to the threat of climate change, sustainable development, and efforts to eradicate poverty* (pp. 3–24). Cambridge, UK and New York, NY, USA: Cambridge University Press. https://doi.org/10.1017/9781009157940.001

IPCC. (2019). Climate change and land: An IPCC special report on climate change, desertification, land degradation, sustainable land management, food security, and greenhouse gas fluxes in terrestrial ecosystems. In P. R. Shukla, J. Skea, E. Calvo Buendia, V. Masson-Delmotte, H.-O. Pörtner, D. C. Roberts, P. Zhai, R. Slade, S. Connors, R. van Diemen, M. Ferrat, E. Haughey, S. Luz, S. Neogi, M. Pathak, J. Petzold, J. Portugal Pereira, P. Vyas, E. Huntley, K. Kissick, M. Belkacemi, & J. Malley (Eds.). In press.

Kabbaj, S., & Alaoui, S. B. (2019). *Le Bio au Maroc. Livre Blanc : Les Filières Bio, Un secteur à fort potentiel*. CEBIO. Février 2019.

Kabbaj, S., & Alaoui, S. B. (2018). *Les Filières Bio: Enjeux et Promesses pour le Maroc et l'Afrique. Avril 2018*. Imprimerie First Paper, ISBN9 789920 354660.

Kassam, A., Friedrich, T., & Derpsh R. (2022). State of the global adoption and spread of conservation agriculture. In A. Kassam (Ed.), *Advances in conservation agriculture* (Vol. 3). Adoption and spread. ISBN-13: 9781786764751
Kassam, A., & Kassam, L. (2020). *Rethinking food and agriculture*. Elsevier.
Laghrour, M., Moussadek, R., Mekkaoui, M., et al. (2015). Impact of no-tillage on physical proprieties of a clay soil in Central Morocco. *Journal of Material and Environmental Science, 6*, 391–396.
Laghrour, M., Moussadek, R., Mrabet, R., et al. (2016). Long and midterm effect of conservation agriculture on soil properties in dry areas of Morocco. *Applied and Environmental Soil Science*, Article ID 6345765. https://doi.org/10.1155/2016/6345765
Laghrour, M., Moussadek, R., Mrabet, R., & Mekkou, M. (2019). Effet à moyen et à long terme du semis direct sur la matière organique, la stabilité structurale et la compaction des sols argileux au Maroc. *Revue Marocaine des Sciences Agronomiques et Vétérinaires, 7*(2), 356–362.
Lal, R. (2016). Feeding 11 billion on 0.5 billion hectare of area under cereal crops. *Food and Energy Security, 5*(4): 239–251. https://doi.org/10.1002/fes3.99.
Lal, R., Mohtar, R. H., Assi, A. T., Ray, R., Baybil, H., & Jahn, M. (2017). Soil as a basic Nexus tool: Soils at the center of the food–energy–water Nexus. *Current Sustainable/renewable Energy Reports, 4*(3), 117–129. https://doi.org/10.1007/s40518-017-0082-4
Lal, R. (2021). Feeding the world and returning half of the agricultural land back to nature. *Journal of Soil and Water Conservation, 76*(4), 75A LP-78A. https://doi.org/10.2489/jswc.2021.0607A.
Lembaid, I., Moussadek, R., Mrabet, R., Douaik, A., & Bouhaouss, A. (2021). Modeling the effects of farming management practices on soil organic carbon stock under two tillage practices in a semi-arid region, Morocco. *Heliyon 7*(1). https://doi.org/10.1016/j.heliyon.2020.e05889.
Lembaid, I., Moussadek, R., Mrabet, R., & Bouhaouss, A. (2022). Modeling soil organic carbon changes under alternative climatic scenarios and soil properties using DNDC model at a semi-arid mediterranean environment. *Climate, 10*, 23. https://doi.org/10.3390/cli10020023
MAPMEFDR. (2022). Moroccan Ministry of Agriculture, Water and Forest, and Rural development website. https://www.agriculture.gov.ma/fr/ [Accessed in February 24, 2022].
Mbow, C., Rosenzweig, C., Barioni, L. G., et al. (2019). Food security. In P.R. Shukla, J. Skea, E. Calvo Buendia, V. Masson-Delmotte, H.-O. Pörtner, et al. (Eds.), *Climate change and land: An IPCC special report on climate change, desertification, land degradation, sustainable land management, food security, and greenhouse gas fluxes in terrestrial ecosystems*. Geneva, Switzerland: Intergovernmental Panel on Climate Change (IPCC).
Mbow, C., Smith, P., Skole, D., Duguma, L., & Bustamante, M. (2014). Achieving mitigation and adaptation to climate change through sustainable agroforestry practices in Africa. *Current Opinion in Environmental Sustainability, 6*, 8–14.
Montagnini, F., & Nair, P. K. R. (2004). Carbon sequestration: An underexploited environmental benefit of agroforestry systems. *Agroforestry Systems, 61*, 281–295.
Moussadek, R., Mrabet, R., Zante, P., et al. (2011a). Impact of tillage and residue management on the soil properties and water erosion of a Mediterranean Vertisol. *Canadian Journal of Soil Science, 91*(4), 627–635.
Moussadek, R., Mrabet, R., Dahan, R., et al. (2011b). Effect of tillage practices on the soil carbon dioxide flux during fall and spring seasons in a Mediterranean Vertisol. *Journal of Soil Science and Environmental Management* 2(11):362–369.
Moussadek, R., Mrabet, R., Dahan, A., et al. (2014). Tillage system affects soil organic carbon storage and quality in Central Morocco. *Applied and Environmental Soil Science*, Article ID 654796, p. 8. https://doi.org/10.1155/2014/654796
Mrabet, R. (2008). *No-tillage systems for sustainable dryland agriculture in Morocco* (153 p.). INRA Publication. Fanigraph Edition.
Mrabet, R., Bahri, H., Zaghouane, O., Chiekh M'hamed, H., El-Areed, S. R. M. & Abou El-Enin, M. M. (2022). Chapter 6: Adoption and spread of conservation agriculture in North Africa. In A. Kassam (Ed.), *Advances in conservation agriculture* (Vol. 3). Adoption and spread. Cambridge,

UK: Burleigh Dodds. ISBN: 978–1786764751. https://shop.bdspublishing.com/store/bds/detail/workgroup/3-190-106341
Mrabet, R., Moussadek, R., Devkota, M. & Lal, R. (2021). Chapter 14: No-tillage farming in Maghreb region: Enhancing agricultural productivity and sequestrating carbon in soils. In R. Lal (Ed.), *Advances in soil science: Soil organic matter and feeding the future: Environmental and agronomic impacts. CRC Press Taylor and Francis Group*. https://doi.org/10.1201/9781003102762-14.
Mrabet, R., Moussadek, R., Fadlaoui, A., & Van Ranst, E. (2012). Conservation agriculture in dry areas of Morocco. *Field Crops Research, 132*, 84–94.
Mrabet, R. (2011). No-tillage agriculture in West Asia & North Africa. In P. G. Tow, I. M. Cooper, I. Partridge, & C. J. Birch (Eds.), *Rainfed farming systems* (pp. 1015–1042). Dordrecht, Netherlands: Springer. ISBN: 978–1–4020–9131–5.
Mrabet, R., Saber, N., El-Brahli, A., Lahlou, S., & Bessam, F. (2001a). Total, particulate organic matter and structural stability of a Calcixeroll soil under different wheat rotations and tillage systems in a semiarid area of Morocco. *Soil and Tillage Research, 57*, 225–235.
Mrabet, R., Ibno-Namr, K., Bessam, F., & Saber, N. (2001b). Soil chemical quality changes and implications for fertilizer management after 11 Years of no-tillage wheat production systems in semiarid Morocco. *Land Degradation & Development, 12*(6), 505–517.
Nichols, V., Verhulst, N., Cox, R., & Govaerts, B. (2015). Weed dynamics and conservation agriculture principles: A review. *Field Crops Research, 183*, 56–83.
Ortiz-Bobea, A., Ault, T. R., Carrillo, C. M., Chambers, R. G., & Lobell, D. B. (2021). Anthropogenic climate change has slowed global agricultural productivity growth. *Nature Climate Change, 11*(4), 306–312. https://doi.org/10.1038/s41558-021-01000-1
Palma, J. H. N., Graves, A. R., Bunce, R. G. H., Burgess, P. J., de Filippi, R., Keesman, K. J., van Keulen, H., Liagre, F., Mayus, M., Moreno, G., Reisner, Y., & Herzog, R. (2007). Modeling environmental benefits of silvoarable agroforestry in Europe. *Agriculture, Ecosystems and Environnement, 119*(2007), 320–334.
Praharaj, S., & Maitra, S. (2020). Importance of legumes in agricultural production system: An overview. *Agro-Economist, 7*(2), 69–71.
Ramachandran, P. K., Nair, B. M., & Kumar Nair, V. D. (2009). Agroforestry as a strategy for carbon sequestration. *Journal of Plant Nutrition and Soil Science, 172*, 10–23.
Roe, S., Streck, C., Beach, R., et al. (2021). Land-based measures to mitigate climate change: Potential and feasibility by country. *Global Change Biology, 27*, 6025–6058. https://doi.org/10.1111/gcb.15873
Sabir, M., & Roose, E. (2010). Zones forestière l'arbre et la GCES (chapitre 10, pp. 277–290). In *Gestion durable de l'eau et des sols au Maroc: Valorisation des techniques traditionnelles méditerranéennes* (p. 343). edt IRD.
Sims, B., Corsi, S., Gbehounou, G., Kienzle, J., Taguchi, M., & Friedrich, T. (2018). Sustainable weed management for conservation agriculture: Options for smallholder farmers. *Agriculture, 8*, 118. https://doi.org/10.3390/agriculture8080118
Smith, P. (2018). Managing the global land resource. *Proceedings of the Royal Society B, 285*, 20172798. https://doi.org/10.1098/rspb.2017.2798
Steensland, A. (2021). Global agricultural productivity report: Climate for sustainable agricultural growth. Edited by T. Thompson. Virginia Tech College of Agriculture and Life Sciences.
Tanji, A., El Gharras, O., Mayfield, A., & El Mourid, M. (2017a). On-farm evaluation of integrated weed management in no-till rainfed crops in semi-arid Morocco. *African Journal of Agricultural Research, 12*(16), 1404–1410.
Tanji, A., El Gharras, O., Ouabbou, H., & Mladen, T. (2017b). Weed dynamics in no-till rainfed crops in Chaouia, Semi-Arid Morocco. *Journal of Agricultural Science, 9*(11), 198–208.
Temani, F., Bouaziz, A., Daoui, K., Werye, J., & Barkaoui, K. (2020). Olive agroforestry can improve land productivity even under low water availability in the South Mediterranean. *Agriculture, Ecosystems & Environment, 307*, 107234, 28 February 2021.

van Noordwijk, M., & Brussaard, L. (2014). Minimizing the ecological footprint of food: Closing yield and efficiency gaps simultaneously? *Current Opinion in Environmental Sustainability, 8*, 62–70. https://doi.org/10.1016/j.cosust.2014.08.008.
The World Bank. (2014a). *Réalisation de l'étude sur le semis direct dans le cadre du projet d'intégration du changement climatique dans la mise en oeuvre du plan Maroc Vert (PICCPMV). Phase 1: Diagnostic analytique de la situation actuelle du système de semis direct (SSD).* Document n°: 11706-N711–13a. Washington, DC, USA: The World Bank.
The World Bank. (2014b). *Réalisation de l'étude sur le semis direct dans le cadre du projet d'intégration du changement climatique dans la mise en oeuvre du plan Maroc Vert (PICCPMV). Phase 2: Feuille de route pour le développement et promotion du semis direct au niveau des exploitations agricoles.* Document n°: 943-N711–14b. Washington, DC, USA: The World Bank.
WCCA. (2021). Declaration. In *The 8th world congress on conservation agriculture*, 21–23 June, Bern, Switzerland. https://8wcca.org/wp-content/uploads/2021/07/8WCCA-declaration.pdf
Willer, H., Schlatter, B., Trávníček, J., Kemper, L., & Lernoud, J. (Eds.). (2020). *The world of organic agriculture statistics and emerging trends 2020* (21st edn.). Research Institute of Organic Agriculture (FiBL) and IFOAM – Organics International, Frick and Bonn. https://shop.fibl.org/chde/5011-organic-world-2020.html
Workman, S. W., & Allen, S. C. (2004). *The practice and potential of agroforestry in the Southeastern United States.* This document is Circular 1446, one of a series through the Center for Subtropical Agroforestry (CSTAF), School of Forest Resources and Conservation.
Yigezu, Y. A., El-Shater, T., Boughlala, M., & Devkota, M., Mrabet, R., & Moussadek, R. (2021). Can an incremental approach be a better option in the dissemination of conservation agriculture? Some socioeconomic justifications from the drylands of Morocco. *Soil and Tillage Research, 212*, 105067. https://doi.org/10.1016/j.still.2021.105067
Zeigler, M., & Steensland, A. (2016). *2016 Global Agricultural Productivity Report. Sustainability in an Uncertain Season.* Washington, D.C.: Global Harvest Initiative. https://globalagriculturalproductivity.org/wp-content/uploads/2019/01/2016_GAP_Report.pdf

Integrating Climate-Smart Agriculture and the Water, Energy, and Food Nexus Approach to Address Sustainability Challenges Under a Changing Climate

Abdul Waheed Bhutto, Sadia Karim, and Aqeel Ahmed Bazmi

Abstract The availability of water, energy, and food (WEF) resources is essential for sustainable well-being. The WEF Nexus directly correlate life-changing Sustainable Development Goals-SGD-2 (zero hunger), SGD-6 (clean water and sanitation), SGD-7 (affordable and clean energy), SGD-9 (industry, innovation, and infrastructure), and SGD-12 (responsible consumption and production). Whereas climate-smart agriculture (CSA) is an integrated approach to address the interlinked challenges of food security and climate change. CSA correlates to SGD-13 (take urgent action to combat climate change and its impacts). Climate change threatens people with food insecurity, water scarcity, flooding, infectious diseases, extreme heat, economic losses, and displacement. Traditionally, global challenges, though interconnected, have been addressed singly, at times reducing one problem while exacerbating the others. In an idealized form, the nexus approach seeks to look at all three elements, i.e., food, energy, and water as an interrelated system. Hence, the nexus approach simultaneously examines interactions and trade-offs and assesses accomplishments. The nexus approach focuses on strengthening synergies and reducing trade-offs among these sectors. SDGs are obliging us to work in an integrated way and not in silos. This chapter aims to hold a discussion on adopting CSA in the context of the progress on multiple sustainable development goals covering the WEF nexus while at the same time mitigating the adverse effects of climate change on agriculture. The chapter also focuses on identifying the knowledge gaps and discusses the interlinked roles of institutions, infrastructure, and information on developing policies by countries and

A. W. Bhutto (✉) · S. Karim
Dawood University of Engineering and Technology, Karachi, Pakistan
e-mail: abdul.waheed@duet.edu.pk

S. Karim
e-mail: skarimdar@hotmail.com

A. A. Bazmi
Process and Energy Systems Engineering Center-PRESTIGE, Department of Chemical Engineering, COMSATS University Islamabad (CUI), Lahore Campus, Defense Road, Off Raiwind Road, Lahore, Pakistan
e-mail: abazmi@cuilahore.edu.pk

M. Behnassi et al. (eds.), *Climate-Smart and Resilient Food Systems and Security*,
https://doi.org/10.1007/978-3-031-65968-3_6

international agencies. It is important to identify an array of tools and resources essential for stakeholders in decision-making. This work has implications for research, innovation, and policy design.

Keywords WEF Nexus · Climate change · Food security · Role of CSA

1 Introduction

An irreversible effect of human activity on the environment is species extinction and reduction in ground cover, which upsets the balance of nature and makes ecosystems more fragile and less resistant to disruptions. Scientists are observing changes in the Earth's climate in every region and across the whole climate system, according to the latest Intergovernmental Panel on Climate Change (IPCC) Report (IPCC, 2021). The scenarios show that if no new policies are adopted, food production and energy generation could further increase by around 60%, and water consumption by around 20% over the period 2015–2050, leading to further degradation of resources and increasing environmental pressure (Van Vuuren et al., 2019).

Facts and Figures (UN-Water, 2021)

- 72% of all water withdrawals are used by agriculture, 16% by municipalities for households and services, and 12% by industries.
- It typically takes between 3,000 and 5,000 L of water to produce 1 kg of rice, 2,000 L for 1 kg of soya, 900 L for 1 kg of wheat, and 500 L for 1 kg of potatoes.
- While almost 800 million people are currently hungry, by 2050 global food production would need to increase by 50% to feed more than 9 billion people projected to live on our planet.
- Water-harvesting and water conservation techniques could boost rainfed kilocalorie production by up to 24% and, if combined with irrigation expansion, by more than 40%.
- In sub-Saharan Africa, irrigated areas are expected to more than double by 2050, benefiting millions of small-scale farmers. However, it has been estimated that 41% of current global irrigation water use occurs at the expense of environmental flow requirements.
- The food production and supply chain accounts for about 30% of total global energy consumption. (FAO 2011).
- 90% of global power generation is water intensive.
- Power plant cooling is responsible for 43% of total freshwater withdrawals in Europe (more than 50% in several countries), nearly 50% in the USA, and more than 10% of the national water cap in China.
- Global water demand is projected to increase by 20 to 30% per year by 2050.
- Global water demand (in water withdrawals) is projected to increase by 55% by 2050, mainly because of growing demands from manufacturing (400% increase).

- By 2035, water withdrawals for energy production could increase by 20% and consumption by 85%, driven via a shift towards higher efficiency power plants with more advanced cooling systems (that reduce water withdrawals but increase consumption) and increased production of biofuel.

Extreme weather—including massive storms such as Typhoon Haiyan and Hurricane Sandy, and severe floods and droughts—is likely to become more frequent and intense as global warming accelerates. Under such dynamic conditions, achieving the SDGs requires all relevant stakeholders to work together and manage the synergies and trade-offs among different management or governance sectors (e.g., food, health, water, and energy). Historically, global challenges, though interconnected, have been addressed singly, at times reducing one problem while exacerbating the others. The nexus approaches simultaneously examine interactions among multiple sectors (Liu et al., 2018). They allow the analysis of interlinkages between sectors, to reap positive synergies and to effectively manage trade-offs. The water-food-energy nexus is central to sustainable development. All three are essential for human well-being, poverty reduction, and sustainable development. The demand for all three will increase significantly over the next decades under the pressure of population growth and mobility, economic development, international trade, urbanization, diversifying diets, cultural and technological changes, and climate change (Hoff, 2011). Agriculture is the largest consumer of the world's freshwater resources, and more than one-quarter of the energy used globally is expended on food production and supply. The inextricable linkages between these critical areas require a suitably integrated approach to ensure water and food security, sustainable agriculture, and energy production worldwide.

Definitions

- **Adaptive capacity**: It is the ability of a human or natural system to adapt, i.e. to adjust to climate change, including to climate variability and extremes; prevent or moderate potential damages; take advantage of opportunities; or cope with the consequences.
- **Resilience**: can be described as the capacity of systems, communities, households or individuals to prevent, mitigate or cope with risk and recover from shocks.
- **Enhanced resilience**: Reduce vulnerability to drought, pests, diseases and other climate-related risks and shocks; and improve capacity to adapt and grow in the face of longer-term stresses like shortened seasons and erratic weather patterns.
- **Financial capital**: This capital refers to the supply of funds and financial services.
- **Food security**: It exists when all people, at all times, have physical and economic access to sufficient, safe and nutritious food that meets their dietary needs and food preferences for an active and healthy life (FAO, 1996).
- **Human capital**: This reflects knowledge and skills, job potential, adaptive capacity (WB/CIAT, 2017).

- **Natural capital**: It signifies resources such as land and production resources, water resources, animals, habitats, natural products and plants, and environmental facilities in terms of how these resources and services are used for people's livelihood.
- **Physical capital**: Different types of facilities and equipment used in agricultural production that help farmers measure the expected profits and/or the extent of climate change incident.
- **Social capital**: it refers to networks and unions, social tools gained by individuals to achieve their livelihood goals.
- **Time of emergence**: It refers to the time when conditions are projected to distinctively differ from the past due to climate change. In other words, it is when the effects of human-induced climate change become apparent.
- **Vulnerability**: can be defined as the propensity or predisposition to be adversely affected.

According to Wheeler and von Braun (2013), it is likely that climate variability and change will exacerbate food insecurity in areas currently vulnerable to hunger and undernutrition, indicating the need for considerable investment in adaptation and mitigation to build climate-smart agricultural systems. Climate-smart agriculture (CSA) is an approach for transforming and reorienting agricultural systems to support food security under the new realities of climate change. Climate related changes threaten agricultural production and increase the vulnerability of people dependent on agriculture for their livelihoods. These threats can be reduced by increasing the adaptive capacity of farmers as well as increasing resilience and resource-use efficiency in agricultural production systems (Lipper et al., 2014). CSA emphasizing on capacity to implement flexible, context-specific solutions, supported by innovative policy and financing actions (Lipper et al., 2014). CSA is a relatively new concept which was launched by FAO in 2010 (FAO, 2010). The background paper also identified that considerable investment is required in filling data and knowledge gaps and in research and development of technologies, methodologies, as well as the conservation and production of suitable varieties and breeds. Institutional and financial support will be required to enable smallholders to make the transition to CSA (FAO, 2010).

2 Climate Change

Human activities are continuing to affect the Earth's energy budget by changing the emissions and resulting atmospheric concentrations of radiatively important gases and aerosols and by changing land surface properties. The climate is changing across our planet, largely because of human activities. The most compelling evidence of climate change derives from observations of the atmosphere, land, oceans, and cryosphere. Unequivocal evidence from in situ observations and ice core records shows that the atmospheric concentrations of important greenhouse gases such as

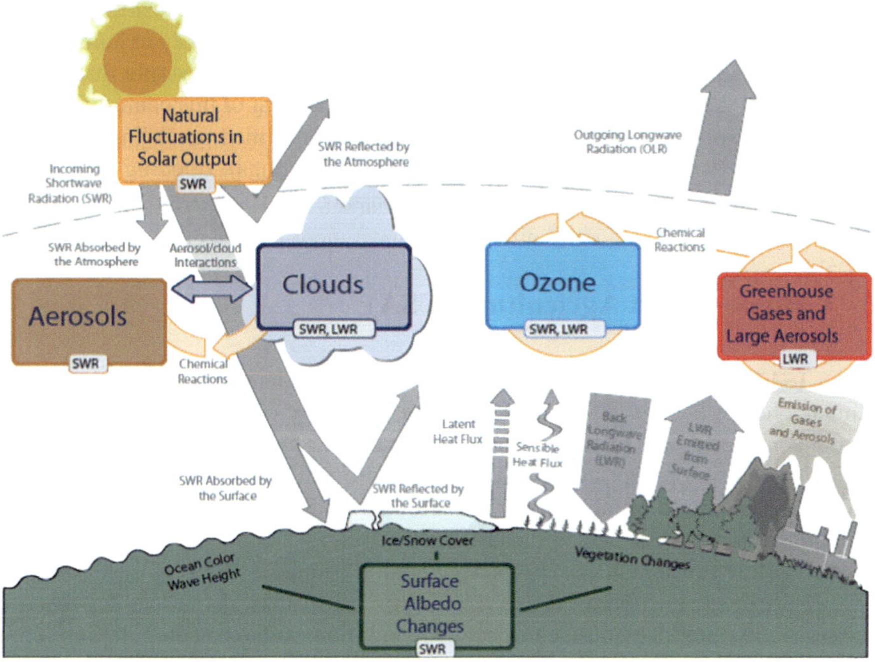

Fig. 1 Main drivers of climate change. *Source* Cubasch et al. (2014)

carbon dioxide (CO_2), methane (CH_4), and nitrous oxide (N_2O) have increased over the last few centuries. Climate change, whether driven by natural or human forcing, can lead to changes in the likelihood of the occurrence or strength of extreme weather and climate events or both. The Earth's climate system is powered by solar radiation (Fig. 1). There are many indicators of climate change. These include physical responses such as changes in the following: surface temperature, atmospheric water vapor, precipitation, severe events, glaciers, ocean and land ice, and sea level:

- Climate change is intensifying the water cycle. This brings more intense rainfall and associated flooding, as well as more intense drought in many regions.
- Climate change is affecting rainfall patterns. In high latitudes, precipitation is likely to increase, while it is projected to decrease over large parts of the subtropics. Changes to monsoon precipitation are expected, which will vary by region.
- Coastal areas will see continued sea level rise throughout the twenty-first century, contributing to more frequent and severe coastal flooding in low-lying areas and coastal erosion. Extreme sea level events that previously occurred once in 100 years could happen every year by the end of this century.
- Further warming will amplify permafrost thawing, and the loss of seasonal snow cover, melting of glaciers and ice sheets, and loss of summer Arctic Sea ice.

- Changes to the ocean, including warming, more frequent marine heatwaves, ocean acidification, and reduced oxygen levels have been clearly linked to human influence. These changes affect both ocean ecosystems and the people that rely on them, and they will continue throughout at least the rest of this century.
- For cities, some aspects of climate change may be amplified, including heat (since urban areas are usually warmer than their surroundings), flooding from heavy precipitation events, and sea level rise in coastal cities.

3 Climate-Smart Agriculture (CSA)

Water use by the agricultural sector along with inefficient irrigation methods and climate change has led to the depletion and insecurity of water resources and the consequent instability of the agricultural system (Zarei et al., 2021). Sustainable agricultural development avoids inefficient traditional and non-economic methods and relies on the use of modern agricultural knowledge and methods and improves economic returns while achieving sustainability.

Agriculture is the major user of scarce natural resources (85% of the developing world's freshwater withdrawal and 42% of its land). It is also a leading cause of underground water depletion, agrochemical pollution, soil exhaustion, loss of biodiversity through deforestation, and an important contributor to global climate change, accounting for up to 30% of greenhouse gas emissions.

Climate change's negative impacts are already being felt, in the form of increasing temperatures, weather variability, shifting agro-ecosystem boundaries, invasive crops and pests, and more frequent extreme weather events. On farms, climate change is reducing crop yields, the nutritional quality of major cereals, and lowering livestock productivity. At the same time, agriculture is a major part of the climate problem. It currently generates 19–29% of total greenhouse gas (GHG) emissions. Without action, that percentage could rise substantially as other sectors reduce their emissions. Additionally, 1/3 of food produced globally is either lost or wasted. Climate change is estimated to have already reduced global yields of maize and wheat by 3.8% and 5.5%, respectively, and several researchers warn of steep decreases in crop productivity when temperatures exceed critical physiological thresholds. Addressing food loss and waste is critical to helping meet climate goals and reduce stress on the environment.

Building resilience to every type of risk is essential to be prepared for uncertainty and change. CSA is an approach to developing the technical, policy, and investment conditions to achieve sustainable agricultural development for food security while adapting to climatic changes. CSA is an approach for transforming and reorienting agricultural systems to support food security under the new realities of climate change (Lipper et al., 2014). CSA systematically considers the synergies and trade-offs that exist between productivity, adaptation, and mitigation. It aims to simultaneously achieve three outcomes:

- Increased productivity: Produce more and better food to improve nutrition security and boost incomes, especially for 75% of the world's poor who live in rural areas and mainly rely on agriculture for their livelihoods.
- Enhanced resilience: Reduce vulnerability to drought, pests, diseases, and other climate-related risks and shocks; and improve capacity to adapt and grow in the face of longer-term stresses like shortened seasons and erratic weather patterns.
- Reduced emissions: Pursue lower emissions for each calorie or kilo of food produced, avoid deforestation from agriculture, and identify ways to absorb carbon out of the atmosphere.

CSA necessitates simultaneous pursuance of efficiency and resilience at various scales. Given that shocks will occur more often due to climate change, being efficient without being resilient will not be helpful in addressing the challenges of climate change over the long term. Measures aiming at building economic resilience at the farm level include but are not limited to increasing income, enabling saving, and promoting diversification. A straightforward model of resilient agricultural systems is increasing diversity of production at the farm and landscape level supported by trade partnerships to overcome the variability of extreme events. Creating co-benefits fills the gap in funding for investment and develops a broader set of trade-offs in remote communities. Improving the performance of existing indigenous technologies and adopting the best practices will also play an important role in developing the resilience of agricultural systems. However, policy interventions are required to address the investment needs of land management schemes and infrastructure, such as local roads and irrigation systems, which are an important source of job creation in rural areas. These public works can be supported by social protection schemes to provide work, food, and income to food-insecure people.

The CSA approach strengthens the livelihoods and food security of the local community by improving the management and use of natural resources at the regional level and adopting indigenous methods and technologies for the production, processing, and marketing of agricultural goods (FAO, 2013). The commonly practiced safety nets for the poorest and most vulnerable communities usually occur in the form of humanitarian relief and food aid, cash payments, agricultural inputs, and public works often after a critical event has occurred. Instead, communities can collectively plan safety-net strategies and resource transfers that are predictable and flexible enough to be scaled up and then scaled down when the crisis subsides (Steenwerth et al., 2014).

CSA promotes coordinated actions from public, private, and civil society stakeholders at the international to local levels in: (1) building evidence and assessment tools; (2) developing coordinated and evidence-based policies; (3) fostering coherence between climate and agricultural policies; (4) strengthening national and local institutions; (5) increasing financing and its effectiveness; and (6) linking climate and agricultural financing (Lipper et al., 2014).

CSA stresses the importance of building evidence to identify viable options and necessary enabling activities. It provides tools for assessing different technologies and practices in relation to their effects on national development and food security objectives under the site-specific effects of climate change.

4 Water, Energy, and Food (WEF) Nexus

We are already aware that direct inputs of water are needed in the production of food and energy, while energy is required for the storage and distribution of food as well as in water extraction, conveyance, and treatment. Any limitation in inputs from one of these components disturbs the quantity, quality, and access of the others to a good nexus balance.

The nexus concept builds on many of these approaches by emphasizing the importance of understanding connections, synergies, and trade-offs. The word 'nexus' refers to approaches that address the linkages between multiple distinct entities. The nexus terminology was first used in the natural resource realm in 1983 under the Food-Energy Nexus Program, which sought integrated solutions to food and energy scarcity (Sachs & Silk, 1990).

The WEF nexus describes interactions among domains that yield gains or trade-offs when analyzed together rather than independently. Hoff (2011) presented initial evidence for how a nexus approach can enhance the WEF security by increasing efficiency, reducing trade-offs, building synergies, and improving governance across sectors. The WEF nexus calls for the integrated study of such systems to improve the resiliency of these natural resources and adapt to a changing world (Kondash et al., 2021). The nexus-focused research can provide new insight into potential trade-offs and synergies of infrastructure investments, policies, or resource management options.

5 Key Principles of the WEFE Nexus

The key principles of the WEFE Nexus can be summarized as follows (Position paper on water, energy, food, and ecosystems (WEFE) nexus and SDGs, JRC Technical Report, JRC114177, 2019):

- Understand the interdependence of resources within a system across space and time and focus on the whole system's efficiency rather than the productivity of individual components. This will provide integrated solutions that contribute to the sustainability of water, energy, and food security policy objectives and to maintaining healthy ecosystems.
- Recognize the interdependence between water, energy, food, and ecosystems and promote rational and inclusive dialog and decision-making processes and efficient use of these resources in an environmentally responsible way.

- Identify integrated policy solutions to optimize trade-offs and maximize synergies across sectors and encourage mutually beneficial responses that enhance the potential for cooperation between all components, and public–private partnership at multiple scales.
- Ensure coordination across sectors and stakeholders to enable synergies and increase solution sustainability.
- Value the natural capital of land, water, energy sources, and ecosystems and encourage governments and business to support the transition to sustainability, for example, using nature-based solutions.

6 Integrating CSA and the WEF Nexus Approach

Increasing efficiency in the use of natural resources and other agricultural inputs is a central element for sustainable food and water cycles. Ecosystem approaches that combine both food and energy production, such as agroforestry or integrated crop-livestock-biogas systems, could substantially mitigate these risks while providing both food and energy to rural and urban populations. The adaptive capacity overwhelmingly depends upon the resources available for adaptation like information, technology, economic resources, institutions, and so on, as well as the ability or capacity of that system to use the resources effectively in pursuit of adaptation (Bogdanski, 2012).

Unless food and energy production are well balanced within the agro-ecosystem, energy remains just another external input for smallholder farming systems. The heavy reliance on wood fuels in developing countries has severe implications for forests and climate change. Climate variability will enhance the pressure on agricultural production systems and forests, and thus making availability of wood fuels important sources of energy for many of the world's poor, more vulnerable. Likewise, the majority of fertilizers are still fossil fuel based. Production of fertilizer in a closed loop substituted portion of demand of fertilizer through the use of residual biomass, e.g., post-harvest residues, residues from livestock production and slaughter, or food processing. The substitution of mineral fertilizers with bio-based alternatives is an important direction in materials and energy recovery (Christel et al., 2014). Hence, bioenergy and food provision cannot be addressed in isolation from each other and the environment on which they depend. They need to be equally addressed to strengthen people's adaptive capacity to climate change.

Changing climatic parameters affect energy generation sources such as hydro, wind, solar, and thermal systems, whereas the sustainability of the energy sector has a direct impact on food security, under a changing global climatic scenario. The possible energy management strategies toward CSA are: replacing a certain portion of synthetic fertilizers with organic manures and biofertilizers; promoting renewable energy sources like solar, wind, biofuels; adopting fuel-efficient engines; fostering efficient heat management of greenhouses; help building energy-efficient food storage strategies; diverting waste-land from crop to biofuel production;

designing hydropower generating dams; large-scale afforestation; and forest conservation program (Bhattacharyya et al., 2020). The integrated production of food and bioenergy may lead to many synergies by adopting different technologies allowing maximum utilization of crops, livestock, and their byproducts. Technologies for small-scale renewable applications are mature and may often provide synergies with agricultural production (Bogdanski, 2012).

7 A Novel Approach

Food systems require urgent transformations. Transformations require knowledge on the multiple dimensions of food systems (e.g., production, trade, consumption, culture, human and animal health, livelihoods and employment, food waste, and environmental sustainability), as well as a mechanism to translate these insights and analyzes into governance processes (Turnhout et al., 2021). Barron-Gafford et al. (2019), while working on hybrid approach to address the challenge, identified novel ideas of explorations toward the resilience of food and energy systems under the future projected increased environmental stress involving heat and drought.

Data is still a major problem globally, especially on the WFE nexus. There is a need to have a regional and country system for monitoring energy, water, and food security as well as CSA while putting in place proper policy environment and institutional arrangements to manage and monitor the exploitation of natural resources (Okumu et al., 2021). Computational modeling is a powerful tool for quantifying the WEF nexus interactions. Most nexus methods are confined to disciplinary silos (Lazaro et al., 2021). Li et al. (2019) come up with a new approach to sustainably manage agricultural water-energy-food nexus from a holistic perspective by proposing an optimization model under uncertainty; the interaction among water, food, and energy nexus had not adequately considered, for example, the production of agricultural new energy based on food production and processing. Bieber et al. (2018) developed a modeling framework to simultaneously optimize investment in power generation and water infrastructure. Endo et al. (2020) reviewed the methodology and the process through which the nexus approach has academically and geographically expanded across, and synthesized more than one discourse, discipline, and domain via its methodology as interdisciplinary and integrated approaches. The study identified that, at present, a well-established nexus methodology that combines interdisciplinary and transdisciplinary approaches, does not exist.

8 Vulnerability of Pakistan's Current Agriculture to Climate Change, Its Repercussions and Way Forward

In Pakistan, Agriculture contributes 19.2% to the GDP and provides employment to around 38.5% of the labor force while 65–70% of the population depends on agriculture for its livelihood (GOP, 2021). Sixty percent of the population is directly or indirectly reliant upon rainfed agriculture that depends on predictable weather patterns (Syed et al., 2022). According to the Global Climate Risk Index 2021, Pakistan was at number 8 in the list of countries most affected from 2000 to 2019 (Eckstein et al., 2021). Pakistan continuously ranks among the most affected countries both in the long-term index and in the index for each respective year.

The major events include: back-to-back floods since 2010; the worst drought episode (1998–2002) as well as more recent droughts in Tharparkar and Cholistan; and the intense heat wave in Karachi (in Southern Pakistan generally) in July 2015; severe windstorms in Islamabad in June 2016; increased cyclonic activity; and increased incidences of landslides and Glacial Lake Outburst Floods (GLOFs) in the northern parts of the country (GOP, 2021). The country often experiences periods of severe droughts, followed by devastating floods (WorldBank, 2017). In the aftermath of the 2010 floods, one-fifth of the country's land area was submerged, damaging the economy, infrastructure, and livelihoods, and leaving 90 million people food insecure.

Pakistan is the world's fifth most populous country, and its population is growing at a rate of approximately 1.8% per year (GOP, 2021).

Agricultural growth rate has been constrained by shrinking arable land, climate change, water shortages, and large-scale population and labor shift from rural to urban areas. Increasing agricultural productivity, therefore, requires the adoption of new approaches. Since most of the arable land is already in use, productivity gains to meet this growing and predominantly urban population will likely be achieved through sustainably increasing cropping intensity. CSA technologies and practices that improve efficiency in livestock systems while simultaneously reducing emissions may include improved animal feed and feeding techniques to reduce methane and nitrous oxide, improved breeding, adapted manure storage and management practices, and improved pastures and management of grazing lands to enhance productivity and create carbon sinks.

The country produces wheat, rice, cotton, sugarcane, and maize and these crops are affected by climate change. Resulting from rising temperatures, northern Pakistan experienced a noted shift in monsoon pattern and increased occurrence of cyclones over the past several years with impacts on the agricultural sector. The country boasts a comprehensive network of irrigation infrastructure including canals and dams, yet water resources for agriculture are expected to decline in coming years. Critical investments in improved seeds, farming technology and techniques, and water infrastructure are needed to tackle the emerging challenges for the agricultural sector, especially in the context of declining water availability and climate change impacts.

The WorldBank (2017) has identified areas to address chronic challenges to Pakistan's agricultural sector, namely drought, flood, and intense heat. Water management strategies, improved crop and livestock varieties, integrated pest management and manure management, and renewable energy technologies for the agricultural sector are among the most widely adopted strategies in Pakistan. The country's new Pakistan Climate Change Act (PCCA), 2017 sets the stage for the establishment of the Pakistan Climate Change Authority and Pakistan Climate Change Fund, which are expected to help mobilize domestic and international funds for mitigation and adaptation interventions in the country, including CSA. The main objectives of Pakistan's National Climate Change Policy (NCCP) include developing climate-resilient agriculture and food systems for all agro-ecological zones in the country. Pakistan has also launched a National Food Security Policy 2018 and one of its goals is to make agriculture more productive, profitable, and climate resilient (GoP, 2018). Figure 2 shows the elements for agriculture and food security in National Food Security Policy of Pakistan (GoP, 2018).

Syed et al. (2022) identified areas like new and advanced feeding technologies, changes in animal feed composition, better and more efficient livestock breeds, improved manure management practices, and the management of pasture and grazing land to enhance agricultural productivity. According to the WorldBank (2017), leveraging existing financing, unlocking new sources of financing, promoting pro-poor policies, and empowering institutions to take action, transformational change toward more sustainable and productive agricultural systems can be achieved in the country. Figure 3 presents a selection of CSA practices with high climate smartness scores

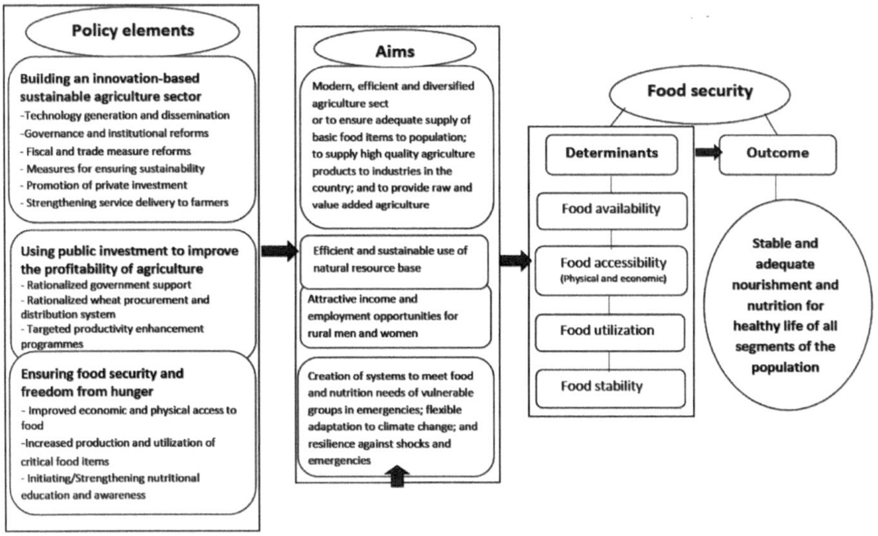

Fig. 2 Elements for agriculture and food security in National Food Security Policy of Pakistan. *Source* GoP (2018)

according to expert evaluations. The major initiatives taken by the Government of Pakistan include (GOP, 2021):

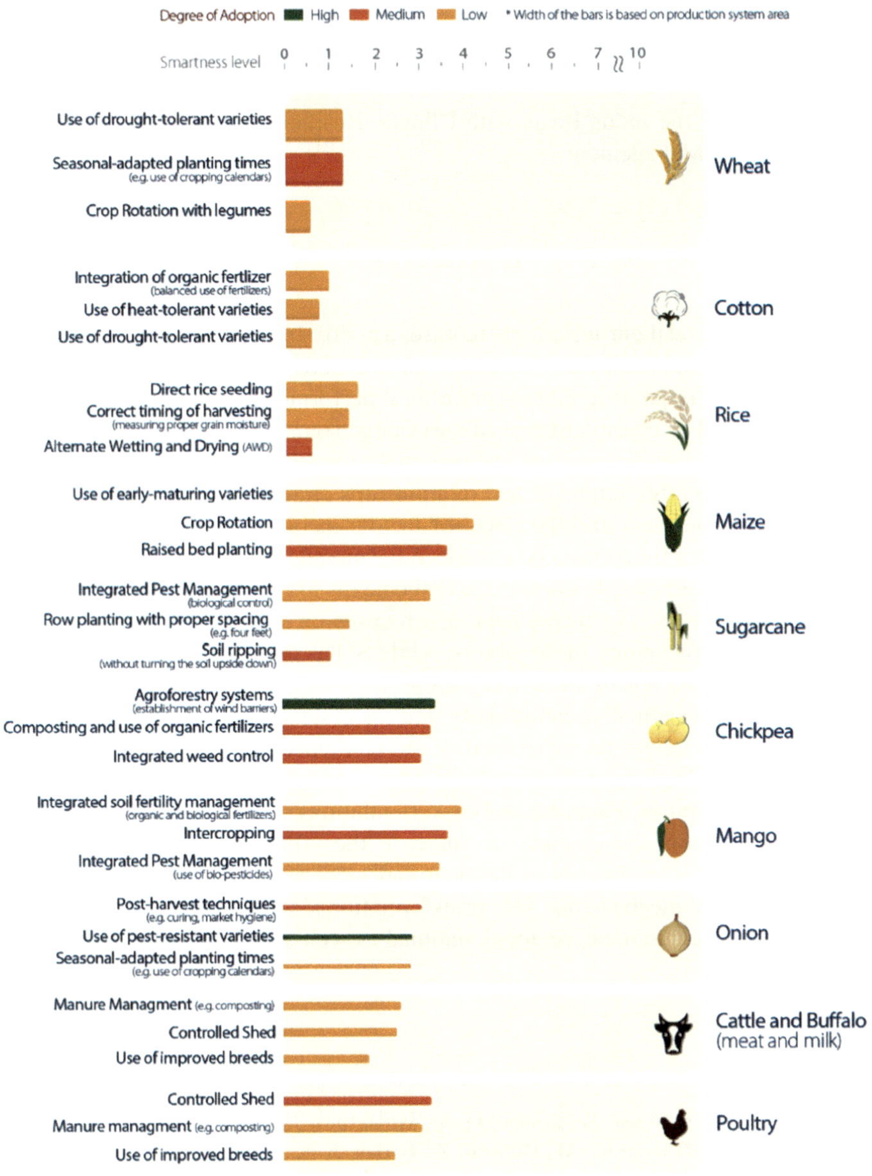

Fig. 3 A selection of CSA practices with high climate smartness scores according to expert evaluations. *Source* WorldBank (2017)

- Ten Billion Tree Tsunami Program (TBTTP);
- Protected Areas Initiative;
- Wildlife Management;
- Participation in Reducing Emissions from Deforestation and Forest Degradation (REDD+);
- Pakistan Climate Resilient Urban Policy Framework;
- Clean Green Pakistan Index and Champions Program;
- Scaling-up of Glacial Lake Outburst Flood (GLOF II) Risk Reduction in Northern Pakistan; and
- Transforming the Indus Basin with Climate Resilient Agriculture and Climate-Smart Water Management.

9 Conclusions and Recommendations

A productive and resilient agriculture requires a major shift in the way land, water, soil nutrients, and genetic resources are managed to ensure that these resources are used more efficiently. Improving global agricultural performance is central to addressing poverty and food insecurity. Improved knowledge significantly influences adoption. Farmers' knowledge of their land and soil quality are also important factors.

The precise configurations of and relationships among WEF nexus components vary by place and time, and the range of factors involved further complicates the ability to develop a functional, systematic WEF model. However, there is a need for co-production of actionable knowledge in the nexus to facilitate the development of sustainable solutions. The nexus approach focuses on strengthening synergies and reducing trade-offs among these sectors, while SDGs are obliging us to work in an integrated way and not in silos. Integrating knowledge of CSA and WEF nexus, when applied conceptually, can identify and understand cross-domain interactions that contribute to long-term sustainability and resilience.

To meet the objectives of CSA, a smarter system is to be developed, with minimum resource consumption. Planning, policies, regulations, and coordination programs across these sectors are essential to increase the efficiency. The transformation requires considerable changes in national and local governance, legislation, policies, and financial mechanisms. This transformation also necessitates improving the access of producers to local, regional, national as well as international markets.

References

Barron-Gafford, G. A., Pavao-Zuckerman, M. A., Minor, R. L., Sutter, L. F., Barnett-Moreno, I., Blackett, D. T., . Thompson, M., Dimond, K., Gerlak, A. K., Nabhan, G. P., Macknick, J. E. (2019). Agrivoltaics provide mutual benefits across the food–energy–water nexus in drylands. *Nature Sustainability, 2*(9), 848–855. https://doi.org/10.1038/s41893-019-0364-5

Bhattacharyya, P., Pathak, H., & Pal, S. (2020). *Energy management for climate-smart agriculture. Climate Smart Agriculture* (pp. 73–84). Springer.

Bieber, N., Ker, J. H., Wang, X., Triantafyllidis, C., van Dam, K. H., Koppelaar, R. H. E. M., & Shah, N. (2018). Sustainable planning of the energy-water-food nexus using decision making tools. *Energy Policy, 113*, 584–607. https://doi.org/10.1016/j.enpol.2017.11.037

Bogdanski, A. (2012). Integrated food–energy systems for climate-smart agriculture. *Agriculture & Food Security, 1*(1), 9. https://doi.org/10.1186/2048-7010-1-9

Christel, W., Bruun, S., Magid, J., & Jensen, L. S. (2014). Phosphorus availability from the solid fraction of pig slurry is altered by composting or thermal treatment. *Bioresource Technology, 169*, 543–551.

Cubasch, U., Wuebbles, D., Chen, D., Facchini, M. C., Frame, D., Mahowald, N., & Winther, J.-G. (2014). *Introduction*. Retrieved from Cambridge University Press.

Eckstein, D., Künzel, V., & Schäfer, L. (2021). Global climate risk index 2021. *Who suffers most from extreme weather events*, 2000–2019.

Endo, A., Yamada, M., Miyashita, Y., Sugimoto, R., Ishii, A., Nishijima, J., Fujii, M., Kato, T., Hamamoto, H., Kimura, M., Kumazawa, T., & Qi, J. (2020). Dynamics of water–energy–food nexus methodology, methods, and tools. *Current Opinion in Environmental Science & Health, 13*, 46–60. https://doi.org/10.1016/j.coesh.2019.10.004

FAO. (1996). Rome declaration on world food security and world food summit plan of action. Retrieved from http://www.fao.org/3/w3613e/w3613e00.htm

FAO. (2010). *"Climate-smart" agriculture: Policies, practices and financing for food security, adaptation and mitigation*. Food and Agriculture Organization (FAO).

FOA. (2013). *Climate-smart agriculture: Sourcebook*. Food and Agriculture Organization of the United Nations (FOA).

GoP. (2018). *National food security policy*. Retrieved from Islamabad.

GoP. (2021). *Pakistan economic survey 2020–21*. Retrieved from Islamabad.

Hoff, H. (2011). Understanding the Nexus. Background paper for the Bonn2011 Nexus conference: The Water, Energy and Food Security Nexus.

IPCC. (2021). *Climate change 2021: The physical science basis. Contribution of working group I to the sixth assessment report of the intergovernmental panel on climate change*. Retrieved from Cambridge University Press.

Kondash, A. J., Herrera, I., Castellanos, E., Baker, J., Leiva, B., Van Houtven, G., Fuentes, G., Alfaro, G., Henry, C., Wade, C., Redmon, J. H., & Hoponick Redmon, J. (2021). Food, energy, and water nexus research in Guatemala—a systematic literature review. *Environmental Science & Policy, 124*, 175–185. https://doi.org/10.1016/j.envsci.2021.06.009

Lazaro, L. L. B., Giatti, L. L., Bermann, C., Giarolla, A., & Ometto, J. (2021). Policy and governance dynamics in the water-energy-food-land nexus of biofuels: Proposing a qualitative analysis model. *Renewable and Sustainable Energy Reviews, 149*, 111384. https://doi.org/10.1016/j.rser.2021.111384

Li, M., Fu, Q., Singh, V. P., Ji, Y., Liu, D., Zhang, C., & Li, T. (2019). An optimal modelling approach for managing agricultural water-energy-food nexus under uncertainty. *Science of the Total Environment, 651*, 1416–1434. https://doi.org/10.1016/j.scitotenv.2018.09.291

Lipper, L., Thornton, P., Campbell, B. M., Baedeker, T., Braimoh, A., Bwalya, M., Caron, P., Cattaneo, A., Garrity, D., Henry, K., Hottle, R., & Torquebiau, E. F. (2014). Climate-smart agriculture for food security. *Nature Climate Change, 4*(12), 1068–1072. https://doi.org/10.1038/nclimate2437

Liu, J., Hull, V., Godfray, H. C. J., Tilman, D., Gleick, P., Hoff, H., Pahl-Wostl, C., Xu, Z., Chung, M. G., Sun, J., & Li, S. (2018). Nexus approaches to global sustainable development. *Nature Sustainability, 1*(9), 466–476. https://doi.org/10.1038/s41893-018-0135-8

Okumu, B., Kehbila, A. G., & Osano, P. (2021). A review of water-forest-energy-food security nexus data and assessment of studies in East Africa. *Current Research in Environmental Sustainability, 3*, 100045. https://doi.org/10.1016/j.crsust.2021.100045

Sachs, I., & Silk, D. (1990). *Food and energy: Strategies for sustainable development*. United Nations University Press.

Steenwerth, K. L., Hodson, A. K., Bloom, A. J., Carter, M. R., Cattaneo, A., Chartres, C. J., . Hatfield, J. L., Henry, K., Hopmans, J. W., Horwath, W. R., Jenkins, B. M., & Jackson, L. E. (2014). Climate-smart agriculture global research agenda: scientific basis for action. *Agriculture & Food Security, 3*(1), 11. https://doi.org/10.1186/2048-7010-3-11

Syed, A., Raza, T., Bhatti, T. T., & Eash, N. S. (2022). Climate Impacts on the agricultural sector of Pakistan: Risks and solutions. *Environmental Challenges, 6*, 100433. https://doi.org/10.1016/j.envc.2021.100433

Turnhout, E., Duncan, J., Candel, J., Maas, T. Y., Roodhof, A. M., DeClerck, F., & Watson, R. T. (2021). Do we need a new science-policy interface for food systems? *Science, 373*(6559), 1093–1095. https://doi.org/10.1126/science.abj5263

UN-Water. (2021). Water, food and energy. Retrieved from https://www.unwater.org/water-facts/water-food-and-energy/

Van Vuuren, D. P., Bijl, D. L., Bogaart, P., Stehfest, E., Biemans, H., Dekker, S. C., Doelman, J. C., Gernaat, D. E., Harmsen, M. (2019). Integrated scenarios to support analysis of the food–energy–water nexus. *Nature Sustainability, 2*(12), 1132–1141. https://doi.org/10.1038/s41893-019-0418-8

WB/CIAT. (2017). *Climate-smart agriculture in Pakistan. CSA country profiles for Asia series.* Retrieved from Washington, D.C.

Wheeler, T., & von Braun, J. (2013). Climate change impacts on global food security. *Science, 341*(6145), 508–513. https://doi.org/10.1126/science.1239402

WorldBank. (2017). *Climate-smart agriculture in Pakistan. csa country profiles for Asia series. International Center for Tropical Agriculture (CIAT)* (28 p.). The World Bank.

Zarei, S., Bozorg-Haddad, O., Singh, V. P., & Loáiciga, H. A. (2021). Developing water, energy, and food sustainability performance indicators for agricultural systems. *Scientific Reports, 11*(1), 22831. https://doi.org/10.1038/s41598-021-02147-9

Food and Water Security Through Climate-Smart Agriculture in Asia and Africa

Saima Baig

Abstract Most countries in Asia and Africa are primarily agricultural economies, dependent on specific climate variables for food production. These agrarian, developing, and poor economies are likely to be affected, perhaps adversely, by a changing climate, which could result in declining national economies. Crucially, the livelihoods of vast numbers of people in Asia and Africa are directly linked to agriculture, which will be influenced by changes in temperature and precipitation. These changes may lead to a decrease in water availability and food security in the concerned regions. Migrations due to droughts, floods, and wildfires, could also increase food insecurity and water scarcity. As a growing population drives up the demand for food, the production of staple food items is struggling to keep up due to deteriorating natural resources that directly impact crop yields, soils, water, and biodiversity. Increasing temperatures, weather variability, shifting agroecosystem boundaries, invasive crops and pests, and more frequent extreme weather events will have a negative impact on crop yields, fisheries, livestock, water availability, and nutritional quality of food items. Agriculture also contributes to climate change and is thus a major part of the climate problem. Agricultural production generates 15 to 25% of total global emissions, which will only rise without action. Climate adaptation and mitigation strategies are therefore required to ensure continued livelihoods, food security, and water availability. Climate-Smart Agriculture (CSA) is an integrative approach that incorporates the management of landscapes, ecosystems, cropland, livestock, forests, and fisheries to raise agricultural productivity, enhance resilience to climate change, increase adaptation, and address the contribution of agriculture to global greenhouse gas (GHG) emissions.

Keywords Africa · Asia · Climate change · ClimateSmart agriculture · Food security · Water scarcity

---

S. Baig (✉)
Chartered Environmentalist, Liverpool, UK
e-mail: saima.baig@aya.yale.edu

M. Behnassi et al. (eds.), *Climate-Smart and Resilient Food Systems and Security*,
https://doi.org/10.1007/978-3-031-65968-3_7

1 Introduction

"*It is unequivocal that human influence has warmed the atmosphere, ocean and land. Widespread and rapid changes in the atmosphere, ocean, cryosphere and biosphere have occurred*", says the IPCC's 6th Assessment Report (IPCC, 2021:5). This was just the latest in a long series of warnings put forth by climate scientists over the last few decades. Unprecedented changes over many hundreds to even thousands of years have occurred "*across the whole climate system, as well as various aspects of climate change*" (IPCC, 2021:9). Climate change and a warming planet due to human activities is "*already affecting many weather and climate extremes in every region across the globe*" (IPCC, 2021:10).

By 2017, anthropogenic climate change had already caused global mean surface temperatures to reach approximately 1 °C above pre-industrial levels (likely range 0.8 °C to 1.2 °C) (IPCC, 2018a, 2018b). At the current rate of increase, warming levels are likely to reach 1.5 °C between 2035 and 2050 (IPCC, 2018a, 2018b). The decade ending in 2015 was 0.87 °C warmer than the average, when compared to 1850–1900. Past and ongoing greenhouse gas (GHG) emissions are resulting in the increase of current global warming at 0.2 °C per decade (IPCC, 2018a, 2018b). Global surface temperature was 1.1 °C higher in the decade between 2011 and 2020 when compared to 1850–1900 (IPCC, 2021). The number of weather-related disasters has increased five-fold, with more than 11,000 such disasters between 1970 and 2019 (WMO, 2020) (Fig. 1).

IPCC's 6th Assessment Report (AR6) has indicated in no uncertain terms that "*Global surface temperature will continue to increase until at least the mid-century under all emissions scenarios considered. Global warming of 1.5* °C *and 2* °C *will be exceeded during the twenty-first century unless deep reductions in carbon dioxide (CO^2) and other greenhouse gas emissions occur in the coming decades.*" (IPCC,

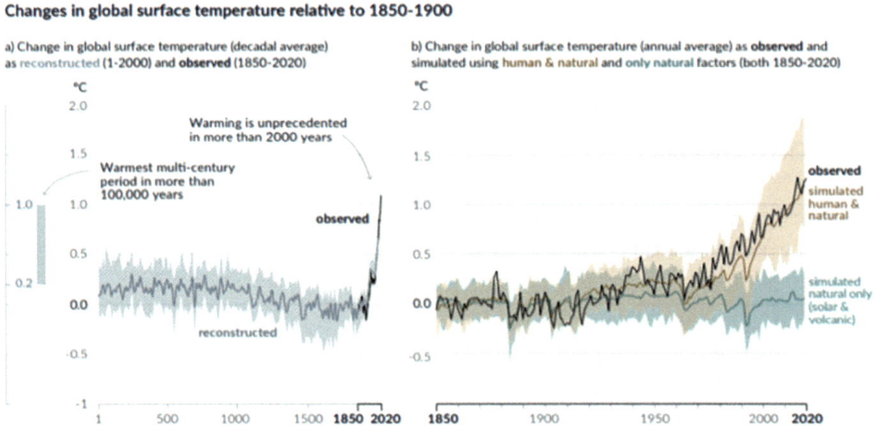

Fig. 1 Human influence has warmed the climate at a rate that is unprecedented in at least the last 2000 years. *Source* IPCC (2021:7)

2021, p: 17). Changes in the climate system will be exacerbated with increased global warming, accompanied by increases in the frequency of hot extremes, intensity of precipitation, cyclones, melting sea ice, permafrost, and snow-cover.

While the impacts of climate change will be felt globally, developing countries, particularly those in Asia and Africa are the most likely to be the hardest hit. These agrarian economies are dependent on climate variables for food production and for the livelihood of local communities. As the climate changes and the world warms, global agricultural production will be affected, in some cases positively but for the agricultural economies of Asia and Africa, these effects will almost certainly be negative. Decreased water availability due to variability in temperature and precipitation will impact agricultural processes leading to losses of livelihoods, food scarcity, and declining economies.

This chapter aims to highlight the impacts of climate change on water and food security in Asia and Africa. It will also consider the implications of Climate-Smart Agriculture (CSA) for food security and water availability in light of climate change. Finally, case studies from Asia and Africa will highlight the potential of CSA in these regions.

2 Climate Change in Asia and Africa

In the coming decades, we are likely to observe increased changes in climate in all regions of the world (IPCC, 2021). Reaching temperatures of 1.5 °C or above will result in varying spatial and temporal impacts (Ebi et al., 2016). For Asia and Africa, an increase in mean temperatures by 1.5 °C or 2 °C will mean that they are likely to see high temperatures, changes in wet and dry seasons, variability in winds and precipitation, sea-level rise, melting glaciers, and ice sheets, and rises in the occurrences and intensity of extreme weather events. Agricultural and ecological droughts[1] will become a norm as the global water cycle is impacted by a warming world (IPCC, 2014a). These effects are already being felt by many Asian and African countries, which are already experiencing increases in hot extremes, heavy precipitation, and agricultural and ecological drought.

Africa

Since 1901, large parts of Africa have already warmed by more than 1 °C, experiencing more heatwaves and hot days (IPCC, 2014a). The year 2019 was one of the three warmest years since records began, and this warming trend is expected to continue (WMO, 2020). According to the IPCC 5th Assessment Report, large areas of Africa will exceed 2 °C of warming above pre-industrial levels by the end of this century (IPCC, 2014a) (Fig. 2).

[1] A period with abnormal soil moisture deficit, which results from combined shortage of precipitation and excess of evapotranspiration, and during the growing season impinges on crop production or ecosystem function in general.

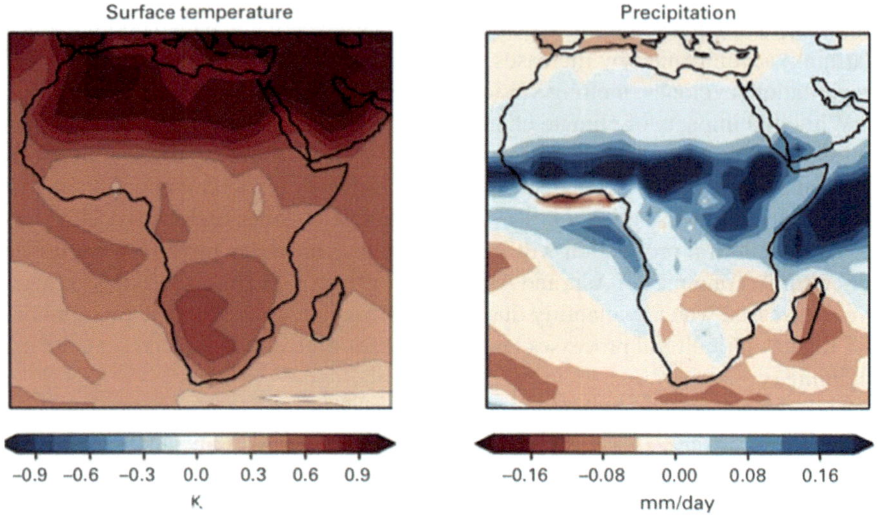

Fig. 2 Multi-model average forecasts of near-surface temperature and precipitation for the five-year period 2020–2024. Colors show anomalies relative to the period 1981–2010 for the average of several international forecasts contributing to the WMO Lead Center for ADCP (https://hadleyserver.metoffice.gov.uk/wmolc/). Forecasts are initialized with observations and start on or after 1 November 2019. *Source* WMO (2020)

In sub-Saharan Africa, a 1.5 °C to 2 °C rise in mean temperature will result in more severe droughts and an increase in arid land, in addition to flooding and shifts in rainfall (World Bank, 2013). With a 2 °C rise in global mean temperatures, there will be significant changes in the frequency and intensity of temperature extremes in all sub-Saharan regions (IPCC, 2014a).

In semi-arid regions of West Africa in the Sahel, an overall shift has been observed toward increased temperatures and decreased annual average rainfall since the 1960s (Kotir, 2011). Projections indicate the strongest drying in the western Sahel and significant increases in the maximum length of dry spells (IPCC, 2014a). Both West and Central Africa will face large increases in the number of hot days under both 1.5 °C and 2 °C scenarios. Central Africa is likely to experience a decrease in the length of wet periods and an increase in heavy rainfall and in 2019, flooding affected the Sahel and surrounding areas (WMO, 2020).

In the Horn of Africa and parts of East Africa, there is likely to be an increase in annual precipitation concentrated in bursts, increasing the risk of flooding (World Bank, 2013). These impacts were observed in late 2019, when there were floods and landslides associated with heavy rainfall, following very dry conditions in 2018 and 2019, in the Greater Horn of Africa (WMO, 2020). Sudan was hit with heavy rains and flash floods in August 2021, following on from 2020, when a wetter than usual rainy season meant that the Nile River and some of its tributaries reached their highest levels in a century (NASA and OCHA, 2021).

Trends indicate continued warming and decreasing rainfall in North and Southern Africa for the period 2020–2024 (WMO, 2020). According to IPCC (2014a), heatwaves and hot days have already spiked, and there are likely to be further decreases in precipitation in North Africa and southwestern regions of South Africa by the end of the century. Southern Africa experienced severe drought in 2019, and its western region is projected to become drier. By the end of the twenty-first century, there is likely to be an increase in the frequency of droughts and the number of heat waves. Projections show that in the 2 °C rise in temperature scenario, precipitation will decrease by approximately 20% in Southern Africa (IPCC, 2014a). Under the same scenario, the number of consecutive dry days will increase in Namibia, Botswana, northern Zimbabwe, and southern Zambia. The Zambezi basin will face reductions in rainfall at five to 10%. Projections for a 1.5 °C scenario also indicate less rainfall over areas in the Zambezi basin, in addition to parts of Western Cape, South Africa. Temperatures will rise faster at 2 °C and the southwestern region of South Africa, as well as parts of Namibia and Botswana, are projected to experience the greatest increases in temperatures (IPCC, 2014a).

Furthermore, sea levels will rise, potentially degrading coastal infrastructure and decreasing the livelihoods of coastal communities. Already, sea-levels increased by 5mm annually in various coastal areas around the continent; more than the global average of 3-4mm per year (WMO, 2020). A critical challenge for the continent is coastal degradation and erosion, especially in West Africa. In Benin, Côte d'Ivoire, Senegal, and Togo, approximately 56% of the coastlines are already eroding (WMO, 2020). This situation is projected to worsen in the future. Although coastal degradation is not considered to be primarily driven by sea-level rise, combined with other environmental and socio-economic factors, an encroaching sea is likely to compound the negative consequences already being felt (UNFCCC, 2020; WMO, 2020).

Asia

Over the past century, escalating temperature extremes and warming trends have been observed across most of Asia, with more warm days and fewer cold days across the continent. During 1900–2009, a particularly strong warming trend was observed during the November to March cold season (IPCC, 2014a).

In East and South Asia, there has been an increasing trend in annual mean temperatures during the twentieth century (IPCC, 2014a). Since the 1960s, Southeast Asia has experienced temperature increases at a rate of 0.14 °C to 0.2 °C per decade, a rising number of hot days, as well as a decline in cooler weather (IPCC, 2014a).

According to the IPCC (2014a), precipitation has also been increasingly variable across the region. In the period 1990–2005, there was a significant rise in precipitation in northern and central Asia but a decrease in southern Asia. After the 1970s, east Asian winter and summer monsoons have experienced weakening, leading to low levels of average summer precipitation in north China, among other areas. Trends indicate future increases in mean and extreme precipitation in east, south, and southeast Asia.

Low lying and crowded coastal cities in many south and southeast Asian countries are particularly under intense stress, and this situation is likely to worsen, with hundreds of millions of people who live there particularly vulnerable. Sea-level rise in many coastal and riverine southeast Asian countries could cause massive flooding and inundation of land (World Bank, 2013). Projections indicate that over one million people living in coastal regions of south and southeast Asia will be at risk from flooding (IPCC, 2014a). Some countries in south Asia, on the other hand, are—and will continue to be—impacted by water scarcity, as well as intense cyclones and flood events, with inconsistencies in the monsoon season and heat extremes (World Bank, 2013).

Glacial melt from the Himalayas will result in the dual issues of increased flooding, on the one hand, and water scarcity due to the reduction of snow melt into river basins of the Indus, Ganges, and Brahmaputra on the other hand (World Bank, 2013), as glaciers shrink and some even disappear. According to projections, a global temperature rise of 1.5 °C in these river basins could result in a temperature increase of 1.4–2.6 °C, while the 2.0 °C scenario could lead to 2.0–3.4 °C of warming (Lutz et al., 2019). This means that the resultant likely increase in monsoon precipitation would lead to more severe and frequent floods in this region, as well as landslides, GLOFs[2], and sedimentation (Lutz et al., 2019).

West Asia experienced rising temperature trends in the last few decades, in addition to a downward trend in precipitation and an increase in extreme weather events (IPCC, 2014a). Rising temperatures, low rainfall, and drought will continue to impact countries across the Middle East and West Asia. This has already come to pass in Iraq and Syria, as more than 12 million people are losing access to agricultural and drinking water as well as electricity, due to a severe water crisis (OCHA, 2021). Drought conditions have arisen because of unseasonably low levels of precipitation across the eastern Mediterranean Basin, since the autumn of 2020 (OCHA, 2021). Record low rainfall has decreased river flow into the Euphrates and Tigris (OCHA, 2021). Syria is facing the worst drought in seven decades and Iraq is also suffering through drought due to an unusually dry winter, its second driest in 40 years (OCHA, 2021).

It is quite clear that this drastic variability in climate will intensify issues around food security and water availability in Asia and Africa, both already impacted by a rising population.

3 Climate Change Impacts on Food Security and Water Availability

Consecutive drought and sandstorms severely impacted Madagascar from April to August 2021, according to the World Food Program (WFP). This news made headlines by the time of writing this chapter, highlighting catastrophic levels of hunger and food insecurity following four years without rain. The famine-like conditions have been attributed to a warming atmosphere, to the extent that people have resorted

[2] Glacial Lake Outburst Floods.

to eating locusts and cacti due to a lack of food (OCHA, 2021). The Grand Sud, an area in the south of Madagascar, has been suffering from one of the worst droughts the region has seen in four decades, culminating in an acute humanitarian crisis, affecting hundreds of thousands of people (UN, 21 August 2021).

On September 8, 2021, the president of Kenya declared the drought affecting various parts of the country a national disaster (Nairobi, 2021). An estimated two million Kenyans face hunger and declining livelihoods as drought reduced access to pastures (FAO Kenya, 2021). Water access has also been impacted with people in 56% of the affected counties having to walk longer distances to obtain water for domestic and livestock use (FAO Kenya, 2021). Poor rainfall during the March to May season caused this acute food insecurity among the populations residing in the north, north-eastern, and coastal areas (FAO Kenya, 2021). The current situation followed locust invasions and flash floods due to heavy rainfall, exacerbated by Covid-19 in 2020 (IFRC, 2020). Food prices in urban areas increased due to this disaster (FAO Kenya, 2021).

The previously mentioned water crisis and drought, currently threatening twelve million people in Syria and Iraq, might result in a total collapse of food production and water availability. The level of the Euphrates River fell from 500 cubic meters per second in January 2020, to 214 cubic meters per second in June 2020 (OCHA, 2021). Large swathes of agricultural land in Syria face severe drought conditions, suggesting an inability to grow crops in addition to a lack of drinking water for at least five million people (OCHA, 2021).

In Iraq, seven million people are affected because thousands of kilometers of farmland, together with fisheries and drinking water resources, have been severely depleted (OCHA Iraq, 2021). In the Nineveh governorate, wheat production is expected to reduce by 70%, and by 50% in the Kurdish region. People have been compelled to spend exorbitant amounts of money to buy drinking water (OCHA Iraq, 2021).

In September 2021, the Indian Meteorological Department issued red alerts as news media[3] reported excessive rains over the coastal city of Odisha, India, impacting almost two million people. The town of Puri experienced 341 mm of rainfall, breaking an 87-year September record, while Bhubaneswar recorded 194 mm in 24 h, beating a 63-year record, all due to a depression in the Bay of Bengal. People in low-lying areas had to be evacuated, houses were damaged and other infrastructure were impacted. The agriculture-based economy of the affected districts could be severely stricken due to floods.

Box 1: Impacts on Natural Resources

Asia and Africa are natural resource rich regions, home to a vast diversity of flora and fauna. These species are also being impacted by increasing climate

[3] https://indianexpress.com/article/india/heavy-rain-lashes-coast-of-odisha-imd-issues-red-alert-7504997/ and https://www.indiatoday.in/india/story/people-dead-heavy-rain-odisha-record-rainfall-bhubaneswar-puri-1852471-2021-09-14.

variability across the two continents, with the result that endemic biodiversity is decreasing at an alarming rate.

People in many communities in these regions supplement their food and livelihoods by relying on biodiversity and natural resources, such as by foraging non-timber forest products, among a host of other strategies. As species decline in quantity and quality, human wellbeing is affected in terms of food availability and supplementary livelihoods strategies.

3.1 Agriculture, Crops, and Food Security

Food security: "A situation that exists when all people, at all times, have physical, social and economic access to sufficient, safe and nutritious food that meets their dietary needs and food preferences for an active and healthy life" (FAO, 2001).

News of drought, crop failure, and food shortages are becoming commonplace as the climate crisis continues. Rising temperatures, changing precipitation, and higher frequency and intensity of extreme events are already resulting in declining food security.

Africa is more vulnerable to climate change than probably any other region. According to WMO (2020), agriculture is the mainstay of its national and regional economies, as well as the livelihoods of most of its households. In addition, a vast majority of its population are small-scale farmers, who depend on seasonal rainfed agriculture. It is, therefore, at considerable risk of reduced agricultural productivity, and by extension, food security, due to heat and drought stress, pest, and disease damage, to say nothing of flood impacts. This, coupled with a rising population, means that the African continent is highly vulnerable to food insecurity and erosion of livelihoods, especially among rural communities. As crops fail and agricultural productivity declines, urban communities will face (and are already facing) rising food prices (FAO 2021a).

As pastoral lands change from grass to savannas due to rising temperatures, there might be an extensive reduction in grazing land for cattle—another important source of livelihoods in Africa (FAO 2021a). Lower animal growth rates and productivity of pastoral systems are being observed across the continent (IPCC, 2019). Furthermore, crop pests and diseases have already shown high levels of variability because of climate change (IPCC, 2019). Food security in the continent's drylands is also declining. Africa is projected to have the highest number of people vulnerable to increased desertification after Asia (IPCC, 2019).

From the 1960s to 1984, there was a decrease in rainfall across the Sahel, which resulted in famine and the deaths of hundreds of thousands (FAO, 1996). In 2013, a World Bank report predicted that warming of 1.5 °C to 2 °C in sub-Saharan Africa will mean a loss of 40–80% of cropland for maize, millet, and sorghum by the 2030s-2040s (World Bank, 2013)—all of these are crucial staple crops for the region.

Food availability deteriorated across Ethiopia, Somalia, Kenya, and Uganda due to poor rainfall during March–May 2019 (WMO, 2020). The number of people suffering from food insecurity in Somalia and Kenya increased from 1.6 to 2.1 million and from 0.7 to 3.1 million respectively, between late 2018 and late 2019 (WMO, 2020). Conversely, heavy rainfall from October to December 2019 caused widespread floods, which again resulted in crop losses and livestock deaths in central and southern Somalia, south-eastern Ethiopia, northern and eastern Kenya, and South Sudan. Indeed, this heavy rainfall created conditions that prompted a favorable environment for the most severe locust outbreak the region had seen in decades (WMO, 2020), which decimated vast areas of cropland. In 2020, almost 60% of all displacements in the East and Horn of Africa were due to disasters related to climate change (IDMC, 2020; WMO, 2020). In Somalia, 650,000 people were displaced in 2020, due to the flooding of the Juba and Shabelle rivers, and faced excessive food insecurity, malnutrition, and starvation (FAO, SWALIM, 2020). This situation continued in 2021 (FAO SWALIM, 2020; UNHCR, 2020; OCHA, 2021). Many states in Sudan were impacted by heavy rains and floods due to the rising levels of the Nile in 2021; 12,700 homes were damaged and 4,800 were destroyed, displacing people, and increasing the need for food assistance (OCHA, 2021).

Poor rains in 2018 in North Africa resulted in a 30% decrease in cereal production in Morocco—below the average of the preceding five years. However, in Algeria and Tunisia, there was a higher-than-average crop yield due to favorable rains (WMO, 2021). This again points to the uncertainty of seasonal and annual climate variability. Countries in the middle east and north Africa (MENA) region are also susceptible to food price volatility due to increase in global food prices (Behnassi et al. eds., 2021), as well as decreased production.

Box 2

Since 2012, the number of undernourished people in sub-Saharan countries has risen by almost 46 precent already, and more than one-third of the world's undernourished (282 million) are found in Africa.

Source: FAO, 2021

The number of people needing food assistance rose to 13.8 million in southern Africa in 2019, almost three million more than 2018. In addition, declining rains resulted in low cereal outputs of seven percent below the five-year average (WMO, 2020. Furthermore, tropical cyclones Idai and Kenneth destroyed an estimated 480,000 hectares of crops mainly in Mozambique, but also to a lesser extent in Malawi and Zimbabwe (FAO, 2019a). Approximately 2.2 million people were displaced in Mozambique, intensifying food security concerns (UNHCR, 2020).

Conflict drives displacement and food insecurity in Central and West Africa. However, four percent of people displaced in this region were due to natural disasters in 2019 (WMO, 2020). In addition, floods and pests compounded the situation with the result that food availability and access declined, and food prices increased (WMO, 2020). The IPCC's Fifth Assessment Report identified West Africa as a climate change hotspot and extensive climate variability is projected to significantly lesson crop production and yields (IPCC, 2014b). There is thus no doubt that the resultant impacts on food security will also be drastic.

Box 3
Gross Domestic Product (GDP) in five African subregions would decline significantly due to global warming. Under scenarios ranging from 1 °C to 4 °C rise in temperature over pre-industrial levels, overall continental GDP is projected to decrease from 2.25% to 12.12%, with west, central and east Africa seeing higher impacts than southern and north Africa.
Source: ACPC (undated)

According to the IPCC (2014b), different warming scenario projections indicate devastating consequences for crop production, which will directly impact not only individual food availability and livelihoods but also regional and national economies. Cereal crops across Africa will experience variable reductions in mean yield under the worst-case scenario. Reduction in yield is projected at 13% in west and central Africa, 11% in north Africa, and eight percent in east and southern Africa. Yields of staples, such as millet and sorghum, will decrease the least by 2050 (by five and eight percent respectively). On the other hand, rice and wheat are likely to suffer the most with yield losses of 12% and 21% respectively by 2050.

As of 2020, hunger impacted 21% of the population in Africa, with the highest levels in sub-Saharan Africa at almost 67%, and one-third of the world's undernourished live in the continent (FAO 2021a; UN SDG, 2021). Stunting[4] in children is among the highest in sub-Saharan Africa at 32.3% (UN SDG, 2021), while the prevalence of food insecurity is higher among women, with approximately 41% of women in the region suffering from anemia[5] (UN SDG, 2021).

In the North Africa and western Asia regions stunting was estimated among 17.8% of children in 2020, while 31.8% of the women were anemic (UN SDG, 2021). The number of hungry people in the Middle East and North Africa (MENA) has doubled from 16.5 million people in 1990–1992 to 33 million people in 2014–2016 (JEBAS, 2017). During the same period the proportion of undernourished people increased from 6.6 to 7.5% in the region (ibid).

---

[4] Children are considered stunted, or chronically malnourished, when they are too short for their age (UN SDG Report, 2021).

[5] Anemia is defined as the prevalence of hemoglobin concentrations below 110 g per liter for pregnant women and 120 g per liter for non-pregnant women (UN SDG Report, 2021).

Box 4

More than half of the world's undernourished are found in Asia (418 million)

Source: FAO (2021)

Agriculture is the dominant source of food and incomes in Asia as well. With 58% of the population living in rural areas, 81% rely on agriculture for their livelihoods. Adverse climate impacts on rice and wheat production, as well as other food crops, will increase food insecurity (IPCC, 2014a), although reduction in agricultural yields will be lower for a 1.5 °C scenario, as compared to a 2 °C one (IPCC, 2018a, 2018b).).

While the impacts of climate change on food production vary for different regions, many areas in Asia are still expected to experience a decline in agricultural productivity. South Asia's Gangetic plains may experience a decrease of almost 50% in high-yielding wheat due to heat stress (IPCC, 2014a).

Rice is the staple food for a large part of the population, and over 90% of global rice production is from Asia. Most scenarios show that higher temperatures and shorter growing periods will result in lower rice yields. Some studies do indicate that increased atmospheric CO2 may result in higher rice yield due to more photosynthesis. However, most projections show an overall downward trend in rice production and yield, especially in Japan, eastern China, southern parts of Indochina, and northern parts of south Asia, primarily due to water shortages for this water intensive crop. Added to the variability in temperature and precipitation are the effects of sea-level rise. This will inundate low-lying areas in rice growing regions, such as those in the Mekong River Delta and Bangladesh (IPCC, 2014a).

Meanwhile, higher frequency of drought in certain regions will impact cotton production (such as western Turkmenistan and Uzbekistan) (IPCC, 2014a), another source of livelihoods. Some areas, however, will benefit. For example, parts of northern and eastern Kazakhstan may experience a longer growing season, warmer winters, and increasing winter precipitation, which may mean better cereal production. However, the trend overall is expected to be more adverse than beneficial (IPCC, 2014a).

Box 5

While the impacts of climate change will be felt globally, experts recognise that Asia will be among the hardest hit. Low lying and crowded coastal cities in many south and southeast Asian countries are most at risk, and the hundreds of millions of people who live there are particularly vulnerable.

*Source: IUCN (*https://www.iucn.org/regions/asia/our-work/climate-change-asia*)*

Crop production will also be negatively affected by extreme weather events (IPCC, 2012). This means that extreme temperatures could decrease rice yields

(Mohammed & Tarpley, 2009; Tian et al., 2010), while higher precipitation may flood agricultural areas, thus lowering crop production (IPCC, 2014a). Rising temperatures will also impact livestock with changes in feed quality, diseases, and water scarcity (IPCC, 2018a, 2018b).

In addition to crops and livestock, Asian communities also rely on fisheries and aquaculture. In fact, Asia is a leader in both global capture fisheries and aquaculture (FAO, 2010). Fish production is a vital source of livelihoods. In 2008, almost 86% of global fishers and fish farmers were from Asia (IPCC, 2014a). Fisheries also provide food security in the Asian Arctic (Zeller et al., 2011). Inland fisheries, an important source of livelihoods and food security, is immensely impacted by a host of threats, such as overfishing, habitat loss, pollution, and dam construction (IPCC, 2014a). Added to these, climate change will influence flow regimes, and increasing water temperatures will affect reproduction of many fish species (IPCC, 2014a). Sea-level rise too will adversely alter fisheries and aquaculture in Asian river basins. Marine fisheries will see huge changes due to massive redistribution of species, as well as general decrease in productivity and fish yields, specifically in the tropics, as coral reefs decline (IPCC, 2014a) and mangroves and seagrasses disappear.

With decreases in crop, livestock, and fish yields across the region, there is likely to be a general increase in food prices and the cost of living in both rural and urban areas. In the Arab region, the gap between domestic food production and consumption has been widening since the global increase in food prices in 2008 (JEBAS, 2017). Urbanization is taking place at a rapid scale in Asia and the urban poor will be impacted by higher cost of food (IPCC, 2014a). Furthermore, drought, floods, and extreme weather events are expected to impact urban populations too, in terms of food security, livelihoods, and asset availability (IPCC, 2014a).

In 2020, nine percent of the population was impacted by hunger in Asia, with more than half of the planet's undernourished found on the continent (UN SDG, 2021). The rate of stunting in children in central and southern Asia was 29.8%, third after Oceania and Africa, while almost half (49%) of the women in these areas were anemic during the same year (UN SDG, 2021). Eastern and south-eastern Asia had 13.4% of children suffering from stunting and 19.5% of women with anemia (UN SDG, 2021).

Clearly, populations across both Africa and Asia are most likely to be impacted by the multiple and compound climate related risks that will drive them toward food insecurity, low nutrition, and poverty. These risks increase between 1.5 °C and 2 °C of global warming. Post Covid-19 pandemic figures, combined with the continuing extensive climate variability, are even more stark. It is highly likely that food availability and access will be even lower as new data comes in—this time not just because of climate impacts and conflict, but also due to the pandemic.

3.2 Water Scarcity

United Nations' sustainable development goal 6 states: "Ensure availability and sustainable management of water and sanitation for all".

Global surface temperatures have reached 1.09 °C in 2021 due to human-induced climate change, increasing faster than any in the other 50-year period for the last 2,000 years. The past five years have been hotter than any other since 1850 (IPCC, 2021). This rise is likely to make existing water availability problems even worse as the water cycle changes, wet areas become wetter, and arid or dry areas become prone to more droughts. As the world continues to warm, each degree rise in temperature will intensify the frequency and severity of rainfall, flooding, and drought (IPCC, 2021). Changes in the hydrological cycle will result in higher levels of evapotranspiration in tropical regions, while some areas around the equator will face higher rainfall—droughts will become worse and will last longer in the first case, flooding will occur in the second (IPCC, 2021).

Box 6

In Africa, especially sub-Saharan Africa, more than ***a quarter of the population*** *spends more than half an hour per round trip to collect water.*

Source: UNDESA

Africa and Asia are going to be hugely impacted by such variability in the water cycle. Already, communities in the two continents face extensive water scarcity, where people in some areas must travel for miles to get potable water.

Sub-Saharan Africa will experience more severe droughts in many locations (IPCC, 2021). As soil moisture decreases in southern Africa, the productivity of agricultural land will be impacted due to hydrological, agricultural, and ecological drought (IPCC, 2021). For example, three years of very low rainfall and overuse of water in Cape Town, South Africa resulted in a severe water crisis and made headlines in 2018. Dams dried up and the government had to take extreme measures to manage water scarcity at the municipal level (Parks et al, 2019). The water crisis in Cape Town was felt the most by its poor residents, many of whom share communal taps. While the situation has become somewhat better there, the urban poor still face the same water availability issues (Parks et al., 2019).

The MENA region experiences some of the most severe water scarcity issues in the world and many countries in the region heavily consume non-renewable water aquifers (JEBAS, 2017), and most have already exhausted their water supply development potential (El Solh, 2014).

In South Asia, changes in the frequency and intensity of monsoon rains will affect crop productivity, livelihoods, and the health of millions of people. Furthermore, floods and then water scarcity will become prevalent as Himalayan glaciers melt, which have been retreating since the 1990s (El Solh, 2014). Ice volume in Asia is likely to decrease by approximately 36% under the 1.5 °C scenario, affecting future water availability in the Indus basin, which is highly dependent on melt water (Kraaijenbrink et al., 2017). Downstream irrigation and potable water availability will become difficult in the major global climate change hotspot of Indus, Brahmaputra, and Ganges basins, to say nothing of the floods and landslides that will subsequently occur (IPCC, 2021).

Some projections indicate that in east, south, and southeast Asia, as well as in the Sahel, precipitation due to monsoons is likely to increase by the end of the century, but overall, the monsoon season will have a lot of variability (IPCC, 2021). It is not necessary that there will be a direct increase in water availability of the Indus, Brahmaputra, and Ganges basins (Lutz et al., 2019). In some regions, such as the Ganges basin, periods of water excess may be followed by more periods of water shortage (Lutz et al., 2019). Any changes in the length of wet and dry periods could result in increasing crop water demand, adding further stress to water resources (Lutz et al., 2019). The Indian sub-continent will experience severe droughts in many regions (IPCC, 2021).

Globally, 2.3 billion people live in water-stressed countries and 2 billion people (26%) lack access to safe drinking water (UN-Water, 2021). Water stress levels in Western and Southern Asia and Northern Africa have been observed more than 60%, 77%, and 108% respectively (UN SDG, 2021). Water stress levels in southeast Asia and sub-Saharan Africa increased from 2017 to 2018 (UN-Water, 2021) and in Eastern Asia they were 45.6% (UN SDG, 2021). In sub-Saharan Africa alone, 76% of the population lacks access to safe drinking water (UN-Water, 2018) and half of those without access to basic drinking water services (387 million) live there (UN SDG, 2021).

The 4.5 billion people who live in Asia use around 65% of the global water supply. Out of this, 30% of the population is already experiencing water scarcity (Satoh et al., 2017). It is estimated that by 2050, 20% of the land area in the Asia–Pacific region, home to 1.2 to 2 billion people, will experience severe water stress (Satoh et al., 2017). The demand for water will increase by 30–40% for agriculture, industry, and household use by 2050, compared to 2010. Projections indicate that by the 2050s, Asia's water demand will be larger than all the other continents of the world put together (Satoh et al., 2017).

Box 7

72 percent of all water withdrawals are used by agriculture

Source: UN-Water (2021)

An array of NASA satellites observed human activities across the globe to map where freshwater resources were changing. This landmark study, using 14 years of observations of the Earth from space (from 2002 to 2016), highlighted 30 global hotspots, where freshwater was experiencing major changes. North-western China, northern and eastern India, and the Middle-East were some of the areas where extensive water overuse had led to serious declines in freshwater availability (NASA, 2018). For example, Saudi Arabia lost 6.1 gigatons annually of stored groundwater due to unprecedented growth of irrigated agricultural land from 1987 onwards, while Xinjiang Province in China lost 5.5 gigatons per year due to an increase in irrigation water use and evaporation (NASA, 2018).

While accessibility to and availability of safe drinking water is essential, there is another important aspect: water is intrinsically linked to agriculture. As increased areas experience water scarcity, crop productivity will be impacted, amplifying food insecurity and lack of nutrition. This is especially so in rice growing Southeast Asian countries (McCartney et al., 2016). Rapidly growing urban areas will place more pressure on water resources, and this will only be compounded by climate change. Clearly, there is a crucial need for agricultural and domestic water management.

4 Climate-Smart Agriculture

Climate-Smart Agriculture is "agriculture that sustainably increases productivity, enhances resilience (adaptation), reduces/removes GHGs (mitigation) where possible, and enhances achievement of national food security and development goals" (FAO, 2013).

The global demand for food is being driven up due to a rapidly rising population. Concurrently, deteriorating natural resources, including soils, water and biodiversity, are resulting in decreased agricultural productivity. In 2020, nearly 690 million people (8.9%) went hungry—an increase of 10 million in one year and 60 million in five years (FAO, 2020). Feeding the world is only going to become more challenging, particularly due to agriculture's vulnerability to climate change. On the flip side, agricultural production also contributes 15–25% of GHG emissions to the global total (Vermeulen et al., 2012). The twentieth century's 'Green Revolution', which degraded soil and water resources and reduced biodiversity through the extensive use of monoculture, fertilizers, and pesticides, adversely impacted human and ecosystem health (CGIAR and CCAFS undated) and contributed to global warming. It is therefore crucial that global agricultural processes are modified and managed sustainably, so that they not only enhance productivity but also increase the resilience of landscapes, ecosystems, and local communities, in addition to mitigating the impacts of climate change.

The challenges of providing food to rising populations as the climate crisis progresses is, and will continue to be, felt most acutely in Asia and Africa. Climate-Smart Agriculture (CSA) is an integrative approach that can help to effectively manage croplands, fisheries, livestock, and ecosystems, as well as mitigate the interlinked issues of climate change, food security, and water scarcity (CGIAR and CCAFS undated) in these regions. According to the Research Program on Climate-Smart Agriculture and Food Security (CGIAR and CCAFS undated), CSA has three objectives:

- Sustainably increasing agricultural productivity, to support equitable increases in farm incomes, food security, and development;
- Adapting and building resilience of agricultural and food security systems to climate change at multiple levels; and
- Reducing greenhouse gas emissions from agriculture (including crops, livestock, and fisheries).

CSA was first defined by FAO in a background document for the 2010 Hague Conference on Food Security, Agriculture, and Climate Change (FAO, 2013). This three-pronged approach ensures that, in addition to increasing crop productivity, CSA also enhances resilience, and adaptation to climate change, while contributing to mitigation by reducing emissions. It adopts a landscape methodology to transform and re-orient agricultural development, with climate change at its center (FAO, 2013; Lipper et al., 2014). Current and future food security forms an integral part of it but with a strong emphasis on adapting to and mitigating climate change (CGIAR CCAFS undated). While it aims to increase productivity, CSA also recognizes the necessity to reduce short-term climate risks to agriculture and increase the resilience of farmers to shocks and long-term stresses to achieve sustainable food security objectives (ibid), particularly in developing countries. At the same time, it seeks to maintain healthy and resilient ecosystems that provide crucial services for human wellbeing (CGIAR, CCAFS undated). Finally, it provides mitigation opportunities through several integrated strategies. It does all of this by seeking synergies and managing trade-offs between productivity, adaptation/resilience, and mitigation (CGIAR CCAFS undated).

Indeed, CSA is an innovative approach for sustainable agriculture and food systems that enhances food security (HLPE, 2019). However, it goes further by combining the goals of sustainable agriculture with resilience, while also aiming to reduce agricultural emissions (CGIAR CCAFS undated). It incorporates ecosystems-based adaptation to assist people in adapting to climate change by maintaining and strengthening ecosystem services, as well as sustainable land management methodologies—integrating land, water, biodiversity, and environmental management—to increase food production. Participatory approaches, involving marginalized groups and women, are used to manage trade-offs and identify synergies between different context-specific options. As this is no "one size fits all" approach, diverse ground realities, differences in landscapes and ecosystems, varying institutional arrangements and policy environments are all crucial components of the methodologies applied (CGIAR CCAFS undated).

CSA uses a systems approach based on synergies, trade-offs, and cost benefit analyses between landscapes, ecosystems, and value chains (CGIAR CCAFS undated). This can provide resource-poor farmers with a framework to achieve sustainable increases in crop productivity in the face of unquestionable climate variability. Incorporating value chains ensures efficient food supply management and decreases in food waste. Such a framework can help to secure local livelihoods and contribute to global food security (CGIAR CCAFS undated). It can also support national economies, decrease poverty, and assist developing countries in Asia and Africa to meet their adaptation and mitigation targets. Importantly, implementing CSA activities can help transform agri-food systems to meet the 17 SDGs (FAO 2021b), by helping countries to advance their climate, food security, and income objectives, as well as bringing about positive environmental and social changes (FAO, 2019b).

CSA approaches are widely being applied globally based on five crucial action points (Box 8) and offer many benefits in the context of Asia and Africa. Nations and communities in these continents will continue to suffer the most as climate impacts accelerate. Implementing CSA approaches and interventions can help to increase food productivity and availability, enhance the resilience of agricultural systems and farmers so that they can adapt to climate change, and devise mitigation strategies to ensure a decrease in the contribution of agriculture to global emissions.

Box 8

Implementation of CSA requires five crucial action points:

(1) Expanding the evidence base for CSA
(2) Supporting enabling policy frameworks
(3) Strengthening national and local institutions
(4) Enhancing funding and financing options
(5) Implementing CSA practices at field level

Source: FAO (2021b)

Here we discuss case studies where CSA approaches have been implemented in Asia and Africa, demonstrating the potential for food and water security and climate goals.

- ***Participatory rangeland management in the Syrian Arab Republic*** (IFAD, 2012)

The Badia are Syrian steppes that have been severely degraded following years of drought and intensive grazing. A participatory rangeland management project was implemented in collaboration with local communities to reduce Badia herders' vulnerability to climate change and restore the rangelands' long-term productivity. Native plants were introduced to help meet fodder needs, fix the soil, and decrease soil encroachment. These steps helped restore ecosystems and decrease the community's vulnerability to climate impacts. Birds and animals returned to the area following two years of resting, reseeding, and replanting. Furthermore, income-generation opportunities for women were also created through truffle growing and collection. According to IFAD (2012), "*In 2010, a community with a 100 000-ha grazing area could earn up to USD one million through the sale of truffles*". Higher household income means additional income-generating opportunities for women could be put in place through literacy classes and training. More economic independence resulted in shifting gender relations, further increasing community resilience.

- ***Integrating traditional practices for CSA into crop and livestock production systems in Botswana*** (FAO, 2021b)

Botswana is an arid country with just five to six percent of arable land suitable for agriculture. Nevertheless, more than half of its population is rural, dependent on rainfed agriculture, with half still relying on subsistence agriculture and livestock farming. The agricultural sector in Botswana is vulnerable to climate change due to low adaptive capacities of its farmers. Lack of education, poverty, and lack of access to assets or capital hinders their capacity to recover from shocks or to enhance their livelihoods. However, Botswanan farmers have extensive traditional knowledge and are implementing CSA practices, but these are isolated and have not been scaled up to a broader level.

A project aimed to develop resilient farming systems across various agro-ecological zones with the primary outcome to build institutional capacity to scale-up these local-level practices. Several CSA options to improve crop and livestock production systems, based on indigenous practices, as well as new technologies and practices were identified and implemented. The project also increased capacities for research and for the provision of extension services, in particular by including indigenous CSA practices in crop and livestock production systems. Case studies on successful implementation of CSA activities and value chain interventions were documented and shared with policymakers and other stakeholders to increase knowledge. Workshops and training were provided at various levels to disseminate knowledge on CSA, as well as on indigenous practices to increase institutional capacity.

- ***Punjab Irrigated Agriculture Productivity Improvement Program, Pakistan*** (World Bank, 2020a)

The main objective of this project is to improve water use in irrigated agriculture in the Punjab, around the Indus River and its tributaries. It contributes to increased agricultural production and livelihoods and supports positive environmental outcomes by improving physical delivery and irrigation practices. Crop diversification and effective application of inputs will translate into greater agricultural output per unit of water used. By 2019, high efficiency irrigation systems had been installed covering 23,500 hectares, and installation in a further 3,677 hectares was in progress. The project improved 11,916 water courses and 621 water ponds were constructed. Water management improved in 5.7 million acres of farmland. The project estimates that 500,000 farm households are directly benefiting from the project and 15,000 full-time jobs were created. All of this contributes to increased food security and improved water management.

- ***Scaling up climate-smart crop systems in Sri Lanka*** (FAO, 2021b)

A recent project in Sri Lanka is supporting the transition to more productive and resilient smallholder farming, as well as reducing GHG emissions in rice production. A combination of heavy rains, soil tillage, and poor nutrient cycling in the uplands results in soil erosion, loss of soil fertility, and loss of productivity. In addition, siltation of water reservoirs causes difficulties for the irrigation system and decreased water productivity in the lowlands.

The project is providing training to farmers to optimize water use, labor, and machinery. Furthermore, it also builds an evidence base for CSA through the assessment of future climate change impacts on rice, maize, green gram, onion, chili, and potato, thus helping to identify the vulnerabilities of farmers to climate change impacts. This evidence base will assist policy makers to make informed decisions regarding effective adaptation strategies. Studies also provide opportunities for future climate-smart activities that could help increase the resilience of the local population.

Farmers cut back rice irrigation requirements by 10 to 20% by adopting new water management practices. Training in alternating wetting and drying techniques helped farmers increase land under irrigation by 15% during the dry season. The whole community had the highest water capacity ever recorded during the dry season, due to implementing various new water management practices. Precise application of fertilizer reduced its application by 27%. All these practices will contribute to the triple goals of increasing productivity, improving adaptation, and supporting mitigation.

- ***Drought-tolerant maize for Africa*** (CCAFS Big Facts)

Maize is an important source of food and livelihood in sub-Saharan Africa. However, 40% of the continent's maize-growing areas face occasional drought stress with the resultant decrease in yield of 10 to 25%. Furthermore, 25% of the maize crop experiences frequent drought and the loss of about half the harvest. A project released more than 100 drought-tolerant maize varieties between 2007 to 2013 to reduce vulnerability, improve food security, and thus increase resilience.

After being tested in laboratories as well on farmers' fields, the maize varieties were released to farmers in 13 countries. The maize could help more than four million people to escape lack of food and poverty in the long run.

- ***Using remote sensing to monitor water productivity—Africa and the Near East*** (FAO, 2021b)

Future food security requires sustainable management of water resources, especially in the face of climate change. Because agriculture consumes over 70% of global freshwater withdrawals, it is imperative to develop a system to monitor and improve water productivity in agriculture.

Project WaPOR (Water Productivity Open-Access Portal) uses remote sensing to monitor water productivity in near real-time. The data is uploaded on a public portal, freely available to countries and a variety of stakeholders, to help them monitor water productivity in agriculture. The information can be used by policymakers to identify water productivity gaps, formulate solutions to reduce these gaps, and sustainably increase agricultural production that can help end-users benefit from participatory, innovative, and informed strategies.

- ***Climate-Smart Staple Crop Production in China*** (World Bank, 2020b)

In China, a recently completed project reduced greenhouse gas emissions by 23,732 tons of CO2 equivalent and increased the soil carbon sink by 71,683 tons of CO^2 due to improved water management and improved use of inputs for wheat production. Higher crop residue retention combined with improved soil water content and soil structure helped increase the carbon sequestration potential. Crop yield increased by over eight% due to the adoption of CSA technologies such as efficient irrigation, fertilizer, and pesticide use practices.

- ***Role of Sustainable Wheat Production to Ensure Food Security in Central and West Asia and North Africa*** (JEBAS, 2017)

Wheat is the basic staple food in the Central and West Asia and North Africa (CWANA) region and wheat cultivation is a means of livelihood for a large proportion of the population. It comprises over 60% of the daily caloric and protein intake. However, wheat productivity is very low due to several environmental stresses, with the result that most countries in the region import wheat. Climate change is expected to further complicate food production shortfalls leading to a food crisis. A wheat breeding Program by the International Center for Agricultural Research in the Dry Areas (ICARDA) has developed high yielding and heat resistant wheat varieties that have been released by many countries in the region. This, accompanied by other crop management practices, has resulted in an average increase of wheat production by 28%. This is another example indicating that the adoption of climate-smart technologies can be an effective way of increasing food security.

According to estimates, the demand for food is going to increase by 50% by 2050, and this demand is unlikely to be met by utilizing the existing food production systems (FAO, 2017). The case studies presented here show the various means by which CSA can increase agricultural productivity while mitigating its climate impacts and enhancing the adaptive capacity of communities.

Using good quality, high yielding, and well-adapted varieties of seeds, integrated and conservation friendly pest management, efficient fertilizer dispensation, and sustainable water management can increase crop production. Soil and rangeland management through sustainable grazing improved feed management and planting of grasses, and native species for fodder can provide benefits to livestock farmers. Enhanced land and water management that incorporates soil management, rainwater harvesting, reusing wastewater, efficient irrigation practices, and watershed management procedures will contribute toward ecosystem conservation and water availability for both agriculture and domestic consumption. Recycling and reusing waste materials will help eliminate pollution and regenerate ecosystems and landscapes. Integrated approaches such as conservation agriculture, agro-forestry, silvopastoralism, crop-fish, and crop-livestock combinations are just some approaches that can be utilized to enhance sustainability and contribute to mitigating GHG emissions.

As mentioned before, CSA is context-specific, therefore greater understanding of landscapes and ecosystem dynamics, use of innovative technologies and indigenous practices will all help achieve the three CSA objectives. These must be accompanied by participatory approaches, involving women and marginalized communities, enabling institutional structures and strong policy environments. Successful, sustainably managed, participatory CSA activities that enhance institutional and policy structures can provide solutions for food and water security, increasing adaptive capacity and mitigating the climate impacts of agricultural systems.

5 Conclusion

CSA can prove to be an important strategy to deal with the vagaries of climate change and to increase people's resilience to its impacts. Asia and Africa—comprised of agrarian economies with rising populations—are going to be the most affected by the ongoing climate emergency. Feeding the populations of these two regions is a significant challenge, as is the availability of sustainably managed water resources.

There is a triple link here: both food and water security are linked to each other through agriculture, and both agriculture and freshwater availability are substantially impacted by climate change. There is a need to therefore increase crop productivity, while managing freshwater systems in the face of climate change.

Populations in Asia and Africa are extremely vulnerable to climate change due to extreme events, sea intrusion, and other environmental issues that are impacting livelihoods, infrastructure, and economies. Such impacts have also decreased food availability, nutrition, and water accessibility. Trends indicate that these issues will intensify with an increase in future temperatures. There is, therefore, a need to decrease vulnerabilities and increase resilience by helping crop systems and people adapt to climate change. Finally, agriculture is also a considerable contributor to global warming, and while most countries in Asia and Africa are not major emitters of GHGs, there is still a need to continue to decrease global emissions if we are to keep temperatures below 1.5 °C according to the goal of the Paris Agreement.

CSA is an approach that can help Asian and African countries achieve these outcomes by implementing innovative and participatory activities that incorporate indigenous practices. Because CSA employs a landscape approach together with ecosystems-based adaptation and sustainable agriculture management, its outcomes will provide environmental and biodiversity benefits, further contributing to increasing adaptive capacities and resilience, while decreasing the contribution of food systems to climate change.

It is important for policymakers, implementers, and farmers to understand and incorporate its objectives of increasing agricultural productivity and adaptation and resilience to shocks, and mitigating climate change. However, not all of these objectives will always be achieved; therefore, synergies and trade-offs will have to be considered. The five action points (Box 8) are equally important when designing and implementing activities, which must be participatory and innovative, incorporate indigenous knowledge wherever possible, and ensure a diversity of stakeholders including women and marginalized groups.

The case studies presented demonstrate the applicability of CSA in Asian and African regions by highlighting how it can increase food security and livelihoods, achieve climate adaptation and mitigation targets, and ultimately contribute to the achievement of SDGs.

References

ACPC. (undated). *Economic growth, development and climate change in Africa, African Climate Policy Centre (ACPC)*, United Nations Economic Commission for Africa, https://archive.uneca.org/sites/default/files/uploaded-documents/ACPC/annex_21b_-_climate_change_impacts_on_africas_economic_growth_-_report.pdf

Behnassi et al. (eds.) (2021). *Emerging challenges to food production and security in Asia, Middle East and Africa—climate risks and resource scarcity*, Springer.

CCAFS Big Facts. *Drought-tolerant maize boosting food security in 13 African countries*. https://ccafs.cgiar.org/bigfacts/#theme=evidence-of-success&subtheme=crops&casestudy=cropsCs2

CGIAR and CCAFS (undated). *Climate smart agriculture 101, research program on climate smart agriculture and food security*, CGIAR, CCAFS. https://csa.guide/

Ebi, K. L., Ziska, L. H., & Yohe, G. W. (2016). The shape of impacts to come: Lessons and opportunities for adaptation from uneven increases in global and regional temperatures. *Climatic Change, 139*(3), 341–349. https://doi.org/10.1007/s10584-016-1816-9

El Solh, M. (2014). The role of science in enhancing food security AFED (2014) Arab environment: Food security. In A. Sadik, M. El Solh, & N. A. Saab (Eds.) *Annual Report of Arab Forum for Environmental Development*, pp. 44–71.

FAO. (2021a). FAO, IFAD, UNICEF, WFP and WHO. *The State of Food Security and Nutrition in the World*. (2021). *Transforming food systems for food security, improved nutrition and affordable healthy diets for all*. Rome, FAO.https://doi.org/10.4060/cb4474en

FAO. (2021b). *Climate-smart agriculture case studies—Projects from around the world*. (2021). Food and Agriculture Organization of the UN, Rome.https://doi.org/10.4060/cb5359en

FAO. (2020). The State of Food Security and Nutrition in the World. *Transforming food systems for food security, improved nutrition and affordable healthy diets*. FAO. http://www.fao.org/3/ca9692en/online/ca9692en.html#chapter-Key_message

FAO. (2019a). *Early warning early action report on food security an agriculture*

FAO. (2019b). *Climate-smart agriculture and the sustainable development goals: Mapping inter-linkages, synergies and trade-offs and guidelines for integrated implementation*. Food and Agriculture Organization of the UN, Rome.

FAO. (2017). *Climate smart agriculture sourcebook, summary*. Food and Agriculture Organization of the UN, Rome.

FAO. (2013). *Climate smart agriculture: Sourcebook*. Food and Agriculture Organization of the United Nations

FAO. (2010). *The State of World Fisheries and Agriculture*. Food and Agriculture Organization of the United Nations (FAO), Fisheries and Aquaculture Department, Rome, Italy, 197 pp.

FAO. (2001). *State of food insecurity*. FAO, http://www.fao.org/3/y1500e/y1500e00.htm

FAO. (1996). Gommes R., & Petrassi, F., *Rainfall Variability and Drought in Sub-Saharan Africa, Environment and Natural Resources Service (SDRN)*, FAO Research, Extension and Training Division.

FAO Kenya. (2021). *FAO and Kenyan Government sign action plan to mitigate drought in ASALs Counties*. http://www.fao.org/kenya/news/detail-events/en/c/1418809/

FAO SWALIM. (2020). *Somalia Facing Risk of Floods in the Juba and Shabelle Basins*, http://www.faoswalim.org/article/somalia-facing-risk-floods-juba-and-shabelle-basins

HLPE. (2019). High Level Panel of Experts on Food Security and Nutrition (HLPE). (2019). *Agroecological and other innovative approaches for sustainable agriculture and food systems that enhance food security and nutrition.* A report by the High-Level Panel of Experts on Food Security and Nutrition of the Committee on World Food Security. July 2019. HLPE Report 14. 162 pp. www.fao.org/cfs/cfs-hlpe

IDMC. (2020). *Global Report on Internal Displacement.* Internal Displacement Monitoring Centre (IDMC). Norwegian Refugee Council (NRC). https://www.internal-displacement.org/global-report/grid2020/

IFAD. (2012). *International Fund for Agricultural Development (IFAD).* Syrian Arab Republic: thematic study on participatory rangeland management in the Badia. http://www.ifad.org/pub/pn/badia.pdf

IFRC. (2020). Red Cross raises the alarm over a "triple menace" of floods, COVID-19 and locusts. Press Release. (2020). *International Federation of Red Cross And Red Crescent Societies.* https://reliefweb.int/report/kenya/red-cross-raises-alarm-over-triple-menace-floods-covid-19-and-locusts

IPCC. (2021). Climate change 2021: The physical science basis. Contribution of Working Group I to the Sixth Assessment Report of the Intergovernmental Panel on Climate Change [Masson-Delmotte, V., P. Zhai, A. Pirani, S. L. Connors, C. Péan, S. Berger, N. Caud, Y. Chen, L. Goldfarb, M. I. Gomis, M. Huang, K. Leitzell, E. Lonnoy, J. B. R. Matthews, T. K. Maycock, T. Waterfield, O. Yelekçi, R. Yu & B. Zhou (eds.)]. Cambridge University Press. In Press.

IPCC. (2019). Summary for Policymakers. In: Climate Change and Land: an IPCC special report on climate change, desertification, land degradation, sustainable land management, food security, and greenhouse gas fluxes in terrestrial ecosystems [P.R. Shukla, J. Skea, E. Calvo Buendia, V. Masson-Delmotte, H.- O. Pörtner, D. C. Roberts, P. Zhai, R. Slade, S. Connors, R. van Diemen, M. Ferrat, E. Haughey, S. Luz, S. Neogi, M. Pathak, J. Petzold, J. Portugal Pereira, P. Vyas, E. Huntley, K. Kissick, M. Belkacemi, J. Malley, (eds.)]. In press.

IPCC. (2018). Allen, M.R., O.P. Dube, W. Solecki, F. Aragón-Durand, W. Cramer, S. Humphreys, M. Kainuma, J. Kala, N. Mahowald, Y. Mulugetta, R. Perez, M. Wairiu, & K. Zickfeld, 2018: Framing and Context. In: Global Warming of 1.5 °C. An IPCC Special Report on the impacts of global warming of 1.5 °C above pre-industrial levels and related global greenhouse gas emission pathways, in the context of strengthening the global response to the threat of climate change, sustainable development, and efforts to eradicate poverty [Masson-Delmotte, V., P. Zhai, H.-O. Pörtner, D. Roberts, J. Skea, P.R. Shukla, A. Pirani, W. Moufouma-Okia, C. Péan, R. Pidcock, S. Connors, J.B.R. Matthews, Y. Chen, X. Zhou, M.I. Gomis, E. Lonnoy, T. Maycock, M. Tignor, & T. Waterfield (eds.)]. In Press.

IPCC. (2018). Summary for Policymakers. In: Global Warming of 1.5 °C. An IPCC Special Report on the impacts of global warming of 1.5 °C above pre-industrial levels and related global greenhouse gas emission pathways, in the context of strengthening the global response to the threat of climate change, sustainable development, and efforts to eradicate poverty [Masson-Delmotte, V., P. Zhai, H.-O. Pörtner, D. Roberts, J. Skea, P.R. Shukla, A. Pirani, W. Moufouma-Okia, C. Péan, R. Pidcock, S. Connors, J.B.R. Matthews, Y. Chen, X. Zhou, M.I. Gomis, E. Lonnoy, T. Maycock, M. Tignor, T. Waterfield (eds.)]. In Press.

IPCC. (2014a). Hijioka, Y., E. Lin, J.J. Pereira, R.T. Corlett, X. Cui, G.E. Insarov, R.D. Lasco, E. Lindgren, & A. Surjan, 2014: Asia. In: Climate Change 2014: Impacts, Adaptation, and Vulnerability. Part B: Regional Aspects. Contribution of Working Group II to the Fifth Assessment Report of the Intergovernmental Panel on Climate Change [Barros, V.R., C.B. Field, D.J. Dokken, M.D. Mastrandrea, K.J. Mach, T.E. Bilir, M. Chatterjee, K.L. Ebi, Y.O. Estrada, R.C. Genova, B. Girma, E.S. Kissel, A.N. Levy, S. MacCracken, P.R. Mastrandrea, & L.L.White (eds.)]. Cambridge University Press, Cambridge, United Kingdom and New York, NY, USA, pp. 1327–1370.

IPCC. (2014b). Climate Change 2014: Synthesis Report. Contribution of Working Groups I, II and III to the Fifth Assessment Report of the Intergovernmental Panel on Climate Change [Core Writing Team, R.K. Pachauri & L.A. Meyer (eds.)]. IPCC, Geneva, Switzerland, 151 pp.

IPCC. (2012). Managing the risks of extreme events and disasters to advance climate change adaptation. A Special Report of Working Groups I and II of the Intergovernmental Panel on Climate Change [Field, C.B., V. Barros, T.F. Stocker, Q. Dahe, D.J. Dokken, K.L. Ebi, M.D. Mastrandrea, K.J. Mach, G.-K. Plattner, S.K. Allen, M. Tignor, & P.M. Midgley (eds.)]. Cambridge University Press, Cambridge, UK and New York, NY, USA, 582 pp.

JEBAS. (2017). Sustainable agriculture and food security in the Arab world (SAFSAW). *Journal of Experimental Biology and Agricultural Sciences, 5*(Spl-1- SAFSAW). https://doi.org/10.18006/2017.5(Spl-1-SAFSAW).S1.S14

Kotir, J. H. (2011). Climate change and variability in Sub-Saharan Africa: a review of current and future trends and impacts on agriculture and food security. *Environment, Development and Sustainability, 13*(3).

Kraaijenbrink, P. D. A., Bierkens, M. F. P., Lutz, A. F., & Immerzeel, W. W. (2017). Impact of a 1.5 °C global temperature rise on Asia's glaciers. *Nature, 549*, 257–260. https://doi.org/10.1038/nature23878

Lipper, L., Thornton, P., Campbell, B., et al. (2014). Climate-smart agriculture for food security. *Nature Climate Change, 4*, 1068–1072. https://doi.org/10.1038/nclimate2437

Lutz, A. F., ter Maat, H. W., Wijngaard, R. R., et al. (2019). South Asian river basins in a 1.5 °C warmer world. *Regional Environmental Change, 19*, 833–847. https://doi.org/10.1007/s10113-018-1433-4

Mohammed, A. R., & Tarpley, L. (2009). High nighttime temperatures affect rice productivity through altered pollen germination and spikelet fertility. *Agricultural and Forest Meteorology, 149*(6–7), 999–1008.

McCartney, M., Johnston, R., & Lacombe, G. (2016). Building climate resilience through smart water and irrigation management systems. *Experiences from Southeast Asia.*

Nairobi (2021). Spokesperson's Office State House, Nairobi 8th September 2021. *Press Release: President Kenyatta declares drought a national disaster*. https://www.president.go.ke/2021/09/08/spokespersons-office-state-house-nairobi-8th-september-2021-press-release-president-kenyatta-declares-drought-a-national-disaster/

NASA. (2020). *Record Flooding in Sudan, NASA Earth Observatory.* https://earthobservatory.nasa.gov/images/147288/record-flooding-in-sudan

NASA. (2018). *NASA Satellites Reveal Major Shifts in Global Freshwater*, Press Release. https://www.nasa.gov/press-release/nasa-satellites-reveal-major-shifts-in-global-freshwater

OCHA. (2021). *Somalia 2021 Gu' Season Floods Update #1.* United Nations Office for the Coordination of Humanitarian Affairs. www.unocha.org/somalia

OCHA. (2021). *Syrian Arab Republic: Euphrates water crisis & drought outlook.* UN Office for the Coordination of Humanitarian Affairs. https://reliefweb.int/report/syrian-arab-republic/syrian-arab-republic-euphrates-water-crisis-drought-outlook-17-june-2021

OCHA Iraq. (2021). *Iraq: Drought.* UN Office for the Coordination of Humanitarian Affairs.

OCHA. (2021). Sudan Flash Floods Update No. 05. [EN/AR], UN Office for the Coordination of Humanitarian Affairs. https://reliefweb.int/report/sudan/sudan-floods-flash-update-no-05-9-august-2021-enar

OCHA. (2021). *Press Release Water crisis and drought threaten more than 12 million in Syria and Iraq.* UN Office for the Coordination of Humanitarian Affairs. https://reliefweb.int/report/syrian-arab-republic/water-crisis-and-drought-threaten-more-12-million-syria-and-iraq

OCHA. (2021). SUDAN: Floods Flash update No. 10.. [EN/AR], UN Office for the Coordination of Humanitarian Affairs. https://reliefweb.int/report/sudan/sudan-floods-flash-update-no-10-2-september-2021-enar

Parks, R., McLaren, M., Toumi, R., & Rivett, U. (2019). *Experiences and lessons in managing water from Cape Town*, Grantham Institute Briefing Paper No 29, Imperial College London.

Satoh, Y., Kahil, T., Byers, E., Burek, P., Fischer, G., Tramberend, S., Greve, P., Flörke, M., et al. (2017). Multi-model and multi-scenario assessments of Asian water futures: The Water Futures and Solutions (WFaS) initiative. *Earth's Future, 5*(7), 823–852.

Tian, X., Matsui, T., Li, S., Yoshimoto, M., Kobayasi, K., & Hasegawa, T. (2010). Heat induced floret sterility of hybrid rice (Oryza sativa L.) cultivars under humid and low wind conditions in the field of Jianghan Basin, China. *Plant Production Science, 13*(3), 243–251.

UN. (2021). *Climate change and hunger in Madagascar: A UN Resident Coordinator blog.* Climate and Environment, United Nations News. https://news.un.org/en/story/2021/08/1098042

UN SDG. (2021). *The United Nations Development Goals Report 2021*: https://unstats.un.org/SDGS/Report/2021 and https://sdgs.un.org/goals

UN. (2021). *Progress towards the Sustainable Development Goals—E/2021/58, Report of the General Secretary, Economic and Social Council Session.* United Nations.

UN-Water. (2021). *Water Scarcity Facts and Figures, UN Water.* United Nations. https://www.unwater.org/water-facts/scarcity/

UNDESA. International Decade for Action 'Water for Life' 2005–2015, UN Water, United Nations Department of Economic and Social Affairs (UNDESA). https://www.un.org/waterforlifedecade/africa.shtml

UNFCCC. (2020). *Climate change is an increasing threat to Africa, article.* United Nations Framework Convention on Climate Change. https://unfccc.int/news/climate-change-is-an-increasing-threat-to-africa

UN. (2018). *Sustainable development Goal 6 Synthesis Report 2018 on Water and Sanitation.* United Nations, New York

UNHCR. (2020). *One year on, people displaced by Cyclone Idai struggle to rebuild*, UNHCR/Hélène Caux. https://www.unhcr.org/news/stories/2020/3/5e6a6e50b/year-people-displaced-cyclone-idai-strugglerebuild.html

Vermeulen S.J., Bruce M. Campbell, and John S.I. Ingram. (2012). *Climate Change and Food Systems, The Annual Review of Environment and Resources.* https://doi.org/10.1146/annurev-environ-020411-130608

WFP Madagascar Country Brief. (2021). https://www.wfp.org/countries/madagascar

World Bank. (2021). *Climate smart agriculture.* The World Bank. https://www.worldbank.org/en/topic/climate-smart-agriculture

World Bank. (2020a). *PK Punjab Irrigated Agriculture Productivity Improvement Program Project.* The World Bank Group. https://projects.worldbank.org/en/projects-operations/project-detail/P125999

World Bank. (2020b). *China—Climate Smart Staple Crop Production.* World Bank Group. http://documents.worldbank.org/curated/en/982891632315957357/China-Climate-Smart-Staple-Crop-Production

World Bank. (2013). *Potsdam institute for climate impact research and climate analytics.* Turn Down the Heat: Climate Extremes, Regional Impacts, and the Case for Resilience. World Bank. https://openknowledge.worldbank.org/handle/10986/14000

WMO. (2021). *WMO ATLAS of Mortality and Economic Losses From Weather, Climate And Water Extremes.* WMO-No. 1267. https://library.wmo.int/index.php?lvl=notice_display&id=21930#.YTIpJo5KhPa

WMO. (2020). *State of the Climate in Africa 2019, World Meteorological Organization.* WMO-No. 1253. https://library.wmo.int/index.php?lvl=notice_display&id=21778#.YTImC45KhPb

Zeller, D., Booth, S., Pakhomov, E., Swartz, W., & Pauly, D. (2011). Arctic fisheries catches in Russia, USA, and Canada: Baselines for neglected ecosystems. *Polar Biology, 34*(7), 955–973.

Energy-Economy-Environment Nexus: Moving Toward an Innovative Framework for Food-Secure Future

Syed Asif Ali Naqvi, Syed Ale Raza Shah, Mustafa Kan, Muhammad Ashfaq, and Mirza Barjees Baig

Abstract Energy is the basic driving force of an economy, and there is a robust linkage between economic development and energy consumption. The industrial revolution, which enhanced the living standards of millions of people worldwide, would not have been possible without energy attained first from fossil combustion. Variations in technology, economic development, ever-stringent environmental protectionism, and the rising worldwide competitiveness have induced a trend of innovative market-based clarifications. This chapter centrally focuses on the 3-E framework (Energy-Economy-Environment) and develops an understanding of the energy world's state and the doctrines of exhaustible resource economics. It provides basic knowledge about judicious decision-making about the food value chain and energy markets, with particular reference to developing nations of the Asian region. Fundamental principles underlying the non-renewable energy system and their impacts on marginal populations, and possible solutions are also discussed. The readers will quickly learn the prevailing food security scenarios, renewable and non-renewable energy, and their allied impacts on the environment.

S. A. A. Naqvi
Department of Economics, Government College University, Faisalabad, Pakistan
e-mail: syedasif_1@yahoo.com

S. A. R. Shah
School of Economics and Finance, Xi'an Jiaotong University, Xian, China
e-mail: syedaaleraza56@gmail.com

M. Kan
Agricultural Faculty, Department of Agricultural Economics, Kirsehir Ahi Evran University, Kirsehir, Turkey
e-mail: mustafa.kan@ahievran.edu.tr

M. Ashfaq (✉)
Faculty of Management Studies, The University of Faisalabad, Faisalabad, Pakistan
e-mail: ashfaq9@hotmail.com; dean.nbs@tuf.edu.pk

M. B. Baig
Water and Desert Research, Prince Sultan Institute for Environmental, King Saud University, Riyadh, Saudi Arabia
e-mail: mbbaig@ksu.edu.sa

M. Behnassi et al. (eds.), *Climate-Smart and Resilient Food Systems and Security*,
https://doi.org/10.1007/978-3-031-65968-3_8

Keywords Energy · Economy · Environment · Food security · Economic development

1 Understanding the State of the Energy World

Accessing energy is essential for economic development and poverty reduction. Guaranteeing that the whole population has adequate energy access is a pressing challenge for global expansion. The energy systems of nations can have significant environmental influences (Ollila, 2007). Historical and current energy systems are dominated by fossil fuels which contribute to greenhouse gas (GHG) emissions considered as the main driver of global climate change. If nations want to meet global climate targets and circumvent dangerous environmental variations, they must have a substantial and concerted transition in energy sources and systems. Harmonizing the difference between economic development and the environment provides the ultimate goal of ensuring that everyone has access to enough sustainable energy to sustain a better living standard.

North America, Europe, and Eurasia jointly absorbed a significant part of the total resources in 1965, comprising more than 80% of the world's energy use (Zhang et al., 2013). These regions' energy consumption has increased since the 1960s, and their relative proportion has decreased considerably. The worldwide energy usage has increased, most notably in the Asia Pacific, where over this span, the total energy use has expanded more than 12 times (Artuc et al. 2019).

Two ways can equate energy use by a source, i.e., primarily energy usage and corrected energy use for fossil fuel conversion inefficiencies. There is also a forward process that indicates their crude form, also referred to as primary energy consumption (Ou et al., 2011). This form of consumption is about how much coal, oil, and gas power are being used as inputs to the economy's network of energy. However, when converted to final animation, the inefficiency of fossil fuels does not take into account this strategy. Therefore, those values should be displayed, which are corrected using the substitution process—which better estimates the final energy demand and considered as a better way of measuring the share in different sources of energy.

While the advent of energy is associated with coal consumption, the population should shift to the substitute for modern fuels by taking a lasting view of human energy system evolution (Gale et al., 2007). The energy dimension also covers the power between humans and animals. The use of petroleum products are eliminating human and animal labor in high-income countries. The human labor force's contribution, particularly in agriculture and manufacturing, is still essential in some countries with low to medium incomes, but the variations in composition are continuing to progress.

Worldwide there is increasing recognition of the value of prioritizing energy productivity to improve national productivity. Over 80 developing economies have agreed to the United Nations' 'Sustainable Energy for All' program, to enhance

Table 1 National energy productivity-related targets

Nation	Target	Baseline	Target Year
USA	Double energy productivity by 2030	2010	2030
Australia	Forty percent more energy production upto year 2030	2015	2030
Indonesia	One percent energy intensity reduction/annum	2005	2025
Singapore	Aims to attain a 35 percent improvement	2005	2030
Japan	Thirty percent energy efficiency improvement	2003	2030
South Korea	Forty six percent energy intensity reduction	2007	2030
EU	Energy efficiency to conserve 27 percent of energy consumption	2012	2030

Source Sustainable Energy for All, http://www.se4all.org/our-vision/our-objectives/

energy efficiency that is one of the goals of United Nations. Sample of National Energy Productivity-Related Targets is given in Table 1.

Although the energy growth of emerging economies is increasing globally, its trend is declining dramatically for many high-income nations; the growing divergence of the twentieth century has already started to be replaced by an ascending convergence in many developed countries. Substantial regional differences in the gross domestic product (GDP) per capita are present and persistent within technologically progressive countries.

The allocation of energy capital may have a considerable effect on worldwide energy markets. Domestic energy consumption is also an essential factor in oil trading. There would be little energy left to export if a nation rich in resources have its high local uses. Similarly, a country with low energy demand can still be a net energy exporter even though its natural resources are comparatively small. Impacts on the trading in the energy sector may be geopolitical: individual nations, for example, may prefer to share fuel resources to preserve energy security rates.

2 Energy Security

Energy is a vital tool for developing goods and services and improving the individual, social, and economic well-being of all economies (Vera & Langlois, 2007). Different industries need this input to manufacture goods and services, transport, and households need energy for heating and cooling their houses, and power the domestic devices. Empirical studies indicate that energy use is well connected to indices of economic growth and well-being, which is why energy supplies are defined as a standard for measuring progress toward sustainable development for most people in the world. Nevertheless, both developed and developing economies face many energy challenges, including insufficient and inadequate energy sources, energy market uncertainty, high and unstable energy prices, and environmental pollution, as well as degradation arising from energy production, distribution, and use. For nations that rest on energy imports from politically unstable regions, the danger of energy supply disturbances and the destabilizing impact pose additional risks.

Food intake is adversely impacted by fuel shortages, which prohibit people from cooking food or disinfecting water by boiling. High prices and a shortage of energy force many households to struggle to protect their cooking fuel and adopt coping strategies that leads to adverse effects on nutrition, such as the trade in food for fuel and skip or undercook meals. This lack of energy leads to the improper care of crops, animals, and the transport of farm products after harvesting.

However, the literature on energy protection has somewhat specific and often contradictory outcome. It has been partially because different authors concentrated on multiple risk sources and carried out research that varies in terms of the impacts of the varied risks. Energy security is characterized as the continued accessibility of energy sources at reasonable prices in the international energy agency, developed in the 1970s to organize a comprehensive response to oil supplies disruption. Likewise, energy insecurity is the loss of welfare that may occur due to a variation in the price of energy (Bohi & Toman, 1993). Energy insecurity is also characterized as a damage of well-being due to a variation in energy prices or availability.

Energy security means assuring suitable, consistent supplies of energy at judicious rates and in ways that do not endanger significant national values and objectives (Andrews, 2005; Jun et al., 2009). Energy security requires maintaining sufficient and secure energy resources at fair costs and not jeopardizing essential national principles and priorities (Mujiyanto & Tiess, 2013). Our concept, therefore, includes three key aspects: physical energy protection, which is access to energy supply; economic energy security that is the sustainable procurement of resources and construction of energy infrastructure; and environmental sustainability, including the use of energy resources in ways that encounter today's requirements without becoming concerned with the development of energy infrastructure.

Energy reserves denote the approximate sum of known and acceptable certainty of energy sources extracted at an economically viable cost with the present technology. A significant determinant of energy security is the nation's energy resources and the degree to which it is established. Countries producing both renewable and non-renewable energy are usually more protected than those relying only on fossil-based energy supplies. Over time, however, improvements alter the energy supplies of a country and thus boost or deteriorate its energy security.

Energy security is a task many economies seek to ensure that their markets function continuously and their populations have contact to modern and renewable energy that is sufficient, reliable, and affordable. It is an urgent concern as demand for energy has proliferated because of economic development, population growth, new energy consumption, and employment. However, energy resources are scarce and, in most cases, non-renewable in terms of energy supplies required to meet these requirements (Guertin et al., 2003). Besides, energy production, transport, and usage are significant sources of GHG that ground global warming. This chapter studies the multidimensional nature of energy security and offers policy measures to improve energy security.

A vibrant conceptualization of future goals is necessary to achieve sustainable energy security. Energy security has multiple dimensions, and risk is one of them. Several studies have concentrated on sources and impacts of risk. Energy security risks that developing nations face are decreasing over time due to better policies. The concept of energy security is quantified by a set of variables; each depicts a specific aspect of the settled definition. This set's arrangement is vital—it should address all the topographies, and it should not include indicators that do not add to the coherent understanding of energy security. The indicator set, taken as a whole, should represent the measurement of energy security and not something else. Risk-related indicators like energy supply resources, market dynamics, energy supply vulnerability, and technological development are estimated by using available data (Mitchell & Watson, 2013).

A huge fraction of the population in many developing nations does not have access to modern energy because of high prices (Reddy et al., 2009). For example, in Kenya, most people use kerosene for cooking and lighting, and kerosene's price has recently increased rapidly, which has reduced its use and switched households to other fuel oil. While liquefied natural gas is available, its high cost limits its use to households with medium and high-income groups. Higher energy prices can be predicted in this scenario due to the rising deregulation of energy markets and reducing energy subsidies. Import requirements and export earnings are more robust food security indicators than primary demand or supply volatility or stock and flow patterns in global cereal markets. Short-term demand and supply price elasticity differentiate fuel and food commodities markets. Therefore, the prices of the market can be volatile and respond vigorously to current or future demand and supply events.

3 Impacts of Energy Price Shocks on the Economies

Since the year 2003 the prices of natural gas, coal, and electricity have steadily risen and price of crude oil has doubled in the USA (Peterson, 2006). Consequently, a rise has occurred in crude oil prices, and this increase in the oil price rise will continue. Economic growth is primarily driven by the increase of oil demand and not a provisional supply decline, and according to market analysts, oil prices will slightly go down over the next ten years on average. The spike in energy prices influences household income and expenditure, and households use their savings to dampen inflation's adverse effect in the short term. The synchronized association among the economy, energy, and environment is revealed in Fig. 1. The environment is the basis of economic development. The environment, as the material state of economic and social development, can both endorse and deter the development of economy.

The mechanism of pricing works efficiently as a policy tool when it is correctly devised and used (Boardman et al., 2017; Dales, 2002). However, if households attempt to restore their savings, their spending could marginally decrease. In particular, energy prices have less impact on firms than on households. Non-energy firms have raised their earnings less quickly than in the absence of a rise in energy costs, but

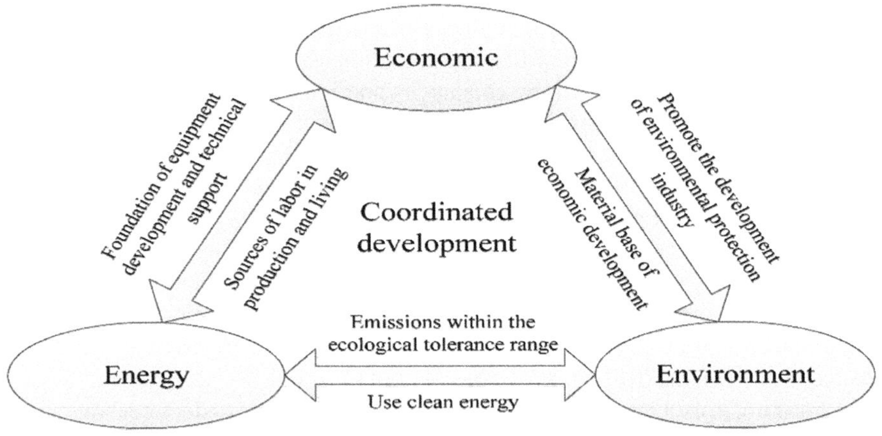

Fig. 1 Relationship among the economy, energy, and environment. *Source* Zuo et al. (2017)

entrepreneurs' income as a whole has risen. Most of this income rise is the continuation of past patterns and gradual decline in the depreciation. The oil-producing firms' income has, of course, risen sharply (Romero, 2008). The majority of industries transfer higher energy costs to their consumers. The level of investment spending on infrastructure and capital has elevated since 2003, and the same trend is prevailing in recent years due to the continuous rise in energy prices (Kesicki, 2010).

4 Doctrines of Exhaustible Resource Economics

Harold Hotelling introduces the term "The economics of exhaustible resources", which carried the topic into contemporary economics; and Solow (1974) raised the question of social interest in the pace of exploitation of the world's endowment of exhaustible resources. Hotelling (1931) did not only study the optimum use of non-renewable resources from a micro-economic point of view but also tried to maximize "the social value of the mine" derived from an exhaustible resource. The concept of resource includes a logical maintenance structure and has played an essential role in fostering the exhaustion myth. Resources may or may not be a tangible entity; it does not represent a list of assets, given the fact that the use of products covers all business ventures. Savers' acts and entrepreneurs' policies primarily generate capital goods. They represent the assets that the entrepreneur has marshaled in pursuing a plan (Vinturella & Erickson, 2003).

The hypothesis of the economic exhaustion of mineral resources is not valid as all extractive resources would be depleted in the long run. This situation has influenced analytical thinking about sustainable public policy and contributed to policy debates about government subsidies for renewable fuels (Kraft and Furlong, 2019). The economics subject has massively perpetuated this theory: first, it continually concentrates on problems related to the global exhaustion of resources; second, it has embraced a traditional but erroneous definition of resources as a physical grouping of produced durable goods.

5 Role of Trade for Food and Energy Secure Future

Managing non-renewable energy reserves and attaining food security are among the most significant challenges for nature conservation, as a balance needs to be struck between different allocations of resources. Food security continues to be a high priority for growth and global interest and it is part of the Sustainable Development Agenda 2030. Food security is also a core component of the human development and capability paradigm, since food access and entitlements are critical for reinforcing essential human capabilities (Lawn, 2003). We are also in a world of economic and social multilevel governance where exhaustible resources are being negotiated in their aptitude to withstand the routines.

In this context, the concept of the green economy, i.e., renewable raw materials, application of research, development, innovation, and industrial biotechnology in sectors such as food, feed, paper and pulp, and biofuels production, is the best option for a sustainable future. Compared to the environmental weight of the green economy, the bio or green economy focuses on new growth openings in traditional and emerging bio-based sectors while allowing for worldwide challenges (e.g., raw material supply insecurity) and resource and environmental restraints (Page, 2018).

6 Role of Environmental Protocols in Reshaping the Eco-Friendly Energy Industry

Economic growth is forecasted to grow mostly from its short-term losses, while energy price fluctuations are likely to keep the GDP level below for the next ten years in developed nations. The standard of living will significantly go down than GDP, as more income would be required to purchase a barrel of oil than if energy cost has not risen, and unfavorable shifts in terms of trade would influence the living standards for the coming years.

The sense of the term eco-friendly is understood to us all. Earth has become a global village with a wide range of pollution, disaster, and diseases. In such chaos, the world and its living things must be preserved. As inhabitants of Mother Earth, we serve a responsibility to safeguard and protect its climate from these incidents.

According to James Smith,[1] environmental sustainability is the need of time, and there are many reasons as follows:

---

[1] https://energycentral.com/c/ec/importance-eco-friendly-environment.

- Environmental problems are one of the significant challenges faced by society today. It can be connected with cutting trees down, limiting supplies, and natural resources exploitation, etc.
- Erosion is another issue and can happen due to several factors and causes much damage to the Earth's environment. It can be instigated due to over-development of land for buildings or homes, abnormal meteorological events, severe weather conditions, etc.
- The problem of pollution is growing over time and creating alarming situations for us all. Smoke from vehicles and industries is the primary source of pollution (Francis et al., 2012). There are three types of pollution, namely, water, air, and soil, and our globe is, unfortunately, the victim of all three. Compared to the above issues, the world must go eco-friendly to go back to its original state. Most industries do not recycle hazardous and waste materials and drop in rivers that is destroying the water's quality and marine life.

Human activities are responsible for the damages in the ozone layer. Promoting self-sustainability and self-sufficiency practices is necessary for the safe future of the world. We have to keep the species that are becoming endangered or extinct due to global warming and the worse changes that the Earth faces every passing day. The goal should be to foster a world that is greener and healthier. Besides, the goods that require renewable energy, like hybrid vehicles, fuel-less generators, solar power generators, and eco-friendly air conditioners, should be promoted to support the very idea of an eco-friendly and sustainable environment. The use of eco-friendly goods enhances efficiency, productivity, and effectiveness and encourages the conservation of the green environment for a lasting change in the following ways:

- Purchasing an eco-friendly product will indeed cost less than buying a normal one. In this way, it will help in utilizing money effectively;
- Eco-friendly products, for that matter, guarantee a healthier lifestyle, advance the environment, and offer several benefits;
- Sustainability is the core element of eco-friendly development. When the environment has been polluted and products are toxic, it can be the right sustainability call. The concept of 'sustainability' includes the environmental, social, and economic aspects;
- Conservation and preservation of the environment increase life quality in terms of mortality, age, and diseases. If one wants to go environmental-friendly, he could get a better shot at living a healthy life with eco-friendly options;
- Financial and economic development is another aspect closely linked to a sustainable environment. It is believed that building, ideas, creativity, and strategies lead to growth and ultimately development, but all goals should be transformed and aligned with the climate (Blakely & Leigh, 2013). Economic development helps not only the nations but also the environment. Governments have to deviate from the traditional mindset of greed and exploitation of natural resources and pursue a better and healthier future to support the environmental sustainability indicators like energy and water consumption, air pollution, and solid and hazardous waste produced to protect Mother Nature. When we take Green as an individuals'

initiative, the areas related to green energy can have a stronger pull for growth. Recycling is also an outstanding practice by which the resource use efficiency can be enhanced and waste can be avoided. Recycling also ensures effective, sustainable, and judicious use of natural resources;

- To put it simply and in a practical situation: millions of tons of materials are being wasted daily, industries release toxic products into the ocean every day, which makes the quality of life harder for marine life, vehicles and industries emit smoke, and thousands of trees are caught up in it. Nature is being exploited without bearing in mind what will possibly happen in the future. Nonetheless, it can still be saved by shifting to alternate courses and using the resources wisely.

7 Strategies for Conserving Augmented Efficiency and Alternative Sources

The growing industrialization is assumed as the prioritized goal for national development. The efficient usage of energy plays a vital role in the sustainable development of industries. Furthermore, the effective execution of energy efficiency technology in the industrial infrastructure and manufacturing chain is a part of the smart emission of GHG. Still, the cost-effectiveness of any plan is always a restriction for its adoption and implementation. GHG emissions in the form of carbon dioxide from mining and the consumption of fossil energy sources cause climate variations (Chevallier, 2011). Climate impacts already have their effects on every sector of life. Conventional sources of energy, such as coal and natural gas, contribute significantly to climate change. More than one-third of global warming emissions are accounted for by energy production; most of these are emitted by coal-fired power plants (Bin Hitam, 1999).

Traditional energy that can also be called dirty energy comes from fossil fuels, and it is often produced from coal, oil, and natural gas. Dependence on traditional energy exposes consumers to price variations and harms human health and the environment. Renewable energy provides substantial benefits for the climate, health, and economy. Confidence in conventional energy sources exposes consumers to price volatility and threatens the environment and human health. Agriculture, forestry, and land use account for 18.4%[2] of total GHG emissions. Food, transport, and direct industrial processes are responsible for approximately 26%, 16.2%, and 5.2% of global GHG emissions (Poore & Nemecek, 2018). Renewable energy delivers significant benefits to climate, human health, food security, and the economy. Many sources of renewable energy do not pollute the environment through gas emissions. While on the other hand, geothermal energy systems emit air contaminants, and overall, these air emissions are much lower than those of coal and gas power plants are. An emphasis on improving both end-use energy efficiency and investment to boost energy productivity can contribute toward global greenhouse gas emission mitigation in the energy

[2] https://ourworldindata.org/co2-and-other-greenhouse-gas-emissions.

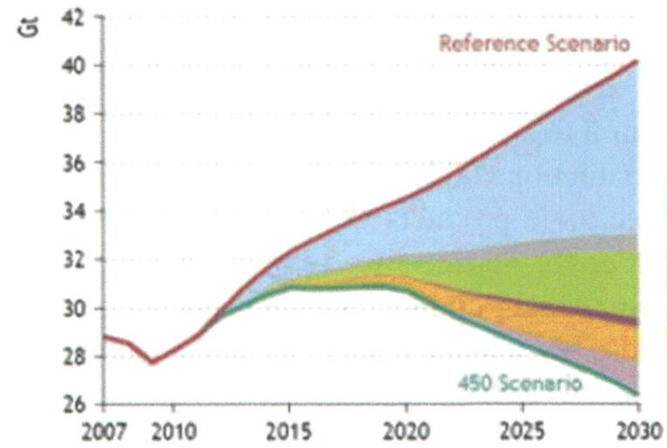

	Abatement (Mt CO_2)		Investment ($2008 billion)	
	2020	2030	2010-2020	2021-2030
Efficiency	2 517	7 880	1 999	5 586
End-use	2 284	7 145	1 933	5 551
Power plants	233	735	66	35
Renewables	680	2 741	527	2 260
Biofuels	57	429	27	378
Nuclear	493	1 380	125	491
CCS	102	1 410	56	646

Fig. 2 World energy sector related CO2 emissions abatement options. *Source* http://www.iea.org/ (Roadmap, A.R.E. 2014 R Emap 2030)

sector by 2030. Energy sector related mitigation options opted by different nations are given in Fig. 2.

However, wind and solar are clean energy sources and do not need water to work. Agriculture and residential users who are getting energy from wind and solar sources do not pollute water supplies (Liang & Liao, 2007). On the contrary, the effect of fossil fuels on water supplies can be substantial. Both coal mining and the extraction of natural gas pollute drinking water supplies. Each year, a significant proportion of total energy production is wasted by transmission, lack of heat, and inefficient technical investment for households and industries, resulting in enlarged carbon emissions and climate change. Enhancement in energy efficiency is the easiest way to tackle climate change, clean up the environment, and save customers and entrepreneurs capital in cost-effective ways.

Energy efficiency means using less energy to perform the same task while reducing the energy waste, and it can be attained with energy-efficient lighting, cooling and heating systems, and ventilation. Better design of houses, landscaping, and other items such as the shaded trees, modifying energy use behavior like switching off the lights when leaving the room can also help in this regard. Such systems may use, for instance, feedback in real-time or automatic controls. Implementing local, state, or federal policies and programs can significantly increase energy efficiency.

The regulations and the system choices include building codes for goods sold or used within the jurisdiction and other requirements that set minimum energy efficiency, for example, requiring light-emitting diode bulbs, using minimum efficiency standards for new homes, the inclusion of energy quality requirements in sustainable growth policies. Several local government agencies have adopted sustainability strategies, and some are incorporating energy conservation programs to minimize consumption levels. Community-based administrations and governments implemented programs that incentivize energy efficiency measures for residential or business consumers to educate consumers about energy efficiency. Companies, governments, or administrations are launching energy efficiency policies or developing internal energy efficiency objectives.

8 Doctrines of Markets and Price Risk Administration

The correlation between the spot and the futures price is a widely studied topic, and it can help investors determine the price they need to consider during decision-making. It can also discover any possible arbitration between location and future rates. The amount of petroleum prices is equally crucial. If oil prices are too high, oil-importing countries' growth rate will fall as energy cost puts downward pressure on industries with their main cost drivers for oil products (Heun & de Wit, 2012).

Energy markets are much more volatile than other commodity markets, so risk mitigation is also a paramount concern. Energy prices, for example, can be affected by weather, turmoil, changes in tax and legal systems, OPEC decisions, transportation issues, and supply and demand. If oil prices are too great, changes in oil markets and higher energy markets tend to reduce economic activity. Literature also argues that current and future supply and demand rates are vital because they affect economic development, both in oil exports and imports. Multiple economic and non-economic factors such as shocks to supply and demand, local and global events, and geopolitical challenges also influence crude oil markets (Stimson et al., 2006). Both spot and futures prices hit by the above-stated events cause abnormal increases and falls in the general price levels.

The nature and relationship between spot and future oil prices pose concerns between these two prices. The future forecasts are higher than current estimates for crude oil (Alquist & Kilian, 2010). It is claimed that the future of crude oil has been traded around the world since 1983, making it the biggest and most active potential market as it leads to global development and international financial stability (Helleiner, 2010). This fact reflects the low transaction costs and the extensive use of short selling mechanisms that lead to a rapid reaction of the future market to new knowledge. This new knowledge causes an increase in the quality of information flowing to the spot market and represents rapid shifts in current prices due to the future's markets. There is no evidence that this price dominates the other, and both forecasts are equally important and sensitive to external factors. Researchers take this fact to define the long- and short-term relationship of crude oil current to future prices.

9 Natural Resources for Sustainable and Food-Secure Future

Over recent decades, food security has been characterized by substantial growth in production and trade increase (Schmidhuber 2007). Production and better terms of trade have met many' dietary needs but by no means all of the global population. The historical background shows a relative increase in the world's population, agriculture, crop field, agricultural water diversion, and farm nitrogen use during the last five decades.

Throughout the value chain, there is a loss, commonly known as food waste. In developed countries, production losses are low, where storage and transportation facilities are more, while in developing countries, production losses are more and storage facilities are low (Bond 2013). Altogether, about one-third of the food harvested is taken away due to food loss and waste. This means that the caloric consumption per capita calculated in developed nations is slightly higher than the actual caloric intake, where food waste has to be deducted from consumers' purchases. It also infers that up to one-third more people could be fed with prevailing production if food losses are expressively reduced. Nevertheless, it is unclear if infrastructure investment to minimize producer food loss is the most cost-effective way to content growing demand, particularly in developing regions.

Worldwide, water resources will be sufficient to produce the food required in 2050, but many parts of the world will face a significant water shortage. Water scarcity will cause more competition, constrain farm productivities, and influence rural and urban communities' incomes and livelihood opportunities. The agricultural sector distributes nitrogen and phosphorous from a few localized mineral sources to soil and water throughout the planet, making it comparatively challenging to recycle (Vaughan et al., 2007). Though mineral reserves are uncertain, some researchers consider that mineral quality's origins could permanently disappear before the end of the twenty-first century. Conversely, it is possible that agricultural demand can still be met if application efficiencies are upgraded and prevailing technologies for recovering and reusing gradually replace mining. The final major crop nutrient contained in fertilizer is potassium, which is obtained from extensive mineral reserves. Most of the researchers believe that these reserves are sufficient for the future.

The debate indicates that by the mid-century, when the world population is projected to rise, there will be ample water, land, and nutrients to satisfy the global food demand. The reserves mentioned above will, in turn, sustain a rise in global water, nutrients, and food production that can readily be moved from place to place. Such free movement, especially for water, is often not feasible. Also, if global supply is above the demand point, food scarcity may be a concern because resources cannot fulfill local food requirements.

In particular, for food production (e.g., agriculture, forests, and aquacultures), both renewable and non-renewable natural resources are essential and consumed to a certain degree in all the food system activities. Sustainable natural resource management is, therefore, crucial for food security. In many cases, however, food system activities considerably deteriorate the natural resources they hang on while contributing to climate changes due to local and regional pollution. Furthermore, population growth and development are exerting pressure on the available stocks of scarce natural resources. If we want to confirm that all people have safe and nutritious food in suitable amounts, we need to be conscious of their stocks used in food system activities to accomplish them efficiently and lessen environmental influences.

10 The Price of a Clean Environment

The environment also affects human and animal health. The effects of air pollution carry essential financial and economic implications. These implications include the societal cost of mortality and morbidity and the household, hospital, public budgets, and decision-making. Additionally, the deleterious influence of air pollution is not confined to human health. Several other influences are worthy of consideration, like the built environment, broader ecological systems, animal and plant health, and other consequential effects on agricultural and forestry resources' productivity. In this viewpoint, air pollution can have essential co-benefits for different policy aims, and air quality may simultaneously be advantageous from interventions that pursue other priorities.

Pricing and investment play a vital role, and the question of pricing is ahead of investment, as the misrepresentation of the former will give the latter a wrong answer and probably a high price to the welfare of future generations. The composition, location, size, and timing of the investment required to meet potential environmental demand would probably be changed if the prices were increased. Over the past 20 years, the world has made a significant and enduring commitment to protect the environment. Overall, the expenses on environmental harms are rising significantly with apparent values for governments' costs at all levels and of industry. Still, if the upward trend continues into the next century, this augmented spending could influence the global markets competitiveness. Besides the level of expenditure, the allocation of resources is changing. The share of costs devoted to land protection is projected to rise relative to air and water protection over the next decade (Watts 2015). Further, the prices of pollution control are growing when unmet environmental needs are still quite large, and people want more environmental improvements and an explicit policy in this regard.

11 Energy Linkage with Food Security

Globally, about two and half billion people lack access to cooking and heating to renewable energy and instead rely on biomass for their home-based energy needs. The achievement of food safety, economic development, and poverty decline remain elusive targets without access to energy (Brew-Hammond 2012). Energy and food security are related to market volatility. Since oil price inflation is harmful to food security, energy use in this sector should be diversified, from excessive dependence on fossil fuels to an appropriate mix of renewable and non-renewable energy resources to promote energy and food health.

Bioenergy production interrelates with food and forestry production in composite ways. It can strive for land, water, and other resources and strengthen conventional food and forestry production by offering new markets for biomass flows that earlier were deliberated waste products. Bioenergy demand can fetch opportunities for cultivating new types of crops and integrating bioenergy production with food and forestry production in ways that improve overall resource management. It can also lead to overexploitation and degradation of resources. Bioenergy development eventually depends on bioenergy products' priority versus other products obtained from land and on how much biomass can be mobilized in total from the farming sector. According to World Business Council for Sustainable Development (WBCSD) greenhouse gas reductions need to be achieved while global energy demand is likely to rise by 40% by 2030. The nexus perspective for developing countries identifies a number of major risks and challenges to a sustainable future (Fig. 3) and these all issues require the attention of the policymakers.

As it is clear from the Fig. 3 petroleum products and derivatives (energy demands) are considered as the significant determinants of economic development for an economy (Kalbasi et al., 2010). Energy is widely used in the manufacturing of petrochemical products to supply various industries, including transportation, agriculture, manufacturing, homes, and as the raw material. This is why energy has excellent value and has an effect on other commodity prices. Since the first oil price shock of 1973, investigating the impacts of energy prices, specifically on macro- and microeconomic levels, has become one of the essential energy economics subjects. Wide ranges of scientists have tried to identify possible causes of these variations in energy prices and their allied impacts on food security in different nations.

Many studies have found that higher oil prices could account for decline in the growth of GDP and a rise in the inflation rate and food security ultimately (Charfeddine et al. 2020; Bergmann 2019). Crude oil price changes impact the GDP and

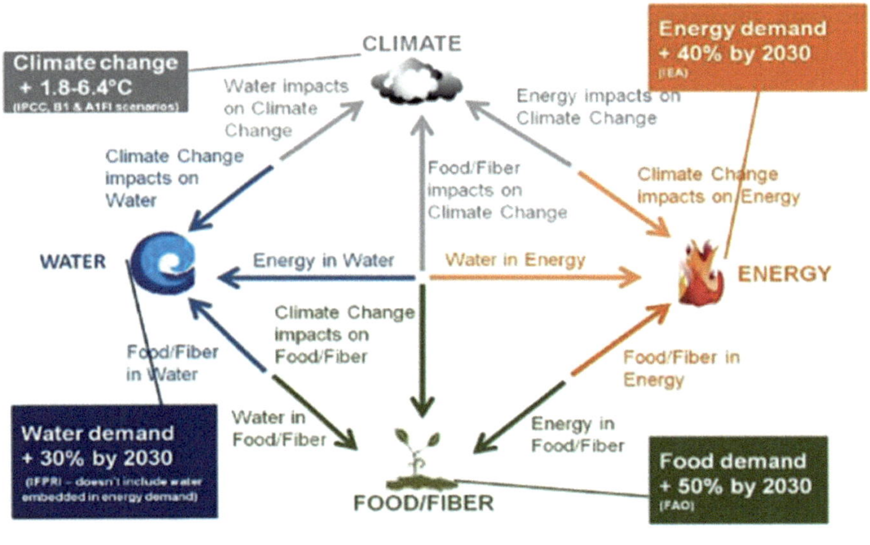

Fig. 3 The Climate Change, energy, water and food production nexus. *Source* http://www.wbcsd.org/work-program/sector-projects/water/waterenergyfood.aspx

consumer price index for both developed and developing economies. Further, this effect is far less on oil-importing countries' petroleum price fluctuations than the emerging economies. Energy protection literature proposes the increase in energy safety by diversifying energy supplies, making fossil fuel imports more flexible, and raising self-sufficiency through the growth of renewable energy projects. Regarding the effect of oil prices on other commodity prices, various studies are being conducted to determine the impact of oil and different energy prices on agricultural commodities.

12 Conclusions and Recommendations

Energy occupies an important place and is considered a cornerstone and vital to realize economic development and combat poverty. Modern world started witnessing developmental by taping and using various energy sources wisely. At the global level, some two and half billion people could not have renewable energy for cooking and heating and had to rely on biomass to meet their home-energy needs. Energy consumption levels and sources (except nuclear and solar) negatively impact the environment and leave a carbon footprint at the places wherever they are being used. For example, industrial development, substantial improvements in the economic sectors, the vast population growth, have necessitated the use of an increased production and consumption of carbon-linked resources causing an imbalance in the relationship between economic growth and its linkage to environmental quality. Therefore, it would be appropriate to revisit use of use of energy resources, reexamine the policies related to governing energy management and economic growth, and explore the new, clean, and green technologies to reduce the environmental impacts resulting from energy consumption. Devoted efforts are needed to enhance energy efficiency in all the applicable sectors to realize higher outputs by applying the same energy input levels. The analyses presented in the chapter indicate that energy efficiency improvement programs are very important particularly for developing countries by improving the productivity of the systems at the same time to witness economic growth and mitigate GHGs emissions. Furthermore, developing countries must explore and identify new renewable sources to bring them into the energy mix. Many potential renewable energy resources do exist like nuclear, hydro, wind, and solar. It is anticipated if harnessed appropriately, they can contribute toward economic growth and reduce carbon footprint as well. However, without creating awareness among the consumers, very technical sound strategies would be of no use. There is a need to launch a public awareness campaign, to spread the knowledge about the usability of renewable energy and anthropogenic implications of conventional energy. This objective can be achieved by designing suitable and proper training and awareness programs for the general public, civil society, and the organizations working to bring behavioral change in society. Media campaigns can make the general public realize that by using the economic use of energy and its proper mix can make the world a better place to live in for the present masses as well the future generations. In order to meet the rising need for energy, scientists, researchers, and the academia are continuously struggling to look for energy resources that are economical, socially acceptable, and don't damage the environment. However, it would not be possible unless all the stakeholders play their due role and accept their responsibilities.

References

Alquist, R., & Kilian, L. (2010). What do we learn from the price of crude oil futures? *Journal of Applied Econometrics, 25*(4), 539–573.

Artuc, E., Lopez-Acevedo, G., Robertson, R., & Samaan, D. (2019). *Exports to jobs: boosting the gains from trade in South Asia*. The World Bank.

Andrews, C. J. (2005). Energy security as a rationale for governmental action. *IEEE Technology and Society Magazine, 24*(2), 16–25.

Bin Hitam, S. (1999). Sustainable energy policy and strategies: a pre-requisite for the concerted development and promotion of the renewable energy in Malaysia.

Blakely, E. J., & Leigh, N. G. (2013). *Planning local economic development*. Sage.

Bohi, D. R., & Toman, M. A. (1993). Energy security: Externalities and policies. *Energy Policy, 21*(11), 1093–1109.

Bond, M., Meacham, T., Bhunnoo, R., & Benton, T. (2013). *Food waste within global food systems*. Swindon, U.K.: Global Food Security.

Boardman, A. E., Greenberg, D. H., Vining, A. R., & Weimer, D. L. (2017). *Cost-benefit analysis: Concepts and practice*. Cambridge University Press.

Chevallier, R. (2011). Towards the climate change Summit in Durban: The role of the South African government as an agent of change in Africa. seminar, African Agency: Transnational security challenges: migration health and climate change, University of Kent June 16th.

Dales, J. H. (2002). *Pollution, property & prices: an essay in policy-making and economics*. Edward Elgar Publishing.

Francis, C. D., Kleist, N. J., Ortega, C. P., & Cruz, A. (2012). Noise pollution alters ecological services: Enhanced pollination and disrupted seed dispersal. *Proceedings of the Royal Society b: Biological Sciences, 279*(1739), 2727–2735.

Gales, B., Kander, A., Malanima, P., & Rubio, M. (2007). North versus South: Energy transition and energy intensity in Europe over 200 years. *European Review of Economic History, 11*(2), 219–253.

Guertin, C., Kumbhakar, S. C., & Duraiappah, A. K. (2003). *Determining demand for energy services: investigating income-driven behaviours* (p. 50). International Institute for Sustainable Development.

Helleiner, E. (2010). A Bretton Woods moment? The 2007–2008 crisis and the future of global finance. *International Affairs, 86*(3), 619–636.

Heun, M. K., & de Wit, M. (2012). Energy return on (energy) invested (EROI), oil prices, and energy transitions. *Energy Policy, 40*, 147–158.

Hotelling, H. (1931). The economics of exhaustible resources. *Journal of Political Economy, 39*(2), 137–175.

Jun, E., Kim, W., & Chang, S. H. (2009). The analysis of security cost for different energy sources. *Applied Energy, 86*(10), 1894–1901.

Kesicki, F. (2010). The third oil price surge–What's different this time? *Energy Policy, 38*(3), 1596–1606.

Kraft, M. E., & Furlong, S. R. (2019). *Public policy: Politics, analysis, and alternatives*. Cq Press.

Lawn, P. (2003). How important is natural capital in terms of sustaining real output? Revisiting the natural capital/human-made capital substitutability debate. *International Journal of Global Environmental Issues, 3*(4), 418–435.

Liang, R.-H., & Liao, J.-H. (2007). A fuzzy-optimization approach for generation scheduling with wind and solar energy systems. *IEEE Transactions on Power Systems, 22*(4), 1665–1674.
Mitchell, C., & Watson, J. (2013). *New Challenges in Energy Security: The UK in a Multipolar World—Conclusions and Recommendations. In New Challenges in Energy Security* (pp. 238–257). Palgrave Macmillan, London.
Mujiyanto, S., & Tiess, G. (2013). Secure energy supply in 2025: Indonesia's need for an energy policy strategy. *Energy Policy, 61*, 31–41.
Ollila, M. (2007). *Department of Industrial Engineering and Management Environmental and Quality Management Unit*. Helsinki University of Technology.
Ou, X., Xiaoyu, Y., & Zhang, X. (2011). Life-cycle energy consumption and greenhouse gas emissions for electricity generation and supply in China. *Applied Energy, 88*(1), 289–297.
Page, K. N. (2018). *Fossils, heritage and conservation: managing demands on a precious resource.* In Geoheritage (pp. 107–128). Elsevier.
Peterson, J. (2006). *Economic Effects of Recent Increases in Energy Prices*. Government Printing Office.
Poore, J., & Nemecek, T. (2018). Reducing food's environmental impacts through producers and consumers. *Science, 360*(6392), 987–992.
Reddy, B. S., Balachandra, P., & Nathan, H. S. K. (2009). Universalization of access to modern energy services in Indian households—Economic and policy analysis. *Energy Policy, 37*(11), 4645–4657.
Romero, R. E. (2008). *Monetary policy in oil-producing economies*. Citeseer.
Schmidhuber, J., & Tubiello, F. N. (2007). Global food security under climate change. *Proceedings of the National Academy of Sciences, 104*(50), 19703–19708.
Stimson, R. J., Stough, R. R., & Roberts, B. H. (2006). *Regional economic development: analysis and planning strategy*. Springer Science & Business Media.
Solow, R. M. (1974). *The economics of resources or the resources of economics*. In Classic papers in natural resource economics (pp. 257–276). Palgrave Macmillan, London.
Vaughan, R. E., Needelman, B. A., Kleinman, P. J., & Allen, A. L. (2007). Vertical distribution of phosphorus in agricultural drainage ditch soils. *Journal of Environmental Quality, 36*(6), 1895–1903.
Vera, I., & Langlois, L. (2007). Energy indicators for sustainable development. *Energy, 32*(6), 875–882.
Vinturella, J. B., & Erickson, S. M. (2003). *Raising entrepreneurial capital*. Elsevier.
Watts, N., Adger, W. N., Agnolucci, P., Blackstock, J., Byass, P., Cai, W., & Cox, P. M. (2015). Health and climate change: Policy responses to protect public health. *The Lancet, 386*(10006), 1861–1914.
Zhang, G. J., Cai, M., & Hu, A. (2013). Energy consumption and the unexplained winter warming over northern Asia and North America. *Nature Climate Change, 3*(5), 466–470.
Zuo, Y., Shi, Y. L., & Zhang, Y. Z. (2017). Research on the sustainable development of an economic-energy-environment (3E) system based on system dynamics (SD): A case study of the Beijing-Tianjin-Hebei Region in China. *Sustainability, 9*(10), 1727.

Climate Change and Innovative Ecological Interventions Through Climate-Smart Agriculture in the Cholistan Desert of Pakistan

Rahmatullah Qureshi, Mirza B. Baig, and Azaiez Ouled Belgacem

Abstract Cholistan Desert is one of the largest deserts of the Pakistan and part of the Great Indian Desert, located in the Sothern part of Punjab province. This desert is spread over 2.6 million hectares measuring 480 km in length and 32 and 192 km in breadth. Human and livestock population in the desert is 0.155 and 2.0 million, respectively. This is a hyper-arid desert that receives an annual mean rainfall ranging from 100–250 mm having temperature fluctuations of 6 to 50 °C. The main land use in the area is livestock rearing. Most of the area (81%) is under small and big sand dunes, while 19% consists of alluvial flats and sandy hummocks. The area is severely affected by desertification due to scanty vegetation cover, high wind velocity and soil salinity and overgrazing. Local communities are largely dependent on livestock keeping for earning their livelihood. The area is facing major issues such as extreme scarcity of water, drought, over-grazing, deforestation, wind erosion, salinity and encroachment for expanding agriculture that results in the decline of natural resources including plant biodiversity. The agriculture is highly vulnerable to temperature and drought in the Cholistan desert. Local inhabitants are chiefly exposed to the impact of climate change on the crop as well as livestock. The prospective upsurge of 2.8 °C in the maximum day temperature and 2.2 °C fall in night temperature by the midcentury is reported in Pakistan. The objective of this chapter is to introduce climate-smart interventions by introducing climate-resilient crops. Initially, it explains the impact of climate change on the current situation in addition to existing natural resources in the Cholistan desert and, then, latent climate-smart interventions to combat food security issues. Climate-smart strategies include reclaiming the desert lands into

R. Qureshi (✉)
Department of Botany, Pir Mehr Ali Shah Arid Agriculture University, Rawalpindi, Pakistan
e-mail: rahmatullahq@uaar.edu.pk

M. B. Baig
Water & Desert Research, Prince Sultan Institute for Environmental, King Saud University, Riyadh, Kingdom of Saudi Arabia
e-mail: mbbaig@ksu.edu.sa

A. O. Belgacem
FAO Office Riyadh, Riyadh, Kingdom of Saudi Arabia
e-mail: a.blegacem@cgiar.org

M. Behnassi et al. (eds.), *Climate-Smart and Resilient Food Systems and Security*,
https://doi.org/10.1007/978-3-031-65968-3_9

productive ecosystems by conserving existing natural resources such as lands, water, and biodiversity along with possible alternatives to improve the livelihood of local communities.

Keywords Cholistan · Hyper-arid desert · Desertification · Livelihood · Biodiversity · Climate change · Climate-smart agriculture · Productivity system

1 Introduction

Pakistan is a subtropical country with an area of 79.6 million hectares (mha) and lies between 24° to 37° N and 61° to 75° E. Approximately 70 mha (87.94%) of the area is arid to semiarid, which receives up to 250 mm annual rainfall (Abd El-Ghani et al., 2017). The deserts of Pakistan cover 11 mha (13.82%) of the land situated in the central and southeastern regions of the country (Abd El-Ghani et al., 2017; Akram & Abdullah, 1990).

The Cholistan Desert, locally known as *Rohi* is the 7th largest in the world and part of the Great Indian Desert, which is located along the south border of Punjab, Pakistan. This desert comprised on three districts namely Bahawalnagar, Bahawalpur, and Rahim Yar Khan. The word "Cholistan" is derived from the Turkish language; *Chol* means a desert (Anonymous, 2002) and *istan*, a Persian suffix that denotes "land of" (Anonymous, 2001). Others believe that it is originated from Iraqi word *Chilistan,* which means waterless wasteland (Ahmad, 2011). The word Cholistan is derived from *Cholna*, which means "moving". This desert is a cradle of the Hakra civilization, once it was a wealthy land, but is now a massive desolate barren land (Abd El-Ghani et al., 2017).

With rapidly increasing human population at about 2% rate in Pakistan, there is a need to increase production of food and fiber to meet growing needs. Desert can sustain the life of local communities for supporting the growing population and livestock if dealt with scientific interventions (Ahmad, 2008; Kahlown et al., 2004; Khan et al., 1996; WWAP, 2009). Water harvesting and its shortage for human and livestock consumption can be overcome the problem of water shortage. Besides, the developmental strategies such as desert aquaculture and cottage industry can enhance the livelihoods of desert dwellers. The desert lands can also be used for reclaiming and turning into green lands through saline agriculture (Adeel & Safriel, 2008).

1.1 Location

Cholistan Desert is located in the southern part of Punjab Province and Eastern Pakistan (Kahlown et al., 2004). It lies between 27° 42′ and 29° 45′ north latitudes and 69° 52′ and 75° 24′ east longitude (Baig et al., 1980) at 112 m above mean sea level (Fig. 1). It spreads over approximately 2.6 million hectares (Ahmad, 2008;

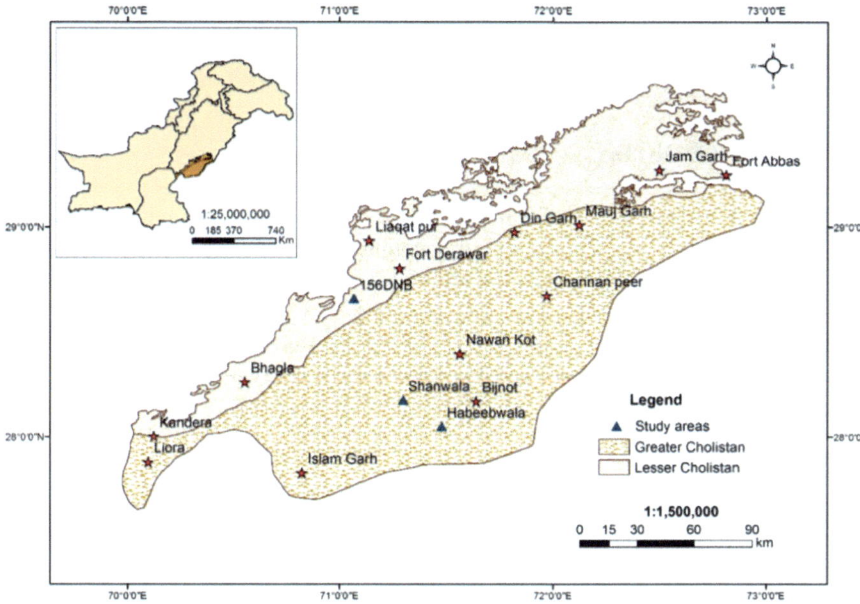

Fig. 1 Location map of Cholistan Desert, Pakistan

Kahlown et al., 2004; WWAP, 2009) with a length of about 480 km and 32 to 192 km breadth (FAO/ADB, 1993; Akbar & Arshad, 2000). It is bounded by the Sutluj River and agricultural lands of Bahawalpur Division in the Northwest; Sukkur district (Sindh) in the south; an international boundary line runs between India and Pakistan in the southeast (Akbar & Arshad, 2000).

The Cholistan desert is an arid land with extreme seasonal fluctuations and a wide variety of edaphic conditions. The human population in this desert consists of over 110,000 nomadic pastorals, which are originally Buddhist and Sikh religions, but now comprises 95% Muslims (Islamic religion) and 5% Hindu communities. The economy of this region is mainly pastoral and livestock keeping with a prominence of nomadic life. The people of Cholistan lead a semi-nomadic life, moving from one place to another in search of water and fodder for their animals. *Saraiki* is the spoken language of this region. More than 81% of the desert is under small and large sand dunes, and 19% consists of alluvial flats and sand hummocks (Akram et al., 2008).

Based on soil, topography and parent material, this desert is divided into two geomorphic regions by the old Hakra River, such as: (a) Lesser Cholistan in the northern region adjoined by canal irrigated areas that cover about 7,770 km^2; and (b) Greater Cholistan that contains 18,130 km^2 (Sial & Arshad, 2003). The Lesser Cholistan is occupied by saline alluvial flat plains (locally called *"Dahars"*) intermingled with homogenized low sandy ridges. These soils are saline or saline-sodic in nature with pH ranging from 8.2 to 8.4 and 8.8 to 9.6, respectively. The Greater

Cholistan is a wind directed sandy desert and comprised of river terraces, large sand dunes, ridges and depressions (Baig et al., 1980).

1.1.1 Lesser Cholistan

This is a northern part of Cholistan which comprised on approximately 7770 km^2 (Akhter & Arshad, 2006). This zone is adjacent to the canal irrigated area that establishes the desert margin, having saline alluvial flat plains alternating with low sandy ridges/dunes (Ahmad, 2008). The area is occupied by stable and somewhat less moving sand dunes that may attain to 30 m of height. Due to its location in low-lying areas, lands are saline, compact and hard and locally known as *Dahars* (Akhter & Arshad, 2006).

1.1.2 Greater Cholistan

This is a large portion of desert located in the south that covers an area of 18,130 Km^2, dominated by large sand dunes. These dunes are of large masses of sand formed by the winds, comprised of a number of old river terraces with large and a wide range of forms of sand ridges and valleys that may attain 100-m height (Akram et al., 2009; Akhter & Arshad, 2006). This region contains 80% large sand dunes and 20% flat plain of hard clay, locally named as *Dahars* (Ahmad et al., 1992; Akhter & Arshad, 2006).

1.2 Topography and Climate

Based on geomorphology, the topography of Cholistan Desert is divided into two regions: the lesser Cholistan covers about 7770 km^2 comprising large saline, hard, and compact areas, alternating with low sandy ridges in the northern region. Greater Cholistan covers an area of 18,130 km^2 in the southern region. The area consists of stabilized, semi-stabilized, or shifting dunes with undulating plains. The climate of the area is subtropical, arid, scorching hot, monsoonic, sporadic long droughts and high summer winds with low relative humidity and high rate of evapo-transpiration (Akram et al., 1986; Arshad et al., 2007).

1.3 Temperature

The mean summer temperature is ranged between 34 to 38 °C; whereas, in winter it varies from 15 to 20 °C. June is the hottest month in which temperature remains 45 °C and may rises up to 51 °C (Ahmad, 2005). Mean winter temperature varies from 14 to 16 °C falling below freezing in December and January. The frost may happen sporadically for a few days once or twice in December. The relative humidity fluctuates from 50 to 65%. Owing to low rainfall, high temperature and high evapo-transportation, this desert turned into extremely arid (Akhter & Arshad, 2006).

1.4 Rainfall

Aridity is the prominent feature of Cholistan with wet and dry years taking place in clusters. The annual rainfall is highly erratic both on temporal and spatial scales. Most of the rains are trapped in monsoon (July to September) with some occasional showers in winter and spring (January to March). This is an arid hot sandy desert that receives 100–250 mm annual rainfall (PCRWR, 2009; Akram et al., 2009) with an average of about 160 mm. The rainfall varies from 100 mm in the north to 200 mm in the south and may reach 250 mm along the Pakistan-India boarder. Most of the rainfall appears during July to September in the monsoon and January to March in winter (Kahlown et al., 2004).

1.5 Wind

Like the rest of deserts of Pakistan, winds drive between southwest and west in hot and rainy season. Sandstorms are very common during summer season, especially during May and June. During post-monsoon and winter, winds are very light and variable (Qureshi, 2012).

2 Natural Resources

The deserts are highly fragile, however sustains precious and venerable resources which typically include flora and/or vegetation, landforms, water and wildlife. An imaginary picture is shown under Fig. 2 that indicates the flowchart of natural resources found in the Cholistan desert. There is overexploitation of the flora/vegetation and wildlife resources in this desert ecosystem; while the land-resources are vulnerable due to the scarcity of water that hampered the production of these resources (Chaudhry et al., 2004).

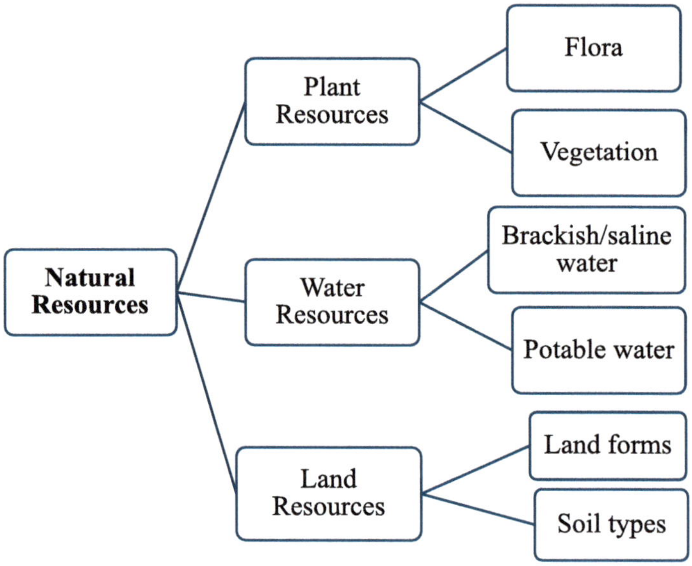

Fig. 2 Flowchart of the natural resources of Cholistan Desert, Pakistan

2.1 Plant Resources

2.1.1 Flora

The flora of Cholistan is composed of Saharo-Sindian elements with some exceptional Indian elements (Ali & Qaiser, 1986). Most of the floristic composition is that of Thar and Nara Desert (Abd El-Ghani et al., 2017; Qureshi & Bhatti, 2008a, 2008b, 2010; Qureshi, 2008a, 2012). No complete record of the plants of this desert exists and a very few studies have been conducted (Arshad & Akbar, 2002; Akhtar & Arshad, 2006; Arshad & Rao, 1994). The pioneer work of Arshad and Rao (1994) indicated the presence of 118 plant species belonging to 82 genera and 32 families. Later on, Wariss et al. (2013) reported 154 species belonged to 106 genera and 38 families. The largest family that contributed large number of species in the Cholistan is Poaceae with 34 species, followed by Papailionaceae and Zygophyllaceae (10 species spp.), Asteraceae (9 spp.) and Aizoaceae, Capparidaceae & Chenopodiaceae (6 spp.). Being conserved area, Lal Suhnra National Park is located in the Cholistan desert and has comparatively good diversity of plants reported by Wariss et al. (2014) indicating 212 species distributed across 162 genera and 50 families. They reported that herbs formed the dominant fraction in terms flora (49%), followed by grasses (22%), shrubs (14%), and trees (6% each). Some photos depicting xeric condition of the Cholistan are given below under Plate 1.

Plate-1

Fig. *Salvadora oleoides*

Fig. *Propospis cineraria*

Fig. *Calligonum polygonoides*

Fig. *Tamarix aphylla*

Plate-1

Fig. *Aerva javanica*

Fig. *Capparis decidua*

Fig. *Zizyphus nummularia*

Fig. *Zizyphus mauritiana*

Plate-1

Fig. *Leptadenia pyrotechnica*

Fig. *Dipterygium glaucum*

Fig. *Haloxylon salicornicum*

Fig. *Crotalaria burhia*

Plate-1

Fig. *Calotropis procera*

Fig. *Cenchrus ciliaris*

Fig. *Citrullus colocynthis*

Fig. *Tribulus longipetalus*

Plate-1

Fig. *Neurada procumbens*

Fig. *Convolvulus prostratus*

2.2 Vegetation

The vegetation of Cholistan desert is xerophytic in natures modified to extreme salinity, temperature, moisture fluctuations and a wide range of edaphic conditions. Greater Cholistan is comparatively rich in vegetation cover owing to a variety of landforms with a number of nutritious and stress-tolerant species of grasses, shrubs, trees and forbs. Although these species are slow growing, but respond quickly to the favorable climatic conditions, especially during rains, and produce plentiful fodder. Important genera of grasses include *Lasiurus, Cenchrus, Sporobolus, Cymbopogon* and *Panicum*; shrubs include *Suaeda, Capparis, Calligonum, Leptadenia, Salsola & Haloxylon*; while. *Prosopis, Tamarix, Zizyphus* and *Acacia* are notable native trees. Each site depicts a typical plant community with xeric genetic adaptations (Akbar & Arshad, 2000; Akbar et al., 1996).

Various studies have been reported the structure and composition of plant communities. Baig et al. (1975) identified six major plant communities with the dominance of *Haloxylon stocksii, Prosopis cineraria, Ochthochloa compressa, Tribulus longipetalus, Dipterygium glaucum* and *Calligonum polygonoides*. Based on adaptability in response to extreme aridity, Khan (1992) reported that *Cenchrus ciliaris* and *Panicum turgidum* are highly adapted grasses, *Calligonum polygonoides, Ziziphus nummularia* and *Haloxylon stocksii* as highly adapted shrubs, and *Prosopis cineraria* and *Acacia jacquemontii* as highly adapted tree and shrub, respectively. Another study conducted by Arshad and Rao (1995) reported dominant plant communities in relation to soil type. According to them, sand dunes were dominated by a *Calligonum polygonoides* community, sandy plains by *Calligonum polygonoides-Prosopis cineraria-Capparis decidua* community, compacted soils by

Fig. 3 Sand dune vegetation associated with *Calligonum polygonoides* and *Aerva javanica*

Capparis decidua-Prosopis cineraria community, and saline soils by *Haloxylon stocksii-Suaeda fruticosa-Tamarix dioica* community.

2.2.1 Sand-Dune Vegetation

Sand dunes are mostly sparse in terms of vegetation cover (Fig. 3); yet, they are dominated by grasses and sedges such as *Stipagrostis plumosa, Aristida adsecnsionis, Panicum turgidum, Lasiurus scindicus*, and *Cyperus conglomeratus* (Qureshi & Bhatti, 2005a) along with some typical shrubs like *Aerva javanica, Calligonum polygonoides, Leptadenia pyrotechnica* and *Dipterygium glaucum* as a common herb (Arshad & Akbar, 2002; Chaudhry, 1992; Rao et al., 1989).

2.2.2 Inter-Dunal Vegetation

Due to enrichment of clay particles and some organic matter in this micro-habitat, the vegetation cover is comparatively thicker in inter-dunal areas (Fig. 4). The grasses such as *Cymbopogon jwarancusa, Cenchrus ciliaris, Sporobolus ioclados, Panicum antidotale* and *Ochthochloa compressa* are very common. Herbs are also frequently found in this area that include and *Crotalaria burhia, Dipterygium glaucum, Tribulus longipetalus*; while, very common shrubs and trees include *Calligonum polygonoides, Capparis decidua, Aerva javanica* and *Haloxylon salicornicum, Prosopis*

Fig. 4 Inter-dunal vegetation dominated by *Panicum turgidum, Aerva javanica* and *Salvadora oleiodes*

cineraria, Salvadora oleoides, respectively (Arshad et al., 2008). The vegetation composition is almost alike as that of other desert habitats (Qureshi & Ahmad, 2010; Qureshi et al., 2009; Qureshi, 2008b).

The desert converts into a green tinge after the rains in the early spring. The ephemerals cannot withstand hot scorching summer and the long drought and escape soon, after producing seed for the next favorable season. The permanent and perennial vegetation in the desert include trees and shrubs such as *Acacia nilotica, Prosopis cineraria, Tamrarix aphylla, Salvadora oleoides, Zizyphus nummularia, Calligonum polygonoides* and *Leptadenia pyrotechnica* (Qureshi, 2012).

2.2.3 Vegetation of Sandy Plains

These are slightly undulated areas encircled by large dunes. The soils of these areas are loose, calcareous with yellow to pale brown color with sandy loam texture. The dominant tree and shrub-layer include *Prosopis cineraria, Salvadora oleoides, Capparis decidua, Ziziphus nummularia, Z. spina-christi, Calotropis procera* and *Leptadenia pyrotechnica* (Arshad et al., 2008). The herb-layer and bushes include *Aerva javanica, Crotalaria burhia, Citrullus colocynthes, Tribulus longipetalus, Limeum indicum, Heliotropium crispum*, etc. (Fig. 5).

The most prominent grasses found in this microhabitat are *Aristida adscensionis, A. funiculata, A. mutablilis, Cenchrus biflorus, C. ciliaris, C. setigerus, Cymbopogon*

Fig. 5 Vegetation of sandy plains

jwarancusa Dactyloctenium aegyptium, Erogrostis ciliaris, E. barreleri, Lasiurus scindicus, Ochthochloa compressa, Panicum antidotale, P. turgidum and *Stipagrostis plumosa* (Qureshi & Bhatti, 2007a).

2.2.4 Vegetation of Saline-Sodic Land

There are certain plain areas which have high pH and EC and dominance of halophytes. The composition of the flora of such areas includes *Tamarix dioica, Haloxylon salicornicum, H. stocksii, Suaeda fruticosa* and *Salsola imbricata* along with grasses such as *Aeluropus lagopoides, Cymbopogon jwarancusa, Desmostachya bipinnata, Lasiurus scindicus, Sporobolus ioclados* and *Ochthochloa compressa,* (Naz et al., 2009; Qureshi, 2012). Figure 6 indicted some dominant plant species of this micro-habitat.

2.2.5 Plants as Food

Other than pasture source, plant resources are being utilized as food in the Cholistan desert. Being a rangeland area, this desert does not sustain proper agriculture due to the water unavailability. Yet, the inhabitants of this area collect seeds, fruits/pods and vegetative parts for preparing curry and food dishes to meet the need of their diet. Though, these plant resources are less palatable and deficient in dietary components, however, wild plants provide instant relief during the drought and famine period. The fruit of *Capparis decidua* (Karar), roots/pods of *Moringa oleifera* (Suhanjna), and pods of *Prosopis cineraria* (Jand) are gathered for food and sold in the local market.

Fig. 6 Vegetation of saline lands dominated by *Cymbopogon jwarancusa* and *Tamarix dioica*

The floral buds of *Capparis decidua* and *Moringa oleifera* are cooked as vegetables; whereas, unripe fruits are cooked as vegetables and pickled. Most expensive pickle have good market in Europe and served in top restaurants. *Calligonum polygonoides* (Phog) is a very common shrub and its flower buds either fresh or sun dried are used to make a luscious dish locally called as *Phugesi*. Fruits of *Salvadora oleoides* (*Peelu*) and *Zizyphus numularia* (*Jhar beri*) growing in Cholistan desert are collected and harvested for human utilization (Fig. 7). During the famine period, seeds of *Panicum turgidium*, and *Cenchrus ciliaris* are mixed with flour for bread making. In addition to this, gums obtained from various plants, such as *Acacia nilotica, Prosopis cineraria.*

2.3 Plants as Folk-Medicine

The nomads of Cholistan desert use medicinal plants as a first source of medicament. They are living in far away from cities and lack basic health facilities, therefore they depend on the available plant resources to meet their healthcare needs (Qureshi, 2018; Qureshi & Bhatti, 2008b; Qureshi et al., 2010). Some of the commercial oriented important medicinal plants are collected and harvested from the desert areas and sold either in local market (*baazaar*) or big cities of the country (Fig. 8). These are collected during various growing seasons for preparing medicines. *Aloe vera* and *Citrullus colocynthes* (Fig. 8) are collected, dried, and sold in the market which

Fig. 7 Collection of wild fruit from *Salvadora oleoides* by the nomadic girls

is a potent drug dealing with digestive problems (Qureshi, 2018). For example, *Corchorus depressus* mixed with seeds of *Papaver somniferum* and candy are ground with water (locally known *Thandai*) and used as demulcent drug to reduce thirst during summer season.

Fagonia cretica is boiled in water and used as bath to treat skin diseases, including itching and irritation. The floral branches and leaves of *Azadirachta indica* are ground in water and used orally as blood purifier and hepato-protective agent. All parts of *Calotropis procera* are used to treat various diseases (Bhatti et al., 2001). For example, the powder of leaves is dusted on wounds to heal. The young twig is forcefully given to cattle to treat colic and indigestion due to overgrazing. *Leptadenia pyrotechnica* was used to treat eczema/ringworm. According to Qureshi (2018), most of the desert dwellers use a single species to treat two or more diseases. For instance, *Achyranthes aspera* was used to treat asthma, cough, cold, pneumonia, joint pain, cuts, and wounds. *Calotropis procera* was used in earache, sciatic pain and pneumonia, painful joints, and chest pain. *Capparis decidua* was reported in asthma, cough, joint pain, inflammation, boils, muscular injuries, and wounds. *Tamarix aphylla* was applied on boils and wounds. *Schweinfurthia papilionacea* is used for typhoid fever, skin impurities, chronic ulcers, and cancerous wounds. Fruits of *Withania coagulens* are soaked in water overnight and used as blood purifier and cooling agent. *W. somnifera* is used in glandular swellings, joint pain, and spermatorrhea and as a nerve tonic (Qureshi, 2018). A list of most important medicinal plants is provided in Table 1.

Fig. 8 Collection and harvesting of *Cirtullus colocynthes* for commercial use

2.4 Plants for Fuel and Construction

Trees are very few in the Cholistan desert and all of these are heavily exploited for fuel-wood collection. The common trees of this desert include *Acacia nilotica, Capparis decidua, Prosopis cineraria, P. juliflora, Tamarix aphylla,* and *Salvadora oleoides*. A large amount of wood is collected by the dwellers for not only their household use but also to sell out in the nearby localities/small cities (Fig. 9). Besides these trees, various shrubs such as *Acacia jacquimontii, Aerva javanica, Calligonum polygonoides, Haloxylon salicornicum, H. stocksii, Salsola imbricata* and *Zizyphus* spp. are also used as fuel-wood source. The collection and digging of large numbers up to the root-level loosens the soil, thus exacerbated the process of desertification. The uprooting of plants imposes many long-lasting harmful effects, not only on the meticulous species composition but also on the desert ecosystem as well.

For constructing house in the desert areas, local people use various plant materials for roof thatching as fiber. These houses, locally called "*Gopas*", are constructed with mud walls, armed with branches of *Capparis decidua, Prosopis cineraria* or *Acacia nilotica*. Usually, a single straight trunk of *Acacia nilotica, Prosopis cineraria* or *Capparis decidua* is kept in center to hold the thatched roof. As framework, various woody branches of these trees are used for contracting these houses and for thatching mostly *Leptadenia pyrotechnica* is used along with *Crotalaria burhia* and *Dipterygium glacum* (Figs. 10 and 11).

Table 1 Medicinal plants of the Cholistan Desert, Pakistan

Sr. No	Family	Plant name	Vernacular name
1	Acanthaceae	*Blepharis scindica* T. Anders.	Utangan
2	Aizoaceae	*Zaleya pentandra* (L.) C. Jeffrey	Itsit
3	Aizoaceae	*Trianthema portulacastrum* L.	Itsit
4	Aizoaceae	*Gisekia pharnaceoides* L.	Buloka sag
5	Amaranthaceae	*Aerva javanica* var. *javanica*	Boi
6	Amaranthaceae	*Amaranthus viridis* L.	Cholai
7	Asclepiadaceae	*Calotropis procera* subsp. *hamiltonii* (Wight) Ali	Ak
8	Asclepiadaceae	*Leptadenia pyrotechnica* (Forsskal.) Decne	Khipp
9	Asclepiadaceae	*Oxystelma esculentum* (L. f.) R. Brown	Dudhi
10	Asclepiadaceae	*Pergularia daemia* (Jacq.) N. E. Brown.	Karial
11	Asteraceae	*Echinops echinatus* Roxb.	Unt katara
12	Asteraceae	*Launaea nudicaulis* Less	Dudhkal
13	Asteraceae	*Oligochaeta ramosa* (Roxb.) Magenitz	Birham dandi
14	Boraginaceae	*Cordial gharaf* Ehrenb. ex Asch.	Lassura
15	Boraginaceae	*Heliotropium crispum* Desf.	Kali lani
16	Boraginaceae	*Heliotropium strigosum*	Gorakh Pan
17	Brassicaceae	*Farsetia hamiltonii* Royle	Lathi or Farid booti
18	Caesalpinaceae	*Cassia italica* subsp. *italica*	Ghora wall or Sana
19	Caesalpinaceae	*Cassia fistula* L.	Amaltas
20	Capparidaceae	*Capparis decidua* (Forsskal.) Edgew	Karir
21	Capparidaceae	*Capparis spinosa* L.	Kubber
22	Capparidaceae	*Cleome brachycarpa* Vahl. ex. DC.	Noli or Kastoori
23	Capparidaceae	*Cleome scaposa* DC.	Khastoori boti
24	Capparidaceae	*Dipterygium glaucum* Decne.	Phel
25	Chenopodiaceae	*Haloxylon recurvum* Bunge. ex. Boiss.	Khar or Sajji
26	Chenopodiaceae	*Haloxylon salicornicum* (Moq.) Bunge	Lana
27	Chenopodiaceae	*Salsola baryosma* (Roem. ex. Scult.) Dany.	Lani
28	Chenopodiaceae	*Suaeda fruticosa* Forssk. ex J. F. Gmelin	Kali lani
29	Compositae	*Pulicaria crispa* (Cass.) Benth. & Hook. f.	Bui
30	Convolvulaceae	*Convolvulus prostratus* Forssk.	Hiran booti
31	Convolvulaceae	*Cressa cretica* L.	Oini
32	Cucurbitaceae	*Citrullus colocynthis* (L.) Schrad.	Tumma/ Kaur tumma
33	Cucurbitaceae	*Cucumis melo* var. *agrestis* Naudin	Chibbarr
34	Cucurbitaceae	*Momordica balsamina* L.	Jangli Karela
35	Cucurbitaceae	*Mukia maderaspatana* (L.) M.J. Roem	Gawala kakri
36	Cuscutaceae	*Cuscuta reflexa* Roxb.	Akash bail

(continued)

Table 1 (continued)

Sr. No	Family	Plant name	Vernacular name
37	Cyperaceae	*Cyperus rotundus* L.	Mootha
38	Euphorbiaceae	*Chrozophora sabulosa* Kar. & Kir.	Nilakari
39	Euphorbiaceae	*Euphorbia granulata* Forssk.	Dudheli
40	Euphorbiaceae	*Euphorbia prostrata* Ait.	Hazar dani
41	Malvaceae	*Abutilon muticum* (Del.ex DC.) Sweet	Kanghi buti
42	Meliaceae	*Azadirachta indica* A. Juss.	Neem
43	Meliaceae	*Melia azedarach* L.	Bakain
44	Mimosaceae	*Acacia jacquemontii* Benth.	Banwli
45	Mimosaceae	*Acacia nilotica* (L.) Delile	Babul or Kikar
46	Mimosaceae	*Albizia lebbeck* (L.) Benth.	Shirin
47	Mimosaceae	*Prosopis cineraria* (L.) Druce	Jand
48	Molluginaceae	*Glinus lotoides* L.	Gandi-booti
49	Molluginaceae	*Mollugo cerviana* (L.) Seringe	Padi or Sarr
50	Molluginaceae	*Mollugo nudicaulis* Lamk.	Gandi-buti
51	Neuradaceae	*Neurada procumbens* L.	Chhapri
52	Nyctaginaceae	*Boerhavia procumbens* Banks ex Roxb.	Biskhipra
53	Papilionaceae	*Alhagi maurorum* Medic.	Jawansa
54	Papilionaceae	*Crotalaria burhia* Buch.-Ham. ex Benth.	Chagg
55	Papilionaceae	*Indigofera argentea* Burm. f.	Neel
56	Poaceae	*Cenchrus biflorus* Roxb.	Mohabat Boti
57	Poaceae	*Cenchrus ciliaris* L.	Daman
58	Poaceae	*Cenchrus setigerus* Vahl.	Chuti Daman
59	Poaceae	*Cymbopogon jwarancusa* (Jones) Schult.	Katrin or Khavi
60	Poaceae	*Panicum antidotale* Retz.	Murrot
61	Poaceae	*Sporobolus ioclados* (Nees ex Trin.) Nees	–
62	Polygonaceae	*Calligonum polygonoides* L.	Phog
63	Polygonaceae	*Polygonum plebejum* R. Br.	Indrani
64	Rhamnaceae	*Ziziphus nummularia* (Burm. f.) Wight & Arn.	Jhar beri
65	Salvadoraceae	*Salvadora oleoides* Decne.	Pilu
66	Solanaceae	*Datura metel* L.	Dhatura
67	Solanaceae	*Solanum nigrum* L.	Mako
68	Solanaceae	*Solanum surattense* Burm. F.	Kateli
69	Solanaceae	*Withania coagulens* (Stocks) Dunal	Paneer
70	Solanaceae	*Withania somnifera* (L.) Dunal	Asgandh
71	Tamaracaceae	*Tamarix aphylla* (L.) Karst	Jhao, Ukan, Frash
72	Tiliaceae	*Corchorus depressus* (L.) Stocks	Boophali

(continued)

Table 1 (continued)

Sr. No	Family	Plant name	Vernacular name
73	Tiliaceae	*Grewia villosa* Willd.	Jalidar
74	Zygophyllaceae	*Fagonia cretica* L.	Dhmasa
75	Zygophyllaceae	*Peganum harmala* L.	Harmal
76	Zygophyllaceae	*Tribulus longipetalus*	Tirkandi or
77	Zygophyllaceae	*Tribulus terrestris*	Bakhara
78	Zygophyllaceae	*Zygophyllum simplex* L.	Lunak

Fig. 9 Collection of fuelwood species from the desert for selling to the nearby villages

2.5 *Fodder Resources and Livestock*

Agro-pastoral plays a vital role in the growth of local economy and accounted for more than half of the total agricultural income and shared 10.6% of the national GDP of Pakistan (Ghazali, 2010). The desert regions of the country are of vital importance to the national economy, since these are rangelands and livestock husbandry is the only source of income for the local communities. Besides, milk and meat production may offset by the climatic fluctuations, especially in those areas which are facing

Fig. 10 Construction of conical thatched huts

Fig. 11 Internal structure of conical huts

food starvation. In Pakistan, 8.1% of buffaloes, 13.5% of cattle, 15.3% of sheep and 14.4% of goats are raised in the desert areas (Farooq et al., 2009). Nevertheless, there are various constraints that limit the local livestock productivity such as shortage of fodder, high rate of diseases, limited availability of veterinary services and poor access to animal vaccination (Farooq et al., 2009). For the sustainable production of livestock in desert habitats, there is a need to ensure the availability of plentiful fodder resources and management (Hussain & Durani, 2009). Since Cholistan is a rangeland

Fig. 12 Grazing in the Cholistan desert

so the vegetation is typically xerophytic in nature adapted to the xeric/arid conditions. The plants are responsive to the extreme temperature fluctuations, moisture, and edaphic conditions. Besides these harsh climatic conditions, this rangeland sustains and supports the livestock by providing palatable plant species with wide variety of growth-pattern.

The first and foremost such interesting species are grasses such as *Cynodon dactylon, Cenchrus ciliaris, C. setigerus, C. pennisetifromis, Eragrostis barrelieri, Lasiurus scindicus, Ochthocloa compressa, Panicum antidotale, P. turgidum* and *Sporobolus* species which provide feed to the livestock and grazers (Fig. 12). The domesticated animals prefer to graze/brows on various species; for instance, camel is fond of halophytic plants such as *Haloxylon/Suaeda* species, *Calligonum polygonoides* and *Salvadora oleoides*. Under the severe drought-conditions, goat and other livestock may browse on young *Leptadenia pyrotechnica* and *Crotalaria burhia,* which are less palatable and usually left due to the higher fiber and silica contents, and due to ample availability of fodder. The poisonous or thorny species such as *Calotropis procera, Tephrosia* spp. and *Dipterygium glaucum* are not browsed by animals; however, in shortage of fodder, these are browsed by goat and camel.

3 Agro-Pastoralism in Cholistan

Three groups of pastoralists exist in Cholistan. These are as follows.

Fig. 13 Transhumance in the Cholistan Desert with temporary settlement

3.1 Transhumant

This group keeps moving along with their livestock (i.e. transhumant and agro-pastoralists) on transhumance during the monsoon season in the desert (Fig. 13). They migrate with their families and herds of animals in the far-flung areas of the desert in search of pastures and water. During extreme dryness and high temperature, they transfer towards the irrigated areas (Chaudhry et al., 2004; Qureshi & Bhatti, 2005b).

During the months of April–May, they migrate to the nearby irrigated areas. The reason for their movement is rising temperatures and the reduction of forage and water in the desert area. During this period, they feed their livestock on weeds/wheat stubbles and drinking water for humans and livestock. The nomads and their herds return to the desert around July–September with the beginning of the monsoon rains (Qureshi & Bhatti, 2005b).

3.2 Nomadic Population

This group of pastoralists lead nomadic life, not adhering the regular timing/schedule or location of their pastures (Fig. 14). There is no limit of traveling distance while migrating from place to another. These are typical some clans who keep large herds of animals mainly camels and cows. They trail the distribution of rainfall and the resulting forage usually within the desert from Toba to Toba and do not prefer to

Fig. 14 Nomadic life in the Cholistan Desert

go to the irrigated areas. During shortage of water and pastures, these clans move to the large area and sometimes ask for the help of other tribe/clan to get water and rangeland. The quantity and quality of available forage is the major concern of pastoralists in the desert; however, there are other factors such as excessive heat, lack of shade, avoiding diseased areas, close to the market, availability of labor and social/ritual relations which are determining their movement-patterns (Qureshi & Bhatti, 2005b).

3.3 *Village Bound*

This group consists of those clans who keep their livestock round the year in the nearby areas of the village (Fig. 15). Normally, they return to their village every night.

Fig. 15 Permanent settlement on the periphery of Cholistan Desert

4 Wildlife Resources

Cholistan desert is the host of specific wildlife adapted to the extreme climatic conditions. The plants of this desert provide food/feed and shelter to the birds, lizards, and wild animals. The Black Buck is very rarely seen in the desert. Hunting coupled with anthropogenic activities by the human being have resulted into declining of 50 to 80% wildlife population over the last decade. Lal Suhnra National Park is established in this desert near the Bahawalpur by the Government of Pakistan for conserving Black Buck and other wildlife. The local communities have less understanding about the behavior of different wildlife species; nevertheless, various hunters/gatherers are well acquaintance about specific wild animals belonged to a particular clan or cast preferring the hunting of different animals. For example, "*Baagri*" tribe has gathering habit to eat porcupine, jackals, snakes and lizards and is familiar with the behavior of these animals. Likewise, another "*Chachans*" are specialized in capturing deer and desert foxes alive (Chaudhry et al., 2004).

Some faunal species such as Blackbuck, Chinkara, Nilgai, and Hubara Bustards have been listed among the most threatened species. Wild Ass can only be cited near Indian border and its population is declining. Swan, jackal, fox, rabbit, taloor, falcon, wolf, deer and blue bull are common fauna found in this desert. Amongst the birds, partridge, quails and sand grouse are common (Akbar & Arshad, 2000; Baig et al., 2007).

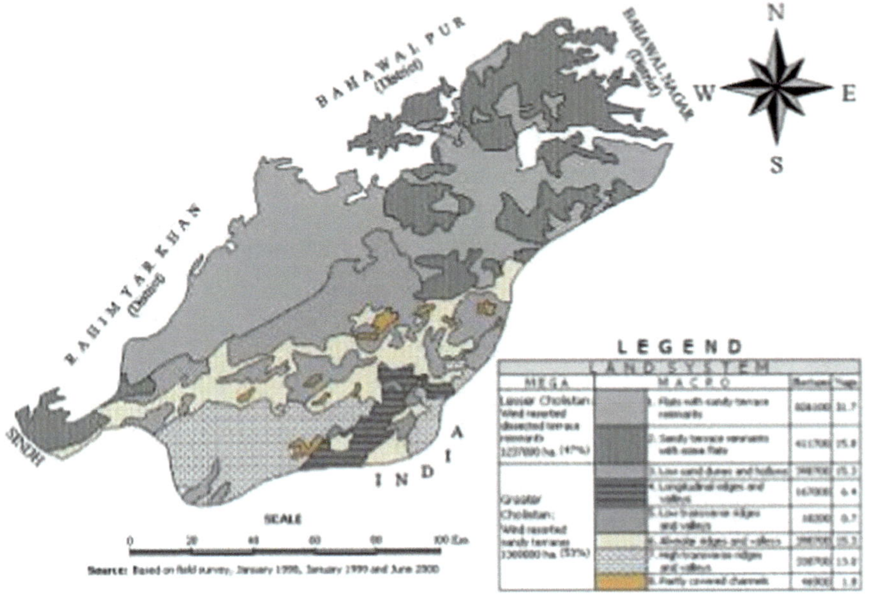

Fig. 16 Landforms of Cholistan Desert. *Source* Ahmad (2008)

5 Land Resources

5.1 Landforms/Microhabitats

Based on geomorphology accountable for soils, moisture and vegetation, eight micro-Landforms/habitats are found in both Lesser and Greater Cholistan (Fig. 16). These microhabitats are an important component of the range ecosystems for pastoralism.

5.2 Soils

The soils of Cholistan are chiefly formed by the deposition of aeolian sands or alluvium deposits. The alluvium deposits are composed of granites, schists, gneiss and slates. The soils of Cholistan are generally saline, alkaline and gypsiferous. Approximately, 1.13 million ha land comprised on stable and un-stabilized dunes, 0.95 and 0.06 million ha sandy and loamy soils, respectively. Topographically, most of the area is undulating to steep dunes lying parallel to each other which are associated by small streamers. These are coarse textured and structure-less soils which are well drained. The gently sloping areas have deep to very deep sandy soils that are calcareous, coarse textured and very well drained (Ahmad, 2008; Baig et al., 1975; Qureshi & Bhatti, 2005a). Loamy soils are found in flat plains/low lying areas by trapping sand and formation of hummocks on the surface. These soils are moderately deep, relatively well drained, calcareous and with a moderately coarse to medium texture (Ahmad, 2008; Baig et al., 1975; FAO/ADB, 1993). About 0.44 million ha are clayey flat plains (locally known as '*dhars*') which are shallow to moderately deep, poorly drained and calcareous. The soils are classified as either saline with pH 8.2–8.4 or saline sodic with pH 8.8–9.6 (Arshad et al., 2008). Due to low in organic matter, the water-holding capacity of these soils is relatively low and reduced (Akhter & Arshad, 2006; Khan et al., 1996; Qureshi & Bhatti, 2005a). Sand is the large sized particle that facilitates the rapid movement of water into deep horizons of soil along with high evaporation rate, which obstruct the natural deposition of surface waters (Ahmad, 2008).

6 Water Resources

6.1 Potable/Rainwater

There are no permanent and natural water bodies found in Cholistan. Rainfall is the chief source of potable/sweet water in Cholistan (PCRWR, 2009; WWAP, 2009) mostly collected in natural depression or manmade ponds called locally "*tobas*". Approximately, 598 *tobas* are found in Cholistan (Ahmad, 2008; CDA, 1996). During monsoon season (i.e. from July to September), most of the rainfall is received with some occasional showering in winter. Rainwater is harvested in flat plains from the catchment areas (Ahmad, 2008; Khan et al., 1990). If an appropriate amount is harvested, that would be available for the human, livestock as well small agriculture practices (Ahmad, 2008; Baig et al., 1980). Rains recharge the groundwater that contributes various fresh water sources (Akram et al., 2008). Yet, a large quantity of the rainwater is lost as runoff due to sandy nature soils and the topography of Cholistan (Akbar & Arshad, 2000; Baig et al., 2007). The inhabitants of desert mainly depend on the rainwater collected during the rainy season, which is collected and stored in depressions or ponds. Two types of ponds such as *Tobas* and *Kunds* are found in Cholistan to meet water requirements as follows:

6.1.1 Tobas

The runoff and rainwater are collected in ponds locally identified as *Tobas* (Fig. 17). The term *Toba* is known as *Kacha* that means earthen/made up of mud. The water pond is not made up of concrete and is dug out in low lying area for collecting rainwater which is utilized by both the human and animals abundantly found in the Cholistan desert (WWAP, 2009). Though, 600 out of 1600 tobas are functional. Due to transportation of pollutants and sediments, these ponds are also unfit for the human and livestock consumption.

Fig. 17 Collection of runoff water in the Tobas

6.1.2 Kunds

These pods locally known as *Dahrs* are either of circular or rectangular in shape with concrete structures used to store rainwater for the household and human consumption. Approximately, 200 Kunds are constructed in the area for obtaining clean water unless (Fig. 18). They are less contaminated than Tobas (PCRWR, 2009; WWAP, 2009). The plain areas encircled by sand dunes serves as catchments for the collection of water or runoff. These catchments may vary from few to several hectares in size.

Fig. 18 Kunds in the desert habitat

6.2 Ground Water

Owing to the high salt contents, the ground water of the area is saline and unsuitable for human and livestock consumption (PCRWR, 2009). Most of the dugout wells have brackish water that is not suitable for both drinking and agricultural purposes (Akbar & Arshad, 2000; Baig et al., 2007). Nonetheless, yet this saline water is being used for domestic and other purposes. These water resources are very deep with almost trifling recharge from rainwater (Abdullah et al., 1990; Ahmad, 2008). Most of the groundwater is located at depths ranging from 30 to 90 m with the salinity levels of 368 to 35,000 mg/l of total dissolved solids (Ahmad, 2008; Baig et al., 1980). There are two major sweet water sources in the Cholistan; however, these are bordered by saline water. The main potable/sweet water extends to 80 km from Fort Abbas towards Moujgarh, which is 10–15 km wide. This aquifer located between 40 to 100 m below the surface and has an anticipated volume of 10 gigaliters. The second source lays 20 km north-west of Derawar Fort. It inhabits an area of 50 km^2 having thickness of 100 m, and occurs about 25 m under the surface surrounded by saline water (Ahmad, 2008; FAO/ADB, 1993). Some of the sweet water is also found in isolated pockets at Phulra, Moujgarh, Dingarh and Derawer Fort along the abandoned Hakra River bed and Bhai Khan, Ghunnianwala, Islamgarh, Lakhewala and Renhal near the Pakistan-India border (Ahmad, 2008; Baig et al., 1975, 1980). Factors like low rainfall, high rate of water penetration into the sands, and high evaporation rate hinder the natural accumulation of surface water.

7 Issues and Challenges of the Cholistan Desert

This is almost a neglected region for its proper development. Despite of low productivity, this desert supports fairly high human and livestock populations. The increasing incidence of prolonged drought, desertification, deforestation, and soil erosion are serious problems of the area. As local pastoralists solely dependent on the livestock for their livelihood, so there is a penchant to only raise the livestock population. The over-exploitation of the natural vegetation by grazing, browsing, and chopping of trees as well as shrubs for fuel purposes caused serious environmental degradation that imperils the natural resource base of this region (Qureshi & Bhatti, 2007b). The region lacks proper infrastructure, communication means, basic health, and educational facilities. Since this is a high potential area for livestock sector, there is a dire need to develop and initiate an appropriate and vibrant livestock marketing system in association with industry related to dairy, wool and other products. The region is also rich in traditional artifacts; however, there is lack of financial and logistic support for the cottage industry and local handicrafts. Natural calamities such as soil erosion and increasing intensity of soil salinity caused severe threats to the vegetation and to this unique ecosystem. Followings are some of the solutions to mitigate these problems and enhance the productivity of this rangeland.

7.1 Need for Sustainable Technologies and Strategies

Many workers like Khan et al. (1996), Akhter and Arshad (2006) and Ahmad (2008) have been stressing the need to develop and launch an integrated and comprehensive strategy that may reduce the adverse effects of drought and desertification. Besides, short- and long-term action plans are urgently required for the development of the area. In the long-term planning, there is a need to develop irrigation sources for the effective saline agriculture in terms of efficient use of land and saline water. Based on the secondary data available in the literature and personal experience of the authors, a comprehensive strategy needs to be developed by adopting a holistic approach.

8 Ecological Interventions Through Climate-Smart Agriculture

8.1 Combating Soil Erosion

Shifting dunes and soil erosion is very common phenomenon of the arid sandy deserts that leads the 2nd most serious issue in Cholistan. Various anthropogenic activities such as chopping/cutting of trees and shrubs for the construction, building/fuel source and overgrazing coupled with natural calamities result in soil erosion and desertification. The vegetation cover is the only protecting means that acts in soil bindings. This issue can be tackled and remedied by reseeding and plantation of grasses such as *Lasiurus scindicus, Panicum antidotale, P. turgidum, Ochthochloa compressa, Saccharum spontaneum* and shrubs *Leptadenia pyrotechnica, Tamarix dioica, Haloxylon salicornicum, Acacia jacquimontii* that are required for the conservation of sandy soils.

8.2 Rehabilitation of Degraded Desert Lands Through Seeding Ball Technology

A sound range management planning including application of mulch and selection of plant species and varieties is required for the rehabilitation of degraded desert lands of Cholistan. Reseeding is a good strategy for the restoration and improvement of desert rangelands. Many techniques have been developed to restore the eroded arid lands. Recently, the use of seed balls (Fig. 19) is a new technique appraised by research and practitioner groups (Gornish, 2019).

Seed balls, seed pellets or seed bombs are typical designed for wrapping seeds in soil materials by mixing clay, compost, and water which are then dried. These structures are made in a way to tolerate harsh climate for the restoration of arid

Fig. 19 Seed ball technology for the rehabilitation of degraded desert areas

land (Madsen et al., 2016) such as drought that aggravates seed drying soil crusts formation and hamper establishment of seedling and seed loss due to predation. This technology also serves to intact with the soil to overcome wind pressure. These seeds are preserved in these structures till sufficient rain eliminates the coverings and the compost part of the seed ball nurtures the seedlings after emergence. Seed balls are inexpensive and easily made that are also capable of increasing germination rate (Pedrini et al., 2017). Some of the important species such as *Acacia nilotica, Azadirachta indica, Calligonum polygonoides, Capparis decidua, Cordia latifolia, Leptadenia pyrotechnica, Moringa oleifera, Prosopis cineraria, Salvadora oloides, Ziziphus mauritiana* are important to the local communities to meet their daily life requirement and can be used for the seed ball technology. This strategy not only meets the need of the dwellers but also sustainably maintains the genetic diversity and rehabilitation of the eroded lands.

8.3 Shelterbelt Technology

Planting trees for wind breaking has demonstrated an effective means in tumbling the wind speed on the soil surface. Such practice has been found successful in combating soil erosion. Some of the perennial trees and shrubs can be used for the protection of settlements, livestock and agriculture lands through shelterbelt technology. In this intervention, trees are planted in closed several rows at the right angle to the wind direction. For this purpose, some of the ecologically well adopted indigenous species such as *Acacia nilotica* (kikar, babul), *Prosopis cineraria* (*Jand*), *Tamarix aphylla* (*Frash*), and *Salvadora oleoides* (*Wanh*) are used (Fig. 20). Compared with

Fig. 20 Use of shelter belt technology by the desert dwellers

the rest of aforementioned species, *Tamarix aphylla* is the fast growing and ideal species that resist harsh climatic and high wind velocity. It is mostly raised through cuttings before monsoon.

8.4 Rangelands Productivity for Sustaining Livestock

Cholistan spread over an area of approximately 2.61 million hectares. According to Akbar and Arshad (2000), 2 decades ago, the livestock population was around 134,798 Animal Units (AU) in Cholistan. Presently, the livestock population in Cholistan has many-fold increased and approximately 1.318 million heads including 667,000 Cholistani cattle, 80,000 camel, 220,000 goat and 351,000 sheep has been recorded (Khan et al., 2010). An adult cow weighing 350 kg (400 kg at international level) approximately consumes about 7–10 kg dry matter forage per day. Akbar and Arshad (2000) anticipated 344,409 tons total annual forage dry matter demand; whereas, the available forage dry matter in the area was about 242,951 tons. According to them, the area has a deficit of 101,457 tons/year dry matter forage. They also noticed that some 39,709 AU/year are an excess to the present carrying capacity. According to Rafay et al. (2016), Cholistan rangelands can support 91,178 animal units of cow, 455,890 sheep, 303,926 goats and 50,654 camels. The average carrying capacity during wet season was estimated to 17.25 ha/AU/y. During the wet season, stocking rate was calculated as 0.057 AU/ha/year. The forage productivity of this rangeland can be enhanced by 10–35 percent by using appropriate seeding techniques and appropriate species (Akram et al., 2008). The higher number of livestock in this region may threaten the natural vegetation that may exacerbate accelerated desertification. There is a need to manage proper stocking rate during the growing season along with the introduction of high yielding fodder crops that may help reduce the grazing pressure and improve productivity.

Rangelands can feed approximately 2.4 million heads of the livestock. Cholistan is a rangeland and livestock sector is the principal source of livelihood of the natives (PCRWR, 2009; WWAP, 2009); therefore the economy of Cholistan Desert is majorly based on livestock keeping. The nomads increase goats, sheep, cows, and camels in their herds. Camels are considered as the main and popular species. Cholistan Desert upholds well adapted plant species according to high xeric conditions. These species not only sustain the livestock population owned by locals, but also decrease the wind velocity and combat soil erosion. But its productivity in terms of vegetation cover has been reduced to some extent of approximately 20% due to various factors such as overgrazing, removal of slow growing shrubs and trees for firewood, and temporary shelters made by inhabitants in the area (PCRWR, 2009; WWAP, 2009). Some of the important genera of grasses such as *Cenchrus, Lasiurus* and *Panicum* and shrubs like *Calligonum polygonoides* and *Holoxylon salicornicum* are preferred by livestock (Qureshi & Bhatti, 2007a). The native trees like *Prosopis cineraria, Salvadora oloides, Ziziphus mauritiana* and *Acacian nilotica* are important source of feed to the domesticated animals of this desert (Akbar & Arshad, 2000; Rao & Arshad, 1991).

8.5 Climate-Smart Agriculture

This is a gigantic sandy desert of barren lands with high water scarcity (Kahlown et al., 2004). The ground water is saline in nature, which is unsuitable for the humans (WWAP, 2009). On the contrary, this desert has great potential for saline agriculture, beekeeping as well as saline fish-farming (PCST, 2005). The chief source of fresh water in Cholistan Desert is overflow due to sporadic rainfall, which is being utilized to meet the need of humans and livestock. The water is gathered in the ponds locally known as Tobas (Ahmad, 2008). These are mostly located inappropriately and most of the water collected does not last for more than 3–4 months (Adeel & Safriel, 2008).

Agro-pastoralism and pastoralism are the two principal land-use systems of the Cholistan Desert. Cotton-wheat crop rotation is followed on areas where sufficient irrigation water is available. Areas with insufficient irrigation water, crop rotations of sorghum (*Sorghum bicolor*), millet (*Pennisetum typhoides*), chickpeas (*Cicer arietinum*), and pulses are followed. About 67% of the farmers of the Cholistan practice farming on irrigated areas of the desert whereas 33% farmers practice dryland agriculture. The livelihood heavily depends on the income generated through animal husbandry. The nomadic tribes throughout the year keep on migrating their livestock from one place to another in search of water and feed, making the pastoral system of the area unique (Akbar & Arshad, 2000; Akram et al., 2008; Baig et al., 2007).

8.6 Saline Water Treatment for Drinking Purpose

There is an extreme shortage of drinking water in Cholistan. The existing groundwater resources are classified into saline according to WHO standards (Akram & Chandio, 1998) and contain high quantities of dissolved salts and minerals (i.e. 9000 to 24,000 Mg/l) in most of the areas (Akbar & Arshad, 2000; Akhtar & Arshad, 2006; Baig et al., 1980). This water can be made suitable for drinking purpose by lowering its salt contents to acceptable limits through the use of the reverse osmosis membrane technology. The desalination plant can treat up to 50 m^3 (50,000 L) per day at a very nominal unit cost. In order to minimize the energy cost, solar energy and thermal energy can be used which is freely available in the desert areas. The use of solar with reverse osmosis membranes for obtaining drinkable water from saline groundwater in the desert has a great potential that can be used on the mass scale to change the life of desert dwellers (PCRWR, 2009; WWAP, 2009).

8.7 Saline Agriculture by Using Salt/Drought Tolerant Plants

Saline agriculture has a great potential for the reclamation of the desert and can be used as an alternate approach. Various researchers screened grasses and crops under drought and osmotic stress in saline soils. Some of the species such as *Acacia ampliceps, Leptochola fusca, Tamarix aphylla, Parkinsonia aculeata, Casuarina equisetifolia* and *Atriplex amnicola* are capable of rapid growth in soils with EC ranging from 20 to 30 dSm^{-1}. Pakistan Council of Research in Water Resources (PCRWR) experimented for planting *Acacia, Asparagus, Eucalyptus,* Iple Iple (*Leucaena leucocephala*), *Parkinsona aculeata,* Date palm (*Phoenix dactylifera*), *Prosopis cineraria,* Pomegranate (*Punica granatum*), *Tamarix aphylla* and *Zizyphus mauritiana* (Beri). The experiment was conducted by using saline water with total soluble salts (TSS) concentration of 2800 part per million (ppm) and sodium adsorption ratio (SAR) of 14 along with harvested rainwater to flush salts at certain intervals (Abdullah et al., 1990). The list of some salt tolerant grasses and forages cultivated in the Cholistan Desert by using saline water is given in Table 2.

Table 2 Salt tolerant plants cultivated in Cholistan Desert by using saline water

Forage Crop/Grass/Bush	Tolerance (ECe106) 4000–18,000
Alkali grass *(Puccinellia airoides)*	High
Bermuda grass *(Cymodon dactylon)*	High
Kallar grass *(Leptochloa fusca)*	High
Salt grass *(Distichlis stricta)*	High
Desert wheat grass *(Agropyron cristatum)*	High
Barley *(Hordium vulgare)*	High
Rape *(Brassica napus)*	Medium
Clover *(Melilotus)*	Medium
Alfalfa *(California common)*	Medium
Oats *(Avena sativa)*	Medium
Atriplex sp.	High

Source PAMDU (1986)

8.8 Saline Agriculture Technology for Soil Reclamation

Moderate to highly saline groundwater remains the secondary source of water in the desert and is not useable for drinking. However, this kind of unfit saline and brackish waters have been utilized for the irrigation of many agricultural crops in several countries under similar physiognomic and environmental conditions. Salt tolerant plants have adaptability to grow in saline soils. They are capable of growing in saline waters due to their certain in-built mechanism and can be used to yield food, fodder and fuel (Qureshi & Barret-Lennard, 1998). There is great potential of such plants, crops, and grasses in reclaiming the salt affected areas. Such areas can be transformed into productive lands by growing the salt-tolerant crops, using brackish water for irrigation. Researchers have successfully grown crops such as wheat and barley without compromising the crop yields by using saline underground waters with EC ranging from 2 to 16 dS m^{-1} (PCST, 2005). Some of the salt tolerant crops, trees and shrubs, equipped to drought and saline environments of Cholistan are given in the Table 3.

The groundwater of the Cholistan desert is saline and can be used for saline agriculture to grow salt tolerant trees, shrubs as well as vegetables and crops in saline to sodic soils (Abdullah et al., 1990). Moreover, these soils can also be used for cultivating palatable fodder grasses having salt tolerance properties. Approximately, 1 million ha land of Cholistan is sandy to loamy soil that can be used for cropping by using underground saline water and harvested rainwater (Ahmad, 2010). Among such grasses, *Leptochloa fusca* is capable of accumulating and excreting salts from the shoot (Abdullah, et al., 1986; Abdullah et al., 1990). This mechanism makes it useful to reduce excessive salt from the soils through root enabling other plants to grow. Grasses like *Leptochloa fusca* possess fibrous roots that can profusely grow in

Table 3 Some potential drought tolerant plants that can grow well in Cholistan Desert

Botanical Name	Local Name	Description
Acacia nilotica	Kikar/Babul	Tolerates high-level of groundwater salinity and can withstand extreme temperatures
Aloe vera	Kunwar Gandal	Tolerates a wide range of conditions. Usually grows in sub-desert scrub where the vegetation consists mainly of an open formation of grasses ephemeral herbs low thorny shrubs etc
Zizyphus mauritiana	Ber	An intolerant tree that has no particular soil requirements and grow on a variety of soil conditions shallow to deep gravel sand to clay. It prefers a warm temperate to sub-tropical to tropical climate
Atriplex canescens	Kallar Jharian	It prefers sandy loam or loam soils to grow. It is recommended on marginal lands subject to prolonged drought or excess salinity
Simmondsia chinensis	Jojaba/Hohoba	It is most suitable to grow in Cholistan where annual rainfall is low and temperature is very high
Capparis decidua	Karir/Karil	It is commonly found on the sandy plains of desert and sometimes occurs on sand dune also
Haloxylon salicornicum	Lana	It is usually found in inter-dunal areas and windward sides of sand dunes throughout Cholistan Desert. It grows in soils with high Na content and high pH and helps reclaiming sodic soils
Suaeda fruticosa	Lani	It is a small shrub that grows in saline areas of Cholistan Desert
Lens culinaries	Masoor	It is a short slender many-branched annual legume with a bushy growth. Due to this low water requirement, it is viewed as drought-resistant crop
Azadirachta indica	Neem/Nim	It grows on soils that vary from rich loams to nutrient deficient soils and prefers an arid hot tropical and sub-tropical climate with a temperature range 1–45 °C
Phoenix dactylifera	Date/Khajoor	Grows successfully in hot arid climates and saline environments; can survive in a temperature range of 10–58 °C
Prosopis juliflora	Mesquite/Kabli Kikar	It is an aggressive species that can grow in highly saline, alkaline and sodic conditions. It can tolerate a wide range of temperature between −10–50 °C
Parkinsonia aculeata	Vilaiti kikar	It can tolerate drought. It is adapted to a variety of soil conditions and survives at the temperatures between −03–38 °C
Prosopis cineraria	Jand	*P. cineraria*grows successfully in harsh climate of Cholistan Desert and in the deserts of Oman and other middle eastern countries; highly saline alkaline soils (pH values up to 9.8. It is adapted to a variety of soil conditions and survives at the temperatures between −06–50 °C

(continued)

Table 3 (continued)

Botanical Name	Local Name	Description
Leucaena leucocephala	Ipil ipil	It is an aggressive species that can grow in highly saline, alkaline and sodic conditions. It can tolerate harsh temperatures and is highly drought resistant
Casuarinas equisetifolia	Casurina	It is an evergreen conifer like tree. It grows on slightly alkaline soils, withstands salinity but not waterlogging. It is an adapted is a heat-loving lowland tree and can tolerate drought for 6–8 months. It has a remarkable ability to stabilize sand dunes and control erosion
Eucylptus camaldulensis	Sufaida	Grows in slightly alkaline soils, where it can withstand some salinity and waterlogging. It prefers a semi-arid subtropical climate (−5– 40 °C) and it is drought tolerant and can stand a dry season up to 7 months
Grewia asiatica	Falsa	It thrives well in tropical climates and other climates as well. It is highly drought resistant. It can grow in alkaline soil conditions. It can survive in saline-sodic soils
Psidium guajava	Guava/Amrood	Optimum temperatures for its growth range from 23–28 °C. It grows well in moderately saline-sodic soils; can grow on alkaline to acidic soil conditions

Source Qureshi and Barret-Lennard (1998), Kahlown and Majeed (2002), and PCST (2005)

these soils, amplify air exchange, organic matter and hydraulic conductivity, reduce rhizospheric pH, enhance biological activity, dissolve $CaCO_3$, improve salts leaching, lower the water table of waterlogged soils, release plant nutrients and increase organic matter, humus and soil mulching, decrease surface evaporation, and progressively improve soil physical properties (Abdullah et al., 1986; Akhtar et al., 1988; Haq & Khan, 1971; Joshi et al., 1981).

The brackish water decreases evapo-transpiration and enhances resistance to drought (Abdullah et al., 1990). Some of the possible suggestions are: (1) identification and selection of species and varieties tolerant of high salinity; (2) the brackish water owing to a degree of salinity is attuned in line with such species suitable to the nature of the soil; and (3) the selection of irrigation in which soils permeable, well drained and rich in calcium and the hydrates of iron and aluminum (Ahmad, 2010).

Pakistan Council of Research in Water Resources (PCRWR) experimented and planted *Eucalyptus, Acacia, Parkinsonia aculeata, Zizyphus mauritiana* (Beri), *Tamarix aphylla, Prosopis cineraria,* Date palm, Pomegranate, Jojoba and Iple Iple (*Leucaena leucocephala)* by using saline water with 2800 ppm saline water and 14 sodium adsorption ratio for irrigation along with harvested rainwater to flush salts at certain intervals (Abdullah et al., 1990).

8.9 Saline Aquaculture

Presently, there is no existence of surface water in the Cholistan desert. The ground water is highly saline and unfit for crop production. This area has the huge potential for saline agriculture, bee keeping and mainly saline fish farming (PCST, 2005). The chief source of water in the Cholistan desert is runoff that meets the water requirements for human and livestock population. The water collected in tobas runs only for 3–4 months (Adeel & Safriel, 2008). For long-term water conservation, construction of kunds (ponds) at suitable points is the best strategy for minimizing water losses and has a sustainable and successful fish harvest.

8.10 Agricultural Crop Resources

Dryland agriculture is generally affected by drought and the areas receiving less than 300 mm cannot receive good yields. Cholistan is hyper-arid to arid region and agriculture is not practiced due to the unavailability of irrigation water and undulating topography (Adeel & Safriel, 2008). Though water is the limiting factor for the crop production in this region, however, by adopting dryland agriculture technologies, some salt tolerant crops can be sown by using saline water. The farmers of Cholistan cultivate some crops such as guar, millet and sorghum on the onset of sufficient showers (Akbar & Arshad, 2000; Baig et al., 2007). By utilizing saline water, dryland agriculture can be practiced on the vast areas of the desert (PCRWR, 2009). Some of the drought and salt tolerant species suitable for this region is provided in Table 3.

9 Framework for the Development of the Cholistan Desert

After the creation of Pakistan, most of the developmental programs were launched to the high potential areas of the Indus Valley blessed with the world's largest canal irrigation system. Chiefly, these developmental programs were aimed at dealing with agricultural issues, linking the yield gaps and increasing the agricultural productivity. Nevertheless, deserts, and especially Cholistan, were deprived mainly due to harsh climate and little inherent potential in agriculture. The following proposed framework if implanted can enhance the livelihoods of the local communities living in this area.

10 Framework for Extension Education

There is a great potential of rural extension that can popularize the desert reclamation and saline farming technologies. The extension services need to be undertaken to focus on the innovative farming technologies, cottage industries, local handicrafts, livestock rearing and its products. Rural Extension has many implications in the overall process of development and a brief account is presented below:

- There is a need to assess the carrying capacities of rangeland per capita required for each animal head.
- Today electronic media has emerged as a very strong and powerful tool in transferring new information that can be combined to disseminate the appropriate technologies on the development and reclamation of the desert.
- Most of the desert dwellers are living below the poverty line. They do not have the television sets for getting new information but they have Radio sets; therefore, Extension Service can be used through Radio for educating them.
- There is a need to develop government and people participatory approach for the development of rural communities. Through this way, Extension Service should adopt this approach to promote the technical skills of farmers by involving professionals, members of the civil society, and NGOs working in the desert areas.
- There is a need to produce skilled and educated human resources through its capacity building and training programs on the integrated desert development and management in order to make Cholistan productive and self-sustainable.
- Rural areas, particularly in the desert of Cholistan, have low literacy rates; therefore, it seems appropriate to educate them by adopting pictorial extension modes for disseminating the desired information.

11 Conclusions and Recommendations

The harsh climate in terms of high temperature and severe drought makes hindrance in agriculture productivity in the Cholistan desert. Besides, since this is fundamentally rangeland, so range management and livestock rearing are the only successful in areas with limited agriculture practice. This desert is facing various problems such extreme water scarcity, high wind erosion, over-grazing, deforestation, salinity, and increase in cultivation to the marginal lands. Followings are some of the remedial measures to address the aforementioned issues:

- Adoption of the saline agriculture practices by using the brackish and saline water to grow crops and grasses.
- Screening and evaluation of indigenous plant genetic resources and drought- and salt-tolerant crops capable of growing in less water requirements.
- Various water harvesting techniques as discussed in this chapter should be implemented on mass-scale for the collection of run-off water.

- For conserving water, various conservation techniques such as use of sprinkler and drip irrigation, etc. should be used.
- Cholistan desert is open and barren land with maximum solar energy. So, the use of solar device is better option to generate and consume solar energy.
- Exploring photovoltaic system can be a good option for the electrification of small villages.
- Promoting traditional cottage industry and local handicrafts can be a good strategy for combating poverty elevation and may uplift the livelihoods of desert dwellers.

References

Abd El-Ghani, M. M., Huerta-Martínez, F. M., Hongyan, L., & Qureshi, R. (2017). *The deserts of Pakistan* (pp. 457–573). Springer International Publishing.

Abdullah, M., Qureshi, R. H., & Ahmad, N. (1986). Responses of *Leptochloa fusca* to various types of substrate salinities. In *PAK.-US. Biosaline Research Workshop* (pp. 115–123). September 22–26 1985, Karachi, Pakistan.

Abdullah, M., Akram, M., Khan, A. D., & Qureshi, R. H. (1990). Internal water resources management by plants under various root environment stresses with special reference to kallar grass *Leptochloa fusca*. In *Proceedings of the National Seminar on Water Resources Development and its Management in Arid Areas*. October 6–8 1990, Quetta, Pakistan.

Adeel, Z., & Safriel, U. (2008). Achieving sustainability by introducing alternative livelihoods. *Sustainability Science, 3*(1), 125–133.

Ahmad, F. (2005). Historical and archaeological perspectives of soil degradation in Cholistan. *Journal of Geophysical Research, 8*, 31–35.

Ahmad, F. (2008). Runoff farming in reducing rural poverty in Cholistan Desert. *Sociedade & Natureza Uberlândia, 20*(1), 177–188.

Ahmad, F. (2010). *Leptochloa fusca* cultivation for utilization of salt-affected soil and water resources in Cholistan Desert. *Sociedade & Natureza, Uberlândia, 22*(1), 141–149.

Ahmad, F. A., Akbar, G., & Tahir, M. B. (1992). Developing Cholistan Desert—A prospective. *Progressive Farming, 12*(6), 35–40.

Ahmad, F. (2011). Soil classification and micro-morphology: A case study ofCholistan Desert. *Journal of Soil Science and Environmental Management, 2*(11), 321–329.

Akbar, G., & Arshad, M. (2000). Developing sustainable strategies for Cholistan Desert: Opportunities and perspectives. *Science Vision, 5*(2), 77–85.

Akbar, G. A., Taj, N. K., & Arshad, M. (1996). Cholistan Desert, Pakistan. *Rangelands, 18*, 124–128.

Akhtar, J., Waheed, R. A., Malik, K. A., & Haq, M. I. (1988). Effect of kallar grass growth on water transmission characteristics of salt—affected lands. In *First National Congress on Soil Sciences* (pp. 393–400). October 6–8 1985, Lahore, Pakistan.

Akhter, R., & Arshad, M. (2006). Arid rangelands in the Cholistan Desert (Pakistan). *Sécheresse, 17*(1), 210–217.

Akram, M., & Abdullah, M. (1990). Wind erosion and sand dune stabilization in the Cholistan desert. In *Proceedings of the International Symposium on Applied Soil Physics in Stress Environments* (pp. 323–334). January 22–26 1989. PARC, Islamabad, Pakistan.

Akram, M., & Chandio, B. A. (1998). Conjunctive use of rainwater and saline groundwater for desertification control in Pakistan through agroforestry and range management. *Journal of Arid Land, 7*(5), 161–164.

Akram, A., Wallyat, A. K., & Bashir, A. S. (1986). *Desertification processes in Cholistan Desert* (pp. 5–16). Technical report, Pakistan Council of Research in Water Resources, (PCRWR).

Akram, M., Kahlown, M. A., & Soomro, Z. A. (2008). Desertification control for sustainable land use in the Cholistan Desert, Pakistan. In C. Lee, & T. Schaaf (Eds.), *The future of drylands*. In *International Scientific Conference on Desertification and Drylands Research* (pp. 483–492). Tunis, Tunisia, June 19–21 2006.

Ali, S. I., & Qaiser, M. (1986). A phytogeographic analysis of the phanerogams of Pakistan and Kashmir. *Proceedings of the Royal Society of Edinburgh, Section b: Biological Sciences, 89*, 89–101.

Anonymous. (2001). What's in a Name? Or -Stan by Your Land. Slate Magazine. Retrieved 21 April 2020.

Anonymous. (2002). Cultural role of Cholistan in South-East, Punjab. *Journal of Asian Civilizations, 25–26*, 111.

Arshad, M., & Akbar, G. (2002). Benchmark of plant communities of Cholistan desert. *Pakistan Journal of Biological Sciences, 5*, 1110–1113.

Arshad, M., & Rao, A. R. (1994). Flora of Cholistan desert (Systematic list of trees, shrubs and herbs). *Journal of Economic and Taxonomic Botany, 18*, 615–625.

Arshad, M., Ashraf, M. Y., Ahmad, M., & Zaman, F. (2007). Morpho-genetic variability potential of *Cenchrus ciliaris* L., from Cholistan desert, Pakistan. *Pakistan Journal of Botany, 39*, 1481–1488.

Arshad, M., Ul-Hussan, A., Ashraf, M., Noureen, S., & Moazzam, M. (2008). Edaphic factors and distribution of vegetation in the Cholistan desert, Pakistan. *Pakistan Journal of Botany, 40*(5), 1923–1931.

Arshad, M., & Rao, A. R. (1995). Phytogeographical divisions of Cholistan desert. In *Proceedings of the sixth all Pakistan geographical conference* (pp. 55–61). (December 26–29, 1993). Department of Geography, Islamia University, Bahawalpur.

Baig, M. S., Khan, E. H., Zaheer, M. R., & Ahmad, M. (1975). *Reconnaissance soil survey of Cholistan* (p. 163). Directorate of Soil Survey of Pakistan, Lahore (Research Report).

Baig, M. S., Akram, M., & Hassan, M. A. (1980). Possibilities for range development in Cholistan desert as reflected by its physiography and soils. *Pakistan Journal of Forestry, 30*(2), 61–71.

Baig, M. B., Akbar, G., & Malik, A. H. (2007). Sustainable strategies to reclaim sandy desert of Cholistan-Opportunities and perspectives. In *Proceedings of ESDev—2007 CIIT* (pp. 1137–1145). Abbottabad: Pakistan.

Bhatti, G. R., Qureshi, R., & Shah, M. (2001). Ethnobotany of Qadan Wari of Nara Desert. *Pakistan Journal of Botany, 33*(SI), 801–812.

CDA. (1996). Master plan for the development in Cholistan, Cholistan Development Authority (CDA), Bahawalpur, Pakistan.

Chaudhry, S. A. (1992). *The Cholistan desert* (p. 34). Cholistan Institute of Desert Studies, Islamia University, Bahawalpur.

Chaudhry, M. S., Sial, N., & Chaudhry, N. A. (2004). Natural resources and their utilization, with special reference to Cholistan Desert, Pakistan. *Science Vision, 9*(1–2 & 3–4), 1–10.

FAO/ADB. (1993). *Cholistan area development project* (Report No. 59/53 ADB-PAK 58 (Final version)). Food and Agriculture Organization (FAO), Rome, Italy.

Farooq, U., Ahmad, M., & Saeed, I. (2009). Enhancing livestock productivity in the desert ecologies of Pakistan: Setting the development priorities. *The Pakistan Development Review, 48*(4), 795–823.

Ghazali, A. (2010). Analyzing the relationship between foreign direct investment domestic investment and economic growth for Pakistan. *International Journal of Finance & Economics, 47*(1), 123–131.

Gornish, E. (2019). Seed ball strategies for gardening and restoration in arid landscapes. The University of Arizona Cooperative Extension (pp. 1–2). University of Arizona College of Agriculture and Life Sciences Tucson, Arizona 85721.

Haq, M., & Khan, M. F. A. (1971). Reclamation of saline and alkaline soil by growing kallar grass. *Nucleus, 8*(4), 139–144.

Hussain, F., & Durrani, M. J. (2009). Seasonal availability, palatability and animal preferences of forage plants in Harboi arid range land, Kalat, Pakistan. *Pakistan Journal of Botany, 41*(2), 539–554.

Joshi, Y. C., Qadir, A., & Sharma, S. K. (1981). *Sequential changes in germination, root and shoot growth and anatomical characters of halophytes and glycophytes* (pp. 94–97). Annual report for 1980, Central Soil Salinity Research Institute, Karnal, India.

Kahlown, M. A., Akram, M., & Soomro, M. (2004). Assessment methodology for marginal drylands case study: Lal-Suhanra Biosphere Reserve, Pakistan. In *Proceedings of the second international workshop for sustainable management of marginal drylands (SUMAMAD)*. Shiraz: Islamic Republic of Iran.

Kahlown, M. A., & Majeed, A. (2002). Water resources situation in Pakistan: Challenges and future strategies. *Science Vision, 7*(3&4), 46–49.

Khan, S. R. A. (1992). *Agricultural development potential of Cholistan desert*. Linkers Publishers.

Khan, N. Z., Ali, K., & Anania, J. R. (1996). Productivity constraints of Cholistani farmers. *The Pakistan Development Review, 35*(4), 549–563.

Khan, A. D., Akram, M., & Abdullah, M. (1990). Rainwater harvesting in Cholistan desert. In *Proceedings of the National Seminar on Water Resources Development and its Management in the Arid Areas* (pp.59–80). October 6–8 1990, Quetta, Pakistan.

Khan, I. A., Younas, M., & Ali, A. (2010). Cholistan: A potential resource for agro-livestock development. http://icdd.uaf.edu.pk/TourReports/001.pdf

Madsen, M. D., Davies, K. W., Boyd, C. S., Kerby, J. D., & Svejcar, T. J. (2016). Emerging seed enhancement technologies for overcoming barriers to restoration. *Restoration Ecology, 24*, S77–S84.

Naz, N., Hameed, M., Ashraf, M., Ahmad, R., & Arshad, M. (2009). Eco-morphic variation for salt tolerance in some grasses from Cholistan Desert, Pakistan. *Pakistan Journal of Botany, 41*, 1707–1714.

PAMDU. (1986). Desertification process in Cholistan desert, PCRWR Publication No. PADMU 7/86, p. 155.

PCRWR. (2009). Cholistan Desert Pakistan: Case study, Pakistan Council of Research in Water Resources, (PCRWR): Islamabad, Pakistan Publication No 130/2004.

PCST. (2005). Water resources of Pakistan. A report of the expert committee, Pakistan Council for Science and Technology (PCST), Islamabad, Pakistan, pp. 1–256.

Pedrini, S., Merritt, D. J., Stevens, J., & Dixon, K. (2017). Seed coating: Science or marketing spin? *Trends in Plant Science, 22*, 106–116.

Qureshi, R. (2008a). Preliminary floristic list of Chotiari wetland Complex, Nawab Shah, Sindh, Pakistan. *Pakistan Journal of Botany, 40*(6), 2281–2288.

Qureshi, R. (2008b). Vegetation assessment of Sawan Wari of Nara desert, Pakistan. *Pakistan Journal of Botany, 40*(5), 1885–1896.

Qureshi, R. (2012). *The flora of Nara Desert, Pakistan*. Nova Science Publishers, Inc.

Qureshi, R. (2018). Medicinal uses of trees and shrubs by the inhabitants of Nara Desert, Pakistan. Ethnobotany and PhysiologyIn M. Ozturk & K. R. Hakeem (Eds.), *Plant and Human Health* (Vol. 1, pp. 391–407). Springer International Publishing AG.

Qureshi, R., & Ahmad, M. (2010). Some notes on the vegetation of Achhro Thar (White desert) of Nara region, Sindh, Pakistan. *Pakistan Journal of Botany, 42*(5), 2985–2994.

Qureshi, R., & Bhatti, G. R. (2005a). Nara Desert, Pakistan: Part I: Soils, climate and vegetation. *Rangelands, 27*(5), 27–31.

Qureshi, R., & Bhatti, G. R. (2005b). Nara Desert, Pakistan: Part II: Human Life. *Rangelands, 27*(5), 32–35.

Qureshi, R., & Bhatti, G. R. (2007a). Nara Desert, Sindh, Pakistan: Part III: Range types and their plant resources. *Rangelands, 29*(1), 26–29.

Qureshi, R., & Bhatti, G. R. (2007b). Nara Desert, Pakistan: Part IV: Destruction of natural habitats and its impact on plant diversity. *Rangelands, 29*(1), 30–33.

Qureshi, R., & Bhatti, G. R. (2008a). Diversity of micro-habitat and its plant resources. *Pakistan Journal of Botany, 40*(3), 979–992.
Qureshi, R., & Bhatti, G. R. (2008b). Ethnobotany of plants used by the Thari people of Nara Desert, Pakistan. *Fitoterapia, 79*, 468–473.
Qureshi, R., & Bhatti, G. R. (2010). Floristic inventory of Pai forest, Nawab Shah, Sindh, Pakistan. *Pakistan Journal of Botany, 42*(4), 2215–2224.
Qureshi, R., Khan, W. A., & Khan, B. (2009). Study of vegetation and smooth coated Otter in Chotiari wetlands complex, Sanghar, Sindh, Pakistan. *Pakistan Journal of Botany, 41*(5), 2507–2516.
Qureshi, R., Bhatti, G. R., & Memon, R. A. (2010). Ethnomedicinal uses of herbs from Nara Desert Pakistan. *Pakistan Journal of Botany, 42*(2), 839–885.
Qureshi, R. H., & Barrett-Lennard, E. G. (1998). *Saline agriculture for irrigated land in Pakistan: A handbook* (pp. 1–142). Monograph No 50, Canberra, Australia: Australian Centre for International Agricultural Research (ACIAR).
Rafay, M., Abdullah, M., Hussain, T., Rubyand, T., & Qureshi, R. (2016). Grass productivity and carrying capacity of the Cholistan desert rangelands. *Pakistan Journal of Botany, 48*(6), 2385–2390.
Rao, A. R., & Arshad, M. (1991). Perennial grasses of Cholistan Desert and their distribution. In *Proceeding National Seminar on People's Participation in the Management of Resources in Arid Lands* (pp. 6–11). Islamia University, Bahawalpur, Pakistan.
Rao, A. R., Arshad, M., & Shafiq, M. (1989). *Perennial grass germplasm of Cholistan desert and their phytosociology* (pp. 84). Cholistan Institute of Desert Studies, Islamia University, Bahawalpur.
Sial, N., & Arshad, M. (2003). Biodiversity in the surface-dwelling fauna Cholistan Desert Pakistan. *Pakistan Journal of Biological Sciences, 6*(3), 1195–1198.
Wariss, H. M., Mukhtar, M., Anjum, S., Bhatti, G. R., Pirzada, S. A., & Alam, K. (2013). Floristic composition of the plants of the Cholistan Desert, Pakistan. *American Journal of Plant Sciences, 4*, 58–65.
Wariss, H. M., Pirzada, S. A., Alam, K., Anjum, S., & Qureshi, R. (2014). Flora of Lal Suhanra National Park, Bahawalpur, Punjab Pakistan. *Pakistan Journal of Botany, 46*(4), 1331–1341.
WWAP. (2009). The United Nations world water development Report 3, facing the challenges. Jointly Published by United Nations Educational Scientific and Cultural Organization UNESCO and Earthscan London, UK.

Enhancing Food Security Through Climate-Smart Agriculture in Kenya

Ann Wamwea and Richard J. Culas

Abstract Achieving sustainable food and nutrition security is one of the major challenges facing countries in Sub-Saharan Africa. Food crisis and chronic hunger reported by African governments and the global international community are a clear affirmation that millions of people in the continent are at risk of economic and climatic shocks that have threatened food security. Even though Kenya continues to enjoy a stable economic growth which is described as the fastest growing economy compared to other African countries, access to adequate quantities of nutritious food remains a challenge for majority of the population, living in the arid and semi-arid regions, which make up 80% of the country's total land area. The situation in Kenya is intensified by the predominance of rain-fed subsistence agriculture being the main source of livelihoods to majority of the population. The impacts of climate change, which include increasing temperatures, shifting precipitation patterns, more severe and frequent extreme weather events and the loss of ecosystem services and biodiversity, undermine agricultural production systems and food systems. Therefore, climate-smart agriculture presents a great opportunity to transform agricultural systems and increase food production in the face of new realities of climate change. This study adopts the food and nutritional security conceptual framework which outlines the linkage between the three pillars of food security. Data analysis in the study presents a trend analysis of food security situation in Kenya based on hunger and food security indicators data derived from FAO. The study explores various policy measures that the Kenyan government has attempted to formulate over the years to achieve Sustainable Development Goals relating to poverty, nutrition, hunger, and environmental sustainability. This study also emphasizes the role of non-governmental organizations (NGOs) in Kenya to transform and reorient agricultural systems towards the efficient adoption of climate-smart practices.

A. Wamwea · R. J. Culas (✉)
School of Agricultural, Environmental and Veterinary Sciences, Charles Sturt University, Orange, NSW 2800, Australia
e-mail: rculas@csu.edu.au

A. Wamwea
e-mail: wamwea.ann@gmail.com

R. J. Culas
Gulbali Institute, Charles Sturt University, Wagga Wagga, NSW 2678, Australia

M. Behnassi et al. (eds.), *Climate-Smart and Resilient Food Systems and Security*,
https://doi.org/10.1007/978-3-031-65968-3_10

Keywords Food security · Climate-smart agriculture · NGOs · Kenya

1 Introduction

According to FAO et al. (2020), one of the most challenging issues is currently how to provide sufficient food to more than 7.9 billion people around the globe. Food security is considered a complex and multidimensional concept linked to several closely related factors, such as poverty, low access to basic social services, and inadequacy of some public policies. Food security, malnutrition, and chronic poverty are interrelated challenges becoming more complex in the current global scenario, especially for the developing countries in Sub-Saharan Africa. In recent times, the desert locust outbreak and the COVID-19 pandemic have undermined the livelihoods, food and nutrition security for many people who are already poor and vulnerable, characterized by fragile and informal economy, subsistence agricultural sector, and weak healthcare systems (FSIN, 2020).

In 2019, it was revealed that close to 690 million people (8.9% of world population) were estimated to be undernourished globally. The 2020 Global Report on Food Crisis reported that about 135 million people in 55 countries and territories are suffering from acute food insecurity while 73 million of this figure are living in 36 countries in Africa (FSIN, 2021). Although it is the desire for all nations to achieve food security regardless of the conditions in their political and socio-economic systems, in the developing countries food insecurity is a persistent challenge.

Schmidhuber and Tubiello (2007) state that the impact of climate variability on crop productivity and food security is a major subject of concern globally, especially in developing countries which are particularly vulnerable to the catastrophic risks of climate change. This is because of their primary reliance on natural resource dependent income sources for their livelihoods such as agriculture, few endowment of resources with which to adapt to the anticipated change in climatic patterns, and lack of planning and poor management at the central level (Thompson et al., 2010). These factors impede or delay recovery from climate-related shocks and, in some cases, even lead to increased economic and social damage.

In Sub-Saharan Africa, achieving sustainable food security is one of the main challenges that continues to be faced by most of the African countries and the international community at large. In Kenya, access to adequate quantities of nutritious food remains a challenge for most parts of the country, especially in arid and semi-arid regions which make up 80% of the country's total land area. Research measures the economic impact of climate on crops in Kenya and found a non-linear relationship between temperature and revenue on one hand and between precipitation and revenue on the other. Climate affects crop productivity, and it was found that the temperature component of global warming is much more important than precipitation (Kabubo-Mariara & Karanja, 2007).

Climate change is one of the major challenges to food security in Kenya affecting food security both directly and indirectly, thus impacting on agricultural production.

The situation is intensified given the predominance of rain-fed agriculture practiced by the majority of smallholder farmers in Kenya, which is highly sensitive to the weather patterns variability (Kabubo-Mariara & Kabara, 2018). Climate change increases the risks of hunger and under-nutrition through extreme weather events such as droughts, floods, and storms. These climate-related disasters have the potential to destroy crops, reduce production in vulnerable regions, rise health and sanitation risks, increase water scarcity, and intensify conflicts over scarce resources.

Schmidhuber and Tubiello (2007) state that climate change is expected to affect all the components that influence food security: availability, access, stability, and utilization. The overall availability of food is affected by changes in agricultural production as well as changes in the existing arable lands. The changes in food production, with other factors, impact food prices, affecting the ability of majority of poor households to access food markets, thereby reducing their dietary diversity. Decreased water availability and quality, caused by droughts in some areas as an impact of climate change, results in increased health and sanitation issues such as diarrheal disease, which has the potential of increasing malnutrition, thus negatively affecting food utilization.

Climate change in Kenya is quite evident indicated by the continuous rise in temperatures, irregular rainfall patterns with intense downpours causing floods in many parts of the country, which appear in cycles with severe droughts. Kenya is characterized by low and declining crop productivity, like many other developing countries, the nation is particularly susceptible to climate change due to its over-reliance on rain-fed agriculture, aridity, inadequate water supply, and degradation of many of its natural lands (Kabubo-Mariara & Kabara, 2018).

Kenya Climate Change Action Plan (2015–2030) recognized agriculture as the main sector with great potential for contributing, in an integrated way, to the achievement of a range of development goals relating to food security, nutrition, poverty reduction through climate change adaptation and mitigation. FAO (2018) proposed Climate-Smart Agriculture (CSA) as an integrative approach to mitigate the ongoing climate change impacts and as an adaptive approach to its consequences without compromising food security. CSA consists of three pillars:

- Sustainably increasing agricultural productivity and incomes (food security)
- Adapting and building resilience to climate change (adaptation)
- Reducing and/or removing greenhouse gas emissions (mitigation)

Climate-informed policies and institutional frameworks governing agriculture and food and nutrition security play a vital role in providing an enabling environment for building farmers' resilience and adaptive capacity within the context of a changing climate and for stimulating sustainable economic growth. There is a need for international, regional, and national research organizations to strengthen their engagement at national and county levels to inform policy development and implementation for enhanced resilience in agriculture and improved food and nutrition security.

Like many other Sub-Saharan African countries, Kenya has been facing frequent incidences of food deficits. In recent times, food shortages have led to hunger being one of the most evident and greatest problem facing the country. According to the

Kenya Food Security Steering Group, at least 1.4 million people in the country are facing acute hunger attributed to low levels of rainfall resulting to poor harvests in agricultural production. This has thereby limited the population to access sufficient food in terms of quantity and quality, rendering them predominantly reliant on food aid at any given time of year (FSIN, 2020).

According to FAO et al. (2020), Kenya is classified as a low-income and food deficit country among the one third of African countries whose food availability shows an average daily caloric availability below the recommended level of 2100 kcal. The country suffers from frequent droughts and famines with as many as 2–3 million people affected every time a famine or drought occurs, with the most of those affected living in the pastoral, semi-arid and arid areas located in the north, northeast and northwest of the country. Therefore, there is an urgent need to build resilience in rural communities and increase farmers' capacity to meet the challenges associated with climate change (FAO et al., 2020). Consequently, the agricultural sector has been identified to have the ability to support the growing food demand globally.

The primary objective of this study is to analyze food and nutrition security situation in Kenya in terms of availability, access, utilization, and stability; and to review the current state of policies and frameworks on climate change, agriculture, food, and nutrition security, including their level of integration. The study also examines the potential entry point of NGOs enhancing food security through CSA.

Section 2 of this chapter presents a conceptual framework in relation to the concepts (elements) of food and nutritional security as outlined by FAO. Section 3 provides a situational analysis of the food and nutrition in Kenya based on the four pillars of food security. Section 4 details an overview of the effects of climate change on agriculture and food security in Kenya. CSA practices in addressing climate change are discussed in Sect. 5. Section 6 describes the role of NGOs in enhancing food and nutrition security through CSA and reviews the policies and strategies stipulated by the Kenyan government in relation to agriculture, food security, and climate change. Conclusion is given in Sect. 7.

2 Concept of Food and Nutritional Security

The term "food security" is a broad concept that includes issues relating to nature, quality, and security of the production, supply as well as access to food which result in many discussions and definitions of food security. The concept is an important phenomenon with a global recognition that was first used in the mid-1970s at the World Food Conference (1974) defining food security in terms of food supply. At this point, food security was used to analyze the concept of food supply in which, at that time, there was a food crisis that led to the emergence of concerns that global food supply shortages would lead to political instability (FAO, 2006).

Furthermore, food security evolved as a concept that included national level with the evolvement of household and individual food supply. This food access at household levels continued to gain relevance with the understanding that household is a key social unit through which people access their food (Webb et al., 2006). The widely accepted definition of food security is derived from the 1996 World Food Summit Plan of Action describing food security to exist "when all people at all times have physical and economic access to adequate/sufficient, safe and nutritious food to meet their dietary needs and food preferences for an active and healthy life" (FAO, 1996). Food and nutrition security is achieved if adequate food (quantity, quality, safety, socio-cultural acceptability) is available and accessible for and satisfactorily used and utilized by all individuals at all times to live a healthy and active life".

Figure 1 shows a Food and Nutrition Security Conceptual Framework (FIVIMS) by FAO. In this conceptual framework, food availability, food access and food utilization determine the situation, referred to as the food and nutrition status of an individual or a household. The First part of the FIVIMS framework highlights vulnerability context that is related to the need of considering the underlying socio-economic, political, institutional, cultural, and natural factors of a country, as they impact on the different dimensions of food security. The second part of the conceptual framework is Food consumption determined by the food access at household level and the care practices. The third component is Food utilization which is the efficient and effective food consumption by the body as understood to be primarily dependent on a person's health status, which in turn is dependent on general health and sanitation conditions. Lastly, we have the Nutritional status as the fourth component in the FIVIMS framework showing the relationships and interactions of the main issues that affect an individual's nutritional status dependent on food consumption and the biological utilization of this food.

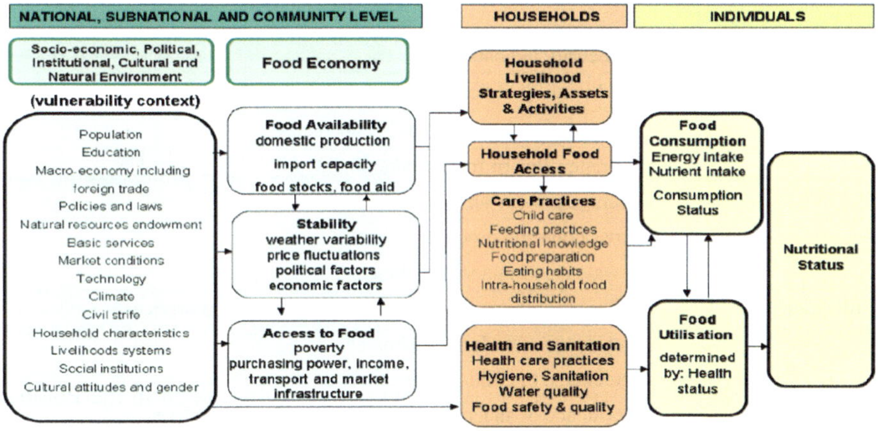

Fig. 1 Food and nutrition security conceptual framework (FIVIMS). *Source* http://www.fao.org/ag/agn/nutrition/docs/Food_and_Nutrition_Security-Strategy_Note.pdf

3 Analysis of Food Security Indicators in Kenya

This section presents a trend analysis of food security situation in Kenya based on hunger and food security indicators data derived from (FAOSTAT, 2021). The comparative data using trend analysis identifies the characteristics that are associated with the prevalence of food security in Kenya. Table 1 provides a summary, but not limited to, of the food security indicators to be analyzed.

3.1 Food Availability

Food availability means the physical availability of food from the farm gate and local markets. Availability is improved by a well-functioning market infrastructure with an adequate road and rail system as well as adequate storage and processing technologies (Gibson, 2012).

3.1.1 Average Dietary Energy Supply Adequacy

Figure 2 represents the Average dietary energy supply adequacy in percentage for a 3-year average period. It expresses the Dietary Energy Supply as a percentage of the Average Dietary Energy Requirement estimated for the population, to provide an index of adequacy of the food supply in terms of calories. The average dietary energy

Table 1 A summary of food security indicators

Dimension	Food security indicators
Availability	Average dietary energy supply adequacy
	Average value of food production
	Average protein supply
Access	Gross domestic product per capita
	Prevalence of undernourishment
Utilization	Access to improved water sources and sanitation services
	Prevalence of undernourishment
	Percentage of child malnutrition
	Prevalence of anaemia among women of reproductive age
	Prevalence of obesity in adult population
Stability	Cereal import dependency ratio
	Value of food imports over total merchandise exports
	Domestic food price level index volatility
	Per capita food production variability

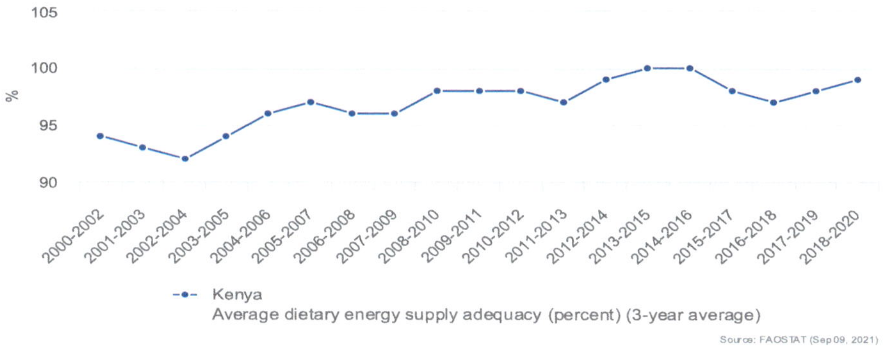

Fig. 2 Average dietary energy supply adequacy (3-year average)

supply (kcal/capita/day) is an indicator calculated at the national level serving as an estimate of the number of calories from foods available for human consumption. This indicator helps determine whether a country's food supply contains enough dietary energy to meet aggregate population needs, and whether measures need to be taken to improve the amount of dietary energy available for the population (FAO, 2011). Kenya dietary energy supply adequacy, although slightly fluctuated over time, has been improved, it was at the level of 99% in 2020, up from 98% the previous year; the upward trend shows the gradual increase in the value of food production.

3.1.2 Average Value of Food Production

Figure 3 expresses the food net production value as estimated by FAO and published by FAOSTAT, in per capita terms. It provides a cross country comparable measure of the relative economic size of the food production sector in the country, in terms of constant 2004–2006 International $ per person, by a 3-year average period. This indicator is calculated in three-year averages, from 2000–02 to 2016–18. The average three-year period reduces the impact of possible errors in estimated production due to the difficulties in properly accounting of stock variations in major food production.

3.1.3 Average Protein Supply

Figure 4 data represents the average protein supply of food in grams per capita per day for a 3-year average providing a national-level estimate of the availability of protein. This indicator offers insight into the nutritional quality of the food supply in the country. Lancet (2008) has shown that consumption of protein from animal source foods protect children against stunting. The average protein supply for Kenya increased from 56.3 g/capita/day in 2004 to 60.7 g/capita/day in 2018 growing at an average annual rate of 0.55%.

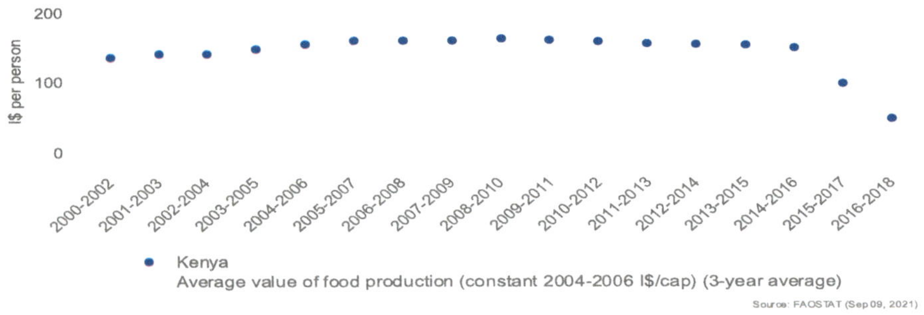

Fig. 3 Average value of food production (constant I$ per person) (3-year average)

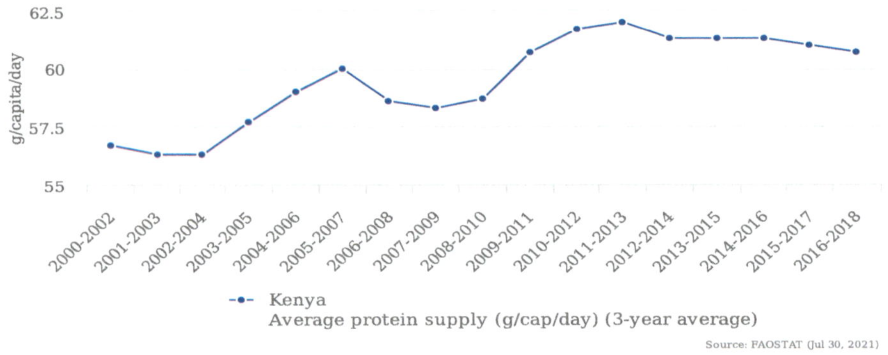

Fig. 4 Average value of protein supply (g/capita/day) (3-year average)

Figure 5 shows the national average protein supply (expressed in grams per capita per day) including the following groups: meat, offal's, animal fats products, milk products, eggs, fish, seafood products and aquatic products. This indicator provides information on the quality of the diet calculated in three-year averages of stock variations in major food. However, Kenya's average supply of protein of animal origin shows a downward trend from 2009 to 2011 period.

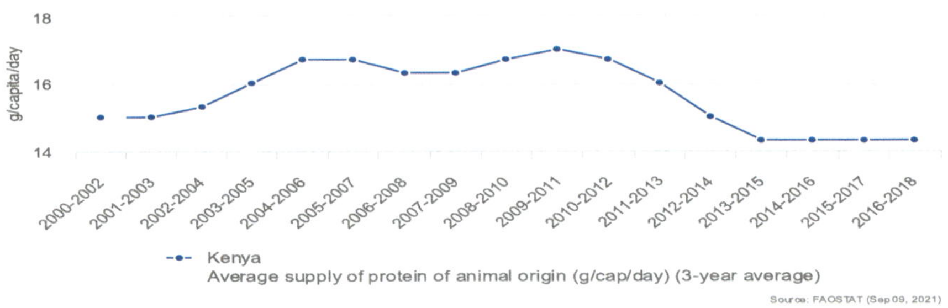

Fig. 5 Average value of protein supply of animal origin (g/capita/day) (3-year average)

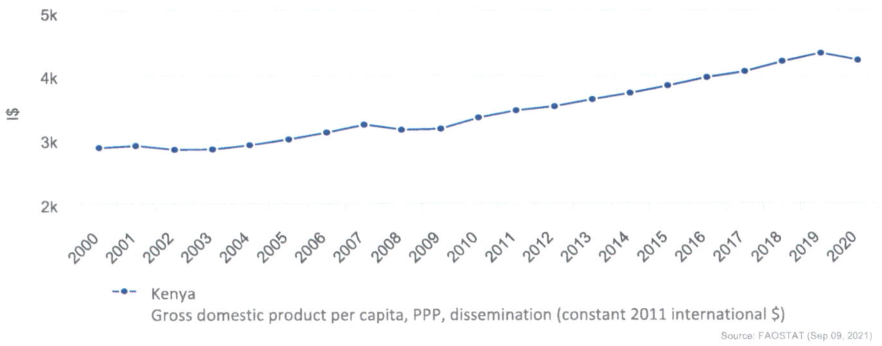

Fig. 6 Gross domestic product per capita (constant 2011 international $)

3.2 *Food Access*

Food access entails ensuring people have adequate access to food, both physical and economic, by growing it, purchasing it, being gifted it, or bartering or trading for it (Gibson, 2012).

3.2.1 Gross Domestic Product Per Capita

Figure 6 shows GDP per capita in terms of the total value of all goods and services produced in a country in a year. It is a very important indicator of the economic strength of a country and a positive change is an indicator of economic growth. The upward trend shown in Fig. 6 from 2009 to 2010 period indicates that the economic growth in Kenya is averaged by 5.4% making the country the fastest growing nation in Sub-Saharan Africa.

3.2.2 Prevalence of Undernourishment

Figure 7 shows the prevalence of undernourishment indicator expressing the probability that a randomly selected individual from the population consumes a few calories that is insufficient to cover her/his energy requirement for an active and healthy life. The indicator is computed by comparing a probability distribution of habitual daily dietary energy consumption with a threshold level called the minimum dietary energy requirement. This indicator is the traditional FAO hunger indicator, adopted as official Millennium Development Goal indicator for Goal 1 of eliminating hunger. In 2020, prevalence of undernourishment for Kenya was 24.8%. Though Kenya prevalence of undernourishment fluctuated substantially in recent years, it tended to decrease through 2006–2020 period ending at 24.8% in 2020.

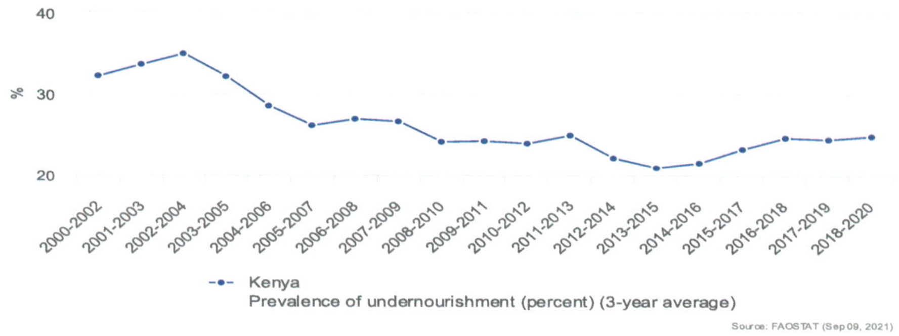

Fig. 7 Prevalence of undernourishment (%) (3-year average)

3.3 *Food Utilization*

Food utilization is simply the ability of a person to optimally or at least effectively absorb the food they eat (Gibson, 2012), and as a result their health status are defined.

3.3.1 Child Malnutrition

Figure 8 shows indicator of child malnutrition exploring several indicators by which we can measure, and track progress related to malnutrition. Inadequate nutrition in children under age five stunts their growth and development leading to undernutrition, which puts children at greater risk of dying from common infections. Three important indicators, stunting, wasting and overweight, under this category have been taken into consideration. Child stunting is the proportion of children under the age of five who suffer from low height for their age (Global Hunger Index, 2019). Child wasting, according to WHO, is the proportion of children under the age of five who suffer from low weight for their height. Child overweight is the proportion of children under the age of five over the weight standards of WHO Child Growth Standards median, reflecting chronic malnutrition.

The analysis of the data between 2000 and 2020 shows a downturn trend of all malnutrition indicators. The year 2008–2009 and 2014 from the figures shows that the stunting rate significantly dropped from 35.2 to 26.0% and the wasting rate decreased from 7.0 to 4.0%. As of 2014, 26.2% of children under five years suffered from chronic malnutrition in Kenya. The lack of adequate nutrients over a long period leads to children experiencing growth failure.

3.3.2 Access to Safe Drinking Water and Sanitation Services

Figure 9 shows basic drinking water services and sanitation services in Kenya. Water is key to human life, food security and nutrition; therefore, safe drinking water and

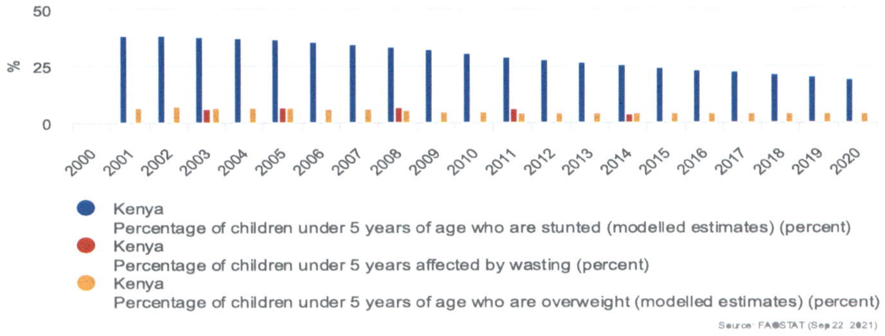

Fig. 8 Percentage of child malnutrition

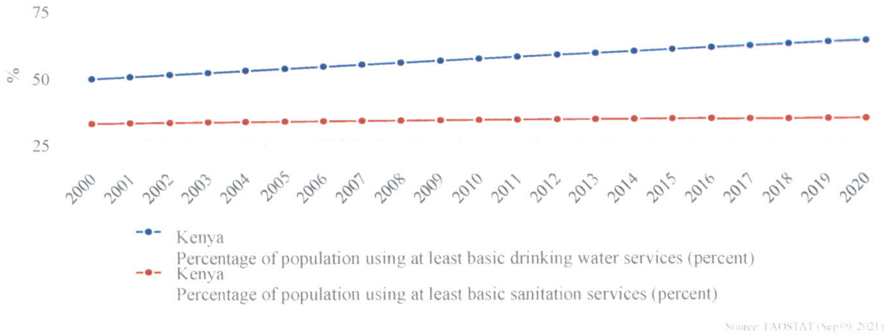

Fig. 9 Percentage of the population using at least basic drinking water and sanitation services

sanitation are fundamental to the good nutrition, health, and dignity of all (FAO, 2015). Although, according to the graph, improvement has been made, access to water and sanitation in Kenya has not been keeping pace with population growth, as only 58% of Kenyans have access to basic drinking water and 30% have access to basic sanitation currently. Since 2000, access to safe drinking water has increased by 12%, while access to basic sanitation has fallen by 5%. In Kenya, 9.9 million people drink directly from contaminated surface water sources and an estimated five million people practice open defecation. Only 25% have hand-washing facilities with soap and water at home.

3.3.3 Anaemia Among Women of Reproductive Age

Figure 10 represents data on the prevalence of anaemia among women of reproductive age (15–49 years). Prevalence of anaemia among pregnant women is the percentage of pregnant women whose haemoglobin level is at low level. Anaemia is a condition in which the number of red blood cells or their oxygen-carrying capacity is insufficient

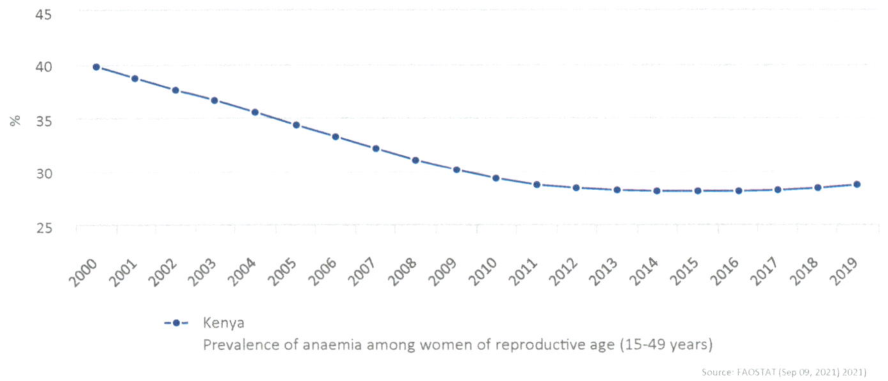

Fig. 10 Prevalence of anaemia among women of reproductive age (15–49 years)

to meet physiologic needs, which vary by age, sex, altitude, smoking status, and pregnancy status. Children under age 5 and pregnant women have the highest risk for anaemia (WHO, 2005). Figure 10 shows that the prevalence of anaemia in women of reproductive age in Kenya from 2000 has been above 25%. WHO has set a global target of achieving a 50% reduction of anaemia in women of reproductive age by 2025 (WHO, 2005).

3.3.4 Obesity in the Adult Population

Figure 11 shows that, in Kenya, the prevalence of obesity in the adult population has increased gradually over the years, from approximately 3.5% in 2000 to 7% in 2016. The higher risk of obesity in Kenya is associated with increased consumption of high-calorie, high-fat diets.

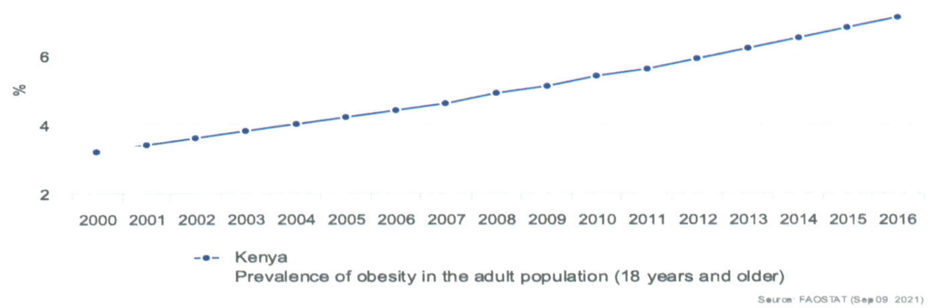

Fig. 11 Prevalence of obesity in the adult population (18 years and older)

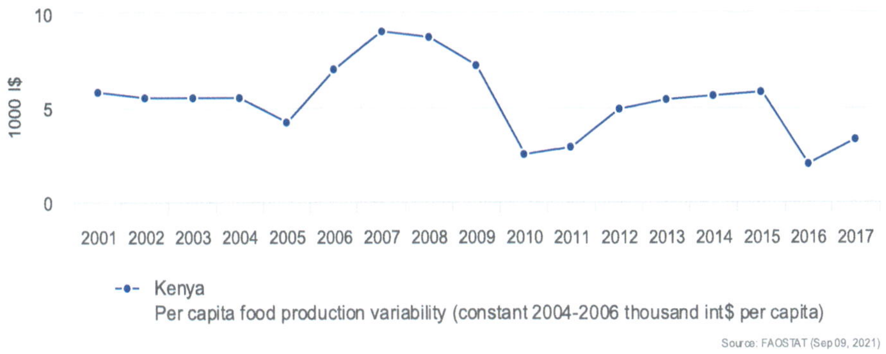

Fig. 12 Per capita food production variability (I$ per person constant 2004–06)

3.4 Food Stability

Food stability refers to household or individual must have access to sufficient food at all times, and is often discussed in the international development literature in association with the term 'vulnerability', which refers in part to the exposure of a population to stresses and shocks of various kinds (DFID, 1999).

3.4.1 Per Capita Food Production Variability

Figure 12 presents data on the per capita food production variability from 2001 to 2017 per thousand $ per capita (where 2004–2006 period is referred as constant). Per capita food production variability corresponds to the variability of the food net per capita production as disseminated in FAOSTAT. This indicator compares the variations of the per capita food production across countries and time. Figure 12 shows that Kenya's mean per capita food production has dropped drastically over the past few years due to increased population growth rates and climatic variability.

3.4.2 Value of Food Imports Over Total Merchandise Exports

Figure 13 details the value of food imports over total merchandise exports. This indicator provides a measure of vulnerability and captures the adequacy of foreign exchange reserves to pay for food imports. This indicator has implications for national food security depending on production and trade patterns. Kenya's food imports have been increasing in recent years.

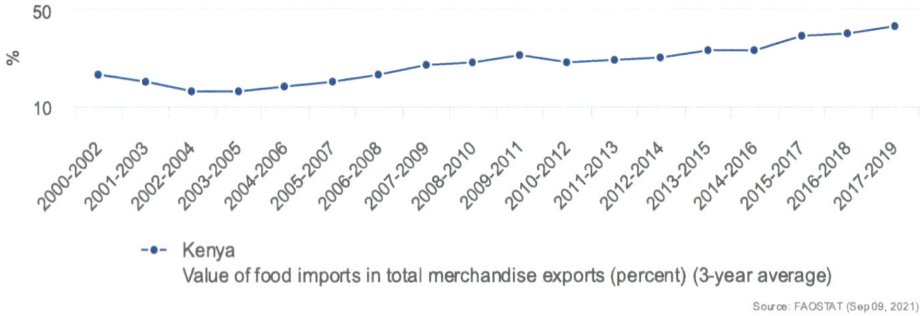

Fig. 13 Value of food imports over total merchandise exports (%) (3-year average)

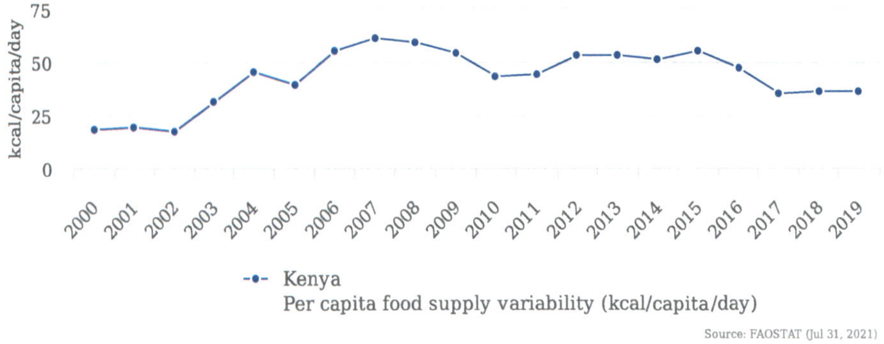

Fig. 14 Per capita food supply variability (kcal/capita/day)

3.4.3 Per Capita Food Supply (kcal/person/day)

Figure 14 displays the per capita food supply measured in kcal per person per day, which compares the variations of the food supply across countries and time. Volatility in the food supply, presumably reflected in price volatility, affects vulnerable households' ability to plan effectively within their resource constraints.

3.4.4 Cereal Import Dependency Ratio

Figure 15 shows that an increase in the total cereal production continues to play a critical role for food security in the country, contributing nearly half of both daily calories and protein intake among the population. Land under cereal production (hectares) in Kenya was reported high in 2018, according to the World Bank collection of development indicators. However, yield (production per unit of land) can be improved to increase the total production.

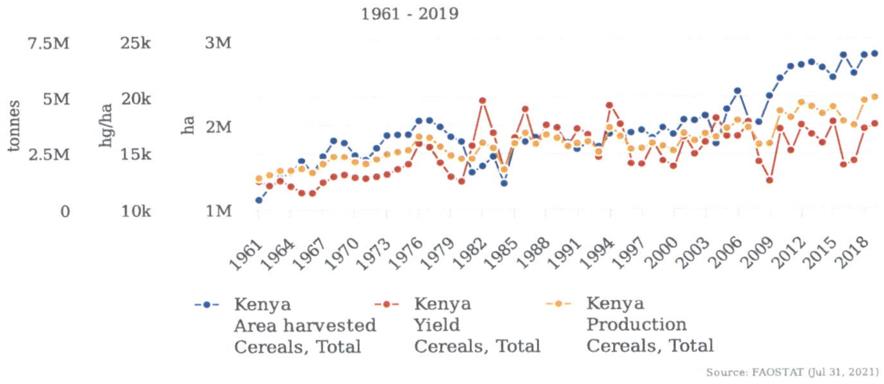

Fig. 15 Cereals total production

4 Effects of Climate Change on Agriculture and Food Security in Kenya

AGRA (2017) reports that agriculture is the mainstay of most African economies contributing up to 55% of the continent's GDP. The sector is the base for food security and acts as the main source of livelihood to approximately 85% of the population that rely on rain-fed food production. However, despite the high dependence on this sector, agriculture remains one of the most vulnerable sectors and is set to continue being exposed to the impacts of climate variability and change (Shah et al., 2009). The low performance of agricultural sector, especially in Sub-Saharan Africa, has been triggered by climate change, low incentives to intensification, poor market access, unfavorable topography, poor soils, and inadequate policy implementation (FAO, 2010).

In Kenya, the agriculture sector is a key contributor to the country's economic growth with about 52% share in total to the gross domestic product. The agricultural sector contributes to employment creation, poverty reduction and food security of majority rural households in the country (GoK, 2017). The sector is among the key priorities in the economic pillar of the Kenya Vision 2030, which aims at achieving an innovative, commercially driven, and modern agricultural sector through institutional reforms, increased productivity, land-use transformation, increased access to markets and development of Kenya's Arid and Semi-Arid Lands (GoK, 2010).

According to the Inter-governmental Panel on Climate Change (IPCC), climate change is expected to have a significant lead in increasing the frequency and intensity of natural disasters and extreme weather events, such as droughts, floods, and hurricanes, rising sea levels and the contamination/salinization of water supplies and agricultural lands. It is also expected that climate variability will cause changes in rainfall patterns, with an expected reduction in agricultural production in already fragile areas of sub-Saharan Africa (IPCC, 2020).

Recent studies show that Kenya is very vulnerable to climate change, with current projections suggesting that the temperature will rise to 2.5 °C between 2000 and 2050. In addition, rainfall will become more intense and less predictable leading to drought occurrence. This will present major challenges for food security and water availability, especially in Kenya's Arid and Semi-Arid Lands (Kabubo-Mariara & Kabara, 2018). The increasing vulnerability of climate change, combined with other underlying factors such as land degradation, declining soil health, high levels of poverty, declining farm sizes, inadequate expertise, low use of technology and poor agricultural farming practices and policies, has greatly hindered the potential for yield increase in Kenya (Morton, 2007).

Climatic variability and change directly affect agricultural production and food security given that most of the population in Kenya live in the rural areas and rely on agriculture as their major source of livelihood. Generally, almost 98% of agriculture practiced in Kenya is rain fed and depends entirely on the bimodal rainfall pattern. It is estimated that only 16% of Kenya's land receives sufficient and consistent rainfall suitable for crop production, whereas the remaining 84% is classified as Arid and Semi-Arid Lands with an average annual rainfall of 400 mm (Shah et al., 2009).

Smallholder farmers are highly vulnerable to climatic and extreme weather conditions as their options for diversifying their resources and income sources are limited. Their vulnerability is worsened by the spread of HIV/AIDS, lack of access to land due to the traditional land tenure systems, lack of adequate water, low levels of technology, and education and institutional mismanagement (Ochieng et al., 2016).

A report by the GoK (2013) states that the situation of food security in Kenya has further been complicated by the high global food prices that are not afforded by many households due to their high levels of poverty. The impacts of higher food prices often result in food crisis and worsened livelihood conditions for the rural poor population. Table 2 gives a summary of the potential impacts of climate change and variability on the four pillars of food security.

The changing temperature and rainfall patterns have profound impacts on Kenya's socio-economic sectors, which are considered climate sensitive including agriculture, livestock, forestry, and energy. These impacts are increasingly worsening and causing an unpredictable future situation if national efforts are not enhanced to make agriculture adaptive to climate changes. This calls for predictability and vulnerability studies on climate variations to reduce food fluctuations probabilities, especially in the rural parts of the country.

Majority of smallholder farmers in Kenya depend on agriculture for survival. These impacts of climate change call for clear response strategies in terms of mitigation and adaptation, particularly to the smallholder farmers in Kenya (Ochieng et al., 2017). It is, therefore, necessary to build their adaptive capacity and resilience to climate change to enable them to protect their livelihoods and achieve food security.

Climate-Smart Agriculture (CSA) presents the opportunity to meet increasing food demands in the face of climate variability. CSA practices have the potential to reverse this situation because of the potential benefits of improving productivity and high income, reducing or removing greenhouse gases, and improving household food security.

Table 2 Impact of climate change on food security

Food security dimension	Potential impacts of climate change
Availability	Decrease in rainfall and rise in evapotranspiration leading to low yields from rain-fed crop production
	Decrease in fertility of soils and land degradation
	Episodes of crop pests and diseases likely to give rise to pre-harvest crop losses
Access	Lower profit margins due to decreased yields and increased cost of inputs
	Increase in demand and competition for available food supplies leading to exorbitant food prices
	Displacement of communities by extreme weather events leading to challenges in accessing food
Stability	Greater uncertainties of food supplies due to erratic weather events
	Unreliable earnings from crop production leading to decreased purchasing capacity
Utilization	Rise in temperature that may alter the safety of food
	Reduced quantities of water supplies, and quality leading to low nutrition
	Climate-related diseases that can affect uptake of some food stuffs

Source Kogo et al. (2021)

5 Addressing Climate Change Through CSA in Kenya

Global population is expected to increase to 9.6 billion with a projected rise in food demand by a minimum of 50–70% by 2050. At the same time, it is forecasted that there will be a 5% decrease in global food production by the same year (Lobell & Gourdji, 2012). The projections show that feeding an increased world population would require raising overall food production. Therefore, by formulating effective adaptation strategies, it is possible to reduce or even avoid some of the negative impacts of climate change on food security.

In a FAO meeting at the Hague conference on Food Security and Climate Change in 2010, the concept of CSA was first presented. FAO defined CSA as agricultural activity that "sustainably and efficiently increases productivity and incomes (adaptation), reduces or removes greenhouse gases (mitigation) and enhances achievement of national food security and development goals" (FAO, 2010). The concept of CSA was generally signifying a balance between food production and environmental stability without compromising any of them. CSA unites numerous issues relating to agricultural and other global development objectives linking to environmental and economic issues.

Achieving the four dimensions of food security availability, access, utilization, and stability of food supply needs to be the overall goal of food production and distribution systems in developing countries (Totin et al., 2018). In the face of climate change,

CSA presents an opportunity to meet the growing world's food demand. According to Campbell et al. (2014), CSA includes farming practices and technologies that have the potential to achieve the "triple-win" of increasing food security and incomes, and climate change adaptation and mitigation through the careful combination of different CSA practices. There are several CSA practices that have been developed globally and among these include irrigation, deep-ploughing, crop rotation, mixed cropping, terracing, mulching, zero or minimum tillage, and cover crops (Lan et al., 2018).

Over the years, CSA is gaining momentum in Kenya, this is attributed to: (i) the fact that agriculture is recognized as a sector with great potential for contributing to the achievement of a range of development goals related to food security and nutrition, poverty reduction, and climate change adaptation and mitigation; and (ii) the recognition of Kenya National Adaptation Plan 2015–2030 on critical CSA practices such as agroforestry, conservation tillage, the limited use of fire in agricultural areas, cultivation of drought-tolerant crops, water harvesting, and integrated soil fertility management, among others (GoK, 2015).

Apart from traditional agricultural techniques, Kenyan farmers have started adopting new, improved technologies, as evident in both crop and livestock production. Some examples include biogas production using biodigesters (especially applied in intensive dairy production), and improved pastures management in pastoral systems in the highlands and sub-humid areas, as well as in intensive and extensive dairy production (mostly through grass–legume associations), among others. For crop production, Kenyan farmers practice terracing and contour bunds for maize, beans, coffee production, use water-efficient irrigation techniques in rice cultivation (mostly in the east), and adopt drought-tolerant crop varieties for cereals and legumes (beans, pigeon peas, cowpeas) in semi-arid areas. Many of these practices help build system's resistance to pests and diseases, such as the case of drought-tolerant varieties for crop and livestock production (Herrero et al., 2010).

Despite the potential benefits presented by CSA practices in the adaptation and mitigation of climate change, Westermann et al. (2018) state that developing countries are not achieving their full potential given the low levels of adoption explained by the low level farm and farmer characteristics, institutional constraints and the complex characteristics of CSA practices (Bernier et al., 2015).

6 Role of NGOs in Enhancing Food Security Through CSA

The term "non-governmental organization" (NGO) was first created by the United Nations to refer to international bodies, which have established headquarters, an executive team, a democratically adopted constitution, and financial independence from the government agencies (Martens, 2002). Over the years, the term NGO has been applied to a wide variety of organizations operating outside the UN framework, such as advocacy organizations, human rights groups, village associations, community environmental groups, faith-based organizations, and farmer cooperatives. A non-governmental organization is an organization that is not part of the government and is not funded by the state. NGOs are therefore typically independent of governments (Willetts, 2002). Although the definition can technically include profit corporations, the term is generally used to mean social, cultural, legal, and environmental advocacy groups having the goals that are primarily non-commercial.

In the last two decades, there has been an increased number of NGOs working in Kenya and other African countries and the various resources that they control have grown significantly (Bingen et al., 2003). This influx was influenced by several factors. One of them being poor standards of living as many African governments failed to address the global environmental issues causing widespread poverty and food insecurity in the countries (Manji & O'Coill, 2002). Second, majority of African governments were faced with poor management, inefficiencies in delivery of services, and rampant corruption. This led to the development of non-profit making organizations for the purpose of handling and managing funds in the process of project implementations (Woods, 2000). Third, NGOs became more popular as change agents of sustainable development addressing issues like climate change, food insecurity, biodiversity, and poverty and demonstrated a comparative advantage over government institutions. Fourthly, Owusu (2003) adds that there was a change of donor policies and international development agencies in 1980's that made several African countries to have adjustment policies for them to be able to receive aid from NGOs.

The extraordinary growth in the number of development-oriented NGOs worldwide is associated with the shift in donor attitudes about development assistance. NGOs are now widely believed to have a comparative advantage over state-led and market-led strategies. In the past two decades, Kenya has witnessed the rapid and unprecedented growth of local NGOs within its borders. This trend reflects similar NGO sector growth throughout the developing world. Scholars have attributed the growth of these key civil society actors to increased policies promoting international donors, an ideology which favors non-state service providers (Michael, 2004).

Over the past two decades, Kenya has experienced significant improvement of food security as a result of CSA related activities promoted by NGOs. Generally, NGOs have played an important role in promoting economic growth, developing society, improving communities, and promoting citizen participation; thus, critically contributing to the socioeconomic development of countries where government capacity and revenues are limited. Specifically, NGOs are scaling up CSA to trigger the desired transformation in agricultural production and food systems with supportive policies, institutions, and financing, that together create an enabling environment for CSA at local, national, and international levels (Totin et al., 2018).

Recently, several important international agreements have been reached that will shape national CSA planning and implementation. The 2030 Agenda for Sustainable Development will serve to guide national development plans over the next years. There are three components of the 2030 Agenda for Sustainable Development: the SDGs, which set the global policy framework; the Paris Agreement on climate change; and the Addis Ababa Action Agenda, which identifies the broad set of financial and non-financial means for implementation of the agenda. In all these components, agriculture is prioritized as a crucial sector to ensure that the needs of both the people and the planet are met.

In this regard, NGOs promote the development of enabling policies and the promotion of an adequate regulatory and legislative environment to promote agricultural production. Increasing policy coherence calls for a systematic assessment of current policies and their intended and unintended effects on the set of development objectives prioritized by a country, including those pertaining to CSA. The NGOs play different roles in this direction: (i) policy implementation and partnership; (ii) information, micro-finance, and insurance access; and (iii) access to agricultural inputs and output markets.

7 Conclusion

Globally, achieving food security is a key policy challenge in many nations more so the developing ones. Measures to address food security need to be multidimensional at all levels and involve different stakeholders. Even though policies targeting food insecurity were developed and implemented since 1980s, the results are not visible. The government has always worked to back up initiatives that aim at increasing the productivity in farms through encouraging utilization of sophisticated technologies. Additionally, national food security has not yet guaranteed household food accessibility and availability.

Enhancing food security requires transition in the systems of food production that are more productive and efficient to preserve the limited natural resources at our disposal. The process of transitioning requires changes in the technical and financial mechanisms. There is a need to set modernized policies on food security that reflect a more unified approach in addressing the problem. The unified approach of the policies will be done to reflect the complicated nature of food security and to try and resolve conflicts and inconsistencies that appear in the current policies, and which are a hindrance to aggressive pursuit of food security in the country.

In Kenya, the majority of population live in the arid and semi-arid parts of the country where agriculture is considered as a key sector due to its significant contribution to the economy, livelihoods, and food security. However, climate change impacts causing seasonal changes in rainfall and temperature has greatly affected the growth in the agricultural sector. It is, therefore, important that drastic measures in adaptation and mitigation are embraced by both the government and NGOs targeting increase in food production, and hence promote the achievement of food security.

There is the need for the national and county governments in Kenya to minimize the climate variability impacts on the agriculture sector by expanding irrigation schemes and promoting the adoption of drought-tolerant crops such as cassava, sorghum, millet, potatoes in the total agricultural production. They should also adopt the use of small dairy stock by all rural households as these animals are potential sources of income and high biological value protein in the form of meat and milk. This will help achieving nutrition security in the country.

References

AGRA. (2017). *Africa agriculture status report: The business of smallholder agriculture in Sub-Saharan Africa*. https://agra.org/wp-content/uploads/2017/09/Final-AASR-2017-Aug-28.pdf

Bernier, Q., Meinzen-Dick, R. S., Kristjanson, P. M., Haglund, E., Kovarik, C., Bryan, E., Ringler, C., & Silvestri, S. (2015). *Gender and institutional aspects of climate-smart agricultural practices: Evidence from Kenya*. CCAFS Working Paper No. 79. CGIAR Research Program on Climate Change, Agriculture and Food Security (CCAFS).

Bingen, J., Serrano, A., & Howard, J. (2003). Linking farmers to markets: Different approaches to human capital development. *Food Policy, 28*(4), 405–419.

Campbell, B. M., Thornton, P., Zougmoré, R., Van Asten, P., & Lipper, L. (2014). Sustainable intensification: What is its role in climate smart agriculture? *Current Opinion in Environmental Sustainability, 8*, 39–43.

DFID (Department for International Development). (1999). DFID sustainable livelihoods guidance sheets. DFID sustainable livelihoods guidance sheets | ENN (ennonline.net).

FAO. (1996). *Rome declaration on food security and world food summit plan of action*. http://www.fao.org/3/w3613e/w3613e00.htm

FAO. (2006). *Food security*. http://www.fao.org/fileadmin/templates/faoitaly/documents/pdf/pdf_Food_Security_Cocept_Note.pdf

FAO. (2010). *Climate-smart'agriculture: Policies, practices and financing for food security, adaptation and mitigation*. http://www.fao.org/3/an177e/an177e.pdf

FAO. (2011). *Dietary protein quality evaluation in human nutrition*. (FAO Food Nutrition Paper). http://www.fao.org/ag/humannutrition/35978-02317b979a686a57aa4593304ffc17f06.pdf

FAO. (2015). *Water for food security and nutrition*. http://www.fao.org/3/a-av045e.pdf

FAO. (2018). *Upscaling climate smart agriculture. Lessons for extension and advisory services*. (Occasional Papers on Innovation in Family Farming). http://www.fao.org/3/I9209EN/i9209en.pdf

FAO, UNICEF, WFP & WHO. (2020). *The state of food security and nutrition in the world 2020. Transforming food systems for affordable healthy diets*.

FAOSTAT. (2021). *FAO country profile*. Retrieved 14 May 2021 from http://www.fao.org/faostat/en/#country/114

FSIN. (2020). *Global report on food crises*. FAO Rome. https://www.wfp.org/publications/2020-global-report-food-crises

FSIN. (2021). *The state of food security and nutrition in the world 2021*. https://www.wfp.org/publications/global-report-food-crises-2021

Gibson, M. (2012). Food security: A commentary: What is it and why is it so complicated? *Foods, 1*(1), 18–27.

Global Hunger Index. (2019). *A closer look at hunger and undernutrition*. Retrieved 24 May 2021 from https://www.globalhungerindex.org/case-studies/2018-kenya.html#fn-03

GoK. (2010). Kenya vision 2030; The popular version. Government Printer.

GoK. (2013). *Ministry of agriculture status report on food security assessment*.

GoK. (2015). *Kenya national climate change action plan 2015–2030*.

GoK. (2017). *Capacity building strategy for agriculture sector*.

Herrero, M. T., Ringler, C., Steeg, J. v. d., Thornton, P. K., Zhu, T., Bryan, E., Omolo, A., Koo, J., & Notenbaert, A. M. O. (2010). Climate variability and climate change: Impacts on Kenyan agriculture. https://cgspace.cgiar.org/bitstream/handle/10568/3840/climateVariability.pdf

IPCC. (2020). *Global warming of 1.5 ºC*. https://www.ipcc.ch/sr15/

Kabubo-Mariara, J., & Kabara, M. (2018). *Climate change and food security in Kenya*. Routledge.

Kabubo-Mariara, J., & Karanja, F. K. (2007). The economic impact of climate change on Kenyan crop agriculture: A Ricardian approach. *Global and Planetary Change, 57*(3–4), 319–330.

Kogo, B. K., Kumar, L., & Koech, R. (2021). Climate change and variability in Kenya: A review of impacts on agriculture and food security. *Environment, Development and Sustainability, 23*(1), 23–43. https://doi.org/10.1007/s10668-020-00589-1

Lan, L., Sain, G., Czaplicki, S., Guerten, N., Shikuku, K. M., Grosjean, G., & Läderach, P. (2018). Farm-level and community aggregate economic impacts of adopting climate smart agricultural practices in three mega environments. *PLoS ONE, 13*(11), e0207700.
Lancet. (2008). *Maternal and child undernutrition.* https://doi.org/10.1016/S0140-6736(07)61869-8
Lobell, D. B., & Gourdji, S. M. (2012). The influence of climate change on global crop productivity. *Plant Physiology, 160*(4), 1686–1697.
Manji, F., & O'Coill, C. (2002). The missionary position: NGOs and development in Africa. *International Affairs (Royal Institute of International Affairs 1944-), 78*(3), 567–583. http://www.jstor.org/stable/3095891
Martens, K. (2002). Mission impossible? Defining nongovernmental organizations. *Voluntas: International Journal of Voluntary and Nonprofit Organizations, 13*(3), 271–285.
Michael, S. (2004). Undermining development: The absence of power among local NGOs in Africa. James Currey Publishers.
Morton, J. F. (2007). The impact of climate change on smallholder and subsistence agriculture. *Proceedings of the National Academy of Sciences, 104*(50), 19680–19685.
Ochieng, J., Kirimi, L., & Makau, J. (2017). Adapting to climate variability and change in rural Kenya: Farmer perceptions, strategies and climate trends. *Natural Resources Forum, 41*(4), 195–208.
Ochieng, J., Kirimi, L., & Mathenge, M. (2016). Effects of climate variability and change on agricultural production: The case of small scale farmers in Kenya. *NJAS - Wageningen Journal of Life Sciences, 77*, 71–78. https://doi.org/10.1016/j.njas.2016.03.005
Owusu, F. (2003). Pragmatism and the gradual shift from dependency to neoliberalism: The World Bank, African leaders and development policy in Africa. *World Development, 31*(10), 1655–1672.
Schmidhuber, J., & Tubiello, F. N. (2007). Global food security under climate change. *Proceedings of the National Academy of Sciences, 104*(50), 19703–19708.
Shah, M., Fischer, G., & Van Velthuizen, H. (2009). Food security and sustainable agriculture: The challenges of climate change in Sub-Saharan Africa. In *Climate change and economic development in Sub-Saharan Africa—AERC Senior Policy Seminar X, Addis Ababa, Ethiopia,* 7–9 April 2008: (Seminar Papers, African Economic Research Consortium, Nairobi, Kenya).
Thompson, H. E., Berrang-Ford, L., & Ford, J. D. (2010). Climate change and food security in sub-Saharan Africa: A systematic literature review. *Sustainability, 2*(8), 2719–2733.
Totin, E., Segnon, A. C., Schut, M., Affognon, H., Zougmoré, R. B., Rosenstock, T., & Thornton, P. K. (2018). Institutional perspectives of climate-smart agriculture: A systematic literature review. *Sustainability, 10*(6), 1990.
Webb, P., Coates, J., Frongillo, E. A., Rogers, B. L., Swindale, A., & Bilinsky, P. (2006). Measuring household food insecurity: Why it's so important and yet so difficult to do. *The Journal of Nutrition, 136*(5), 1404S-1408S. https://doi.org/10.1093/jn/136.5.1404S
Westermann, O., Förch, W., Thornton, P., Körner, J., Cramer, L., & Campbell, B. (2018). Scaling up agricultural interventions: Case studies of climate-smart agriculture. *Agricultural Systems, 165*, 283–293.
WHO. (2005). *Worldwide prevalence of anaemia 1993–2005.*
Willetts, P. (2002). What is a non-governmental organization. http://www.staff.city.ac.uk/p.willetts/CS-NTWKS/NGO-ART.HTM#Part9
Woods, N. (2000). The challenge of good governance for the IMF and the World Bank themselves. *World Development, 28*(5), 823–841.

Ergonomics for Sustainable Agriculture and Food Security

M. G. Mohamed Thariq, Mohamed Mujithaba Mohamed Najim, and Rahul Jain

Abstract Agricultural development is essential to minimize poverty, enhance broader prosperity, and supply food to an increased population in the future. Labor-intensive agriculture is common in low and middle-income countries. Small and medium-scale farm operations are predominant in those countries and the diffusion of technologies is slower when compared to countries with higher income levels. Farmworks usually involve a higher level of physical activities and manual labor, which push farmworkers to a posture with a high risk of causing work-related musculoskeletal disorders (MSDs). The workforce is the most valuable asset of any production operation. Therefore, serious attention should be given to working conditions for improving productivity and output quality. The MSDs can cause severe long-term negative consequences such as reduced workers' productivity, a decline in farm income, other health problems such as depression or stress, and poor quality of life. Such negative consequences lead to less productive farming activities, impacting food security. Several ergonomic interventions in the agriculture and livestock production sector indicate improvements in productivity and performance by sustaining workers' health. Ergonomic interventions create a common environment wherein farmers and their employers collaborate; this would benefit employees from enhanced safety, health, comfort, improved user-friendliness of tools, equipment, and software, and improved quality of work life, whereas employers can benefit from reduced costs and improved productivity. Sustainable agriculture requires a healthy workforce with sustainable human well-being, which is a vital part of a profitable and sustainable agricultural industry. The sustainability of farm workers' well-being requires significant input from ergonomics. Plenty of research is conducted on agriculture ergonomics

M. G. M. Thariq (✉)
Department of Biosystems Technology, Faculty of Technology, South Eastern University of Sri Lanka, Oluvil, Sri Lanka
e-mail: mgmthariq@seu.ac.lk

M. M. M. Najim
Faculty of Agriculture, Sultan Sharif Ali Islamic University, Sinaut Campus, Sinaut, Brunei
e-mail: najim.mujithaba@unissa.edu.bn

R. Jain
Department of Mechanical Engineering, Rajasthan Technical University, Kota, India
e-mail: rjain.npiu.me@rtu.ac.in

M. Behnassi et al. (eds.), *Climate-Smart and Resilient Food Systems and Security*,
https://doi.org/10.1007/978-3-031-65968-3_11

to address the health, well-being, and effectiveness of farm workers. However, more research on the micro-ergonomics level may be needed when considering the workforce's contribution to agriculture and agriculture's contribution to the country's economy and food security.

Keywords Farmworkers · Comfort · Health · Agricultural production · Sustainability · Food security

1 Introduction

Crop domestication and subsequent agriculture started primarily to meet food needs. Now agriculture has evolved not only as a primary source of worldwide food supply but also as the source of raw materials for other industries and medicine. In addition, agriculture plays a major role in economic development; it contributed to 4% of the Gross Domestic Product (GDP) in 2018, whereas it contributes more than 25% of the GDP in some developing nations. Further, agricultural expansion is considered one of the most important requirements to minimize poverty, enhance prosperity at various levels, and supply food to 9.7 billion people by 2050 (The World Bank, 2021). Agriculture is one of the most significant sectors contributing to world employment. According to a World Bank report (2021), 26.7% of the total workforce is employed in agriculture. It was reported that the agriculture sector employs half the world's total workforce (ILO, 2003), and further, the proportion of the workforce in the agriculture sector is higher in developing countries where labor-intensive agriculture is common with predominant small and medium-scale farm operations.

Agriculture, being an occupation, is considered one of the most physically demanding and hazardous sectors in the world (Benos et al., 2020). It has been reported that farmers have higher prevalence rates of Musculoskeletal Disorders (MSDs) than non-farmers, which indicates that they are at a higher risk of developing MSDs compared with other occupations (Osborne et al., 2012). This is due to the strenuous nature of the farm work, which is connected with a higher level of physical activities and high levels of manual labor leading to the development of MSDs (Rosecrance et al., 2006; Walker-Bone, 2002). Previous studies have identified prolonged trunk flexion, moving heavy loads, repetitive lifting, awkward postures of the trunk and wrist, and intensive hard work as ergonomic risk factors associated with farm work (Fathallah et al., 2008; McCurdy et al., 2003; Meyers et al., 1997).

Farmers may suffer from severe long-term pain because of MSDs. In addition to the physical bodily sufferings, the worker can undergo other negative consequences such as a decline in worker productivity, other health problems such as stress or depression, lower farm income, and poor quality of life. For example, farmers working under unfavorable working conditions are likely to be less productive since the workers must cope with several hazards and other risk factors that jeopardize their health and safety. It was reported that, in the Netherlands, MSDs were the leading reason for sick leave in farmers who were self-employed (Hartman et al., 2006). Hence, the MSDs could have negative impacts on labor productivity, ultimately affecting farming activities, which can end up either in low production or a high cost of production; in whatever the case, the impact on food security would be unavoidable.

The workforce is the most valuable asset of any business or production operations. Therefore, serious attention should be given to working conditions for improving productivity and output quality. Hence, it is essential to maintain the health and safety of the workforce in agriculture to produce sufficient food at an affordable cost to meet the growing demands for food in the world. Under this scenario, the need for applications of ergonomics in agriculture, mainly concerning food crop production, becomes crucial. This involves the identification of the risk factors of MSDs and their root causes. And to this effect, the development, implementation, and evaluation of ergonomic solutions are crucial in ensuring the health and safety of the workforce in agriculture. Food security means that all people, at all times, have physical, social, and economic access to sufficient, safe, and nutritious food which meets their dietary needs and food preferences for an active and healthy life (FAO, 2003). Sonnino et al. (2014) emphasized that a sustainable food security program needs to move the focus away from the individual components of the food system (e.g., supply and demand) and should have a holistic approach towards the complex relationships between its different stages and actors. The current study tries to explain a conceptual model for the agriculture workforce's health and safety, applying ergonomics, which is essential for sustainable agriculture and food security (Fig. 1).

2 Ergonomic Hazards Among the Agricultural Workforce

The fundamental principle of ergonomics is always referred to as "fitting the task to the person". According to Helander (1997), ergonomics is the science of design to improve safety, productivity, and operator satisfaction. Schuman (2002) tried to define agricultural ergonomics as part of the art and science of producing a crop (output) safely for the benefit of the consumer and the worker's health. The varieties and complexities in agricultural practices across countries, cultures, and economies and resulting MSDs indicate the need for context-specific ergonomic interventions in farming. Poorly designed or unfitting workstations and work practices slowly over months or years lead to injuries and diseases among workers. Some signs and symptoms appearing in an employee for a long duration may indicate that something is wrong in the workplace and the meantime, and the worker is unknowingly and slowly adapting to injuries. For example, a farmer may feel discomfort while performing his/her tasks or may feel pain in the muscles or joints at home after leaving work. In addition, the farmer may have a few minor muscle strains for some time. Therefore, it is necessary to investigate these issues among farmers with an understanding of what may be perceived as discomfort in some cases that may lead to severe injuries or MSDs (Singh & Arora, 2010) impacting the quality of work and the life of a farmer.

Fig. 1 Conceptual model of ergonomics for food security

Table 1 summarizes the major ergonomic hazards among agricultural workers based on the available literature.

All of the recent studies on ergonomic risks indicate that MSDs are prevalent among farmworkers across countries. Though the MSDs are not lethal, they can ultimately affect farm production and the cost of production. According to Kirkhorn et al. (2010), MSDs increase production costs because of decreased work capacity, medical and insurance costs, worker absenteeism, and loss of employees to turnover, in addition to competition for the workforce from other industries which are less physically demanding. According to Jafry and O'Neill (2000), the European Union (EU) advocates that musculoskeletal injury developed from work may become disabling and result in loss of income because of worker absenteeism, which may disrupt family life.

Table 1 Ergonomic hazards among farmworkers

MSD risk factor/nature of work/task	Major MSD	References
Farmwork (Irish farmers)	Some types of MSDs were experienced by 56% of farmers in the previous year. The most common MSD was back pain.	Osborne et al., 2010
Farm work (Cambodian farm workers) Orchard work	The prevalence of musculoskeletal symptoms with one symptom (at least) was found to be 80.40%. The affected body parts were identified as the lower back, upper back, neck, wrist, hand, and thigh.	Thetkathuek et al., 2018
Thai rice farmers	MSDs were most prevalent in body regions such as pain in the lower back, neck, and shoulders during the last six months.	Sombatsawat et al., 2019
Farmwork (Nepal farmers)	Some kinds of MSDs were found in more than 70% of farmers in the body regions of the neck, shoulder, elbow, lower back, knee, and ankle.	Mahto & Gautam, 2018
Farmwork (Saskatchewan farming population)	Musculoskeletal pain during the past year was experienced by 85.6% of farmers at least in one body region. Pain in the lower back was (57.7%), shoulders (44.0%), and neck (39.6%). 27.9% of farmers were prevented from carrying out regular activities.	McMillan et al., 2015
Land preparation, sowing, irrigation, weeding, and harvesting (Indian agricultural workers)	Pain in the lower back was reported by all male and female agricultural workers.	Vyas, 2012
Farmers in Rajasthan performed manual harvesting	Musculoskeletal disorders in the lower back, fingers, shoulders, and wrists/hands were reported by more than 50% of farm workers.	Jain et al., 2018

(continued)

Table 1 (continued)

MSD risk factor/nature of work/task	Major MSD	References
Lifting/pulling/pushing activities (Farmers in Ireland)	Most of the farmers (around three-quarters) stated that lower back pain is due to either a specific farm activity or repeated work activities, usually lifting.	Osborne et al., 2013
Farmwork (Farmers)	Lower back pain was experienced by 86% of the farmers.	Tomczyszyn et al., 2018
Harvesting (vegetables - tomato and lettuce) prolonged and iterative stooping, lifting loads of great weight Intensive and repetitive cutting	Workers experienced discomfort in the hands, wrists, and lower back.	Fathallah, 2010
Agricultural work (agriculture workforce)	The most prevalent MSD is lower back pain both in developed and developing countries.	Fathallah et al., 2008 Gangopadhyay et al., 2005 Thetkathuek et al., 2018

3 Ergonomics Improvements for Comfort, Health, and Safety of the Agriculture Workforce

Though agriculture is associated with high ergonomic risks for its workforce in both developed and developing countries, ergonomics research and its applications reduce MSDs, improve health, and prevent MSDs among farmworkers (Table 2). Realizing the importance of integrating ergonomics in agricultural tasks to improve work conditions, ILO (2014) published ergonomic checkpoints in agriculture in collaboration with the International Ergonomics Association (IEA). This report is a comprehensive global document with 100 checkpoints that outline practical solutions for ergonomically sound workplace improvements to reduce work-related accidents and illnesses among agricultural workers and increase living conditions and productivity. The US National Institute for Occupational Safety and Health (NIOSH) in 2001 came out with simple ergonomic solutions for farmworkers with the help of researchers and scientists. This makes farm work safer through the science of ergonomics. NIOSH recognizes that making work practices and tools better using ergonomic principles will reduce the MSDs associated with farming practices.

According to Dombeková (2016), an unfitting workplace environment ultimately leads to rising costs by associating with occupational diseases and work-related injuries. The poor working conditions consequently worsen the worker's health and have an impact on the economy at microeconomic and macroeconomic levels. Lange-Morales (2014) in an article on a sustainable world through human factors and ergonomics, stated, from the perspective of green ergonomics as follows: over

Table 2 Ergonomics improvements for health and safety of agriculture workforce

Ergonomic Intervention	Problems solved/ benefits	Reference
Ergonomic application and improvement in workstation design	MSDs associated with harder agricultural activities were reduced.	Singh & Aror, 2010
Standardization in machine design, agricultural equipment, and legislation for health and safety (safer workstation design and safety guidelines)	MSD hazards became minimal in agricultural and forestry sectors in Europe and North America. Occurrence of accidents and injuries reduced among farm workers.	Jafry & O'Neill, 2000
Ergonomic interventions in the workplace	The number of MSD injuries was reduced by 50% on average.	Goggins et al., 2008
Ergonomic interventions for youths/ adults for handling water/feed buckets on the farm (Ergo Bucket Carrier and Easy Lift)	A lower back disorder was reduced by 50% by the use of an Easy lift when compared to manual lifting. The lower back risk was the least with the use of the Ergo Bucket Carrier. The lower back risk was reduced by 60% with the use Ergo Bucket. Carrier for both adults and youths in performing the dumping task when compared with the manual activities.	Fathallah et al., 2016
Educational intervention and knowledge (improvement in agriculture equipment, application of safe work methods, and correct postures)	Many MSD problems were reduced among the farm workers.	Vyas, 2012
Work tool design and development integrating ergonomics	Ergonomic risk factors were reduced and occupational well-being overall was improved in the agro-based industry-related development activities.	Chakrabartia & Bhattachheriya, 2012
Locally manufactured model long-handle hoe compared with a traditional hoe	Local long-handled hoes take 22% less time as compared to traditional hoes to fill the standard plots.	Vanderwa et al., 2011
Ergonomic intervention: use of mesh bags and standard containers for aiding in crop handling among those of small and fresh market vegetable operations	The level of use increased for containers (38% vs. 54%) and bags (8% vs. 17%).	Chapman et al., 2004

(continued)

Table 2 (continued)

Ergonomic Intervention	Problems solved/ benefits	Reference
Modifications to a shoulder strap for apple picking. Addition of a section to the strap which is wider and padded, the section lies on the shoulder, and also it incorporates an elasticized suspension which can minimize the impact of sudden downward forces	The modified bag was preferred by 78.6% of the workers. A difference in pain-related feelings in the back, neck, or shoulder was felt by 71.4% of workers. 64.3% of workers reported that regular use of modified bags may slow down their work.	Earle-Richardson et al., 2005
Intervention: a 5 min. rest break was added to every hour of work in addition to the other scheduled break (e.g., lunch time) to the task of strawberry harvest and the cultivation of citrus-bearing fruit trees	The severity of symptoms among workers was significantly lesser. Variations in productivity were observed.	Faucett et al., 2007
A hip belt was added to a traditional apple harvesting bucket – an ergonomic hip belt	Loading on the shoulder trap was reduced. The load was transferred greatly to the hips.	Freivalds et al., 2006
Workers used a smaller tub and carried cut grapes, by which the average load was reduced from a mean of 26 kg to 21kg	The pain reporting of workers dropped from 61% to 26%. Reduced risk for WMSDs.	Janowitz et al., 2000
Ergonomically improved tools for weeding for hill farm women in India	Postural stress and severity of pain were reduced in different body regions.	Kishtwariaa & Rana, 2012
Addon handles for shovels	Comfort and health effects on farmworkers were not convincing.	Kotowski et al., 2009
Agricultural safety interventions to reduce farm hazards; educating youth, community partnership, self-audit, pre/ post control, or post-only control group	A 20% reduction in hazard scores was obtained through self-audit and it was the most effective intervention method.	Landsitte et al., 2001

time, the natural environment degrades and depletes, and sustainable human well-being and its effectiveness will become a big challenge. Harmonizing a particular task with humans to achieve the production goals occurs within a working system, which ultimately operates within a natural environment. While operating in harmony with the environment, the tasks and systems need to fit humans. The suitable work environment to fulfill corporate needs: maximum output at minimum cost. There are plenty of research studies conducted to reduce MSDs among farm workers (Table 2).

Naeini et al. (2014) identified farming as a high-risk activity associated with various kinds of occupational health problems, especially work-related musculoskeletal disorders (WMSD). They emphasized that ergonomic assessments and improvements will have a significant effect on the enhancement of farmers' health.

4 Ergonomic Interventions and Agricultural Productivity

Several ergonomics interventions indicate improvements in productivity and performance in addition to sustaining workers' health in agriculture and other production sectors. According to Hendrick (1996), ergonomic interventions create a common platform for workers and management to work together where managers can benefit in terms of reduced costs and improved productivity and employees in terms of improved safety, health, comfort, the usability of tools and equipment, including software, and improved quality of work life. Duraj et al. (2000) tested a smaller picking tub with ergonomic improvements to aid harvesting in grapes harvesting activities in California. They found a significant decline in symptoms of pain, a slight change in productivity, overall worker preference, and widespread worker and management acceptance. May et al. (2008) also conducted another study and evaluated ergonomically improved blueberry harvesting rakes and they found that the productivity was significantly higher with rakes attached with the long handle than the traditional short handle rakes, and further, the rakes with long handles were found with the highest preference by users. Earle-Richardson (2005) evaluated ergonomically modified apple buckets for worker acceptance and worker productivity and they found that worker acceptance was extremely high with no negative impacts on productivity. According to Faucett et al. (2007), their ergonomic interventions in agriculture demonstrated that different patterns of rest breaks, together with short rest breaks at the beginning of the work shift, might reduce workers' musculoskeletal discomfort and fatigue over the day with a modest impact on production. Schuman (2002), based on the review on ergonomics in agriculture, stated that the milking system available in the market with a rail system overhead to carry equipment causes a significantly lower level of repetitive strain on the neck, shoulders, arms, and the lower back. This is an example of an application of ergonomics in the dairy industry to enhance productivity and workers' health and safety. Singh and Arora (2010) stated that women-friendly tools and equipment based on their ergonomic characteristics will lower musculoskeletal disorders and improve efficiency and ultimately improve the productivity of the worker. Borah and Kalita (2012) after evaluating the traditional tool for rice pounding concluded that redesigning the traditional tool incorporating ergonomics in the pounding of rice improves work performance and enhances the productivity of women.

Though the enhancement of productivity by ergonomic interventions in agriculture was not estimated, the existing research work shows improved productivity by improving comfort and easiness in carrying out tasks and eliminating the ergonomic risk factors that cause musculoskeletal pain disorders among farm workers.

5 Ergonomics for Sustainable Agriculture

Siemieniuch et al. (2015) identified food security as one of the global drivers and the contribution of sustainability engineering in abating the likely negative effects of the global driver. They further emphasized that sustainability requires significant inputs from ergonomics. However, no studies were found on how ergonomics directly contributes to the food supply (Thatcher et al., 2017). Further, Thatcher et al. (2017) agree that there is plenty of research conducted on agriculture ergonomics to address the issues of health, well-being, and effectiveness of individual farmworkers. Implementing ergonomics at the micro level for sustainable agriculture will help maintain a healthy workforce with sustainable human well-being.

According to Hurst et al. (2007), it is hard to assign a monetary value to human suffering, but it is possible for economic losses incurred because of the issues concerning weaker occupational health and safety levels and deterioration in environmental standards. Further, Hurst et al. (2007) highlight that accidents, ill health resulting in lost working days, deaths, and increased medical expenses create considerable negative impacts on agricultural productivity and have vast social and financial burdens on agricultural enterprises. They estimated that 4% of the gross domestic product is lost due to fatalities, accidents, and work-related diseases. Therefore, a safe, healthy, and environmentally conscious workforce is an essential element of a profitable and sustainable agricultural industry; this becomes much more important when considering the workforce's contribution to agriculture and the contribution of agriculture to GDP and food security in the world (Table 3).

Vyas and Bajpai (2016) suggested that lowering the incidence of workplace injuries and illness by incorporating ergonomics will keep workers employable and cut down the avoidable costs to the healthcare systems; this may ultimately contribute to the sustainability of social security systems. The fundamental principle

Table 3 Percentage of the workforce in agriculture out of total employment and GDP share of agriculture in 2019

Country	Percentage of the workforce in agriculture	The GDP share of agriculture (%)
India	42.6	16.68
Sri Lanka	24.98	7.54
Malaysia	10.28	7.26
Indonesia	28.5	12.71
China	25.33	7.14
Nigeria	34.97	21.91
Kenya	54.34	34.15
Pakistan	36.92	22.04
Thailand	31.43	8.14
Bhutan	55.78	15.82

Source The World Bank (2021)

of ergonomics is to improve the performance of the work system by improving the compatibility between the workers and the components of the system with no risk to workers. Therefore, ergonomics applications by improving the health and safety of workers would contribute to improving the performance in farming, ultimately contributing to the improvement of agricultural sustainability and food security.

6 Conclusion

Ergonomic interventions are essential prerequisites to sustain the agricultural workforce's comfort, health, and safety. To sustain agricultural production and food security for the growing world population and in a broader sense, it is essential to maintain the health and safety of existing and incoming young farmworkers where there is a decline in the agricultural labor force since the younger generation moves away from agriculture for various reasons. Sustaining agricultural production ensures food security. Though it is not much discussed, ergonomic interventions in agriculture play a considerable role in ensuring food security. Reduction in incidences of agricultural activities-based illnesses and injuries may contribute to the sustainability of agricultural production, leading to food security.

Farm work associated with higher ergonomic risk levels causes the development of work-related MSDs in farm workers, which can cause severe long-term negative impacts leading to further negative consequences such as reduced workability, lower farm income, poor quality of life, and the onset of other health problems such as stress or depression. As such, strenuous work conditions with poor ergonomic interventions in agriculture negatively influence farmworkers' health and safety, leading to less productivity and impacting food security. The fundamental principle of ergonomics is to enhance the performance of the work system by improving the compatibility between the workers and the components of the system with no risk to workers. Several ergonomic interventions indicate improvements in productivity and performance. Therefore, ergonomics applications by improving the comfort, health, and safety of workers would contribute to improving the performance in farming, ultimately contributing to the improvement of agricultural sustainability and food security.

References

Benos, L., Tsaopoulos, D., & Bochtis, D. (2020). A review on ergonomics in agriculture Part i: Manual Operations. *Applied Sciences, 10*(6), 1905.

Borah, R., & Kalita, M. (2012). Ergonomic evaluation of pounding of rice with a traditional tool. *Work, 43*(4), 411–416.

Chakrabarti, D., & Bhattachheriya, N. (2012). Ergonomic design intervention strategy for work tools development for women agro-based workers in Northeast India. *Work, 41*(Supplement 1), 1330–1334.

Chapman, L. J., Newenhouse, A. C., Meyer, R. H., Taveira, A. D., Karsh, B. T., Ehlers, J. J., & Palermo, T. (2004). Evaluation of an intervention to reduce musculoskeletal hazards among fresh market vegetable growers. *Applied Ergonomics, 35*(1), 57–66.

Dombeková, B. (2016). Increasing corporate and government spending: Can ergonomics help? *International Advances in Economic Research, 22*(4), 469–470.

Duraj, V., Miles, J.A., Meyers, J.M., Faucett, J.A., Janowitz, I.L., Tarter, M.E., Tejeda, D.G., Smith, R.H. and Weber, E.A. (2000). Harvesting aids in reducing ergonomics risk factors in wine grape hand harvesting. In *Proceedings of the American Society of Agricultural Engineers Meeting.*

Earle-Richardson, G., Jenkins, P., Fulmer, S., Mason, C., Burdick, P., & May, J. (2005). An ergonomic intervention to reduce back strain among apple harvest workers in New York State. *Applied Ergonomics, 36*(3), 327–334.

Fathallah, F. A., Miller, B. J., & Miles, J. A. (2008). Low back disorders in agriculture and the role of stooped work: Scope, potential interventions, and research needs. *Journal of Agricultural Safety and Health, 14*(2), 221–245.

Fathallah, F. A. (2010). Musculoskeletal disorders in labor-intensive agriculture. *Applied Ergonomics, 41*(6), 738–743.

Fathallah, F. A., Tang, S. C., & Waters, T. (2016). Development and evaluation of ergonomic interventions for bucket handling on farms. *Human Factors, 58*(5), 758–776.

Faucett, J., Meyers, J., Miles, J., Janowitz, I., & Fathallah, F. (2007). Rest break interventions in stoop labor tasks. *Applied Ergonomics, 38*(2), 219–226.

Freivalds, A., Park, S., Lee, C., Earle-Richardson, G., Mason, C., & May, J. J. (2006). Effect of belt/bucket interfaces in apple harvesting. *International Journal of Industrial Ergonomics, 36*(11), 1005–1010.

Food and Agriculture Organization of the United Nations (2003). Trade reforms and food security conceptualizing the Linkages. *Commodity Policy and Projections Service Commodities and Trade Division.*

Gangopadhyay, S., Das, B., Das, T., & Ghoshal, G. (2005). An ergonomic study on posture-related discomfort among preadolescent agricultural workers of West Bengal, India. *International Journal of Occupational Safety and Ergonomics, 11*(3), 315–322.

Goggins, R. W., Spielholz, P., & Nothstein, G. L. (2008). Estimating the effectiveness of ergonomics interventions through case studies: Implications for predictive cost-benefit analysis. *Journal of Safety Research, 39*(3), 339–344.

Hartman, E., Oude Vrielink, H. H., Huirne, R. B., & Metz, J. H. (2006). Risk factors for sick leave due to musculoskeletal disorders among self-employed Dutch farmers: A case-control study. *American Journal of Industrial Medicine, 49*(3), 204–214.

Helander, M. G. (1997). Forty years of IEA: Some reflections on the evolution of ergonomics. *Ergonomics, 40*(10), 952–961.

Hendrick, H. W. (1996). The Ergonomics of Economics is the Economics of Ergonomics. *Proceedings of the Human Factors and Ergonomics Society Annual Meeting, 40*(1), 1–10. https://doi.org/10.1177/154193129604000101

Hurst, P., Termine, P. and Karl, M. (2007). Agricultural workers and their contribution to sustainable agriculture and rural development. *International Union of Food, Agricultural, Hotel, Restaurant, Catering, Tobacco and Allied Workers' Associations.* pp. 102. Accessed on 17.12.2021.

ILO. (2014). *Ergonomic checkpoints in agriculture: Practical and easy-to-implement solutions for improving safety, health and working conditions in agriculture (2nd ed.).* Available at https://www.ilo.org/wcmsp5/groups/public/---ed_protect/---protrav/--safe_work/documents/instructional_material/wcms_176923.pdf. Accessed 22 February 2022.

ILO. (2003). *Facts on Agriculture.* Retrieved December 17, 2021, from: http://www.ilo.org.

ILO. (2011). *Safety and health in agriculture.* Labor Protection Department. Accessed on 16.12.2021. https://www.ilo.org/wcmsp5/groups/public/@ed_protect/@protrav/@safework/documents/publication/wcms_110193.pdf.

Jain, R., Meena, M. L., Dangayach, G. S., & Bhardwaj, A. K. (2018). Risk factors for musculoskeletal disorders in manual harvesting farmers of Rajasthan. *Industrial Health, 56*(3), 241–248.

Jafry, T., & O'Neill, D. H. (2000). The application of ergonomics in rural development: A review. *Applied ERgonomics, 31*(3), 263–268.

Janowitz, I., Stern, A., Morelli, D., Vollowitz, E., Hudes, M. and Rempel, D. (2000). *Ergonomics Society Annual Meeting.*

Kirkhorn, S. R., Earle-Richardson, G., & Banks, R. J. (2010). Ergonomic risks and musculoskeletal disorders in production agriculture: Recommendations for effective research to practice. *Journal of Agromedicine, 15*(3), 281–299.

Kishtwaria, J., & Rana, A. (2012). Ergonomic interventions in weeding operations for drudgery reduction of hill farm women of India. *Work, 41*(Supplement 1), 4349–4355.

Kotowski, S.E., Davis, K.G. & Waters, T.R. (2009). Investigation of select ergonomic interventions for farm youth. Part 1: shovels. *Journal of agromedicine, 14*(1), pp.33–43.

Lange-Morales, K., Thatcher, A., & García-Acosta, G. (2014). Towards a sustainable world through human factors and ergonomics: It is all about values. *Ergonomics, 57*(11), 1603–1615. https://doi.org/10.1080/00140139.2014.945495

Landsittel, D. P., Murphy, D. J., Kiernan, N. E., Hard, D. L., & Kassab, C. (2001). Evaluation of the effectiveness of educational interventions in the Pennsylvania Central Region Farm Safety Pilot Project. *American Journal of Industrial Medicine, 40*(2), 145–152. https://doi.org/10.1002/ajim.1082

Mahto, P. K., & Gautam, B. B. (2018). Prevalence of work-related musculoskeletal disorders in agricultural farmers of Bhaktapur District. *Nepal. International Journal of Occupational Safety and Health, 8*(1), 3–7.

May, J., Hawkes, L., Jones, A., Burdick, P., Ginley, B., Santiago, B., & Rowland, M. (2008). Evaluation of a community-based effort to reduce blueberry harvesting injury. *American Journal of Industrial Medicine, 51*(4), 307–315.

McCurdy, S. A., Samuels, S. J., Carroll, D. J., Beaumont, J. J., & Morrin, L. A. (2003). Agricultural injury in California migrant Hispanic farm workers. *American Journal of Industrial Medicine, 44*(3), 225–235.

McMillan, M., Trask, C., Dosman, J., Hagel, L., Pickett, W. and Saskatchewan Farm Injury Cohort Study Team. (2015). Prevalence of musculoskeletal disorders among Saskatchewan farmers. *Journal of Agromedicine, 20*(3), 292–301.

Meyers, J. M., Miles, J. A., Faucett, J., Janowitz, I., Tejeda, D. G., & Kabashima, J. N. (1997). Ergonomics in agriculture: Workplace priority setting in the nursery industry. *American Industrial Hygiene Association Journal, 58*(2), 121–126.

Naeini, H. S., Karuppiah, K., Tamrin, S. B., & Dalal, K. (2014). Ergonomics in agriculture: An approach in the prevention of work-related musculoskeletal disorders (WMSDs). *Journal of Agriculture and Environmental Sciences, 3*(2), 33–51.

Osborne, A., Blake, C., McNamara, J., Meredith, D., Phelan, J., & Cunningham, C. (2010). Musculoskeletal disorders among Irish farmers. *Occupational Medicine, 60*(8), 598–603.

Osborne, A., Blake, C., Fullen, B. M., Meredith, D., Phelan, J., McNamara, J., & Cunningham, C. (2012). Prevalence of musculoskeletal disorders among farmers: A systematic review. *American Journal of Industrial Medicine, 55*(2), 143–158.

Osborne, A., Finnegan, G., Blake, C., Meredith, D., McNamara, J., Phelan, J., & Cunningham, C. (2013). An evaluation of low back pain among farmers in Ireland. *Occupational Medicine, 63*, 53–59.

Rosecrance, J., Rodgers, G., & Merlino, L. (2006). Low back pain and musculoskeletal symptoms among Kansas farmers. *American Journal of Industrial Medicine, 49*(7), 547–556.

Schuman, S. H. (2002). Ergonomics in agriculture: Commentary and literature review. *Journal of Agromedicine, 8*(1), 9–18.

Siemieniuch, C. E., Sinclair, M. A., & Henshaw, M. D. (2015). Global drivers, sustainable manufacturing, and systems ergonomics. *Applied Ergonomics, 51*, 104–119.

Singh, S., & Arora, R. (2010). Ergonomic intervention for preventing musculoskeletal disorders among farm women. *Journal of Agricultural Sciences, 1*(2), 61–71.

Sombatsawat, E., Luangwilai, T., Ong-Artborirak, P., & Siriwong, W. (2019). Musculoskeletal disorders among rice farmers in Phimai District, Nakhon Ratchasima Province. *Thailand. Journal of Health Research., 33*(6), 494–503.

Sonnino, R., Faus, A. M., & Maggio, A. (2014). Sustainable food security: An emerging research and policy agenda. *The International Journal of Sociology of Agriculture and Food, 21*(1), 173–188.

Thatcher, A., Waterson, P., Todd, A., Moray, N., Thatcher, A., Waterson, P., Todd, A. & Moray, N. (2017). State of Science: ergonomics and global issues State of Science: ergonomics and global issues.

The World Bank. (2021). Agriculture and Food. Accessed on 09.11.2021. https://www.worldbank.org/en/topic/agriculture/overview#1

Thetkathuek, A., Meepradit, P., & Sa-Ngiamsak, T. (2018). A cross-sectional study of musculoskeletal symptoms and risk factors in Cambodian fruit farm workers in Eastern Region. *Thailand. Safety and Health at Work, 9*(2), 192–202.

Tomczyszyn, D., Solecki, L., & Pańczuk, A. (2018). Assessment of the type of farmers' low back pain. *Medycyna Pracy, 69*(4), 355–364.

Vanderwal, L., Rautiainen, R., Kuye, R., Peek-Asa, C., Cook, T., Ramirez, M., Culp, K., & Donham, K. (2011). Evaluation of long-and short-handled hand hoes for land preparation, developed in a participatory manner among women vegetable farmers in The Gambia. *Applied Ergonomics, 42*(5), 749–756.

Vyas, R. (2012). Mitigation of musculoskeletal problems and body discomfort of agricultural workers through educational intervention. *Work, 41*(Supplement 1), 2398–2404.

Vyas, R., & Bajpap, N. K. (2016). *Ergonomics in Agriculture: An Approach to Quality Life of Farm Communities., 8*(1), 92–95.

Walker-Bone, K., & Palmer, K. T. (2002). Musculoskeletal disorders in farmers and farm workers. *Occupational Medicine, 52*(8), 441–450.

Precision Agriculture as a Coping Strategy for Climatic Challenges

Maqsood Ahmad, Zahid Ali Butt, Imran Ali Sani, Muhammad Farhan, Sayed Ali Raza, Shahjahan Shabbir Ahmed, and Nazeer Ahmed

Abstract The agriculture sector is suffering from changing climatic factors. Although it is a global phenomenon, poor regions are more at risk. In coming years, livestock and health will be the most affected sectors and they may get damaged even at 1.5 °C rise in temperature during global warming. This deteriorating situation, already happening in many vulnerable countries, can be coped with by incorporating precision agriculture concepts, improving technology, adopting new policies, and utilizing indigenous knowledge. In the present scenario, precision agriculture is one of the leading management strategies to deal with unexpected climatic challenges. Precision agriculture strategies rely on technologies and emphasize on digitalization and Internet of Things (IoT) for rapid and precise information to improve productivity and farm sustainability. These modern and state-of-the-art technologies are available in developed countries but still lack in many developing countries, and without these

M. Ahmad (✉)
Department of Environmental Science, Balochistan University of Information Technology, Engineering and Management Sciences, Quetta, Pakistan
e-mail: maqsood.ahmed@buitms.edu.pk

Z. A. Butt
Department of Botany, Government College Women University Sialkot, Sialkot, Pakistan
e-mail: zahid.ali.butt@gcwus.edu.pk

I. A. Sani · S. S. Ahmed · N. Ahmed
Department of Biotechnology, Balochistan University of Information Technology, Engineering and Management Sciences, Quetta, Pakistan
e-mail: imran.ali@buitms.edu.pk

S. S. Ahmed
e-mail: Shahjahan.shabbir@buitms.edu.pk

N. Ahmed
e-mail: Nazeer.Ahmed@buitms.edu.pk

M. Farhan
Sustainable Development Study Center, Government College University Lahore, Lahore, Pakistan
e-mail: Dr.muhammadfarhan@gcu.edu.pk

S. A. Raza
Water Resources Zone, Irrigation Department, Government of the Punjab, Lahore, Pakistan
e-mail: chemstone@yahoo.com

M. Behnassi et al. (eds.), *Climate-Smart and Resilient Food Systems and Security*,
https://doi.org/10.1007/978-3-031-65968-3_12

sophisticated technologies agriculture production in developing countries cannot be increased. Developing regions like South Asian nations need to increase agriculture production up to 60–70% by 2050 to ensure food security. Initial investment is required to achieve the target, through adopting recommended practices and technologies. As a result, it is estimated that there will be significant economic effects in the form of cost and crop production by adopting the concept of precision agriculture.

Keywords Precision agriculture · Agriculture practices · Climate change · Internet of Things

1 Introduction

An increase of approximately 1.0 °C of global warming above the pre-industrial level has been observed due to anthropogenic activities. If follows the same pace, global warming is expected to reach 1.5–2 °C between 2030 and 2052. Greenhouse gas emissions from the pre-industrial period to the present are expected to remain in the atmosphere for centuries and will have long-term impacts on the climate system. Climate-related risks will be dependent on geographical location, intensity and rate of warming, development level, and implementation of mitigation measures (Masson-Delmotte et al., 2018). Climate change is the most serious threat that affects agricultural productivity. Some analysis revealed that severe weather events and changes in temperature and precipitation can reduce agriculture production, especially in Africa and some parts of Asia (Enete & Amusa, 2010; Gornall et al. 2010). In addition, countries with higher initial temperatures, degraded lands, and weak economies with poor adaptation capacity will face more severe impacts of climate change (Keane et al., 2009). In low-income countries, the agriculture sector contributes more greenhouse gases due to the large number of cattle, inadequate manure management, maladministration of land; and the dependency on agriculture sector makes climate change more serious. For instance, in Ethiopia, agriculture is providing 80% employment, and contributing 43% to GDP (Praveen & Sharma, 2019). This is one example from the developing world that shows how climate change will affect developing regions.

According to IPCC (2007), adaptation, mitigation, and technology options can be used to reduce the impact of climate change on agriculture sector. Private and public organizations are now working to facilitate the farmers' coping capacities with future climate change by educating them about climate-smart agriculture strategies (Nyang'au et al., 2021; Waaswa et al., 2021). Despite the impact of climate change will impact on a global scale, farmers of poor countries with limited resources will be more at risk because of unaffordability and lack of adequate technology (Archer et al., 2007). Although, agriculture is on the priority list of different countries to meet the food requirements with affordable prices, agriculture is still at risk due to climate change. If the projected impact of climate change on agriculture is not addressed timely, a serious food insecurity will be the outcome, especially in developing countries (Nyang'au et al., 2021). For instance, 1.5 °C increase in temperature will affect different sectors such as livestock, health, water supply, and consequently economic growth (Bongaarts, 2019).

Different studies depicted the altering effect of climate change on cropping calendar that causes soil erosion (Savo et al., 2016), and as a consequence, it is placing pressure on the food production system (Tripathi et al., 2016). A prominent example was the drought summer of 2018 in northern Europe in which only Sweden faced 50% reduction in the yield of cereal crops and a drop in livestock production due to shortage of fodder and feed (Grusson et al., 2021). For the Scandinavian region, the increase in average temperature and rainfall is predicted along with variations in the distribution pattern of rainfall throughout the year and within each month; this will also change the balance between evapotranspiration and accordingly the amount of rainfall (Rummukainen et al., 2004; Strandberg et al., 2015).

At the present rate of climate change, there are evidences that the change is affecting crop production globally. Major parameters responsible for food scarcity and consequently affecting societies are temperature, rainfall, and water availability (Kummu et al., 2021). Reliable information is available about developed countries, while it is scarce about developing ones, but it has been reported that Asia, Africa, and Latin America are already facing the impacts of climate change in various sectors, especially in agriculture and fisheries (Matias et al., 2021). Agriculture and ecosystems have been negatively impacted globally in general, and particularly in developing regions. Therefore, to deal with these adverse implications there is a need to develop and disseminate resilience strategies and technologies to farmers to counter the climatic impacts (Talanow et al., 2021; Traore et al., 2015). The concept of precision agriculture can be used to minimize the impact of climate change on agriculture, because it is based on site-specific crop management using data analysis to find primary factors responsible for yield reduction (Ponnusamy & Natarajan, 2021).

2 Smart Agriculture

To cope with present and anticipated adverse climate impacts on food security, we should start work simultaneously on required technologies and policies at national and regional levels. Smart agriculture is the most discoursing concept to deal with the uncertain future climate (Autio et al., 2021). Smart agriculture can be defined as agriculture activities having the capacity to reduce greenhouse gas emissions, enhance the adaptive capacity of farmers; and increase agriculture production sustainably to fulfill food demand and support the economy (Lipper et al., 2014). To enhance the production and quality of agriculture products, smart or precision agriculture is based on data collection and analysis for useful information. Integration of agriculture system with data analysis and machine learning is a basic demand of smart or precision agriculture. The aim of precision agriculture is to develop an integrated system that guides the decision making for farming management (Ponnusamy & Natarajan, 2021). So, climate-smart agriculture works with a political framework and focuses on technical approaches (Chandra et al., 2018).

Although the concept has some acceptance issues in the scientific community, its positive impacts on agriculture, economy, and total environment are well recognized (Neate, 2013; Saj et al., 2017). A number of recent studies demonstrated that different smart agriculture-based technologies such as integrated soil fertility practices for maize and wheat production (Agegnehu et al., 2014; Chivenge et al., 2011), improved nitrogen uses efficiency and alternate drying and wetting for rice production (Gaihre et al., 2015; Huda et al., 2016), and no-tillage especially for maize and wheat (Erensteinand & Laxmi, 2008; Erenstein et al., 2012) may be used to reduce the damage expected due to the climate change.

Several studies on climate-smart agriculture (CSA) emphasize on developing procedures to analyze agricultural systems from plot scale to region. These studies examined the options of new systems in agriculture, combined local practices with scientific knowledge, and recommended multiple level changes in social and organizational systems to cope with anticipated climatic changes (Andrieu et al., 2019). Empirical evidence provided by a pilot-scale project in western Kenya reported cost benefit in implementation due to the involvement of local community. It was also concluded that cost estimates by the involvement of local communities ('ground-based') are much cost efficient and acceptable than the expert driven desk-based estimates (Chaudhury et al., 2016). If summarized, CSA demands the introduction of new technologies, adoption of dynamic policies, and involvement of all key stakeholders to reduce the impact of climate change on the agriculture sector (Branca et al., 2021; Issad et al., 2019).

3 Role and Challenges of Precision Agriculture in the Current Era

The rise in human population demands irresistible food supply for their survival (Mumtaz et al., 2017). It becomes a big challenge in the post-Covid 19 era. Globally, many interventions are adopted with conventional techniques of agriculture system during the digitalization era of precision agriculture (PA) and Internet of Things (IoT). The role of remote sensing technology is of pivotal nature in PA, which aids in monitoring crop conditions as desired. Various parameters are vital in plant health and the PA allows farmers to have precise information on parameters mandatory for improved crop yields quantitatively (Shafi et al., 2019; Sishodia et al., 2020). Timely information about the related factors like temperature, humidity, weed growth, chlorophyll contents, etc., can play a key role in increasing crop productivity. Not a single source can provide this information, but it can be collected from different sources to provide farmers with useful data for timely actions to protect and enhance crop productivity (Deichmann et al., 2016). This can only be done using modern concepts in technology such as IoT. IoT connects different technologies like remote sensors, radars, programs, and association of small objects like smartphone to transfer information among them to generate useful information for humans (Alam, 2018; Jindal et al.,

2018). This information helps to decide about timing and type of fertilizer and suitable irrigation timing, which are key to enhance crop productivity (Snyder et al., 2009). The provision of information to farmers well in time is a helping tool to decide about right decision to cope with the target which is only possible if modern technologies are easily accessible to farmers in an affordable manner. An important reason is the installation of some basic equipment to collect precise information is mandatory (Cuesta et al., 2020; Tseng et al., 2006).

The use of modern technologies as per IoT concept targets better productivity, sustainability, and efficiency of farming with recognized technologies that decrease prices and enhance yields (Porter & Kramer, 2019). For example, it has been estimated that it can reduce lime costs by up to 60% while attaining a target pH rate managing to lower the costs and higher yields; decrease phosphorus rates up to 35% while sustaining production target; decline in Nitrogen application by 40% while sustaining production targets, thus showing encouraging results. This was achieved by computing data from IoT to get useful information and act accordingly (Precision Agriculture, 2021; Shafi et al., 2019).

However, IoT based technologies still have many limitations like cloud blockage and latency, which causes delay in data transfer to concern for timely processing of useful information to consumers (Firouzi et al., 2021; Pramanik et al., 2018). In developing countries, additional issues with IoT exist, some of these are economic limitations, lack of technologies and expertise likely because of technology transfer hesitation from advanced countries and lack of appropriate policies and vision (Liu, 2018; Mishra & Chakraborty, 2020; Ugwuanyi et al., 2021). Pakistan is also lacking this emerging regime of remote sensing due to the lack of resources and awareness, which is impacting the crop yields badly leading to an overall reduction in average produce per unit area. Pakistan is an agricultural country due to its huge arable land and climatic changes, which make it appropriate to grow multiple types of crops. Even with all these natural resources, the farmer's community in Pakistan is still impotent to produce immense yields (Mumtaz et al., 2017).

The major hurdle in good crop yields is the substandard and conventional agronomical practices which lack the proper and in-time examination of crop health influenced by environmental factors. Techniques emphasized by the PA concept are entirely based on farmer's awareness and experiences. Farmers simply used to visit field areas to screen the crop condition, and this practice is time consuming and laborious and results in wastage of resources. For example, in case of insect or pest attack, only visual observation without scientific knowledge may consume a lot of insecticidal spray which not only disturbs the crop nutrition (Noman et al., 2020) but also harms the environment. Many studies have proved that it is possible to obtain significantly high crop yields with minimum expenditure by utilizing modern technologies (satellite, drones, soil sensors, multispectral cameras, etc.) to get appropriate information as required by the PA concept (Dlodlo & Kalezhi, 2015; Lee et al., 2013; Mekala & Viswanathan, 2017; Zhao et al., 2010). Developing regions like South Asian nations need to increase the agriculture production up to 60–70% by 2050 to ensure food security (Sharma et al., 2020). To achieve this target, radical changes in policies are demanded to adopt modern technologies (Tester & Langridge, 2010).

The average cost of adaptation to increase wheat and rice production ranges from 18.7 to 40.51 US \$ ha^{-1}, but the result net return can be increased from 92.73 to 209.61 US \$ ha^{-1}, so can improve the livelihood and food security (Khatri-Chhetri et al., 2016).

4 Conclusion and Recommendations

The PA concept is easy to implement in developed countries because of economic resources, available technologies, skills, vision, and political stability. In developing countries, it is difficult to adopt mainly because of high literacy rates, higher installation and operational costs, inappropriate farm structures, lack of vision, and political instability. Climate change and agriculture crises will be a global issue in the near future (Shukla et al., 2019), thus developed nations should transfer technologies and skills to developing regions irrespective of geopolitical interests. In addition, decision and law makers in developing regions should reconsider both long- and short-term policies for adaptation of recommended practices and technologies to cope with expected agriculture productivity issues due to climate change in the near future.

References

Agegnehu, G., Vanbeek, C., & Bird, M. I. (2014). Influence of integrated soil fertility management in wheat and tef productivity and soil chemical properties in the highland tropical environment. *Journal of soil science and plant nutrition, 14*(3), 532–545.

Alam, T. (2018). A reliable communication framework and its use in internet of things (IoT). *CSEIT1835111| Received*, 10, 450–456.

Andrieu, N., Howland, F., Acosta-Alba, I., Le Coq, J. F., Osorio-Garcia, A. M., Martinez-Baron, D., Gamba-Trimiño, C., Loboguerrero, A. M., & Chia, E. (2019). Co-designing climate-smart farming systems with local stakeholders: a methodological framework for achieving large-scale change. *Frontiers in Sustainable Food Systems, 3*, 37.

Archer, E., Mukhala, E., Walker, S., Dilley, M., & Masamvu, K. (2007). Sustaining agricultural production and food security in Southern Africa: An improved role for climate prediction? *Climatic Change, 83*(3), 287–300.

Autio, A., Johansson, T., Motaroki, L., Minoia, P., & Pellikka, P. (2021). Constraints for adopting climate-smart agricultural practices among smallholder farmers in Southeast Kenya. *Agricultural Systems, 194*, 103284.

Bongaarts, J. (2019). Special report on climate change and land use. *Intergovernmental Panel on Climate Change, 2018.*

Branca, G., Braimoh, A., Zhao, Y., Ratii, M., & Likoetla, P. (2021). Are there opportunities for climate-smart agriculture? Assessing costs and benefits of sustainability investments and planning policies in Southern Africa. *Journal of Cleaner Production, 278*, 123847.

Chandra, A., McNamara, K. E., & Dargusch, P. (2018). Climate-smart agriculture: Perspectives and framings. *Climate Policy, 18*(4), 526–541.

Chaudhury, A. S., Helfgott, A., Thornton, T. F., & Sova, C. (2016). Participatory adaptation planning and costing. Applications in agricultural adaptation in western Kenya. *Mitigation and adaptation strategies for global change, 21*(3), 301–322.

Chivenge, P., Vanlauwe, B., & Six, J. (2011). Does the combined application of organic and mineral nutrient sources influence maize productivity? A meta-analysis. *Plant and Soil, 342*(1), 1–30.
Cuesta, M. A., Castillo-Calzadilla, T., & Borges, C. E. (2020). A critical analysis on hybrid renewable energy modeling tools: An emerging opportunity to include social indicators to optimise systems in small communities. *Renewable and Sustainable Energy Reviews, 122*, 109691.
Deichmann, U., Goyal, A., & Mishra, D. (2016). Will digital technologies transform agriculture in developing countries? *Agricultural Economics, 47*(S1), 21–33.
Dlodlo, N., & Kalezhi, J. (2015). The internet of things in agriculture for sustainable rural development. In *2015 international conference on emerging trends in networks and computer communications (ETNCC)* (pp. 13–18). IEEE.
Enete, A. A., & Amusa, T. A. (2010). Challenges of agricultural adaptation to climate change in Nigeria: A synthesis from the literature. *Field Actions Science Reports. The Journal of Field Actions, 4.*
Erenstein, O., & Laxmi, V. (2008). Zero tillage impacts in India's rice–wheat systems: A review. *Soil and Tillage Research, 100*(1–2), 1–14.
Erenstein, O., Sayre, K., Wall, P., Hellin, J., & Dixon, J. (2012). Conservation agriculture in maize- and wheat-based systems in the (sub) tropics: Lessons from adaptation initiatives in South Asia, Mexico, and Southern Africa. *Journal of Sustainable Agriculture, 36*(2), 180–206.
Firouzi, F., Farahani, B., & Marinšek, A. (2021). The convergence and interplay of edge, fog, and cloud in the AI-driven Internet of Things (IoT). *Information Systems*, 101840.
Gaihre, Y. K., Singh, U., Islam, S. M., Huda, A., Islam, M. R., Satter, M. A., Sanabria, J., Islam, M. R., & Shah, A. L. (2015). Impacts of urea deep placement on nitrous oxide and nitric oxide emissions from rice fields in Bangladesh. *Geoderma, 259*, 370–379.
Grusson, Y., Wesström, I., & Joel, A. (2021). Impact of climate change on Swedish agriculture: Growing season rain deficit and irrigation need. *Agricultural Water Management, 251*, 106858.
Gornall, J., Betts, R., Burke, E., Clark, R., Camp, J., Willett, K., & Wiltshire, A. (2010). Implications of climate change for agricultural productivity in the early twenty-first century. *Philosophical Transactions of the Royal Society B: Biological Sciences, 365*(1554), 2973–2989.
Huda, A., Gaihre, Y. K., Islam, M. R., Singh, U., Islam, M. R., Sanabria, J., Satter, M. A., Afroz, H., Halder, A., & Jahiruddin, M. (2016). Floodwater ammonium, nitrogen use efficiency and rice yields with fertilizer deep placement and alternate wetting and drying under triple rice cropping systems. *Nutrient Cycling in Agroecosystems, 104*(1), 53–66.
IPCC. (2007). *Climate change 2007: The physical science basis: Summary for policymakers* (pp. 104–116). IPCC.
Issad, H. A., Aoudjit, R., & Rodrigues, J. J. (2019). A comprehensive review of Data Mining techniques in smart agriculture. *Engineering in Agriculture, Environment and Food, 12*(4), 511–525.
Jindal, F., Jamar, R., & Churi, P. (2018). Future and challenges of internet of things. *Aircc's International Journal of Computer Science and Information Technology, 10*(2), 13–25.
Keane, J., Page, S., & Kennan, J. (2009). Climate change and developing country agriculture. *International Centre for Trade and Sustainable Development.*
Khatri-Chhetri, A., Aryal, J. P., Sapkota, T. B., & Khurana, R. (2016). Economic benefits of climate-smart agricultural practices to smallholder farmers in the Indo-Gangetic Plains of India. *Current Science*, 1251–1256.
Kummu, M., Heino, M., Taka, M., Varis, O., & Viviroli, D. (2021). Climate change risks pushing one-third of global food production outside the safe climatic space. *One Earth, 4*(5), 720–729.
Lee, M., Hwang, J., & Yoe, H. (2013). Agricultural production system based on IoT. In *2013 IEEE 16Th international conference on computational science and engineering* (pp. 833–837). IEEE.
Lipper, L., Thornton, P., Campbell, B. M., Baedeker, T., Braimoh, A., Bwalya, M., Caron, P., Cattaneo, A., Garrity, D., Henry, K., Hottle, R., & Torquebiau, E. F. (2014). Climate-smart agriculture for food security. *Nature climate change, 4*(12), 1068–1072.
Liu, Z. (2018). Research on the Internet of Things and the development of smart city industry based on big data. *Cluster Computing, 21*(1), 789–795.

Masson-Delmotte, V., Zhai, P., Pörtner, H. O., Roberts, D., Skea, J., Shukla, P. R., Pirani, A., Moufouma-Okia, W., Péan, C., Pidcock, R., & Waterfield, T. (2018). Global warming of 1.5 C. *An IPCC Special Report on the impacts of global warming, 1*(5).

Matias, D. M. S., Kim, H., Anticamara, J. A., Matias, K. M. S., & Tibig, L. V. (2021). Emerging response options and scenarios of slow onset events related to climate change in Southeast Asia. *Current Opinion in Environmental Sustainability, 50*, 175–184.

Mekala, M. S., & Viswanathan, P. (2017). A survey: Smart agriculture IoT with cloud computing. In *2017 international conference on microelectronic devices, circuits and systems (ICMDCS)* (pp. 1–7). IEEE.

Mishra, K. N., & Chakraborty, C. (2020). A novel approach toward enhancing the quality of life in smart cities using clouds and IoT-based technologies. In *Digital twin technologies and smart cities* (pp. 19–35). Springer.

Mumtaz, R., Baig, S., & Fatima, I. (2017). Analysis of meteorological variations on wheat yield and its estimation using remotely sensed data. A case study of selected districts of Punjab Province, Pakistan (2001–14). *Italian Journal of Agronomy, 12*(3).

Neate, P. J. (2013). Climate-smart agriculture: Success stories from farming communities around the world. Retrieved from https://cgspace.cgiar.org/handle/10568/34042

Noman, A., Aqeel, M., Qasim, M., Haider, I., & Lou, Y. (2020). Plant-insect-microbe interaction: A love triangle between enemies in ecosystem. *Science of the Total Environment, 699*, 134181.

Nyang'au, J. O., Mohamed, J. H., Mango, N., Makate, C., & Wangeci, A. N. (2021). Smallholder farmers' perception of climate change and adoption of climate smart agriculture practices in MasabaSouth Sub-county, Kisii, Kenya. *Heliyon, 7*(4), e06789.

Ponnusamy, V., & Natarajan, S. (2021). Precision agriculture using advanced technology of IoT, unmanned aerial vehicle, augmented reality, and machine learning. In *Smart sensors for industrial internet of things* (pp. 207–229). Springer.

Porter, M. E., & Kramer, M. R. (2019). Creating shared value. In *Managing sustainable business* (pp. 323–346). Springer.

Pramanik, P. K. D., Pal, S., Brahmachari, A., & Choudhury, P. (2018). Processing IoT data: From cloud to fog—it's time to be down to earth. In *Applications of security, mobile, analytic, and cloud (SMAC) technologies for effective information processing and management* (pp. 124–148). IGI Global.

Praveen, B., & Sharma, P. (2019). A review of literature on climate change and its impacts on agriculture productivity. *Journal of Public Affairs, 19*(4), e1960.

Precision Agriculture. (2021). https://precisionagriculture.com.au/. Tweeter.

Rummukainen, M., Bergström, S., Persson, G., Rodhe, J., & Tjernström, M. (2004). The Swedish regional climate modelling program, SWECLIM: a review. *AMBIO: A Journal of the Human Environment, 33*(4), 176–182.

Saj, S., Torquebiau, E., Hainzelin, E., Pages, J., & Maraux, F. (2017). The way forward: An agroecological perspective for climate-smart agriculture. *Agriculture, Ecosystems & Environment, 250*, 20–24.

Savo, V., Lepofsky, D., Benner, J. P., Kohfeld, K. E., Bailey, J., & Lertzman, K. (2016). Observations of climate change among subsistence-oriented communities around the world. *Nature Climate Change, 6*(5), 462–473.

Shafi, U., Mumtaz, R., García-Nieto, J., Hassan, S. A., Zaidi, S. A. R., & Iqbal, N. (2019). Precision agriculture techniques and practices: From considerations to applications. *Sensors, 19*(17), 3796.

Sharma, P., Singh, C., Jasrotia, P., Singh, G. P., & Sarial, A. K. (Eds.). (2020). *Improving cereal productivity through climate smart practices*. Woodhead Publishing.

Shukla, P. R., Skea, J., Calvo Buendia, E., Masson-Delmotte, V., Pörtner, H. O., Roberts, D. C., Zhai, P., Slade, R., Connors, S., Van Diemen, S., Ferrat, M., & Malley, J. (2019). IPCC, 2019: Climate change and land: An IPCC special report on climate change, desertification, land degradation, sustainable land management, food security, and greenhouse gas fluxes in terrestrial ecosystems.

Sishodia, R. P., Ray, R. L., & Singh, S. K. (2020). Applications of remote sensing in precision agriculture: A review. *Remote Sensing, 12*(19), 3136.

Snyder, C. S., Bruulsema, T. W., Jensen, T. L., & Fixen, P. E. (2009). Review of greenhouse gas emissions from crop production systems and fertilizer management effects. *Agriculture, Ecosystems & Environment, 133*(3–4), 247–266.

Strandberg, G., Bärring, L., Hansson, U., Jansson, C., Jones, C., Kjellström, E., Kupiainen, M., Nikulin, G., Samuelsson, P., & Ullerstig, A. (2015). *CORDEX scenarios for Europe from the Rossby Centre regional climate model RCA4*. SMHI.

Talanow, K., Topp, E. N., Loos, J., & Martín-López, B. (2021). Farmers' perceptions of climate change and adaptation strategies in South Africa's Western Cape. *Journal of Rural Studies, 81*, 203–219.

Tester, M., & Langridge, P. (2010). Breeding technologies to increase crop production in a changing world. *Science, 327*(5967), 818–822.

Traore, B., Van Wijk, M. T., Descheemaeker, K., Corbeels, M., Rufino, M. C., & Giller, K. E. (2015). Climate variability and change in southern Mali: Learning from farmer perceptions and on-farm trials. *Experimental Agriculture, 51*(4), 615–634.

Tripathi, A., Tripathi, D. K., Chauhan, D. K., Kumar, N., & Singh, G. S. (2016). Paradigms of climate change impacts on some major food sources of the world: A review on current knowledge and future prospects. *Agriculture, Ecosystems & Environment, 216*, 356–373.

Tseng, C. L., Jiang, J. A., Lee, R. G., Lu, F. M., Ouyang, C. S., Chen, Y. S., & Chang, C. H. (2006). Feasibility study on application of GSM–SMS technology to field data acquisition. *Computers and Electronics in Agriculture, 53*(1), 45–59.

Ugwuanyi, S., Paul, G., & Irvine, J. (2021). Survey of IoT for developing countries: Performance analysis of LoRaWAN and cellular NB-IoT networks. *Electronics, 10*(18), 2224.

Waaswa, A., Nkurumwa, A. O., Kibe, A. M., & Ng'eno, J. K. (2021). Understanding the socioeconomic determinants of adoption of climate-smart agricultural practices among smallholder potato farmers in Gilgil Sub-County, Kenya. *Discover Sustainability, 2*(1), 1–19.

Zhao, J. C., Zhang, J. F., Feng, Y., & Guo, J. X. (2010). The study and application of the IOT technology in agriculture. In *2010 3rd international conference on computer science and information technology* (Vol. 2, pp. 462–465). IEEE.

Contribution of Precision Agriculture to Drought and Food Security

Mehmet Metin Ozguven

Abstract Agriculture is one of the most important fields of activity where the essential and basic needs of humanity are produced. Many difficulties and uncertainties such as diseases, pests, weeds, drought, economic crises, pandemic, climatic changes, drought, hail, flood, tornado, decrease in agricultural lands and natural resources, and fragmentation of agricultural lands, are struggled during agricultural activities. These difficult and laborious activities also carry some risks such as the inability to obtain an efficient product for various reasons or a sufficient income to meet the expenditures. Therefore, the aim in agricultural production is to increase productivity and product quality in agriculture, to use minimum inputs, to ensure food safety, to protect natural resources and the environment, and to provide economic, sustainable, and productive management. For such a production, it is necessary to use quality seeds, irrigation, fertilization, plant protection, etc., and applications should be done duly on time and with machinery. The use of precision agriculture technologies not only makes productivity more effective, but also provides an economical production by preventing environmental pollution, protecting natural resources, food safety, and preventing unnecessary input use. In this study, the benefits of variable rate applications that cause the emergence of precision agriculture technology, and the effects of precision irrigation applications on drought are explained. In addition, the potential of precision agriculture technologies to contribute to food security is explained.

Keywords Precision agriculture · Precision irrigation · Water scarcity · Drought · Food safety

M. M. Ozguven (✉)
Department of Biosystems Engineering, Faculty of Agriculture, Tokat Gaziosmanpaşa University, Tokat, Turkey
e-mail: metin.ozguven@gop.edu.tr

M. Behnassi et al. (eds.), *Climate-Smart and Resilient Food Systems and Security*,
https://doi.org/10.1007/978-3-031-65968-3_13

1 Introduction

Agriculture is a sector of vital importance as it enables people to continue their lives, contributes to national income and employment, and provides raw materials to other sectors. For this reason, the agricultural sector has a significant impact on all segments of society with its economic, social, and environmental dimensions. Economically, there are topics such as increasing agricultural production and producer incomes, minimum usage of production inputs, increasing marketing options, etc. Socially, the issues of food quality and safety, agricultural employment problem, support for the socio-economic sustainability of rural areas, ensuring animal welfare, etc., are dealt with. Areas that need to be protected environmentally are biodiversity, wildlife, forests, meadow-pasture, ground and surface waters, soil resources, etc. Taking advantage of the opportunities offered by modern technologies is becoming increasingly mandatory in order to achieve success in studies on all these comprehensive subjects, due to the importance of the issues and its difficulties (Ozguven, 2018b).

In addition to the fact that there are many fields of study that require expertise in agricultural science, agriculture consists of complex events involving biological, physical, and chemical processes. Furthermore, it can be adversely affected by many factors such as weather, climate, soil characteristics, diseases and pests, and environmental pollution that cannot be controlled during agricultural production. For this reason, for an efficient agricultural production, not only expertise is sufficient, but also cultivation conditions must be suitable for demand. The most efficient production will be achieved if the growing conditions meet the optimum demands, provided that the species and breed of the plant and animal grown are also suitable. Providing optimum conditions can be very difficult and costly. Therefore, for a successful agricultural business, the changes in each component should be determined accurately and in a timely manner, and the applications to be made should be decided by considering the possible interactions (Ozguven, 2022). Precision agriculture (PA) technology offers us solutions for all these issues.

PA is the idea of doing the right thing, in the right place, at the right time. This idea is as old as agriculture. However, during the mechanization of agriculture in the twentieth century, there was strong economic pressure to cultivate large fields with uniform agricultural practices. PA provides a way to automate site-specific applications using information technology, thus, making the site-specific application practical in commercial agriculture. PA includes all agricultural production practices that use information technologies, e.g., variable rate application (VRA), yield monitors, and remote sensing (RS) to either tailor the use of inputs to achieve desired results or monitor those results (Bongiovanni & Lowenberg-Deboer, 2004).

PA technologies, combined with control, electronics, computer, and database with the account data, present an advanced system approach. Using global positioning system, geographic ınformation system, VRT, and RS technologies, PA technologies, contrary to common fixed-level application methods, which are applied at all same to the whole land, use the variable-level application methods (based on the application of fertilizer and chemicals to each section to its own needs, tillage at different levels, planting at different norms, irrigation, and drainage at different levels) determining land and plant characteristics of small sections (soil moisture, nutrient level of soil, soil structure, product requirements, yield, etc.). As a result, PA technologies are agricultural production and management methods whose targets are more economic and more environmentally precision production (Özgüven & Türker, 2010).

PA practices are not an agricultural practice that is applied in a single year and left the next year. For accurate analysis in PA, it is recommended to use a yield map of at least 3 years. In PA, for example, with yield maps, regions with low yields and the factors limiting yield in these regions can be determined. In the production to be made in the following year, the management decisions to be taken by considering these factors also allow an increase in productivity. If there are land deficiencies such as low soil pH or soil compaction that limit the yield, they are tried to be eliminated first. In cases where the deficiency cannot be eliminated, the amount of input that cannot be used by the plant can be reduced, thus contributing to economic production by preventing the use of unnecessary inputs and preventing these inputs from harming the environment. PA practices begin with the acquisition of data by using various sensors and RS technologies and continue with soil tests of the production area and determination of soil properties. Application maps are created by associating all information such as yield values, fertilizer and pesticide application norms, climate data, topographic data, weed density, and disease status of previous production seasons with their actual location in the production area. Then, by using appropriate hardware and software, variable-level applications are made in the field (Özgüven, 2018a).

The concept of PA can be expressed in terms such as precision farming, smart farming, VRT, site-specific farming, site-specific management, computer aided farming, and prescription farming. With the recent use of artificial intelligence applications in agriculture, the term Digital Agriculture has been used more widely. The concepts of digital agriculture and PA serve agriculture for the same purpose. But there is a fundamental difference between digital farming and precision farming. PA is based on the concept of variability and the application of different norms suitable for variability. There are also different types of applications in digital agriculture. In addition, PA is applied not only in plant production but also in animal production. This practice is called Precision Livestock Farming (PLF).

In PLF, animals can be continuously monitored individually or as a herd. Areas of monitoring are performance, physiological parameters, welfare status, behavior, housing, product, reproductive and health conditions, etc., and this information is recorded. Warning systems work in case of sudden changes in the general conditions of animals. For this purpose, sensors are used to automatically obtain computerized imaging technologies, RFID tags, movement activity, rumination number, ruminal pH, pulse signal, body temperature, respiratory rate, body weight, body condition score, and location and acoustic information.

2 Variable Rate Application and Its Benefits

The concept of variability is the most important feature that led to the emergence of PA. Understanding this concept will make it easier to understand PA. Photographs of the same cotton field are given in Fig. 1. When looking at Fig. 1, it is seen that the yield is high in some areas and the yield is low in some areas in the top photos; and in the bottom photos, the cotton yields in the field are not the same and the yield is variable in different areas. With PA practices, first of all, the areas where these variabilities in agricultural areas and grown plants are determined. Then, a VRA is made depending on this variability. Therefore, VRA is the most important phase of precision agriculture. In VRA, only as much input is applied as necessary by changing the application norms. VRA systems are classified as variable rate fertilization, variable rate spraying, variable rate planting, variable rate irrigation, and variable rate soil tillage according to the type of application to be made. Although various benefits are obtained from all these applications, variable rate fertilization and spraying applications are the most important VRA applications in providing financial gain and reducing environmental pollution (Özgüven, 2018a). The soil map of a field prepared according to the results of soil analysis is shown in Fig. 2.

Variability in potassium, phosphorus, magnesium, and pH values is seen in Fig. 2, as a result of soil analysis. When the scales on the maps given as examples are examined, variability in different parts of the field can be seen. In PA, variability

Fig. 1 Variability in cotton yield. *Source* Özgüven (2018a)

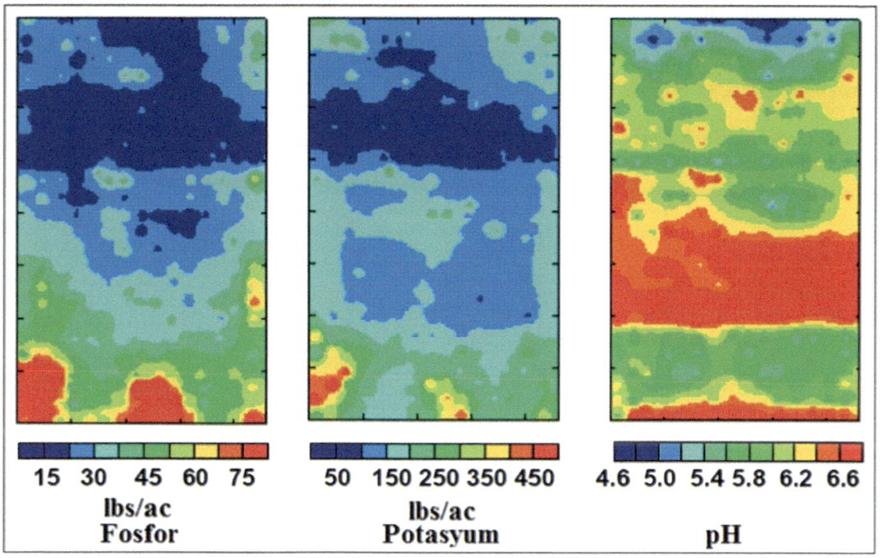

Fig. 2 Varying soil maps. *Source* Davis et al. (1998)

values are taken into consideration during fertilization, and fertilizer is applied at a variable rate needed by field-specific applications. Thanks to the VRA, fertilization, spraying, and planting can be done in certain amounts needed in the field; thus, contributing to economic production by preventing the use of unnecessary inputs and preventing these inputs from harming the environment. In the VRA, the inputs are powder, granule, or liquid. Motors, valves, control, and measuring mechanisms can be used to provide variability. The Current Trimble electric row clutch used for variable rate seeding or input application is shown in Fig. 3, and the apron chain used for solid manure distribution is shown in Fig. 4.

More complex finer adjustments in decision making and input practices do not necessarily increase farm profits or wealth in overly vague, variable, and uncontrolled activities, such as agriculture. The impact of risk factors, such as the weather and price, may exceed the potential benefits of more precise production decisions and methods (Knight & Malcolm, 2009). PA has many potential benefits, such as increased crop yield, cheaper input costs through improved process control, and reduced environmental impact (Weiss, 1996). As a result of the advantages provided by PA technology, an agricultural enterprise will be able to establish a registration system by recording detailed information in GIS, such as yield, disease, climate information, the types of agricultural products produced, the processes such as spraying, fertilization, types of inputs such as seeds, pesticides, fertilizers, etc. In this way, efficiency will be increased by ensuring that more effective business decisions are taken. The income from agricultural production is increased through increased crop yield, higher quality products, and reduced input expenses, such as fertilizer, pesticide, seeds, and water. Increased income makes the agricultural business more stable,

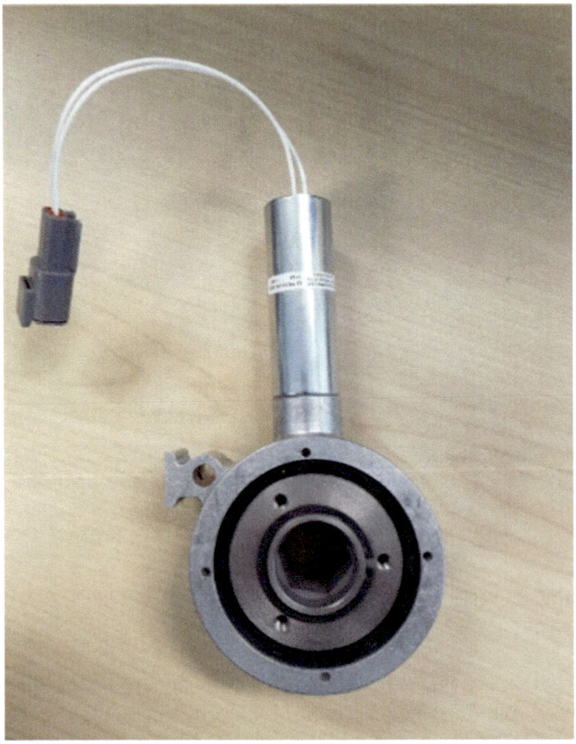

Fig. 3 Current Trimble electric row clutch. *Source* Image provided by Trimble Inc

Fig. 4 Controlling the feed rate of the apron chain. *Source* Fulton et al. (2009)

leading to the opportunity to increase employment. In this way, it contributes to rural development (Ozguven, 2022).

3 Precision Irrigation Applications and Its Effect on Drought

It has been observed that there has been a significant increase in the number and severity of meteorological natural disasters because of global warming and greenhouse effect, which have been more felt in the world in recent years. In the agricultural sector, which is defined as the 'open-top factory', the great financial losses caused by the effect of natural risks threaten the world economy to a great extent (Dinler, 2003). First of all, in the management of natural risks in agriculture; technical protection measures. In addition to early warning systems, the use of systems that provide minimum water consumption against the risk of drought, the tendency to plant pattern that consumes less water, and dwarf fruit growing are some of them (Dinler et al., 2005).

Water scarcity, defined in terms of access to water, is a critical restriction for agriculture in many parts of the world. More than 1.2 billion people, one-fifth of the world's population, live in areas of physical water scarcity and are unable to find enough water for their demands. About 1.6 billion people live in watersheds with water shortages, where human capacity or financial resources will be insufficient to develop adequate water supplies (Fig. 5). Today's water scarcity will possibly cause complexity in the population in the coming years. A growing population is an important factor, but the main causes of water problems are due to other reasons. These include lack of interest in water and poverty, inadequate and incomplete investments, inadequate human resources, ineffective institutions, and mismanagement (IWMI, 2007). Increasing water scarcity is now one of the leading challenges for sustainable development. This challenge will become more urgent as the world population continues to increase, their living standards rise, diets change, and the effects of climate change intensify (FAO, 2019). Internal renewable water resources (IRWRs) from rivers and aquifers amount to 44,000 km^3/year, while withdrawals exceed 4000 km^3/year, which is almost 10% of IRWRs. The local effects of physical water scarcity and freshwater pollution are spreading and accelerating. In many cases, the first sign of scarcity from increased withdrawals is falling groundwater levels (FAO, 2021).

Plants take water from the soil through their roots in order to maintain their normal development. Nutrients in the root environment are transported to the extreme points with this water taken. Since the continuity of the cycle is provided by the removal of water from the leaves of the plant by transpiration, sufficient moisture level must be constantly provided in the root zone of the plant. The water needs of plants vary according to various conditions. These factors are the type of plant, soil properties such as soil salinity, penetration resistance, electrical conductivity, pH values, etc., and climate characteristics. In addition, soil water is absolutely necessary for the

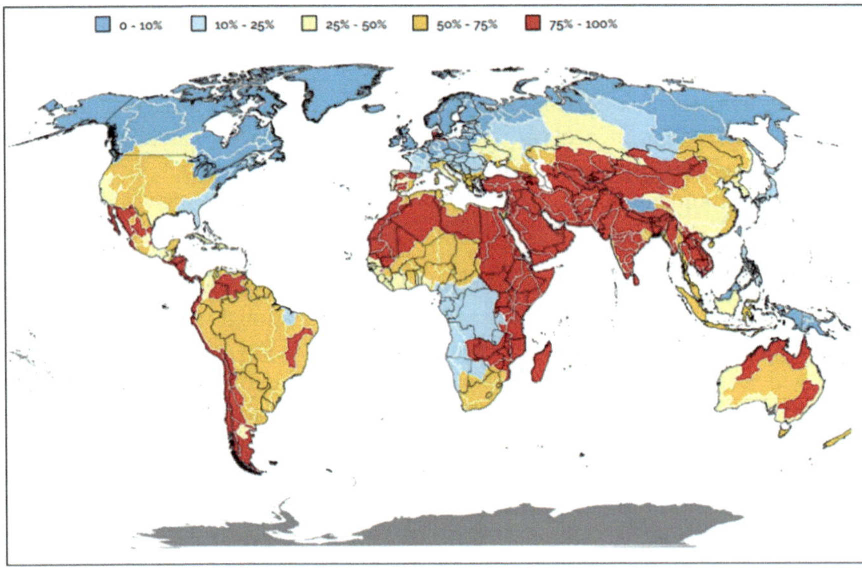

Fig. 5 Water stress level for agricultural sector. *Source* FAO and UN-Water (2021)

conversion of fertilizers into a form that can be taken by plants. Nowadays, when water scarcity is increasing, precision irrigation (PI) practices that meet the specific needs of each plant at the optimum level should be expanded. The only target in PI is not only productivity increase but also includes applications that will allow savings in input usage without causing loss of productivity. PI practices have the potential to minimize their negative environmental impacts, increase optimum water use, and increase economic efficiency. With this system, the data obtained from the field continuously and effectively are analyzed by using advanced hardware and software, more accurate decisions can be made regarding plant breeding, and managers can be facilitated at the decision making stage in the effective management of irrigation networks.

Irrigated agricultural areas constitute 47% of the total agricultural land. Moisture status of the soil, seasonal fluctuations, precipitation-use efficiency, availability of stored water, and the irrigation methodology adopted are important parameters that affect the crop growing systems preferred by the farmers, the grain types that dominate an agricultural region, and the yield targets envisaged by the farmers (Krishna, 2018). Irrigation for intensive agricultural activities can lead to unsustainable water use. Land drainage can lead to lower groundwater levels and also to the destruction of wetlands, which are the habitat of many species. In addition, pesticides and, nitrates from fertilizer use may leak into groundwater, and similarly, nitrogen and phosphorus resulted from fertilizer use may mix with surface water. Irrigation and drainage can affect groundwater levels and cause salinity in the soil (Dişbudak, 2008).

Advanced irrigation management and application technologies combined with advanced detection, modeling, and control technologies are used to ensure the best

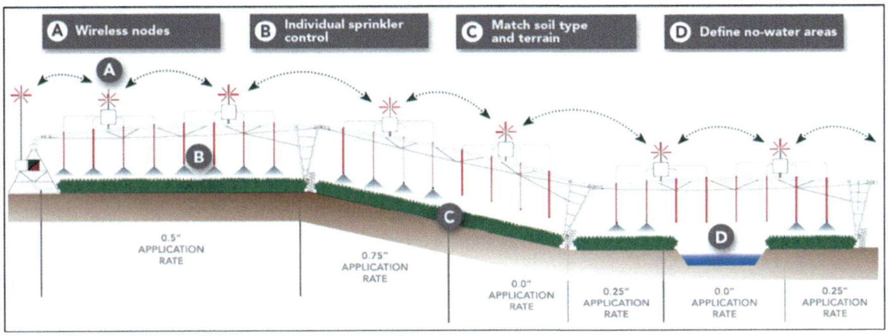

Fig. 6 Individual sprinkler control allowing PI. *Source* Anonymous (2011)

performance in the PI. With PI, plant yield is optimized through systematic information collection and processing about the plant and area, and field-specific targets are set. In this way, reducing the use of inputs prevents waste of resources, minimizing adverse environmental impact. In PI, the goal is not only the yield increase, but also the practices that will allow savings in the use of inputs so as not to result in yield loss. With this system, data obtained continuously and effectively from the field is analyzed using advanced hardware and software, and more accurate decisions are made regarding farming. Furthermore, by using these techniques, it is possible to facilitate managers' decision making in the effective management of irrigation networks (Fig. 6) (Özgüven & Karaman, 2012).

Plant water requirement is determined by neutronmeter, TDR, or humidity sensors. The development and use of humidity sensors have increased in recent years due to the need for calibration, measurement method, and high cost of the device in neutronmeter and TDR measurements. With the humidity sensors, measurements are made by placing them in predetermined places in the field and at the desired soil depth. In addition, these placed sensors can consist of a sensor fusion that determines different soil properties such as electrical conductivity and soil temperature, apart from the humidity sensor. Then, the measured values are transmitted to a central unit by wireless communication methods such as Radio frequency, ZigBee, Wi-Fi, etc. These sensors are usually fixed in the soil and in this way, the moisture status of the soil can be monitored regularly throughout the entire production season. Figure 7 shows the wireless measurement system developed for humidity measurement.

While preparing irrigation maps, the variation in humidity in the field is determined and the data showing these variations are digitized. The geostatistics method can be used to transform the data obtained with humidity sensors into maps. In the geostatistics method, the spatial structure and spatial dependence of the measured moisture values such as soil properties are digitized and using the relationships obtained therefrom, the value of moisture values at the unsampled point is estimated. While creating the map, 3–5 different ranges and colors are generally chosen. The red colored areas on the map show the regions with the lowest humidity, and the green colored areas show the regions where the humidity is higher. In addition, with

Fig. 7 Wireless humidity measuring system. *Source* Pan et al. (2010)

the increase in the use of drones in agriculture in recent years, studies on this subject have begun.

Wu et al. (2019) developed a new drone-mounted ground penetrating radar (GPR) for soil moisture mapping. The entire radar system weighs 1.5 kg and includes a handheld vector network analyzer working as frequency domain radar, a mild hybrid horn-dipole antenna covering a wide frequency range (250–2800 MHz), a GPS for positioning, a microcomputer with the controlling application, and a smartphone for remote control. It is stated that the radar signal is processed using full-wave inversion and the inversion is performed in the time domain and focused on the surface reflection, soil moisture mapping results in the 3 test regions of the developed radar show the accuracy of the technique. In this study, the 500–700 MHz range was used to avoid soil surface roughness effects and to focus on the top 10–20 cm of soil. These areas present a range of landform conditions that lead to specific soil moisture distributions. Soil moisture maps were generated from local measurements using kriging. Obtained soil moisture maps are compatible with topographic conditions of the fields and aerial orthophoto observations (Fig. 8). Finally, these results have been reported to demonstrate the potential and benefits of drone-GPR for rapid, high-resolution mapping of soil moisture at land scale and to support, for example, precision agriculture and environmental monitoring.

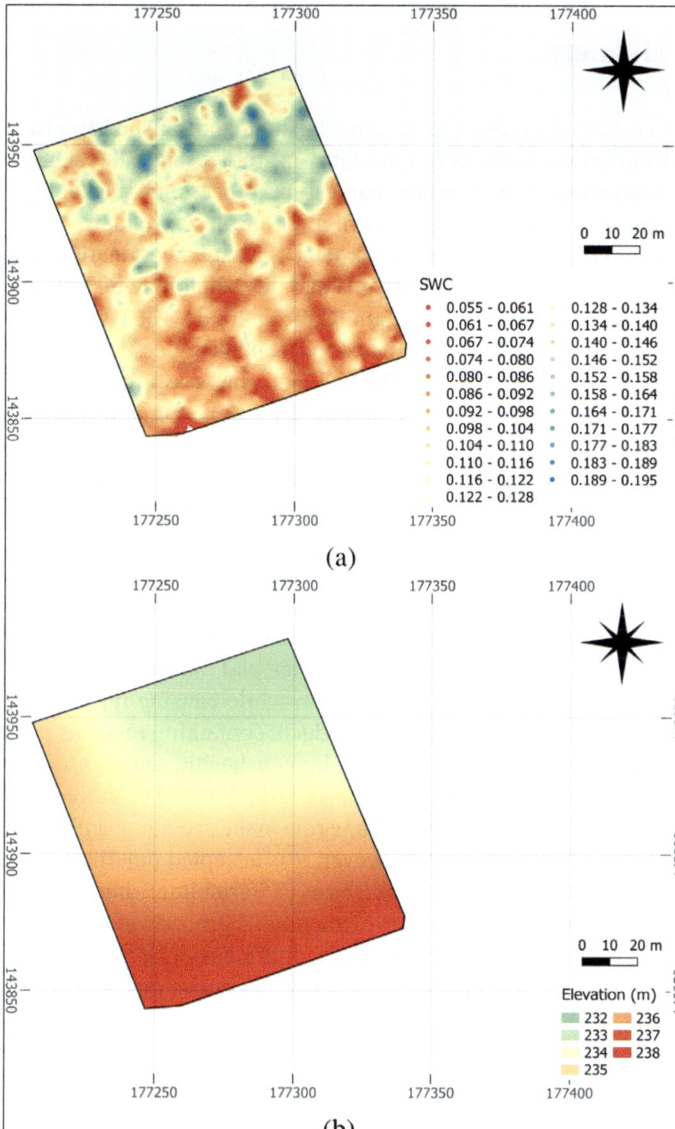

Fig. 8 The studied agricultural area **a** soil moisture map and **b** digital elevation map. *Source* Wu et al. (2019)

4 Potential of Precision Agriculture Practices to Contribute to Food Safety

Food safety covers all kinds of raw, semi-finished, and finished foodstuffs used as human food and animal feed, and all production, storage, and sales stages such as the cultivation, production, maintenance, harvesting, processing, packaging, classification, transportation, preparation and distribution of these products. Food safety is a set of actions and practices designed to protect against contamination and mixing of biological, chemical, and physical hazards and pests that may be harmful to human health throughout the entire supply chain (Cebeci, 2006). The issue of food security is generally examined in four dimensions: food availability, access to food, utilization and use of food, and food stability. These dimensions form the general framework of the definition established by the Food and Agriculture Organization of the United Nations (FAO) (United Nations, 2017). Food security exists when everyone, always, has physical, social, and economic access to adequate, safe, and nutritious food that meets their dietary needs and nutritional preferences for an active and healthy life (FAO, 2016).

After the Second World War, with the agricultural techniques called 'Green Revolution', which started to be applied in agriculture, increases of up to 100% were achieved in yields. The use of chemical fertilizers and pesticides in plant production and the use of feed additives such as hormones and antibiotics in animal production increased the yield and reduced the cost, while causing rapid deterioration of the ecosystem. Some plants and animal products containing residues produced with these techniques have adversely affected human health, and caused a significant increase in health problems, especially cancer (Ak, 2004). The increased use of soil, irrigation, and agrochemicals played a large role in the growth of agricultural production during the Green Revolution. However, it is accepted that the gains often have adverse effects on the natural resource base of agriculture, including land degradation, salting of irrigated lands, over-extraction of groundwater, formation of pest resistance, and erosion of biodiversity. Agriculture has done a lot of damage to the environment through deforestation, greenhouse gas emissions, and nitrate pollution in water resources (FAO, 2011).

Climate-PA aims to increase food security and revenues in a sustainable way, reduce potential risks, adapt to climate change while capturing benefits, and gain flexibility. Protected agriculture links other innovations such as agroecology, agroforestry, and the development of crop varieties with greater tolerance for pests, diseases, drought, flooding, and salinity (FAO, 2013). In climate-PA contributes to sustainable livestock production with mixed plant-livestock systems that fully integrate environmental goals and production goals. Examples include the rotation of pasture and forage plants to improve soil quality and reduce erosion, and the use of livestock manure to maintain soil fertility (FAO, 2017). Today, different approaches are applied to increase food access, such as plant breeding, genetically modified foods, in vitro planting, and the spread of closed ecological systems. In addition to these methods, the use of modern technologies known as PA is also recommended as a way to ensure food safety. Using various methods such as RS and GIS, early intervention is provided to prevent unwanted events through monitoring vegetation at every stage of production from soil processing to the harvest. Different methods of PA can provide the following benefits in terms of food safety (Talebpour et al., 2015):

- Crop quality,
  - Increasing crop quality using spatially variable fertilizer management,
  - Reducing the impact of climate change,
  - Timely control of pests,
  - Increasing irrigation efficiency when water resources are scarce,
  - Protection of crop quality with precision post-harvest application,
- Sustainability and rural development,
- Risk management,
- Traceability.

Managing productivity risk is one of the most difficult decisions for farmers and farm managers. Produced items can be sold at the best prices, but the result can be disastrous if they have little or nothing to sell. Production or yield risk arises from a variety of uncertain factors such as weather events (inadequate rainfall, hail, etc.), diseases, insects, crop variety selection, soil characteristics, or production techniques. PA can enable a producer to manage the spatial component of return risk, but it does not always help him cope with weather risk (Gandonou, 2005). Adopting PA to make risk management decisions is beneficial for getting more and better information and controlling crop growth conditions. Thus, it is possible to address problem areas specific to the area for a net return from season to season, while reducing the possibility of low yields and taking into account temporal variability in yields and spatial variability. Despite the high risks of using PA, such as the possibility of crop loss, financial risk due to investment in bad crop season, and human and technological risks, early estimation of yield using RS data leads farmers to develop self-confidence in early marketing. It also simplifies contracting by bringing agriculture closer to 'production to specification'. This means that farmers will have more control over input application and yield quality, and will be more willing to contract early to earn higher incomes (Lowenberg-DeBoer, 1999).

PA offers opportunities to monitor all practices for the production of agricultural products. This monitoring capability is becoming one of the main subjects of research on farm management monitoring (McBratney et al., 2005). By using the GIS as a component in PA, traceability can be achieved by associating it with the agricultural environment such as soil condition, local characteristics, and basin areas. In this way, direct and indirect environmental characteristics can be determined to document the history of events in the production area from sowing-planting to harvest (Oger et al., 2010). Another important application for traceability in PA is radio frequency identification (RFID) technology. The widespread use of RFID in PA makes it possible to increase the efficiency, productivity, and profitability of agricultural systems, while avoiding undesirable effects on the environment. Obtaining real-time information allows farmers to adjust strategies at any time, providing a solid basis for recognizing differences and changing management action accordingly (Ruiz-Garcia & Lunadei, 2011).

Development has continued since the emergence of PA technology and this development continues today. While it was focused on the field-specific determination of variability before, systems with sensor-based applications were developed later. Today, agricultural machines are being converted into ISOBUS compatible smart machines. Agricultural machinery and production or cultivation areas are equipped with sensors, and even more than one agricultural machine can communicate with each other. Apart from agricultural machines, it is possible to communicate with agricultural robots or drones and agricultural applications can be made in coordination (Ozguven, 2022). However, environmental conditions such as humidity, temperature, and abrasive factors in agricultural areas and technical difficulties such as communication problems at long distances are some of the obstacles to the application of agricultural robots and smart machinery (Özgüven & Közkurt, 2021). In PA technology, all necessary controls are made with electronics. Therefore, machines and tools have intelligent processing units that can communicate with each other, with other units and with the farm operating unit. As seen in Fig. 9, the ISO11783 standard, called ISOBUS, provides a manufacturer-independent plug-and-play solution for each combination of tractors and equipment. In this way, intelligent machines can collect multi-purpose information at high spatial and temporal resolution. Integrated into the farm business, this technology makes a significant contribution to the documentation and traceability of products and processes, including changes in field operations (Auernhammer & Speckmann, 2006).

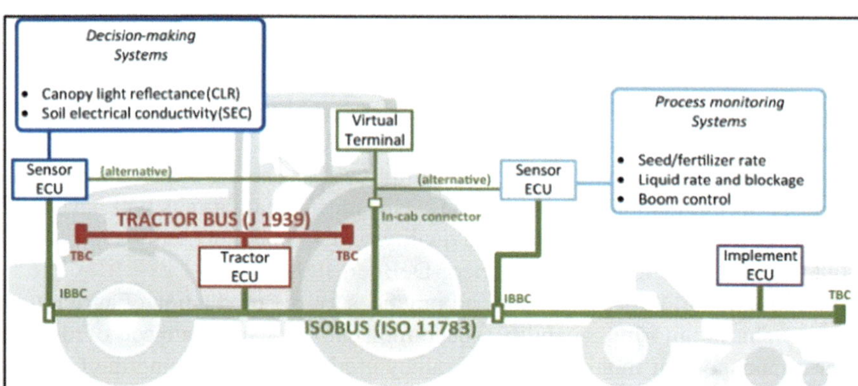

Fig. 9 ISOBUS setup of a tractor and a towed implement. *Source* Paraforos et al. (2019)

5 Discussion

By using remote sensing satellites, drones and sensors in PA, crops, plants and their characteristics, diseases and pests, land use, vegetation indexes, and the physical and chemical properties of soils can be determined. In addition, fertilizer applications and field water conditions can be monitored. In addition to today's PA applications, technologies such as artificial intelligence, robotics, IoT, autonomous vehicles, drones, advanced computers, cloud computing, and big data can be used in agriculture. By using these technologies together in agriculture, it has become possible to take more effective management decisions for better quality production. In addition, the impact of climate change on the world and agriculture is frequently discussed today. In this regard, PA offers significant opportunities to minimize the negative impact of climate change on agricultural production and food supply. For example, using drones, it is possible to monitor irrigation in a very short time and to determine plant water status and drought or flood effects. In addition, the adoption of precision irrigation techniques helps improve water use and maximize plant productivity. These practices help both reduce the impact of drought and increase food production.

6 Conclusion

PA provides very important benefits for agricultural production, which is adversely affected by global market fluctuations, economic crises, climate changes, drought, etc. The VRAs provided by PA ensure the effective use of inputs, decrease production costs and negative environmental effects, and increase product quality and productivity. In addition, PA applications help field or enterprise management, yield estimation, and environmental management. Thus, more competitive, efficient, and sustainable agricultural practices are possible with the use of PA technologies in agriculture. In recent years, good agricultural practices have come to the fore in food safety. However, all notifications and documents created, including this application, are based on the declaration of individuals. In precision agriculture, all application information such as fertilizer and pesticide application norms can be obtained from the PA system. Here, it would be appropriate to give an example of a commercial application that can keep real-time records. By using Trimble's Precision-IQ™ field application with Trimble Farmer Core and Trimble TMX-2050 or GFX-750 displays with Farmer Pro software, real-time and location-based records can be kept of spraying information and amount of spraying in a field. This application gives the ability to instantly collect and manage the information and data from infield activities and then analyze that data back in the office using farm management software1.[1] For this reason, these applications in PA are a very important feature that provides digital recording of all applications in the production and distribution chain of agricultural

[1] See the https://www.trimble.com/en/ website for details.

products. It will bring confidence in the agricultural products produced in this way and will result in higher profits.

References

Ak, İ. (2004). Apolyont Doğal Tarım ve Hayvancılık Projesi. I. Uluslararası Organik Hayvansal Üretim ve Gıda Güvenliği Kongresi. 28 Nisan–1 Mayıs, s. 144 (Turkish).

Anonymous. (2011). Irrigation advances conserving water, energy and labor, *20*(1), Spring 2011.

Auernhammer, H., & Speckmann, H. (2006). Section 7.1 Dedicated Communication Systems and Standards for Agricultural Applications, pp. 435–452 of Chapter 7 Communication Issues and Internet Use, in CIGR Handbook of Agricultural Engineering Volume VI Information Technology. Edited by CIGR-The International Commission of Agricultural Engineering; Volume Editor, Axel Munack. St. Joseph, Michigan, USA: ASABE. (Çevirmenler: Yaşar, E., Tarhan, S. ve Özgüven, M.M.; Çeviri Editörleri: Tarhan, S. ve Özgüven, M.M.).

Bongiovanni, R., & Lowenberg-Deboer, J. (2004). Precision agriculture and sustainability. *Precision Agriculture, 5*, 359–387.

Cebeci, Z. (2006). Gıda İzlenebilirliğinde Bilgi Teknolojileri. Ulusal Tarım Kurultayı, 15–17 Kasım 2006, Çukurova Üniversitesi, Adana. Bildiriler s. 189–195 (Turkish).

Davis, G., Casady, W. W., & Massey, R. E. (1998). Precision Agriculture: An Introduction. University Extension, University of Missouri-System.

Dinler, T. (2003). Tarımda Meteorolojik Karakterli Doğal Afetler ve Risk Yönetim Teknikleri, III. Atmosfer Bilimleri Sempozyumu, İstanbul (Turkish).

Dinler, T., Yaltırık, A., Çetin, B., Özkan, B., Gülçubuk. B., Sürmeli, E., Ekmen, E., Saner, G., Akçaöz, H., Karahan Uysal, Ö., Karaaslan, S., & ve Kıymaz, T. (2005). Tarımda Risk Yönetimi ve Tarım Sigortaları. ZMO Türkiye Ziraat Mühendisliği VI. Teknik Kongresi, Cilt.2, s.1209–1232, Ankara (Turkish).

Dişbudak, K. (2008). Avrupa Birliği'nde Tarım-Çevre İlişkisi ve Türkiye'nin Uyumu. AB Uzmanlık Tezi. T.C. Tarım ve Köyişleri Bakanlığı Dış İlişkiler ve Avrupa Birliği Koordinasyon Dairesi Başkanlığı. Ankara (Turkish).

FAO. (2011). Land Tenure, Climate Change Mitigation and Agriculture. Mitigation of Climate Change in Agriculture (MICCA) Program, June 2011. Rome.

FAO. (2013). Climate-Smart Agriculture Sourcebook. Rome.

FAO. (2016). Food Security Indicators. http://www.fao.org/economic/ess/ess-fs/ess-fadata/en/

FAO. (2017). The Future of Food and Agriculture—Trends and Challenges. Rome. ISBN: 978-92-5-109551-5.

FAO. (2019). Water Scarcity—One of the Greatest Challenges of Our Time. Rome. https://www.fao.org/fao-stories/article/en/c/1185405/

FAO. (2021). *The state of the world's land and water resources for food and agriculture—Systems at breaking point*. Synthesis Report 2021. Rome. https://doi.org/10.4060/cb7654en

FAO and UN-Water. (2021). Progress on Level of Water Stress: Global Status and Acceleration Needs for SDG Indicator 6.4.2. Rome. 96 pp. http://www.fao.org/3/cb6241en/cb6241en.pdf

Fulton, J., Brodbeck, C., Winstead, A., & Norwood, S. (2009). Overview of Variable-Rate Technology. Alabama Cooperative Extension System. Precision Agriculture Series Timely Information. Agriculture, Natural Resources & Forestry.

Gandonou, J. M. A. (2005). *Essays on precision agriculture technology adoption and risk management*. University of Kentucky Doctoral Dissertations. Paper 227. http://uknowledge.uky.edu/gradschool_diss/227

IWMI (International Water Management Institute). (2007). A Comprehensive Assessment of Water Management in Agriculture. Edited by David Molden. Earthscan, UK and USA.

Knight, B., & Malcolm, B. (2009). A whole-farm investment analysis of some precision agriculture technologies. *AFBM Journal, 6*(1), 41–54.
Krishna, K. R. (2018). *Agricultural drones: A peaceful pursuit*. Apple Academic Press. ISBN: 978-1-315-19552-0.
Lowenberg-DeBoer, J. (1999). Risk management potential of precision farming technologies. *Journal of Agricultural and Applied Economics, 31*(2), 275–286.
McBratney, A., Whelan, B., Ancev, T., & Bouma, J. (2005). Future directions of precision agriculture. *Precision Agriculture, 6*(1), 7–23.
Oger, R., Krafft, A., Buffet, D., & Debord, M. (2010). Geotraceability: An innovative concept to enhance conventional traceability in the agri-food chain. *Biotechnologie, Agronomie, Société Et Environnement, 14*(4), 633–642.
Özgüven, M. M., & Türker, U. (2010). Application of precision farming in Turkey, comparative analysis of wheat, cotton and corn production. *Journal of Agricultural Machinery Science, 6*(2), 127–135.
Özgüven, M. M., & ve Karaman, S. (2012). Hassas Sulama Teknolojileri. 27. Tarımsal Mekanizasyon Ulusal Kongresi, 5–7 Eylül Samsun, S:288–297 (Turkish).
Özgüven, M. M. (2018a). Hassas Tarım. Akfon Yayınları, Ankara. ISBN: 978-605-68762-4-0 (Turkish).
Ozguven, M. M. (2018b). The newest agricultural technologies. *Current Investigations in Agriculture and Current Research, 5*(1), 573–580. https://doi.org/10.32474/CIACR.2018.05.000201
Özgüven, M. M., & Közkurt, C. (2021). Agricultural robots and smart agricultural machinery. In *International Symposium of Scientific Research and Innovative Studies*, 22–25 February 2021 (pp. 81–85). Bandırma-Turkiye. ISBN: 978-625-44365-8-1.
Ozguven, M. M. (2022). *The digital age in agriculture*. CRC Press Taylor & Francis Group LLC.
Pan, L., Adamchuk, V. I., Martin, D. L., Schroeder, M. A., & Ferguson, R. B. (2010). Analysis of water use efficiency using on-the-go soil sensing and a wireless network. In R. Kholsa (Ed.), *Proceedings of the Tenth International Conference on Precision Agriculture*, Denver, Colorado, 18–21 July 2010, Colorado State University, Fort Collins, Colorado, USA (p. 13). CD publication.
Paraforos, D. S., Sharipov, G. M., & Griepentrog, H. W. (2019). ISO 11783-compatible industrial sensor and control systems and related research: A review. *Computers and Electronics in Agriculture, 163*, 104863. https://doi.org/10.1016/j.compag.2019.104863
Ruiz-Garcia, L., & Lunadei, L. (2011). The role of RFID in agriculture: Applications, limitations and challenges. *Computers and Electronics in Agriculture, 79*(1), 42–50.
Talebpour, B., Türker, U., & Yegül, U. (2015). The role of precision agriculture in the promotion of food security. *International Journal of Agricultural and Food Research, 4*(1), 1–23. ISSN: 1929-0969.
United Nations. (2017). The Role of Science, Technology and Innovation in Ensuring Food Security by 2030. United Nations. Economic and Social Council. 27 February 2017.
Weiss, M. (1996). Precision farming and spatial economic analysis: Research challenges and opportunities. *American Journal of Agricultural and Applied Economics, 78*, 1275–1280.
Wu, K., Rodriguez, G. A., Zajc, M., Jacquemin, E., Clément, M., De Coster, A., & Lambot, S. (2019). A new drone-borne GPR for soil moisture mapping. *Remote Sensing of Environment, 235*, 111456. https://doi.org/10.1016/j.rse.2019.111456

Climate Change Impacts on Agricultural Sector in Sri Lanka: Necessity of Precision Agricultural Technologies

Lahiru Udayanga and Mohamed Mujithaba Mohamed Najim

Abstract The impacts of climate change and increasing concerns on environmental sustainability have imposed numerous challenges on ensuring the agricultural productivity of Sri Lanka. The recent policy decisions of the country on environmentally friendly agricultural strategies have further influenced this issue. Therefore, maintaining higher productivity with minimum environmental impacts and using optimum inputs (fertilizer, irrigation, labor, and other agro-chemicals) remain a challenge. Adoption of Precision Agricultural Technologies (PAT), which is an emerging discipline that integrates several technologies, such as Automation, Drone Technology (DT), Internet of Things (IoT), Geo-Informatics, Sensor Technology, and Variable-Rate Technology (VRT), remains the most viable solution. Different PAT combinations could effectively support monitoring and respond to variability in soil, environment, and crops, ensuring optimal resource utilization. The combined use of the aforesaid technologies would enable Sri Lankan farmers to detect and monitor soil and crop conditions, enabling customized automation of irrigation, fertilizer application, and pest and disease management. Adequately calibrated crop modeling applications can also support decision-making and operational control in agriculture. Ultimately, accurate and reliable decision support systems could be developed and used for Sri Lankan scenarios by integrating PAT, respecting edaphic, crop, operational, and meteorological parameters relevant to agriculture. However, due to numerous barriers such as a lack of technological expertise and physical resources, poor socio-economic conditions of farmers, gaps in policies and implementation processes, poor coordination among government and other entities, and lack of government support, PAT adoption in Sri Lanka remains relatively low. Therefore, the Sri Lankan government and other key agricultural entities have to take efforts to promote PAT in Sri Lanka, through enhancement of the physical infrastructure facilities, dissemination

---

L. Udayanga
Department of Bio-Systems Engineering, Faculty of Agriculture and Plantation Management, Wayamba University of Sri Lanka, Makadura, Sri Lanka
e-mail: udayanga@wyb.ac.lk

M. M. M. Najim (✉)
Faculty of Agriculture, Sultan Sharif Ali Islamic University, Sinaut Campus, Km 33, Jln Tutong, Kampong Sinaut, Tutong TB1741, Brunei
e-mail: najim.mujithaba@unissa.edu.bn

M. Behnassi et al. (eds.), *Climate-Smart and Resilient Food Systems and Security*,
https://doi.org/10.1007/978-3-031-65968-3_14

of knowledge and technology, empowering the farmers and Agricultural Instructors (AI), while elevating the Technology Readiness (TR) of farmers. In addition, drafting and implementation of long-term policies aiming to promote PAT are vital to ensure agricultural productivity in Sri Lanka with minimum environmental impacts.

Keywords Agriculture · Climate change · Precession agricultural technology · Productivity · Sri Lanka

1 Agriculture Sector of Sri Lanka

Sri Lanka has a unique climatic system and a rich biodiversity, due to its location between 5° 55′ and 9° 5′ North latitudes and 79° 42′ and 81° 53′ East longitudes. Throughout civilization, agriculture has played a critical role in Sri Lanka. At present, agriculture acts as a key pillar of the Sri Lankan economy and the livelihood of its people. The contribution of the agriculture sector to the National Gross Domestic Production (GDP) in 2018 was around 7% (672,923 million LKR), with a growth rate of 4.8%, when compared to 2017 (CBSL 2019). Regardless of the impacts caused by the COVID 19 epidemic, the agriculture, forestry, and fisheries sectors have denoted 5.8% (686,412 million LKR) and 6% (669,802 million LKR) contributions to the GDP in 2019 and 2020, respectively (CBSL 2021). In 2019, the agricultural sector has catered for 25.3% of the labor force in Sri Lanka (DOA 2020). Based on the relative contribution to the GDP, paddy (11.7%), marine fishing and marine aquaculture (11.3%), spices, and other pharmaceutical crops (10.2%) and fruits (9.6%) could be recognized as the leading sub-sectors contributing to the overall income earned by this sector, as shown in Fig. 1 (CBSL 2021).

Three major agricultural production systems can be identified in Sri Lanka, namely, food crops, plantation crops, and minor export crops. Paddy, maize, pulses, fruits, and vegetables are considered as the major food crops in Sri Lanka (CBSL 2021). Paddy is an important subsector in Sri Lankan agriculture, due to its preference as the staple food by the nation. With 983,550 ha under paddy cultivation, Sri Lanka has produced 4,592,056 t of paddy with an average yield of 4,795 kg/ha in 2019. The "Maha" season (the major rainy season in the drier part of the country) accounts for the highest share of paddy production (66.9%). Mostly, high-yielding varieties such as AT362, BG352, and BG360 (3.5-month varieties) are predominantly cultivated, contributing to around 77% of the annual paddy production, followed by 3-month varieties (18%).

Around 3.5% of the total land area in Sri Lanka is under tea cultivation, followed by coconut (3.2%) and rubber (2.8%). Tea and rubber plantations have shown a 9.2% and 3.6% share of the GDP, respectively (Fig. 1). Meanwhile, oleaginous fruits (such as coconut, king coconut, and oil palm) accounted for 9.1% of the GDP in 2020 (CBSL 2021). Among the minor export crops, clove, cinnamon, cardamom, nutmeg, and pepper play a major role in securing export income. In the case of fruit crops, banana, mango, papaw, and pineapple account for the highest land cover. Meanwhile, cabbage

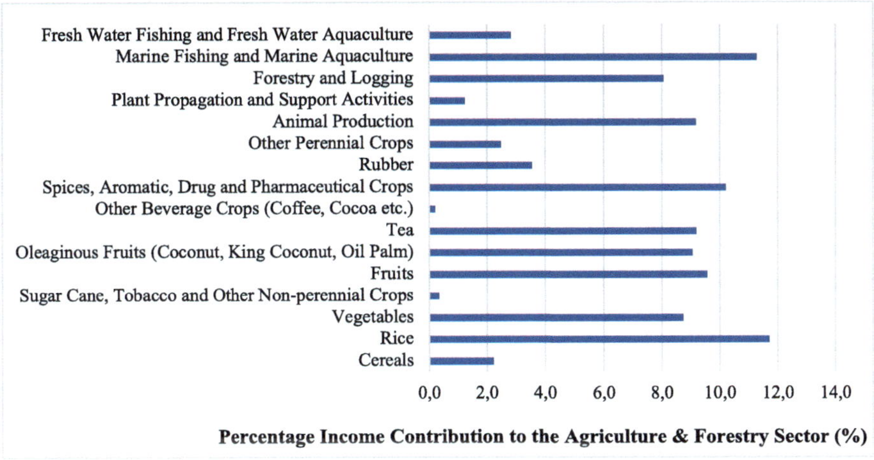

Fig. 1 Economic contribution of different sectors to the Agriculture & Forestry sector. *Source* CBSL (2021)

(116,577 t), carrot (80,767 t), tomato (77,916 t), and radish (44,244 t) have played a dominant role among up-country vegetables in terms of production in 2019. Brinjal (134,863 t), red pumpkin (97,473 t), ash plantain (63,460 t), and bitter gourd (40,970 t) could be recognized as the major low-country vegetables based on production (DOA 2020). Coastal and offshore fishing activities have contributed by 48.0% and 34.2% to the national fish production (505,830 t) in 2019, respectively. Meanwhile, intensive culture-based fisheries have accounted for 12.2% (DOA 2020).

2 Climate Change: The Sri Lankan Perspective

2.1 Climate in Sri Lanka

Due to its geographical location, Sri Lanka is currently experiencing tropical monsoonal climate conditions. Based on the annual cumulative rainfall, effectiveness of monsoon rainfall, soil types, topography, and land use features, Sri Lanka is classified into three major climatic zones: the dry zone, the intermediate zone, and the wet zone. The wet zone is characterized by a relatively higher mean annual rainfall > 2500 mm, while the intermediate zone receives an annual average rainfall of 1750–2500 mm. Meanwhile, the average rainfall received by the dry zone remains < 1750 mm per year (Panabokke 1996). The dry zone accounts for around 64% of the total land area of Sri Lanka (6.56 million ha), followed by the wet zone that accounts for 23%. The intermediate zone extends over an area of 0.85 million ha, accounting for 13% of the total land available in Sri Lanka (Kudagamage, 1998). Sri Lanka has been divided into 46 agro-ecological regions, based on climatic, soil, and topographic conditions to respect rich agro-ecological diversity (Punyawardena et al., 2003). Different agro-ecological conditions in Sri Lanka have provided ideal conditions for cultivation of many crops and animal husbandry.

The climate of Sri Lanka is characterized by two distinct monsoons and two inter-monsoon seasons with convectional activity. The South-West Monsoon (SWM), which extends from May to September, and the North-East Monsoon (NEM), from December to February, are the two main monsoons. Among the two inter-monsoon seasons, the First Inter-Monsoon (FIM) extends from March to April, followed by the Second Inter-Monsoon (SIM), which ranges from October to November (Marambe et al. 2014). The major growing season of Sri Lanka ("Maha" season) begins with the arrival of the SIM rains and continues up to NEM (October to February). Meanwhile, the minor growing season ("Yala" season) extends from May to September, corresponding to the SWM rainfall (Zubair, 2002).

The mean annual temperature of the lowlands is homogeneous and remains at 27 °C up to an altitude of 100–150 m. The temperature decreases rapidly with the increasing altitude in the highlands. At an altitude of around 1,800 m, the mean annual temperature of the city of Nuwara Eliya is around 15 °C. Generally, the lowest temperature conditions in Sri Lanka can be experienced in January, while the highest temperature conditions occur in April and August (Punyawardena et al., 2003). Meanwhile, the Relative Humidity (RH) tends to range from 70 to 90% during the mornings to 55–80% during the late afternoons, based on the geographical location. The period from June to August often corresponds to relatively lower RH levels (40–60%), especially in dry lowlands. On the contrary, higher RH levels can be experienced from December to January, in Sri Lanka (Marambe et al. 2014). The average pan evaporation rate ranges between 3 and 8 mm per day during the "Yala" season, while it ranges between 2 and 5 mm per day in the "Maha" season (Punyawardena, 2007).

2.2 *Climate Change: The Sri Lankan Perspective*

Changes in climate over time, whether due to natural variability or as a consequence of anthropogenic activities, could be considered as climate change (IPCC 2007). A recent study conducted by Premalal (2009) has evidenced that the average temperature of Sri Lanka is increasing at a rate of 0.01–0.03 °C per year. The nighttime minimum temperature has been predicted to increase more than the daytime maximum temperature (Marambe et al., 2015). In another study, increments of 2.5–4.5 °C and 2.5–3.25 °C have been projected to be experienced in the annual average temperature by 2080 under the B2 scenario, based on the downscaling of General Circulation Models (Punyawardena et al., 2013a). Meanwhile, studies based on regional climate model projections and General Circulation Models have also forecasted an increase in temperature of 2–3 °C and 2.84 °C, respectively (Ahmed & Suphachalasai, 2014; Cline, 2007). An increase in temperature in the South-Asian region by more than 2 °C has been predicted to occur by the mid-twenty-first century, while the occurrence of hot days will increase (IPCC 2014). Further, a recent study conducted by Zubair et al. (2015) has predicted an increment of 1.5–2.8 °C for "Yala", followed by 1.1–2.4 °C during "Maha" seasons within the North-Western Province of Sri Lanka, based on 20 General Circulation Models.

A few recent studies have denoted that those apparent variations in rainfall, in terms of season and quantity, couldn't be identified in many parts of the country based on long-term data, due to the high inter-annual variability of rainfall in Sri Lanka (Nissanka et al., 2011; Marambe et al. 2014). Nevertheless, the occurrence of drought and flood events has increased in recent years (2001 onwards), when compared to past years throughout the country. The notable seasonal variations in rainfall have been identified as the cause for this trend (Punyawardena & Premalal, 2013). Further, rainfall is expected to become more erratic, along with an increase in the frequency of extreme rainfall events in the South-Asian region (Kaur & Kaur, 2017). Based on the General Circulation Models, the dry zone of Sri Lanka is expected to become drier, while the wet and intermediate zones would become wetter, under the A2 scenario (Marambe et al., 2015). Under the projections of the B2 scenario, both the dry zone and the central highlands of the island are expected to become drier, while the remaining areas will become wetter at a relatively slow rate (Punyawardena et al., 2013b). In both scenarios, the date of onset and quantity of monsoon rainfall have been predicted to indicate notable variations (Premalal, 2009). While many studies have predicted an increase in the mean annual temperature in Sri Lanka (Cline, 2007; De Silva et al. 2006), the mean annual rainfall is expected to increase by 14% and 5% under the A2 and B2 emission scenarios, respectively (Ahmed & Suphachalasai, 2014; De Silva et al., 2007). According to De Silva et al. (2007), the SWM rainfall could denote an increase characterized by apparent spatial and temporal variations, while the precipitation received during NEM would decline. In general, Sri Lankan temperature is predicted to increase by around 3 °C, while rainfall denote slight increments with apparent spatial and temporal variations (Marambe et al. 2014). An increase in the overall dryness within the DL_{1b} Agro-Ecological Zone has been suggested by a recent study (Udayanga & Najim, 2020). Further, notable increases in minimum and maximum temperatures have been predicted for 70% and 55% of the meteorological stations in Sri Lanka, along with an increase in nighttime temperatures, respectively. Therefore, a narrower diurnal temperature difference could be expected over many regions of Sri Lanka (Naveendrakumar et al., 2018).

3 Impacts of Climate Change on Agriculture

Apparent changes in the climate have led to increasing attention being paid to the impacts of climate change on different sectors, such as agriculture, tourism, health, and industry. Among them, agriculture is a key sector, which is heavily influenced by the changing climatic patterns. An increase in the global annual mean temperature, variability in rainfall, and increased frequency of extreme weather events are expected to occur due to climate change, which have led to extensive disasters in recent years (Marambe et al., 2015).

3.1 Crop Production

In general, the frequent occurrence of intensive rainfall events (> 25 mm h^{-1}) is capable of causing topsoil erosion in many arable lands, especially in the wet zone. Further, such scenarios could lead to nutrient leaching conditions, leading to a decrease in yield and many environmental issues (De Costa & Sangakkara, 2006; Marambe et al., 2015). In addition, consecutive rainfall events could damage the harvest of crops. On the contrary, raising temperatures may enhance biodegradation, thereby degrading the fertility and productivity of topsoil.

With the enhancement of growth and development of different pests, an increased occurrence of pests, pathogens, and diseases could also be expected due to the increasing temperature and humidity conditions, which will further influence the yield (Eeswaran, 2018). In addition, an increase in viral diseases and intensified weed occurrences are also expected in different regions of the country, which may cause severe impacts on agricultural productivity (Fand et al., 2012). The expansion of invasive alien fauna and flora has also imposed an alarming condition on securing high yields (Iqbal et al. 2014). Therefore, the majority of Sri Lankan farming communities tend to utilize different synthetic agro-chemicals (fertilizer, pesticides, weedicides, and fungicides) at intensive rates, leading to serious public and environmental health hazards (Eeswaran, 2018). Degradation of farming communities' socio-economic levels, as well as increased susceptibility to both physical and psychological health issues, remain serious concerns (Berry et al., 2011).

Being the staple food crop of Sri Lanka, rice has a strong bond with the nation that dates back to ancient history. Due to its priority as the main food, paddy is cultivated in all three climatic zones in Sri Lanka. For paddy cultivation, around 1500–2000 mm of water per season is required for establishment, growth, and weed control (Bouman, 2001). Sri Lankan paddy cultivation is mainly based on rainfall and irrigation systems. Therefore, any variability in the magnitude, distribution, and seasonality could have a direct impact on the yield of paddy, especially in rain-fed systems. Due to the influence on catchment recharging, the irrigated rice systems may also experience negative impacts from rainfall variability (Marambe et al., 2015). On the contrary, floods caused by extreme rain events could have severe impacts on paddy cultivation, especially in low-lying areas in Sri Lanka (Nguyen, 2002).

Paddy, as a C3 plant, produces higher yields under higher CO_2 levels due to increased net photosynthetic rate and crop growth rate (De Costa et al., 2003; Delpitiya et al., 2014; Tanaka, 1976). Therefore, elevated CO_2 levels in the atmosphere would favor rice production. However, the increasing ambient temperature acts as the limiting factor, since the optimum vegetative growth of rice could be experienced from 25 °C to 30 °C (Yoshida, 1978). Induction of pollen sterility is a serious concern that can be expected due to global warming. In rice, exposure to elevated temperature conditions (around 36 °C), even for a short duration, could induce a notable level of pollen sterility, especially during flowering (Abeysiriwardena et al., 2002). An increment of 2 °C, in maximum and minimum temperatures, has been predicted to result in an average yield loss of 12% in rice (Delpitiya et al., 2014). Due to seawater intrusion, the salinity levels in ground and surface water bodies in the coastal lowlands of Sri Lanka are being elevated. Lowering of the groundwater table and sea-level rise have been recognized as the main causes of this. Such an increase in salinity could result in a significant strain on rice cultivation systems, particularly in coastal areas (De Costa, 2010; Marambe et al. 2014).

An increase in minimum and maximum temperatures has resulted in narrower diurnal temperature ranges, causing undeniable detrimental impacts on a variety of field crops, including vegetables, fruits, legumes, and tuber crops. According to Marambe et al. (2014), a narrow diurnal temperature range is known to retard the assimilation and tuber bulking in potatoes, leading to a yield loss. The fiber content of fruits and sugary vegetables may rise, reducing flavor and edible quality (Marambe et al., 2015; Titumil and Basak 2010). A rise in evapotranspiration rates, a decrease in soil moisture, and the induction of temporary soil salinity variations caused by rising temperatures could all combine to create harsh conditions for plant growth. Recently, yield losses have been reported from maize (Malaviarachchi et al. 2014) and mung beans (Malaviarachchi et al., 2015) in Sri Lanka, due to an increase in temperature. Frequent occurrences of flood conditions could also act as a stressor for vegetables, fruits, legumes, and tuber crops, leading to notable yield losses (Esham et al. 2017).

Similar to annual or cash crop growers, perennial plantation growers can also suffer the significant impacts of climate change (Esham et al. 2017). Coconut (*Cocos nucifera* L.) is a rain-fed perennial crop that plays a major role in the Sri Lankan economy (Pathiraja et al. 2015). Around 57% of the total coconut extent in Sri Lanka is accounted by the coconut triangle (Kurunegala, Gampaha, and Puttalam), mainly due to the mean and annual rainfall of > 1500 mm and the presence of a mean temperature of around 28 °C (Eeswaran, 2018). The productivity of perennial crops such as coconut could be significantly influenced by high temperatures and water stress conditions (Esham et al. 2017). Prolonged drought conditions have been found to induce yield losses in coconut (Fernando et al. 2007), while aggravating pest attacks (Eriyagama et al., 2010). Variability in rainfall and increasing temperatures are known to cause low production rates in the tea sector, especially in the low-country region (Wijeratne, 2011; Karunaratne et al. 2015). The majority of tea plantations in Sri Lanka are expected to face higher temperatures and dry weather conditions, except those at high elevations (> 1200 m). However, increasing ambient CO_2 levels could increase the tea yield notably (Wijeratne et al. 2007). Meanwhile, periods of high temperature and humidity could elevate the susceptibility of tea to blister blight (*Exobasidium vexans*) occurrence, influencing tea production (De Costa, 2010).

In the case of rubber, rainfall variability and frequent droughts have been identified to cause significant impacts on young plants, immediately after field planting (Jessy et al., 2011; Wijesuriya & Samarappuli, 2009). In addition, rainfall variability has been identified to aggravate the susceptibility of rubber to certain fungal diseases such as *Oidium* leaf fall (*Oidium hevea*). Further, intensive rainfall and drought events could affect the tapping days and disrupt the harvesting process. However, rubber trees can withstand drought conditions better than other plantation crops, and a short dry spell is often advantageous for rubber to avoid certain leaf diseases (Gitz et al., 2020). Home gardens also account for a notable level of agricultural produce in Sri Lanka. A recent study conducted by Weerahewa et al. (2012) has evidenced that high crop diversity, spending more time, paying more attention, and continuous adoption of smart farming technologies have made home gardens more resilient to the changes in climate.

3.2 Farm Animal Production, Fisheries, and Forestry

Heat stress conditions caused by increasing temperature could directly influence the productivity of the livestock and poultry sectors of Sri Lanka, via retarding growth, fertility, and production performance (Wijayagunawardane, 2009). In addition, the limited availability of feed and water due to impacts imposed by rainfall variability and drought events could exert a notable pressure on the livestock farms, especially in dry regions. High-yielding temperate breeds are expected to suffer the most. The majority of the livestock systems located in the dry region mostly depend on rain-fed pasture systems and other foliage crops for feed, along with natural water bodies. Therefore, direct impacts caused on those could indirectly influence the productivity of the livestock sector, especially in extensively managed livestock systems (Eeswaran, 2018). On the contrary, prolonged droughts and flood events would also influence the performance of both the livestock and poultry industries in Sri Lanka (Marambe et al., 2015). However, the present level of understanding on the interactions between climate change and farm animal production is not adequate enough to face the future challenges of the sector (Eeswaran, 2018).

Recent trends in sea-level rise, ocean acidification, increase in oceanic temperature, and degradation of coastal ecosystems have been influencing the productivity of the coastal fisheries sector in Sri Lanka. Unexpected technical disasters, such as the sinking of X-Press Pearl within the territorial waters of Sri Lanka, have further intensified the vulnerability of the coastal fisheries sector. Variations in rainfall patterns, the frequent occurrence of prolonged floods, droughts, and other extreme weather conditions have caused notable impacts on the inland fisheries, especially in seasonal tanks (Senaratne et al., 2009). These factors could induce changes in fish species distribution, growth, and reproduction of fish stock and may lead to alteration of species composition. Therefore, both the inland and coastal fisheries sectors of Sri Lanka remain highly vulnerable to expected changes in climatic patterns.

Forests play a key role in climate change mitigation as a major sink of CO_2. Therefore, negative impacts on forests and plantations could critically influence both the economy and the environment of Sri Lanka. Even at present, a notable fraction of rural communities still relies on non-timber forest products, such as medicinal plants, honey, mushroom types, and fruits as a dominant contributor for livelihoods

(Senaratne et al., 2003). Recent intensification in deforestation, arising due to human settlements, infrastructure development, and expansion of agricultural lands have exerted a high pressure on the sustainability of forest ecosystems in Sri Lanka. At such a time, climate change could cause both direct and indirect impacts on the forests and plantations of Sri Lanka (Marambe et al. 2014).

4 Role of Precision Agricultural Technologies for Resilience and Mitigation

With undeniable evidences for climate change, Sri Lanka is also facing devastating impacts from climate change on agriculture. Therefore, adaptation of climate-smart practices and integration of climate change response policies remain critical in enhancing the overall climate change resilience of Sri Lanka. According to the National Policy Framework (NPF) of Sri Lanka, the modernization of agriculture and adoption of novel technologies have been recognized as a priority area to enhance agricultural productivity, while ensuring environmental sustainability (MOF 2019). Further, after realizing the detrimental impacts of intensive agricultural practices on both ecosystems and humans, the Sri Lankan government has taken strong initiatives to promote eco-friendly agriculture within the country. Under this, restricting the use of inorganic fertilizers and other agro-chemicals (pesticides, weedicides, and plant nutrient supplements) was implemented as a policy decision since May 2021. Therefore, a growing need to ensure sustainable farming while facing changes in the climate and limitations in soil and other resources, with limited use of inorganic agro-chemicals, has emerged in Sri Lanka.

Precision Agriculture Technologies (PAT), also termed as Digital Agriculture, is an emerging discipline that aims to utilize data-intensive approaches to ensure agricultural productivity, with minimum impacts on humans and the environment. Precision Agriculture Technologies (PAT) could provide ideal solutions for farmers to enhance their farming productivity by optimizing the use of inputs and human labor, with minimum impacts on ecosystems (Lindblom et al., 2017). Precision Agriculture Technologies are an integration of several technologies, including Sensor Technology, Internet of Things (IoT), Geo-Informatics, Variable-Rate Technology (VRT), and Unmanned Aerial Vehicles (UAVs), that support monitoring, measuring, and responding to variability in crops, with the goal of optimizing resource utilization and agricultural decision support (Pierpaoli et al., 2013). Different combinations of these technologies could be used to enhance the accuracy and efficacy of farming operations. Precision Agriculture Technologies enable farmers to develop a decision support system for agricultural operations with process automation, timely response to crop and environmental variations, and input optimization (fertilizer, irrigation, and pesticides) of agricultural systems. Therefore, the adoption of PAT would enable farmers to enhance crop monitoring, manage variability in farmlands, and tailor inputs to secure desired outputs (Lambert & Lowenberg-De Boer, 2000).

The concept of PAT initially came into play in early 1990, with the incorporation of Global Positioning Systems (GPS). Since then, a variety of novel technologies such as Sensor Technology, Internet-of Things (IoT), Geo-Informatics, Variable-Rate Technology (VRT), and Remote Sensing (RS) have nurtured the PAT to its current state (Pierpaoli et al., 2013; Robert, 1999). Precision Agriculture Technologies enable farmers to use a vast amount of data generated through a variety of different sensors, models, and other secondary sources to understand the agricultural environment (soil, crops, weather) and operations (farm machinery), thereby allowing farmers to make more effective, timely and accurate decisions. Incorporation of technologies like Machine Learning (ML) and big data analysis could further enhance the speed and accuracy of the above process.

Prediction of the potential yield is a vital requirement for agricultural decision-making, especially in the case of crop management, input optimization, and marketing of crop produce. A variety of approaches in PAT have been successfully used by different countries for yield prediction of different crops. A recent study conducted in China has demonstrated the efficacy of PAT for prediction of rice development and yield based on soil, crop, and meteorological parameters through a Support Vector Machines (SVM) approach (Su et al., 2017). A study conducted by Pantazi et al. (2016) has developed a yield prediction model for wheat production using online predicted soil parameters and Normalized Difference Vegetation Index (NDVI). In another study, multi-temporal Remote Sensing data and Artificial Neural Networks (ANN) have been applied successfully to estimate grassland biomass (Ali et al., 2016). However, a limited focus has been laid on the use of PAT for yield prediction in Sri Lanka. Few recent studies have attempted to incorporate meteorological data and crop parameters through ANN, machine learning, Recurrent Neural Networks (RNN), and SVM for paddy yield prediction (Amaratunga et al., 2020; Muthusinghe et al., 2018; Wickramasinghe et al., 2021). In addition, another study conducted by Shanmugam and Dammalage (2018) has utilized Remote Sensing approaches for paddy yield estimation with high accuracy. However, such PAT has been rarely used for crop yield prediction of other crops in Sri Lanka.

Timely irrigation and application of fertilizers in appropriate quantities are crucial to secure higher agricultural productivity. Often, Sri Lankan farmers exhibit over irrigation and fertilizer applications, which have led to serious environmental concerns with elevated economic costs. The predicted changes in climate have suggested notable impacts on rainfall variability and irrigation water availability, signifying the necessity of optimum irrigation management. Irrigation scheduling and application of appropriate volumes are key factors that are vital for hydrological, climatological, and agronomical balance (Liakos et al., 2018). Under PAT, the frequency, timing, and quantity of irrigation can be set based on three major approaches: weather-based sensing, plant-based sensing, and soil–water-based sensing. The first approach considers the meteorological parameters such as rainfall and evapotranspiration over the considered area for scheduling irrigation. A few recent studies have utilized different ML approaches such as ANN, Extreme Learning Machines (ELM), and

Generalized Regression Neural Network (GRNN) for estimation of evapotranspiration requirements, and thereby define the irrigation requirements of crops (Feng et al., 2017; Mehdizadeh et al., 2017; Patil & Deka, 2016).

In Plant-based Sensing, a wide range of sensing technologies could be utilized to detect the onset and severity of plant stress as a response to the lack of water availability. For this reason, different electromagnetic reflectance results from crops have been used to derive associations between NDVI and crop coefficients through remote and proximal sensing approaches for a broad range of crops (Bastiaanssen et al. 2002; Chavez et al., 2009). Meanwhile, soil–water-based sensing attempts to capture the spatial and temporal variations in soil moisture content, especially within the crop root zone, to optimize crop irrigation requirements (Hossain et al. 2008; Padhi & Misra, 2009). Incorporation of all these three approaches, with IoT and sensor technology has provided a reliable and effective approach to ensure large-scale, low-cost, and site-specific crop irrigation management for many crops in different parts of the world (Smith, 2011). However, only a limited number of studies have attempted to incorporate the above technologies for irrigation management in Sri Lanka, mainly using IoT technology (Laksiri et al. 2019) and Remote Sensing (Bandara, 2003).

Elevated rates of pest and disease incidences are a key challenge that could be expected by the agricultural sector due to climate change. Spraying of pesticides and fungicides is the most widely used practice in controlling such pest and disease outbreaks. Despite being effective, such applications of pesticides are often costly and may lead to serious environmental concerns (Liakos et al., 2018; Marambe et al., 2015). A number of studies have already evidenced the reliability and efficacy of IoT, ML, Proximal Sensing and Remote Sensing technologies for early detection of the onset of pest and diseases in a wide array of crops such as cabbage (Ferentinos, 2018), rice (Chung et al., 2016), wheat (Moshou et al., 2014; Pantazi et al., 2017), onion (Ferentinos, 2018), and strawberry (Ebrahimi et al. 2017). Precision Agricultural Technologies often support farmers by discriminating against plant stress due to pest and disease onset, thereby enabling them to make site-specific applications of pesticides and fungicides (Ferentinos, 2018). The "Permanent Crop Clinic" program, which was initiated as a pilot project in 2009, and subsequently expanded in 2012, is a primary initiative toward applications of PAT in pest and disease management of crops in Sri Lanka (Jayasundara et al. 2016). However, this program is still at its infancy and should be strengthened further by enhancing the expertise knowledge of Agricultural Instructors (AI), Technology Readiness (TR), and attitudes of farmers.

Early detection of weed and management is another significant challenge faced in agriculture, since weed can act as a competitor for space, light, and nutrients, restricting the availability of them to crops (Liakos et al., 2018). Therefore, timely weed management remains crucial to secure higher yields. Similar to pesticides, often broad and specific weedicides are applied by farmers for weed management, in addition to mechanical or manual removal. Difficulties in weed discrimination from the crops have made this task challenging, tedious, and difficult. Proximal and Remote Sensing platforms can effectively aid farmers in detecting and discriminating weeds from desired crops, thereby enabling the use of mechanical tools and methods for the elimination of weeds from croplands with a minimum use of herbicides (Binch &

Fox, 2017; Mogili et al. 2018; Pantazi et al., 2017). At present, many countries are using partial or fully autonomous vehicles, equipped with different sensors, GPS, and IoT technology for early detection and timely elimination of weeds, pests, and diseases with higher accuracy and minimum resources (Lindblom et al., 2017).

A high number of powerful, accurate, and reliable crop modeling applications and platforms have been developed by different entities in the world. These crop models are often used to model the crop suitability, irrigation requirements, fertilizer requirements, and success of different crop management practices, and ultimately for yield prediction (Boote, 2019; Corbeels et al., 2018; France & Thornley, 1984). Such applications may range from simple forms to more complex applications, which utilize numerous data such as crop parameters, soil parameters, meteorological parameters, and other operational conditions (Liakos et al., 2018). Once calibrated to fit the local cropping systems, these models can be effectively used to support agricultural decision-making. As a whole, these systems will combine farmers' field data acquired from soil and crops with real-time weather to provide a decision support system for farmers, enabling them to optimize inputs in terms of quantity and timing, while maintaining higher productivity.

Information and Communication Technologies (ICT) have played a key role in many sectors, i.e., agriculture, in many countries, including Sri Lanka (Jayathilake et al., 2010; Nikkila et al. 2010). Applications of ICT in agriculture, commonly termed as Digital Agriculture or e-Agriculture, have strengthened the productivity and expansion of the agricultural sector (Anoop et al., 2015). Accurate information collection and effective information use continue to be the lifeblood of any industry. Many studies have demonstrated the capacity of ICT applications to promote healthy competition and improve market performance in the purchasing of inputs (fertilizer, pesticides, and other agro-chemicals), farm machinery, and most importantly, for the marketing of agricultural produce (Thompson & Sonka, 1997). Digital Agricultural applications can enhance the degree of information shared among the stakeholders in the agricultural sector and decrease the cost of information dissemination (Jayathilake et al., 2010). Further, through ICT, farmers will gain access to virtual markets (online markets), expanding their potential customers.

In addition, Digital Agricultural platforms can effectively connect farmers with agriculture instructors and extension officers, enabling them to acquire their services easily at a lower cost. Ultimately, e-agricultural applications would enhance the quality of decision-making of farmers, especially on which crops to grow, when to take the harvest to market, and regarding the market rate, while eradicating unnecessary intermediate costs in the supply chain (Narmilan et al., 2020; Rahman, 2009). In the Sri Lankan context, several e-Agricultural platforms—such as the official website of the DOA (www.doa.gov.lk), Wikigoviya website (www.goviya.lk), "Krushilanka" agriculture portal (www.krushilanka.gov.lk), website of the Rice Knowledge Bank, Agriculture Advisory Service (Call extension 1920), and e-SMS Service and Govi Mithuru project—have been launched by the Department of Agriculture (DOA) of the Ministry of Agriculture, under the Sri Lanka e-Agriculture Strategy (Narmilan et al., 2020).

5 Barriers Against Adoption of Precision Agricultural Technologies in Sri Lanka

A diverse interplay of factors such as availability of infrastructure, adequate technical knowledge, attitudes, Technology Readiness (TR), and socio-economic factors of farmers could determine the success of adopting PAT (Duang et al. 2019). With recent advancements in ICT, accessing knowledge, sharing of experience on novel PAT, and marketing of crop produce have become flexible and easier. Therefore, the TR of farmers, in terms of knowledge in ICT, availability of physical resources, and perceptions, would change the fate of Smart Farming in Sri Lanka. However, limited research has been conducted to assess the level of TR, along with drivers and barriers for adopting Smart Farming technologies at the national level. Limited access to up-to-date technology among farmers and Agricultural Instructors (AI) could be identified as the major setback factor in promoting PAT in developing countries (Katke, 2019). Inadequate knowledge of novel technologies, as well as limitations in technology usage, can frequently stymie PAT adoption. Because some PAT require a certain level of ICT literacy, low level of education and the complexity of novel technologies and e-Agricultural applications can act as limitations (Munasinghe and Kanchanamala 2018; Namara et al., 2003). This remains a key constraint, especially for adoption of RS and sensor-based systems.

Limitations in the availability of physical resources such as smartphones, tabs, or laptops can also be identified as a major barrier in promoting PAT among Sri Lankan farmers. Weaknesses in the telecommunication facilities such as low network coverage, insufficient bandwidth, and poor customer care, also inhibit the adoption of Digital Agriculture in Sri Lanka. Several recent studies have highlighted the influence of socio-economic conditions of farmers, such as age, education level, awareness and attitudes toward ICT and e-Agriculture, and other socio-behavioral factors on the preference toward PAT (Jayathilake et al., 2010; Munasinghe and Kanchanamala 2018; Narmilan et al., 2020). Government support and motivation, as well as the implementation of legislation and policy frameworks, are critical for the sustainability of PAT adoption activities in any country (Jayathilake et al., 2010; Katke, 2019). Poor coordination among government entities and other stakeholders, gaps in the legislation and national policies, limitations in transparency of government actions, and lack of funds and technical expertise in Sri Lanka act as significant barriers that influence the promotion of PAT.

6 Conclusions and Way-Forward

Maintaining higher agricultural productivity with minimum environmental impacts has become a challenge in many developing countries, including Sri Lanka. The evident changes in climate and barriers to the use of inorganic agro-chemicals in Sri Lanka have further aggravated this issue. Therefore, adoption of PAT remains the most viable solution. Application of Geo-Informatics and Sensor Technology in monitoring crop health and real-time variability in crops is a key approach that could be initiated under PAT. This will facilitate timely use of fertilizers and other agro-chemicals (pesticides and fungicides) in optimum dosages at the onset of any stress-induced symptoms. Further, the combined use of different sensors, IoT technology, and machine learning technologies will enable Sri Lankan farmers to detect and monitor soil and crop conditions, enabling customized automation of irrigation, fertilizer application, and pest and disease management. Adequately calibrated crop modeling applications can also support decision-making and operational control in agriculture. Ultimately, accurate and reliable decision support systems could be developed and used for agricultural decision-making by integrating the aforesaid technologies with respect to edaphic, crop, operational, and meteorological parameters relevant to agriculture.

Therefore, the Sri Lankan government and other key agricultural entities have to take efforts to promote PAT in Sri Lanka. Improving Internet Service Providers' physical infrastructure facilities in terms of network coverage and bandwidth remains a critical requirement. Further, empowering farmers, AI, and Extension Officers (EO) with physical resources should also be a focus. Dissemination of technology, elevating TR, and capacity building of farmers are vital elements in introducing these Digital Agricultural systems to local farmers. Therefore, different platforms such as workshops, seminars, and demonstrations should be conducted for Training of Trainers (ToT) and farmers on different technologies and applications of Digital Agricultural technologies. Through this, continuous efforts have to be taken to empower the AI and local farmers with up-to-date technical knowledge on precision agriculture technologies.

Establishment of regional crop monitoring and modeling stations with the collaborative efforts of government entities, academics, and researchers will also play a valuable role in PAT promotion among Sri Lankan farmers. The designing of adequate policies and implementation of those with prevailing legislation to support the expansion of PAT are also important. In addition, the government can introduce soft loan schemes and insurance facilities for farmers, who take initiatives to adopt Digital Agricultural technologies. Government and Non-Governmental Organizations should focus on the actual requirements of the local farming communities, while understanding their socio-economic and behavioral factors, in designing intervention programs to promote novel PAT. Adoption of precision agricultural technologies requires technical expertise and capital funds. Therefore, implementation of certain platforms will not be economical for local farmers due to the above requirements. Therefore, on-site advisory services should be provided, opening the skies to farmers, who want better ways to monitor and manage their cropping systems, at discounted rates. These services will enable the farmers to plan and strategize their cropping activities to secure the highest possible productivity in a sustainable manner. Various PAT combinations could be used to improve crop monitoring, manage variability in farmlands, tailor inputs to achieve desired outputs, and finally, market the agricultural produce in Sri Lanka.

References

Abeysiriwardena, D. D. Z., Ohba, K., & Maruyama, A. (2002). Influence of temperature and relative humidity on grain sterility in rice. *Journal of the National Science Foundation of Sri Lanka, 30*(1–2), 33–41.

Ahmed, M., & Suphachalasai, S. (2014). *Assessing the costs of climate change and adaptation in South Asia.* Asian Development Bank.

Ali, I., Cawkwell, F., Dwyer, E., & Green, S. (2016). Modeling managed grassland biomass estimation by using multitemporal remote sensing data—A machine learning approach. *IEEE Journal of Selected Topics in Applied Earth Observations and Remote Sensing, 10*(7), 3254–3264.

Amaratunga, V., Wickramasinghe, L., Perera, A., Jayasinghe, J., & Rathnayake, U. (2020). Artificial neural network to estimate the paddy yield prediction using climatic data. *Mathematical Problems in Engineering*, 2020.

Anoop, M., Ajjan, N., & Ashok, K. R. (2015). ICT based market information services in Kerala–determinants and barriers of adoption. *Economic Affairs, 60*(1), 117–121.

Bandara, K. M. P. S. (2003). Monitoring irrigation performance in Sri Lanka with high-frequency satellite measurements during the dry season. *Agricultural water management, 58*(2), 159–170.

Bastiaanssen, W. G., Molden, D. J., & Makin, I. W. (2000). Remote sensing for irrigated agriculture: Examples from research and possible applications. *Agricultural Water Management, 46*(2), 137–155.

Berry, H. L., Hogan, A., Owen, J., Rickwood, D., & Fragar, L. (2011). Climate change and farmers' mental health: risks and responses. *Asia Pacific Journal of Public Health, 23*(2_suppl), 119S-132S.

Binch, A., & Fox, C. W. (2017). Controlled comparison of machine vision algorithms for Rumex and Urtica detection in grassland. *Computers and Electronics in Agriculture, 140*, 123–138.

Boote, K. (2019). *Advances in crop modelling for a sustainable agriculture.* Burleigh Dodds Science Publishing.

Bouman, B. A. M. (2001). Water-efficient management strategies in rice production. *International Rice Research Notes, 26*(2), 1–1.

Central Bank of Sri Lanka (2019) *Annual Report 2018, Central Bank of Sri Lanka.* Central Bank of Sri Lanka (CBSL), Colombo, Sri Lanka.

Central Bank of Sri Lanka (2021) *Annual Report 2020, Central Bank of Sri Lanka.* Central Bank of Sri Lanka (CBSL), Colombo, Sri Lanka.

Chavez, J. L., Gowda, P. H., Howell, T. A., & Copeland, K. S. (2009). Radiometric surface temperature calibration effects on satellite based evapotranspiration estimation. *International Journal of Remote Sensing, 30*(9), 2337–2354.

Chung, C. L., Huang, K. J., Chen, S. Y., Lai, M. H., Chen, Y. C., & Kuo, Y. F. (2016). Detecting Bakanae disease in rice seedlings by machine vision. *Computers and Electronics in Agriculture, 121*, 404–411.

Cline, W. R. (2007). *Global warming and agriculture: Impact estimates by country.* Peterson Institute.

Corbeels, M., Berre, D., Rusinamhodzi, L., & Lopez-Ridaura, S. (2018). Can we use crop modelling for identifying climate change adaptation options? *Agricultural and Forest Meteorology, 256*, 46–52.

De Costa, W. A. J. M. (2010). Adaptation of agricultural crop production to climate change: A policy framework for Sri Lanka. *Journal of the National Science Foundation of Sri Lanka, 38*(2).

De Costa, W. A. J. M., & Sangakkara, U. R. (2006). Agronomic regeneration of soil fertility in tropical Asian smallholder uplands for sustainable food production. *The Journal of Agricultural Science, 144*(2), 111–133.

De Costa, W. A. J. M., Weerakoon, W. M. W., Herath, H. M. L. K., & Abeywardena, R. M. I. (2003). Response of growth and yield of rice (Oryza sativa) to elevated atmospheric carbon dioxide in the subhumid zone of Sri Lanka. *Journal of Agronomy and Crop Science, 189*(2), 83–95.

De Silva, C. (2006). Impacts of climate change on water resources in Sri Lanka. In *32nd WEDC international conference*, November 13–17, 2006, Colombo, Sri Lanka.

De Silva, C. S., Weatherhead, E. K., Knox, J. W., & Rodriguez-Diaz, J. A. (2007). Predicting the impacts of climate change—A case study of paddy irrigation water requirements in Sri Lanka. *Agricultural Water Management, 93*(1–2), 19–29.

Delpitiya, D. G. P. S., Weerakoon, W. M. W., Nissanka, S. P., Punyawardena, B. V. R., Karunarathne, M. A., & Hearth, R. M. (2014). Impact of climate change on rice yields in Sri Lanka: A crop modeling approach using decision support system for agro technology transfer (DSSAT4. 5). *Ann Sri Lankan Dept Agric, 16*, 71–81.

Department of Agriculture. (2020) *AgStat: pocket book of agricultural statistics*. Department of Agriculture, Peradeniya.

Duang-Ek-Anong, S., Pibulcharoensit, S., & Phongsatha, T. (2019). Technology Readiness for Internet of Things (IoT) Adoption in Smart Farming in Thailand. *International Journal of Simulation: Systems, Science and Technology, 20*, 1–6.

Eeswaran, R. (2018). Climate change impacts and adaptation in the agriculture sector of Sri Lanka: What we learnt and way forward. In *Handbook of Climate Change Communication* (Vol. 2, pp. 97–110). Springer, Cham.

Eriyagama, N., Smakhtin, V., & Chandrapala, L. (2010). *Impacts of climate change on water resources and agriculture in Sri Lanka* (p. 35). IWMI Research Report. Colombo: International Water Management Institute.

Esham, M., Jacobs, B., Rosairo, H. S. R., & Siddighi, B. B. (2018). Climate change and food security: A Sri Lankan perspective. *Environment, Development and Sustainability, 20*(3), 1017–1036.

Fand, B. B., Kamble, A. L., & Kumar, M. (2012). Will climate change pose serious threat to crop pest management: A critical review. *International Journal of Scientific and Research Publications, 2*(11), 1–14.

Feng, Y., Peng, Y., Cui, N., Gong, D., & Zhang, K. (2017). Modeling reference evapotranspiration using extreme learning machine and generalized regression neural network only with temperature data. *Computers and Electronics in Agriculture, 136*, 71–78.

Ferentinos, K. P. (2018). Deep learning models for plant disease detection and diagnosis. *Computers and Electronics in Agriculture, 145*, 311–318.

Fernando, M. T. N., Zubair, L., Peiris, T. S. G., Ranasinghe, C. S., & Ratnasiri, J. (2007). *Economic value of climate variability impacts on coconut production in Sri Lanka*. AIACC working Paper No. 45, International START Secretariat, 2000, Florida Avenue, USA, pp. 1–7.

France, J., & Thornley, J. H. (1984). *Mathematical models in agriculture*. Butterworths.

Gitz, V., Meybeck, A., Pinizzotto, S., Nair, L., Penot, E., Baral, H., & Xu, J. (2020). *Sustainable development of rubber plantations in a context of climate change: Challenges and opportunities* (Vol. 4). CIFOR.

Hossain, M. B., Lamb, D. W., Lockwood, P. V., & Frazier, P. (2010). Field determination of soil moisture in the root zone of deep vertosols using EM38 measurements: calibration and application issues. In *Proximal Soil Sensing* (pp. 255–263). Springer, Dordrecht.

Intergovernmental Panel on Climate Change (IPCC). (2007). *Climate Change 2007: Synthesis Report*, Contribution of Working Groups I, II and III to the Fourth Assessment Report of the Intergovernmental Panel on Climate Change. Geneva; 2007.

Intergovernmental Panel on Climate Change (IPCC). (2014). *Contribution of working group II to the fifth assessment report of the intergovernmental panel on climate change*. Intergovernmental Panel on Climate Change.

Iqbal, M. C. M., Wijesundera, D. S. A., & Ranwala, S. M. W. (2014). Climate change, invasive alien flora and concerns for their management in Sri Lanka. *Ceylon Journal of Science (Bio. Sci.), 43*(2), 1–15.

Jayasundara, M.U.P., Herath, Y.M.C.K., Abeykoon1, A.N. & Cameron K. (2016). The "Permanent Crop Clinic Program" in Sri Lanka: making use of data to resolve crop health problems. In *Annual Symposium of Department of Agriculture-2016*. Department of Agriculture, Sri Lanka.

Jayathilake, H. A. C. K., Jayaweera, B. P. A., & Waidyasekera, E. C. S. (2010). ICT adoption and its' implications for agriculture in Sri Lanka. *Journal of Food and Agriculture, 1*(2), 54–63.
Jessy, M. D., Krishnakumar, R., Annamalainathan, K., & Jacob, J. (2011). Climate uncertainties and early establishment of young rubber plants in traditional rubber growing regions of India. *Natural Rubber Research, 24*(1), 145–147.
Karunaratne, A. S., & Wheeler, T. (2015). Observed relationships between maize yield and climate in Sri Lanka. *Agronomy Journal, 107*(1), 395–405.
Katke, K. (2019). Precision Agriculture Adoption: Challenges of Indian Agriculture. *International Journal of Research and Analytical Reviews, 6*(1), 863–869.
Kaur, H., & Kaur, S. (2017). Sustaining agricultural production in South Asia in the face of climate change: A panel data study. *Asian Journal of Research in Social Sciences and Humanities, 7*(9), 73–94.
Kudagamage, C., & Gannoruwa, P. (1998, June). Present status of horticulture research in Sri Lanka. In *World Conference on Horticultural Research*. pp. 17–20.
Laksiri, H. G. C. R., Dharmagunawardhana, H. A. C., & Wijayakulasooriya, J. V. (2019, December). Design and optimization of IOT based smart irrigation system in Sri Lanka. In *2019 14th Conference on Industrial and Information Systems (ICIIS)* (pp. 198–202). IEEE.
Lambert, D., & Lowenberg-De Boer, J. (2000). *Precision agriculture profitability review* (pp. 1–154). Purdue University.
Liakos, K. G., Busato, P., Moshou, D., Pearson, S., & Bochtis, D. (2018). Machine learning in agriculture: A review. *Sensors, 18*(8), 2674.
Lindblom, J., Lundström, C., Ljung, M., & Jonsson, A. (2017). Promoting sustainable intensification in precision agriculture: Review of decision support systems development and strategies. *Precision Agriculture, 18*(3), 309–331.
Malaviarachchi, M. A. P. W. K. M., De Costa, W. A. J. M., Fonseka, R. M., J. B. D. A. P. K., & Suriyagoda, L. D. B. (2015). Can seasonal temperature extremes alter the growth and yield of mung bean (*Vigna radiata* (L.) Wilczek)?. An investigation with different agronomic management packages conducive for climate adaptation. *Ann Sri Lanka Dept Agric, 17*, 92–96.
Malaviarachchi, M. A. P. W. K., De Costa, W. A. J. M., Fonseka, R. M., Kumara, J. B. D. A. P., Abhayapala, K. M. R. D., & Suriyagoda, L. D. B. (2014). Response of maize (*Zea mays* L.) to a temperature gradient representing long-term climate change under different soil management systems. *Tropical Agricultural Research, 25*(3), 327–344.
Marambe, B., Punyawardena, R., Silva, P., Premalal, S., Rathnabharathie, V., Kekulandala, B., Nidumolu, U. & Howden, M. (2015). Climate, climate risk, and food security in Sri Lanka: the need for strengthening adaptation strategies. *Handbook of climate change adaptation*, 1759–1789.
Mehdizadeh, S., Behmanesh, J., & Khalili, K. (2017). Using MARS, SVM, GEP and empirical equations for estimation of monthly mean reference evapotranspiration. *Computers and Electronics in Agriculture, 139*, 103–114.
Ministry of Finance. (2019). *National policy framework vistas of prosperity and splendour*. Ministry of Finance, Sri Lanka.
Mogili, U. R., & Deepak, B. B. V. L. (2018). Review on application of drone systems in precision agriculture. *Procedia Computer Science, 133*, 502–509.
Moshou, D., Pantazi, X. E., Kateris, D., & Gravalos, I. (2014). Water stress detection based on optical multisensor fusion with a least squares support vector machine classifier. *Biosystems Engineering, 117*, 15–22.
Munasinghe, P. G., & Kanchanamala, H. D. C. (2018). Factors Affecting for the Usage of Mobile Applications among Farmers in Bulathsinhala Area in Sri Lanka. In *Munasinghe, PG and Kanchanamala, HDC (2018). Factors affecting for the usage of mobile applications among farmers in Bulathsinhala area in Sri Lanka. International Research Conference (JUICe2018), Towards a Safer Future, Conference Proceedings, 27th & 28th.*

Muthusinghe, M. R. S., Palliyaguru, S.T., Weerakkody, W.A.N.D., Saranga, A.H., & Rankothge, W.H. (2018). Towards smart farming: accurate prediction of paddy harvest and rice demand. In *2018 IEEE Region 10 Humanitarian Technology Conference (R10-HTC)* (pp. 1–6). IEEE.

Namara, R.E., Weligamage, P., & Barker, R. (2003). *Prospects for adopting system of rice intensification in Sri Lanka: A socioeconomic assessment* (Vol. 75). IWMI.

Narmilan, A., Niroash, G., & Puvanitha, N. (2020). Assessment of current status on smart farming technologies in Batticaloa District, Sri Lanka. *Sri Lanka Journal of Technology, 1*(1), 14–20.

Naveendrakumar, G., Vithanage, M., Kwon, H. H., Iqbal, M. C. M., Pathmarajah, S., & Obeysekera, J. (2018). Five decadal trends in averages and extremes of rainfall and temperature in Sri Lanka. *Advances in Meteorology*, 2018.

Nguyen, N.V. (2002). *Global climate changes and rice food security*. FAO: Rome, Italy, 24–30.

Nikkilä, R., Seilonen, I., & Koskinen, K. (2010). Software architecture for farm management information systems in precision agriculture. *Computers and Electronics in Agriculture, 70*(2), 328–336.

Nissanka, S. P., Punyawardena, B.V.R., Premalal, K.H.M.S., & Thattil, R.O. (2011, December). Recent trends in annual and growing seasons' rainfall of Sri Lanka. In *Proceedings of the International Conference on the Impact of Climate Change in Agriculture. Faculty of Agriculture, University of Ruhuna, Mapalana, Sri Lanka* (pp. 249–263).

Padhi, J., & Misra, R.K. (2009). Monitoring spatial variation of soil moisture in crop fields with EM38. In *Irrigation Australia 2009: Irrigation Australia Irrigation and Drainage Conference: Proceedings*. Irrigation Australia Ltd.

Panabokke, C.R., & Punyawardena, B.V.R. (2010). Climate change and rain-fed agriculture in the dry zone of Sri Lanka. In *Proceedings of the national conference on water, food security and climate change in Sri Lanka. Volume 2: water quality, environment and climate change. Colombo, 9–11 June, 2009* (pp. 141–146). International Water Management Institute.

Pantazi, X. E., Moshou, D., Alexandridis, T., Whetton, R. L., & Mouazen, A. M. (2016). Wheat yield prediction using machine learning and advanced sensing techniques. *Computers and Electronics in Agriculture, 121*, 57–65.

Pantazi, X. E., Tamouridou, A. A., Alexandridis, T. K., Lagopodi, A. L., Kashefi, J., & Moshou, D. (2017). Evaluation of hierarchical self-organising maps for weed mapping using UAS multispectral imagery. *Computers and Electronics in Agriculture, 139*, 224–230.

Pathiraja, P.M.E.K., Griffith, G.R., Farquharson, R.J., & Faggian, R. (2015). The Sri Lankan coconut iIndustry: Current status and future prospects in a changing climate. *Australasian Agribusiness Perspectives, 23.*

Patil, A. P., & Deka, P. C. (2016). An extreme learning machine approach for modeling evapotranspiration using extrinsic inputs. *Computers and Electronics in Agriculture, 121,* 385–392.

Pierpaoli, E., Carli, G., Pignatti, E., & Canavari, M. (2013). Drivers of precision agriculture technologies adoption: A literature review. *Procedia Technology, 8*, 61–69.

Premalal, K. H. M. S. (2009). Climate Change in Sri Lanka. *Global Climate Change and its Impacts on Agriculture, Forestry and Water in the Tropics*, 21.

Punyawardena, B. V. R. (2007). *Agro-ecology (map and accompanying text), National Atlas of Sri Lanka*. Department of Survey, Colombo.

Punyawardena, B. V. R., & Premalal, K. H. M. S. (2013). Do trends in extreme positive rainfall anomalies in the Central highlands of Sri Lanka exist. *Annals of the Sri Lanka Department of Agriculture, 15*, 1–12.

Punyawardena, B. V. R., Bandara, T. M. J., Munasinghe, M. A. K., Jeyaratne, B. M. & Pushpakumara, S. M. V. (2003) *Agro-ecological regions of Sri Lanka (Map)*. Natural Resources Management Centre, Department of Agriculture, Peradeniya, Sri Lanka

Punyawardena, B. V. R., Mehmood, S., Hettiarachchi, A. K., Iqbal, M., De Silva, S. H. S. A., & Goheer, A. (2013a). Future climate of Sri Lanka: An approach through dynamic downscaling of ECHAM4 General Circulation Model (GCM). *Tropical Agriculturist, 161*, 35–52.

Punyawardena, R., Dissanaike, T., & Mallawathanthri, A. (2013v). *Vulnerability of Sri Lanka to climate change-monograph*. Department of Agriculture, Peradeniya.

Rahman, A. M. A. (2009). *Use of information and communication technologies (ICTs) by agricultural extensionists in the Gezira state* (Sudan). Faculty of Agricultural Sciences, University of Gezira.

Robert, P. C. (1999). Precision Agriculture: research needs and status in the USA. In *Precision agriculture'99, Part 1 and Part 2*. In *2nd European Conference on Precision Agriculture, Odense, Denmark, 11–15 July 1999* (pp. 19–33). Sheffield Academic Press.

Senaratne, A., Abeygunawardena, P., & Jayatilake, W. (2003). Changing role of non-timber forest products (NTFP) in rural household economy: The case of Sinharaja World Heritage site in Sri Lanka. *Environmental Management, 32*(5), 559–571.

Senaratne, A., Perera, N., & Wickramasinghe, K. (2009). *Mainstreaming climate change for sustainable development in Sri Lanka: Towards a national agenda for action*. Institute of Policy Studies of Sri Lanka.

Shanmugam, T., & Dammalage, T. L. (2018). Rice Yield Estimation Using Free Satellite and Field Data. *11th International Research Conference of the General Sir John Kotelawala Defence University (KDU IRC)* (pp. 7–13). General Sir John Kotelawala Defence University.

Smith, R. (2011). *Review of precision irrigation technologies and their applications*. University of Southern Queensland.

Su, Y. X., Xu, H., & Yan, L. J. (2017). Support vector machine-based open crop model (SBOCM): Case of rice production in China. *Saudi Journal of Biological Sciences, 24*(3), 537–547.

Tanaka, I. (1976). Climatic influence on photosynthesis and respiration of rice. *Climate and rice*, 223–247.

Thompson, S., & Sonka, S. T. (1997). Potential effects of information technologies on the economic performance of agricultural and food markets. *American Journal of Agricultural Economics, 79*(2), 657–662.

Titumir, R. A., & Basak, J. K. (2010). *Agriculture and Food Security in South Asia: A Historical Analysis and a Long Run Perspective*. Unnayan Onneshan, The Innovators, Centre for Research and Action on Development.

Udayanga, N. W. B. A. L., Najim, M. M.M. (2020). Climate Change Induced Variations in Rainfall Patterns & Potential Adaptation Options of DL_{1b} Agro-Ecological Zone of Sri Lanka: A Case Study. In: Caldera, H. I. U. & Perera, S. A. C. N., editors. *Adapting to Climate Change: A Sri Lankan perspective*. Institute of Biology Sri Lanka, Colombo. pp. 167–180.

Weerahewa, J., Pushpakumara, G., Silva, P., Daulagala, C., Punyawardena, R., Premalal, S., Miah, G., Roy, J., Jana, S., & Marambe, B. (2012). Are homegarden ecosystems resilient to climate change? An analysis of the adaptation strategies of homegardeners in Sri Lanka. *APN Science Bulletin, 2*(2), 22–27.

Wickramasinghe, L., Weliwatta, R., Ekanayake, P., & Jayasinghe, J. (2021). Modeling the relationship between rice yield and climate variables using statistical and machine learning techniques. *Journal of Mathematics*, 2021.

Wijayagunawardane, M.P. (2009). Impacts of global warming on livestock reproduction. *Global Climate Change and its Impacts on Agriculture, Forestry and Water in the Tropics*, 263.

Wijeratne, M. A. (2011). Adaptations to the threats of climate change: Strategies for plantation crops with special reference to tea. *Journal of National Institute of Plantation Management, 25*(1), 23–31.

Wijeratne, M. A., Anandacoomaraswamy, A., Amarathunga, M.K.S.L.D., Ratnasiri, J., Basnayake, B.R.S.B., & Kalra, N. (2007). Assessment of impact of climate change on productivity of tea (Camellia sinensis L.) plantations in Sri Lanka. *Journal of the National Science Foundation of Sri Lanka, 35*(2).

Wijesuriya, W., & Samarappuli, L. (2009). Impact of climate change on rubber plantations. *Journal of National Institute of Plantation Management, 23*(2), 1–17.

Yoshida, S. (1978). *Tropical climate and its influence on rice*. IRRI Research Paper Series 20, International Rice Research Institute, Los Baños, Philippines.

Zubair, L. (2002). El Nino–southern oscillation influences on rice production in Sri Lanka. *International Journal of Climatology, 22*(2), 249–260.

Zubair, L., Nissanka, S. P., Weerakoon, W.M.W., Herath, D. I., Karunaratne, A. S., Prabodha, A.S.M., ... & McDermid, S. (2015). Climate change impacts on rice farming systems in northwestern Sri Lanka. In *Handbook of Climate Change and Agro-ecosystems: The Agricultural Model Inter-comparison and Improvement Project Integrated Crop and Economic Assessments, Part 2* (pp. 315–352).

Climate Change Challenges and Its Preparedness Toward Agriculture Using Climate-Smart Agriculture in Potohar Plateau of Pakistan

Shahzada Adnan, Mirza Barjees Baig, Qamar uz Zaman Chaudhry, Sahibzad Khan, Muhammad Latif, Kalim Ullah, Amna Bibi, Azmat Hayat Khan, and Sajjad Haider

Abstract Climate change has become one of the most threatening challenges to our planet. There is clear evidence about the global temperature increase accompanied by changes in weather and climate. Climate variability and its associated extremes have become more frequent and vigorous in many places of the world. The rapid changes

S. Adnan (✉) · A. Bibi
Pakistan Meteorological Department, National Drought Monitoring and Early Warning Centre, Islamabad, Pakistan
e-mail: shaz_adnan@pmd.gov.pk

A. Bibi
e-mail: aamna.met@gmail.com

M. B. Baig
Prince Sultan Institute for Environmental, Water and Desert Research, King Saud University, Riyadh, Kingdom of Saudi Arabia
e-mail: mbbaig@ksu.edu.sa

Q. Z. Chaudhry
International Climate Change Specialist ADB, Islamabad, Pakistan
e-mail: dgmetpak@hotmail.com

S. Khan
Director General, Pakistan Meteorological Department, Islamabad, Pakistan
e-mail: sahibzad@gmail.com

M. Latif · K. Ullah
Department of Meteorology, COMSATS University, Islamabad, Pakistan
e-mail: muhammad_latif@comsats.edu.pk

K. Ullah
e-mail: kalim_ullah@comsats.edu.pk

A. H. Khan · S. Haider
Pakistan Meteorological Department, Islamabad, Pakistan
e-mail: chiefmetpmd@gmail.com

S. Haider
e-mail: sajjadhaider1391976@gmail.com

M. Behnassi et al. (eds.), *Climate-Smart and Resilient Food Systems and Security*,
https://doi.org/10.1007/978-3-031-65968-3_15

have been observed in floods, droughts, torrential rain, cyclone, tornadoes, heat, cold waves, etc. Temperature is one of the main triggering factors that directly affect the climate of our globe. A small temperature change may impact large and dangerous potential shifts in weather and climate. The oceans are becoming more acidic as a result of global warming, glacier and polar ice caps are melting and sea levels are rising. These changes have become prominent and will likely impact our ecosystem and society in the coming decades. Agriculture activity particularly in rainfed areas (Potohar plateau) has always been a big challenge and it may become more severe under climate change scenarios in the coming future. To address the future challenges of this threat, new advancement in technology particularly Climate-Smart Agriculture (CSA) is the key to make the crops sustainable in the region. The CSA technique could benefit climate change mitigation, reduced soil erosion, drought buffering, stable yields, and reduced field preparation costs.

Keywords Climate change · Climate variability · Climate-smart agriculture · New advancement · Sustainable

1 Introduction

The recent global temperature increase gives a clear indication of global warming which is caused by the constant increase in the greenhouse gases concentrations in the atmosphere (IPCC, 2014). This warming is altering weather and climate patterns all over the world, which is one of the most significant aspects of climate change. The world has witnessed significant changes in temperature, precipitation, wind pattern as well as other climate extremes (flood, droughts, cyclone, snowstorm, heat, and cold wave, etc.) over the recent two to three decades. Moreover, the projected global and regional models show that the intensity and severity of extreme events will continue to rise in the future.

According to National Oceanic and Atmospheric Administration (NOAA), 2020 annual climate report, the average combined temperature (ocean and land) has increased at the average rate of 0.08 °C per decade since 1880 and this increase is also twice 0.18 °C since 1981. Similarly, World Meteorological Organization (WMO) identified the period of 2015–2019 is about to become the hottest period of all time. Moreover, there have been 10 warmest years occurred since 2005, whereas 7 out of 10 experienced since 2014. The climate is breaking its record in each coming year and is ultimately replaced as the climate shifts forward in time. These phenomena are becoming more persistent and more probable which is another indication of climate change.

2 Geography and Climate of Potohar Plateau

There are four districts in the northern Punjab province (Attock, Chakwal, Jhelum, and Rawalpindi) and Islamabad (Pakistan's capital), which encompass the Potohar Plateau of Pakistan (PPP (Rashid & Rasul, 2011). With elevation ranging from 181 to 2232 m, the region comprises an area of around 22,305 km^2 (Nadeem et al., 2012). The Geographical location of PPP is 32°5–34°N latitude and 72°–74°E longitude (Fig. 1).

Due to the high altitude, low temperature, and abundant average annual precipitation (1000–2000 mm), the climate of the PPP region is classified as semi-arid to humid (Adnan & Khan, 2009 Adnan et al., 2020). Two weather systems dominate the region: the southwest summer monsoon (July to September) and eastward propagating western disturbances (December to March), which bring rain during the Kharif (May–September) and Rabi (October–April) crop growing seasons, respectively (Adnan et al., 2020; Ahmed et al., 2019; Hussain & Lee, 2016; Latif & Syed, 2016).

The average inter-annual temperature range lies between 13 °C and 26 °C, with a maximum of 40 °C (June) and a minimum of −5 °C (January) (Adnan et al., 2017). May and June are known as the hottest months, whereas January is the coldest month

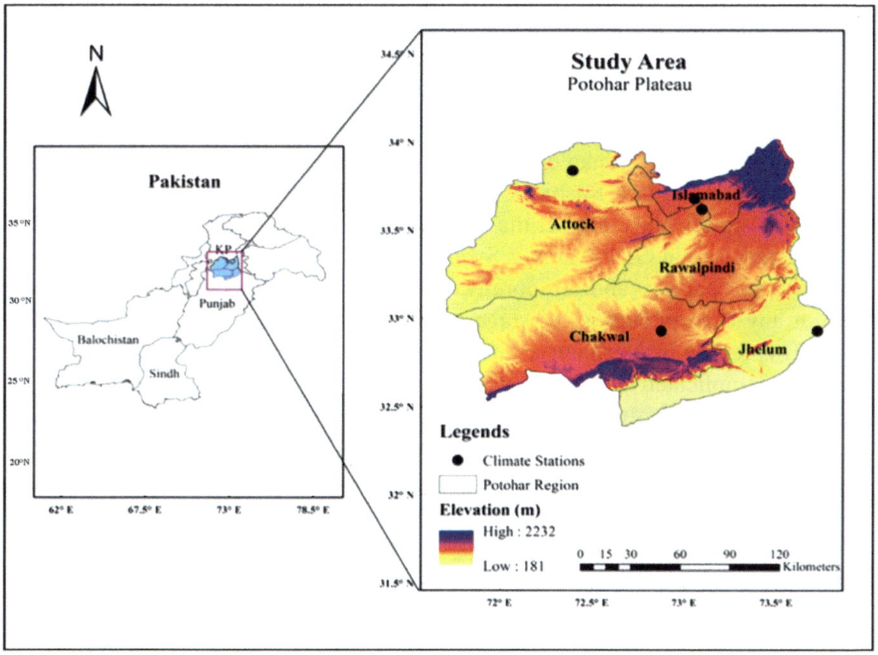

Fig. 1 Geographical location and topography of the Potohar Plateau of Pakistan (the study area). Included districts are Attock, Chakwal, Jhelum, Rawalpindi, and Islamabad (capital territory)

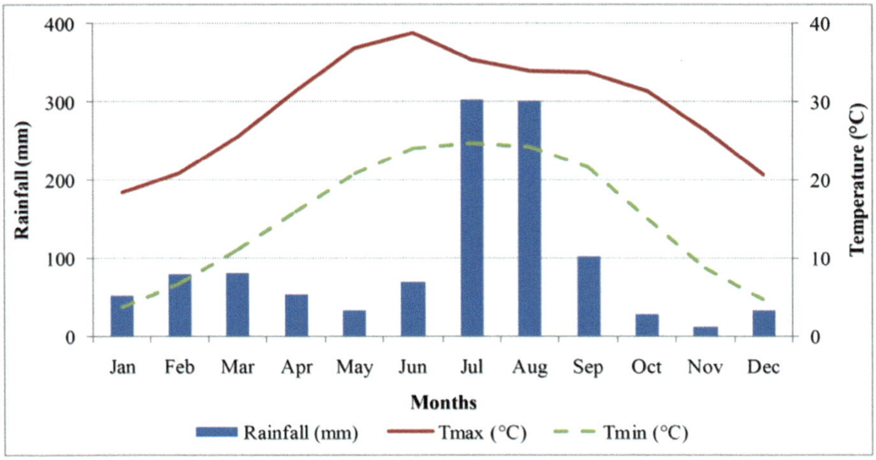

Fig. 2 Annual climate cycle of Potohar Plateau of Pakistan (the study area)

of the year. The minimum temperature tends to be below zero in some of the north eastern side of PPP during winter. Similarly, the rainfall increases latitudinally from south toward north and the highest in the northeastern side of PPP. In the Kharif and Rabi seasons, key crops such as maize and wheat are grown, respectively. The annual climate cycle of the Potohar plateau is shown in Fig. 2.

Around 10% of the country's overall agroecology comes from this region (Ashraf, 2016). Precipitation from the southwest monsoon is the main dominant weather system that significantly provides relief in moisture stress to fulfill the water need for both crop seasons (Adnan et al., 2018). Since agricultural activity in Pakistan is highly reliant on rainfall (Rashid & Rasul, 2011). Small seasonal variations in total rainfall can significantly impact the agricultural industry, which accounts for 24% of the country's total GDP (Latif & Syed, 2016). Rainfall in the study area is highly variable both spatially and temporally. Slight variations in the rainfall and water stress at any crop phonological stage can potentially alter the grain yield (Cakir, 2004). This region lacks canals as an irrigation system due to shallow groundwater levels and complicated terrain. Droughts are common in the region, and they have an impact on crop yields. In addition, more than 70% of the annual rainfall falls in the monsoon season. Delayed start and earlier retreat of monsoons can reduce soil water supply during the summer growing season, reducing crop yield and increasing the risk of drought. Flooding is more likely a result of unpredictable changing weather patterns brought about by climate change unless steps are taken to control run-off at the source.

The Potohar Plateau's cultivation area is approximately 110,600 ha (GOP, 2002–2003). The Potohar region is mostly rain-fed with supplementary canal irrigation in some parts. The amount and intensity of rainfall in such places determine crop productivity. Only 4% of the farmed area in the Potohar Plateau is irrigated, with the remaining 96% relying on rain (Majeed et al., 2010). Wheat-maize/millet cropping systems and wheat-ground-nut cropping systems are the two types of cropping systems in the region (Arif & Malik, 2009). Aside from shrubs, plantings of fast-growing trees have been made along the field edges for fodder and browse (Hussain et al., 2009; Nizami et al., 2004).

The majority of agricultural soils are made up of losses, mountain out-wash, old alluvial deposits, and recent stream valley deposits that have been transported by wind and water. From sand through silt loam to clay loam, they cover a wide range of terrain types, from poor to productive. Gullies and minor hill ranges disrupt the plateau's flat to gently sloping landscape. However, 60% of the land area has been severely eroded, leaving the remaining flat ground primarily cultivated. Only 0.77 million hectares of the total area is cultivated; the rest is largely pastureland.

Mini dams and tube wells are used for irrigation in today's smaller-scale farming. Namal Lake is a natural lake found in the Potohar region's extreme southwest. Although the majority of the water from this lake is sent via the tunnel from the Salt Range to irrigate crops near Mianwali, it is pumped to surrounding locations for irrigation.

Wheat, gram, peanut, millet, sorghum, oilseeds, and fodders are the main crops farmed in the Potohar region. Maize and sunflower are also cultivated in locations with more rainfall. It is possible to grow vegetables and orchards near cities, as long as they have access to irrigation water from dams and tubewells. Only a few protected and inaccessible locations retain natural flora, such as remnants of overthrown thicket savanna, while dense forests can be found in scattered pockets in higher precipitation areas.

Livestock production is another important economic activity in the Potohar region, which accounts for more than a quarter of the total livestock population in Punjab's Barani tract. The most common species are sheep and goats, followed by cattle, camels, and donkeys. Buffaloes are typically kept in subtropical climates or in areas with plenty of water. This area is home to the Dhani breed of draught cow and the Potohar breed of goat. The land covers the distribution of Potohar districts are shown in Table 1.

Table 1 List of land cover distribution (in percentage) in districts of Potohar in the year 2017

Land cover types	Islamabad	Attock	Chakwal	Jhelum	Rawalpindi
Agriculture	37.32	69.17	68.65	37.35	38.52
Grasses	16.42	7.88	7.46	14.38	16.13
Trees/Forests	19.85	16.11	17.03	30.78	29.45
Shrubs/Tall herbs	13.81	2.69	3.1	5.69	10.94
Bare soil/Rocks	3.01	1.05	1.35	5.79	1.58
Built-up	8.94	2.76	2.25	5.55	3.29
Water	0.66	0.35	0.17	0.45	0.09

Source Amir et al. (2019)

Approximately one-quarter of Punjab's total cropland is located in the Potohar Plateau. The area farming is distinguished by a diverse variety of activities and a complex structure. The region's farm productivity is low, which contributes to the region's substantial food shortfall. Livestock is a vital aspect of the farming system, but its interactions with crops and technological ramifications are poorly understood. Based on rainfall distribution patterns, the barani Potohar was stratified topographically into three zones, i.e., low (<500 mm), medium (500–750 mm) and high rainfall zone (>750 mm).

Adnan et al. (2018) reviewed the important agro-climatic aspects of Pakistan. The most important agro-climatic variable affecting agriculture in the area is rainfall. Farmers, even in the high-rainfall zone, face significant year-to-year variability. The intensity of moisture stress in a specific zone is also influenced by land and soil types. Rabi crops have more variability in both rainfall and agricultural production than Kharif. The list of land covers distribution (in percentage) in districts of Potohar for 2017 is shown in Table 1.

3 Agriculture

Wheat is the staple food crop cultivated in all rainfall zones of Potohar. Most crops require a specific terrain type or rainfall zone to thrive. Mustard, maize, and Kharif pulses are more important in areas with abundant rainfall, whereas groundnut, sorghum/millet, and rabi pulses in areas with little or less rainfall. Maize is mainly grown in the Potohar area while pulses, groundnuts, and sorghum/millet crops are also cultivated. As a result, farmers have developed very effective systems for assigning land based on specific crops' fertility and moisture requirements.

Rainfall, terrain type, farm size, power constraints, and livestock ownership all have an impact on cropping intensities. In low and high rainfall areas, cropping intensity has grown from 108 to 129. In all rainfall zones, Rabi season cropping intensity is much higher than Kharif season. Cattle are also an integral part of livestock systems in all rainfall zones. The increased percentage of milk animals in the wet zone, mainly buffaloes, is owing to the closeness to metropolitan markets and the guaranteed availability of green fodder. Draught animals, sheep, and goats are more common in drier areas.

Farmers have met their fodder needs in two ways: (1) by intercropping fodder with grain and (2) by growing sole fodder crops. Intercropping mustard in wheat and maize fodder (from thinning) in maize grain crops is used to manage fodder from limited land resources. Farmers' main solitary fodder crops are sorghum and millet, as well as barley and oats to a lesser extent. During seasons of low fodder, farmers have supplemented their home-grown feed by purchasing concentrates.

Barani farmers produce a slight excess over subsistence needs in grains. They're close to self-sufficiency in pulses, but they're short on feed. All three subsistence staples are in short supply for small farmers, particularly in the arid zone, with fodder being the most severe shortage. In each zone, animal products were ranked as the main source of cash revenue from the farm by the majority of farmers. In general, cash crop sales have outperformed food crop sales. Farmers, particularly small farmers, rely heavily on off-farm revenue.

Small farm sizes are common in the Barani regions. The average farm size is 4 ha in medium and heavy rainfall zones, yet nearly half of the farmers sampled cultivate less than 2 ha.

4 Climate-Smart Agriculture (CSA)

The UN Food and Agriculture Organization (FAO) estimates that feeding the world's growing population will require a 60% increase in total agricultural production. The food security issues are enormous, as many of the resources required for long-term food security are already stressed. On the other hand, climate change is already having a negative impact on agricultural production both globally and locally. Cropping, livestock, and fisheries are all predicted to be more vulnerable to climate change in the next decades, particularly in low-income nations with little adaptation potential. Agriculture has a negative impact on food security as well as rural livelihoods and rural development. To make matters worse the agriculture sector accounts for approximately half of global greenhouse gas emissions when land-use change is included (Lynch et al., 2021).

As a result, climate-wise agriculture practices such as rainwater collecting for agriculture, home, and animal use, as well as managing flash floods in an economically and environmentally sustainable manner, are getting international attention, including in Pakistan. Climate-Smart Agriculture (CSA) ideas and practices can help with both climate change and economic growth and development. If a practice enhances food security while simultaneously addressing at least one of the other CSA goals (adaptation and/or mitigation), then it is classified as CSA.CSA includes hundreds of new ideas and practices from all over the world. Among these operations are water management strategies, improved crop and animal varieties, integrated pest management, manure management, and renewable energy.

4.1 Aims of Climate-Smart Agriculture (CSA)

The CSA aims to increase agriculture productivity while also improving farmers' incomes and livelihoods. Climate-Smart Agriculture (CSA) is an integrated method to addressing the interconnected concerns of food security and climate change, with three specific objectives:

- a sustained improvement in agricultural productivity to support equitably higher farm incomes, food security, and development;

- adapting and strengthening agriculture and food security systems' resistance to climate change on multiple levels; and
- minimizing agriculture's contribution to global warming (including crops, livestock, and fisheries).

While considering national and local interests, the CSA invites individuals to think about these three goals on a variety of scales—from farm to landscape—and at various levels—from local to global. To promote climate change resilience and adaptation, as well as to reduce and/or eliminate GHG emissions whenever practicable, the CSA has been designed.

4.2 CSA Practices

- Agroforestry
- Intercropping
- Crop Rotation
- Organic Composing
- Improved Grazing
- Improved Water Management
- Early Warning Systems.

4.3 Benefits of CSA

- **Stable yields**: CSA helps to stabilize yields against weather extremes by conserving water and soil. In most cases, CSA increases average yields over time.
- **Drought buffering**: By infiltration and reducing runoff and evaporation, CSA increases soil water quality. Increased infiltration increases water quality, while also protecting crops from drought. Mulch also protects the soil from temperature extremes. In the rainfed semi-arid highlands of Mexico, for example, soil water content was 10–20 mm higher in CSA maize fields than in those with traditional tillage and residue removal during dry periods. When CSA fields in southern Africa were compared to traditional tilled plots, infiltration was 24–38 mm per ha higher on average (Richards et al., 2014).
- **Reduced field preparation costs**. CSA lowers the cost of tillage, whether it is done by hand or by machine. Field operating costs in mechanized rice–wheat systems in India were 15% lower under CSA. CSA fields needed 20% less labor than traditional ridge and furrow fields in Malawi's manual maize systems. With CSA, field preparations are reduced, allowing for earlier planting and more fruitful harvests.

- **Reduced soil erosion**: Erosion can be reduced by up to 80% by reducing tillage and preserving soil cover with crop residues. In addition, CSA increases topsoil organic matter, as well as soil biological activity and biodiversity.
- **Climate change mitigation**: In some circumstances, carbon sequestration and reduced GHG emissions may help to mitigate climate change, but adaptation rather than mitigation should be the key policy driver for its promotion.

4.4 Climate-Smart Agriculture Packages for Potohar Plateau of Pakistan

4.4.1 PARC Solar Pumping Irrigation

In Chakri, Punjab's Rawalpindi district, PARC is testing "Solar Pumping Irrigation Systems Coupled with Climate Smart Agriculture Packages." Potohar's rain-fed Soan River Basin farmers are taking part in many climate wise interventions to increase farm output and resilience at this pilot site. As a result of scaling up this pilot, Pakistan's dry and desert regions will benefit from improved food security and livelihoods. As a result of the undulating terrain and occasional rainfall, crop productivity is negatively affected. Due to a lack of crop diversity, mono-cropping dominates agriculture, and farmers' well-being is at the mercy of the rains. The effects of climate change on such populations will further increase future unpredictability. In addition to improving water management, the pilot program aims to improve farmers' livelihoods and raise their resistance to the effects of climate change.

4.4.2 Rainwater Harvesting (RWH)

The Punjab government is building mini dams and ponds to irrigate the arid land in the Potohar area and boost agricultural production. The amount of rain that can be stored is completely dependent on RWH systems. To address the increased demand for food and agriculture activities, it is necessary to investigate the finest RWH sites using good methodologies, i.e., Dugwells, runoff harvesting, and rooftop water harvesting, etc. Improvement of site suitability analysis by improving the requirements for RWH site recognition. Established Land Use Land Cover (LULC), drainage network, drainage density, slope, and rainfall data criteria can also be used to choose the best RWH sites. Other socioeconomic elements can be utilized to refine the site selection criteria, such as distance from archaeological sites, recreational sites, and highways. It is also possible to use the distance from seismic faults as a factor for choosing a mega-dam location. Percolation tanks, farm ponds, and check dams have plenty of room in the Potohar area. The criteria for selecting the best RWH site could be established by using GIS and remote sensing.

4.4.3 Climate Smart Model Farms (CSMF)

In 28 villages, PODA (Potohar Organization for Development Advocacy) created a technological solution called Climate Smart Model Farms (CSMF) and began climate adaptation smart farming. CSMF can adapt to changing climate patterns, severe weather events, and is resistant to high temperatures and long droughts. Sub-plots are divided into one CSMF. (1) Damage Probability Risk CSMF of Fruits/Vegetables (2) CSMF Planting Scheduler for Staple Food Crops (3) Water Deficit Calculator and CSMF of Rainwater Harvesting (4) Composting/Biological Disease Control and Zero Tillage CSMF (5) Livestock/Poultry Animals CSMF. Farm green and dry waste are collected in a composting ditch to make natural fertilizer, and farmers are encouraged to use minimum/zero tillage techniques to preserve soil moisture.

The Punjab government recognizes the project and its practices as the province's best social and economic project. The CSMF project has a long-term social, technological, and financial outlook. PODA is still managing this project and replicating it in other remote areas with the support of local communities and stakeholders for long-term climate adaptation agriculture farming in the country. Now, the Punjab government, local research stations, and other stakeholders are requesting PODA to initiate CSMF project replication in other climate-vulnerable districts as well. Unfortunately, the organization has not had enough funds to start replications in other districts.

4.4.4 The Potohar Climate Smart Irrigated Agriculture Project (PCIAP)

In the Potohar Plateau, a maintenance and rehabilitation program will be carried out to ensure the protection, integrity, and continued serviceability of existing small dams, as well as the identification of new sites and development of more small dams, irrigation systems, and command areas. Water scarcity for domestic and livestock purposes, lack of irrigation facilities, rainwater depletion and degradation of topsoil, recurrent droughts and dry spells, and the low profitability of rain-fed agriculture are all major issues on the plateau. The PCIAP will not only aid in the resolution of these issues but will also boost agricultural productivity in the region.

4.4.5 Use of Fertilizers

According to Sarwar et al. (2016), 72% of the 94 farmers interviewed said that they use chemical fertilizers for their crops. Two commonly used fertilizers are urea ($CO(NH_2)_2$) and DAP, or diammonium phosphate ($(NH_4)_2HPO_4$). These fertilizers are primarily used at different growth stages of the wheat and groundnut crops. Some farmers use DAPs before planting crop seeds and urea after germination. As long as local inputs come from the right source (organic), are applied at the right time, in the right quantity, they will be more effective. Balanced Organic fertilization lowers the GHG carbon footprint and pollution associated with synthetic fertilizer usage.

4.4.6 Farmers' Report on Pests, Diseases, and Weeds

The most common vertebrate pests causing damage to groundnut crops were reported to be rats, while some farmers also reported damage from porcupines and wild boars. Weeds must be removed as soon as possible. During droughts, it reduces crop losses and raises yields.

4.4.7 Pest and Disease Control

Weed removal by hand was discovered to be a normal practice. Rodents were reportedly killed by hand. Some farmers have reported using burrow smoking as a rodent control tool, especially in groundnut crops. None of the farmers reported using frightening devices, e.g., scarecrows and nets to control bird infestations as per Sarwar et al. (2016). Pesticides and herbicides were the most commonly used chemical control methods by farmers, with 49% of farmers using them. To control rodent pests in their fields, more than half of the farmers polled indicated they employed poisoned baits coated with zinc phosphide.

4.4.8 Crop Rotation

Crop rotation helps to keep weeds, pests, and diseases at bay. Intercropping may be used by farmers who do not have enough land to rotate crops. Because of their nitrogen-fixing capacity, legumes are recommended as rotational crops.

4.4.9 Holistic Planned Grazing

According to Sarwar et al. (2016), the goat was the most commonly reared animal by 70% of farmers, followed by cattle, which were reared by 63%. Sheep were owned by 41% of the farmers, while buffaloes were owned by 29% of the farmers. Soil quality and landholding capacity are also improved by rotational grazing. It conserves pastures and biodiversity.

4.4.10 Conservation Measures

Some of the suggestions for maintaining profitable agriculture while preserving faunal biodiversity are

- To boost the floral diversity of these agroecosystems, multiple cropping should be encouraged.
- Unwanted human activities in wild areas should be reduced to the greatest extent possible by effective management and enforcement of existing laws.
- Rather than growing agricultural land, improved crop varieties should be implemented to maximize crop yields on existing cultivable land.
- Conservation relies largely on environmental education and community awareness. It should be made a requirement in primary and secondary schools' formal education. Print and electronic media, on the other hand, can reach out to locals.
- To create successful sustainability initiatives in this understudied Potohar Plateau ecosystem, more research into biodiversity-environment relationships is needed.

5 Conclusions and Recommendations

Climate change has threatened the world as its impacts are more devastating in various sectors especially in agriculture. The frequency of extreme weather events has been increased over the years and those areas which are highly dependent on rainfall have become more vulnerable. The Potohar region of Pakistan is known to be the rainfed region and agriculture activity is highly dependent upon the amount of rainfall. Although this region receives a maximum rainfall, however, the deficiency in the seasonal rainfall is one of the major reasons for crop failure; however, the CSA techniques could be useful to make the crops sustainable in the region. Moreover, these CSA techniques will ensure productivity, sustainability, resilience and minimize the climate change impacts on the crops.

There are three key research implications for the Barani tract that have been identified. To begin with, land type plays a significant role in farmers' crop management decisions. Crop management research, such as fertility and tillage, as well as crop rotations and cropping patterns study, should constantly be considered. As a result, stratification by land type is just as essential as stratification by rainfall zone, if not more so. Second, fodder production must be prioritized in Barani areas to alleviate livestock productivity constraints. Finally, an increase in crop intensity is a great opportunity in these areas where little research has been done so far.

Rainfed agricultural guidelines in Punjab were only disaggregated by rainfall zone till recently. The Potohar Plateau receives a very high amount of rainfall both in summer and winter, so small water storage dams should be built in these areas that will latterly be used for supplementary irrigation. Water harvesting is another option for increasing water per unit farming area, reducing drought impact, and allowing beneficial use of runoff. Water harvesting is effective in recharging groundwater aquifers. Crop models may be able to aid in the long-term analysis of some of these relationships. In terms of socioeconomics, a deeper knowledge of farmers' choices is required. Because these are frequently communal decisions, a study of group decision-making at the village level and the possibility of influencing these decisions is required. Further research into the role of risk aversion in crop or fallow land decisions is also needed.

CSA must be part of a comprehensive set of farm cultural practices to gain the most benefits. Supplemental irrigation, which is optimized by on-farm water management regulations and timely socio-economic interventions, is critical for the long-term utilization of limited water resources, especially groundwater. A water management approach that increases yield or water production is not always the best option in water-scarce places. The best plan is one that falls halfway in the middle. Water usage and valuation policies should be focused on controlling water consumption, lowering water demand, ensuring safe water use and disposal, and encouraging users to take a community approach to utilize and managing water. These policies must be balanced, reasonable, and feasible; otherwise, implementing and/or enforcing them will be challenging. Integrated and participatory research and development programs are the most effective method to understand the full potential of technologies and processes. Supplemental irrigation investment opportunities could include altering policies and regulations governing groundwater development and operation, strengthening or founding water-user associations, and funding water resource development.

References

Adnan, S., & Khan, A. H. (2009). Effective rainfall for irrigated agriculture plains of Pakistan. *Pakistan Journal of Meteorology, 6*(11), 61–72.

Adnan, S., Ullah, K., & Ahmed, R. (2020). Variability in meteorological parameters and their impact on evapotranspiration in a humid zone of Pakistan. *Meteorological Applications, 27*(1), e1859.

Adnan, S., Ullah, K., Gao, S., Khosa, A. H., & Wang, Z. (2017). Shifting of agro-climatic zones, their drought vulnerability, and precipitation and temperature trends in Pakistan. *International Journal of Climatology, 37*, 529–543.

Adnan, S., Ullah, K., Shuanglin, L., Gao, S., Khan, A. H., & Mahmood, R. (2018). Comparison of various drought indices to monitor drought status in Pakistan. *Climate Dynamics, 51*(5), 1885–1899.

Ahmed, R., Latif, M., Adnan, S., & Abuzar, M. K. (2019). Thunderstorm frequency distribution and associated convective mechanisms over Pakistan. *Theoretical and Applied Climatology, 137*(1), 755–773.

Amir, S., Saqib, Z., Khan, A., Khan, M. I., Khan, M. A., & Majid, A. (2019). Land cover mapping and crop phenology of Potohar region, Punjab, Pakistan. *Pakistan Journal of Agricultural Sciences, 56*(1).

Arif, M., & Malik, M. A. (2009). Economic feasibility of proposed cropping patterns under differentsoil moisture regimes of Pothwar Plateau. *International Journal of Agriculture and Biology, 11*, 27–32.

Ashraf, M. (2016). Managing water scarcity in Pakistan: Moving beyond rhetoric. In *Proceedings of AASSA Regional Workshop, Challenges in Water Security to Meet the Growing Food Requirement* (pp. 2–14).

Cakir, R. (2004). Effect of water stress at different development stages on vegetative and reproductive growth of corn. *Field Crops Research, 89*(1), 1–16.

GOP. (2002–2003). Crops under production. Food, Agriculture and Livestock Division, Economic Wing, Islamabad, pp. 2–12.

Hussain, A., Mirza, S. N., Khan, T. A. & Naem, M. A. (2009). Determination of relative species composition and seasonal plant communities of Nurpur reserved forest in scrub rangeland of district Chakwal. *Pakistan Journal of Agricultural Sciences, 46*(1), 34–39.

Hussain, M. S., & Lee, S. (2016). Investigation of summer monsoon rainfall variability in Pakistan. *Meteorology and Atmospheric Physics, 128*(4), 465–475.

Intergovernmental Panel on Climate Change (IPCC). (2014). *AR5 synthesis report: Climate change 2014.*

Latif, M., & Syed, F. S. (2016). Determination of summer monsoon onset and its related large-scale circulation characteristics over Pakistan. *Theoretical and Applied Climatology, 125*(3), 509–520.

Lynch, J., Cain, M., Frame, D., & Pierrehumbert, R. (2021). Agriculture's contribution to climate change and role in mitigation is distinct from predominantly fossil CO_2-emitting sectors. *Frontiers in Sustainable Food Systems, 4*, 300.

Majeed, S., Ali, I., Zaman, S. M. & Ahmad, S. (2010). Productivity of mini dams in Pothwar Plateau: A diagnostic analysis. Research Briefings, Natural Resource Division, Pakistan Agricultural Research Council, Islamabad, 2(13), 1–19.

Nadeem, S. M., Naz, R., Shah, S. I., Beg, M. A., Kayani, A. R., & Mahmood, T. (2012). Season and locality related changes in the diet of Asiatic jackal (*Canis aureus*) in Potohar, Pakistan. *Turkish Journal of Zoology, 36*(1), 1–8.

Nizami, M., Shafiq, I. M., Rashid, A., & Aslam, A. (2004). The soils and their agricultural development potential in Potohar. WRRI and NARC, ISL. Pakistan, pp. 5–7.

Rashid, K., & Rasul, G. (2011). Rainfall variability and maize production over the Potohar Plateau of Pakistan. *Pakistan Journal of Meteorology, 8*(15), 63–74.

Richards, M. B., Sapkota, T. B., Stirling, C. M., Thierfelder, C., Verhulst, N., Friedrich, T., & Kienzle, J. (2014). Conservation agriculture: Implementation guidance for policymakers and investors.

Sarwar, M., Hussain, I., Anwar, M., & Mirza, S. N. (2016). Baseline data on anthropogenic practices in the agro-ecosystem of Pothwar Plateau, Pakistan. *Journal of Animal and Plant Sciences, 26*, 850–857.

Dynamics of Agricultural Extension for Nigeria's Rural Areas Under the Current Scenario of Climate Change

Mohammed Bashir Umar and Usman Haruna

Abstract Dynamics of agricultural extension for Nigeria's rural areas have been discussed in this chapter under the current scenario of climate change. Efforts have been made to highlight different issues affecting agricultural extension scenario in response to dynamics of climatic changes in the country. The agricultural extension dynamics showed shifts in extension paradigms as of colonial era to date from advisory to facilitation or empowerment. Therefore, the vulnerability of agriculture and rural life in Nigeria is compelling extension to think beyond the technology transfer and perform better under pressure of any change. In this context, a paradigm shift is required to be designed from 'technology' to 'knowledge', from 'research centric' to 'farmer centric', and from 'top-down' to 'bottom-up' solutions. In order to revitalize agriculture in the developing world, it is time to develop mechanism of extension pluralism. By knowledge empowerment of key players of extension system in rural communities, many of the agrarian challenges can be addressed better in the developing world. In the context of climate change, available information needs assessment, availability, access and use of climate and weather information, agricultural extension and climate change adaptation and mitigation. Potential synergies between adaptation and mitigation need to be revisited in the near future as the ultimate goal of Nigeria, which is eager to advance to higher stages of development and competitive power. It was concluded that agricultural extension has a role to play in informing farmers about the changing sets of incentives for different choices in agricultural management. Therefore, agricultural extension organizations have an important role to play in advising farmers and other actors on this issue. The following recommendations may be put in place for agricultural extension system to perform effectively and efficiently in Nigeria: climate change scientific research should have a real impact

M. B. Umar
Department of Agricultural Economics and Extension, Faculty of Agriculture, Federal University, Yobe State, Gashua, Nigeria
e-mail: mbumar1966@gmail.com

U. Haruna (✉)
Department of Agricultural Economics and Extension, Faculty of Agriculture, Federal University, Jigawa State, Dutse, Nigeria
e-mail: uharuna77@gmail.com

M. Behnassi et al. (eds.), *Climate-Smart and Resilient Food Systems and Security*,
https://doi.org/10.1007/978-3-031-65968-3_16

on farm productivity and livelihoods; new methodologies for dissemination of information have to be developed or adapted; reforms in agricultural extension should be oriented towards learning rather than teaching paradigm; roles and responsibilities of researchers, extensionists and community should be re-defined and shared. Socio-economic analysis for each methodology should be performed before adapting any method, approach, or strategy. An integrated approach (comprising different strategies) is recommended in diverse socio-cultural, economic, and political situations in order to achieve desired goals. As a conclusion, a sound agricultural extension policy is indispensable to achieve success in transferring knowledge to farmers.

Keywords Dynamics · Extension · Climate change · Rural · Vulnerable

1 Introduction

In Nigeria, the vulnerability of farming is quite evident due to different factors, such as climate change; changes in natural resources quality (desertification over large tracts in the north and constant loss of forest cover and biodiversity in the south); and lack of coping strategies at micro-and macro-levels of decision-making coupled with globalization, emerging market forces like commodity markets, and sustainability constraints. Impliedly, the vulnerability of countries, localities, and societies to the effects of climate change depends not only on the magnitude of climatic stress but also on the sensitivity and capacity of affected people to adapt to or cope with such stresses (Madu, 2012).

Thus, developing countries, and in particular less developed countries, are among the most severely impacted by climate change because of their greater social and economic vulnerability to changing circumstances (World Bank, 2010). Besides, Nigeria remains grouped among the 43 'Low-income food-deficit countries' (LIFDCs) in Africa. Rural dwellers in developing countries like Nigeria are especially affected by local conditions which intensify their climate vulnerability. Ranger and Garbett-Shiels (2012) observed that a number of these conditions are geographical in nature and include remote locations and paucity of socio-economic infrastructure that place considerable limitations on the pursuit of activities in rural communities. As such, Dasgupta et al. (2014) claim that the conditions of rural areas in developing countries that make them vulnerable to climate change are (1) dependence on agriculture and natural resources, high prevalence of poverty, isolation, marginality, neglect by policymakers, and lower human development; and (2) the distinctive characteristics of rural areas make them uniquely vulnerable to the impacts of climate change.

Therefore, Rishi et al. (2010) averred that there is pressing needs to address issues related to climate change adaptation, vulnerability, and coping, particularly in developing countries. Meanwhile, a major challenge for the scientific community is to provide relevant information to policymakers on vulnerability, impacts, and adaptation (VIA) in the context of a changing climate, and to do so in a coherent and coordinated way (UNDP, 2013). Notably, it has been argued that despite the growing knowledge based on climate change and its impacts, much remains to be understood about the linkages between climate change and rural development (Unique Forestry & Land Use, 2013).

Consequently, an assessment of the influence of rural characteristics on vulnerability to climate change is not only important for defining the risks posed by climate change in rural areas, but will also provide information to help identify adaptation measures. Besides, it will enable rural development practitioners and decision-makers to identify the most vulnerable rural areas which will in turn enable them tackle climate change problems with precision. This is because by understanding, planning for, and adapting to a changing climate that individuals and communities can reduce risks and seize opportunities (USAID, 2007).

In this milieu, rural areas still account for almost half the world's population and about 70% of the developing world's poor people. In addition, rural areas are viewed as a dynamic, spatial category, which remains important for assessing the impacts of climate change and the prospects for adaptation (Dasgupta et al., 2014). This is particularly important for Nigeria because it is the most populous country in Africa and 7th in the world with a total population of 160.2 million people, out of which 50% resides in rural areas (Population Reference Bureau, 2013).

In the current scenario of rapidly changing agricultural technologies in the world, Jones (1997) observed that agricultural extension has been recognized as the most important mechanism for delivering knowledge (information), which is an essential input in modern farming. Therefore, it has to reorient itself beyond the narrow transfer of technology models. In its place, it has to invigorate its dynamism for 'transferring knowledge (or) information packages' as the input in modern farming. As such, extension should be more diversified, knowledge intensive, and demand-driven. Presently, challenges for rural extension in the country are more heightened.

Even though, agricultural and rural developments, and thus rural extension, continue to be in transition, these are happening because of the forces that are today driving agriculture in the world. In this chapter, agriculture is referred to the traditional activities (e.g., crop protection, production, processing, marketing, and livestock husbandry) and the natural resource management activities associated with agricultural activities (e.g., water resource management, soil fertility, agroforestry, and fishery). The above scenarios, such as high rural population and changing agricultural technology, lead to in-depth review of the current dynamics of agricultural extension in Nigeria in reference to a changing climate.

2 Vulnerability as a Concept

'Vulnerability' is a concept that entails many understandings and interpretations depending on the disciplines and/or the perspectives of the organizations. In a general definition, vulnerability links the exposure of people, individuals, or groups to threats, their capacity of reaction, and the consequences in terms of a decline in well-being.

The intensity of vulnerability depends in great part on people's capacities to cope with external factors/events, sometimes difficult, and also on the social, economic, political, and environmental systems in which they live.

The above definitions show that the concept of 'vulnerability' is generally used in development work so as to ILO (2015) describe exposure to risks of being harmed, suffering a loss, or being affected by the realization of an adverse event. This can affect entire communities, countries, certain individuals, households, or groups of people. Risks are not equally distributed and certain individuals or groups are more exposed than others due to a wide range of factors, namely, geographic location, status, age, socio-demographic characteristics, and economic situation. These risks may result from natural events or from human actions, some may occur at the individual level (associated with health such as injuries, accident, or social risks like violence, etc.) while others may occur at a wider level (for instance, natural disasters, epidemics, political or economic crises) (Bonilla & Gruat, 2003). The concept of 'vulnerability' may as well guide normative analysis of actions to enhance resilience through risk management; vulnerability is considered as having limited capacity to deal with the prevailing risks (Adger, 2006).

The concept of resilience is closely related and can be defined as 'the ability of households, communities and systems to *ex ante* anticipate, withstand, adapt to and *ex post* recover from the aftermaths of shocks, stresses, and threats (such as natural disasters, epidemics, socio-economic instability, or conflict) in ways that support economic and social development and reduce vulnerability' (ILO, 2015).

3 Key Actors in the Agricultural Extension Practice in Nigeria

One of the relevant factors is to identify and analyze the key actors or stakeholders. Any of the following may be key actors contributing in the extension system in Nigeria (Madukwe, 2008):

- Farms' size (small, medium, or large).
- Input/service providers (such as seed or feed, agrochemicals, machinery/equipment, transport, credit, and insurance).
- Agro-processing enterprises (small, medium, or large).
- Marketers, markets and consumers, wholesalers, retailers, supermarkets, and commodity boards.
- Policymakers and legislators, organizations that influence policy and provide resources.
- Government agencies, ministries of agriculture, science and technology, education, industry and trade, finance and regulatory agencies (IP, ISO).
- Research organizations (national, regional or international whether public, quasi-governmental, and private), training institutions/universities, and other institutions of higher education.

- Extension service, organizations that provide information, and agro-technology transfer services.
- Farmer's organizations: associations, cooperatives, or other NGOs (public, private, quasi-governmental) that facilitate networking.
- Financial and business institutions: outfits that provide business services—such as feasibility studies and business plans—and help in the development of marketing strategies.

4 Dynamics of Agricultural Extension in Nigeria

Colonial and Internal Self-Government characterized the 'diffusionist' period and an approach emerged in the context of what was named as Technology Transfer (persuasive and paternalistic). This paradigm was prevalent in colonial times and reappeared in the 1970s, 1980s, and 1990s when the Training and Visit system in extension work was established. Technology transfer involves a top-down approach that delivers specific recommendations to farmers about the practices they should adopt.

Therefore, from the colonial era up till the present, the agricultural extension service in the public sector has remained dominant as a public service good and employed a variety of systems/approaches. From the early colonial time to the immediate post-independence period (1893–1968) rigorously characterized an export commodity extension approach. The post-independence period through the oil boom era (1969–1975), when the issue of food production was becoming a major concern, was exemplified by the 'diffused/general Ministry of Agriculture' extension system/ approach. Primarily, the policy of the Colonial Government for agricultural extension service was educational and regulatory; consequently, there was little attention paid to food crops and livestock development during the colonial and the immediate post-independence era.

However, in the 1980s and 1990s rural extension institutions faced a deep crisis. In Nigeria, the fiscal and economic crisis in addition to the criticism against the diffusionist extension approach which aimed at supporting agricultural modernization and the progressive consolidation of the neoliberal agenda, led to a period characterized by pressures toward the privatization and pluralistic extension services. Beyond the differences in approach, the crisis of the traditional extension systems was typified by an abrupt reduction in the public funding for extension and the outsourcing of extension work to the private sector.

As such, the post-World Bank-assisted Agricultural Development Project (ADP) period was set apart by the initiation of several donor-supported interventions using various levels of participatory extension approaches within the ADP structure. But, in the main time, the ADPs use all variants/modified versions of the Training (T) and Visit (V) in 'non-project areas' of the states to the level which their inadequate resources and limited staff would allow. The launch of these donor projects marked the foundation of a paradigm shift in agricultural extension service delivery from a top-down, supply-driven, and technology-centered extension system to bottom-up, demand-driven, and client-centered participatory extension system. The period has also been characterized by the introduction of pluralistic extension systems (multiple advisory service providers in public and private sectors) and multilateral funding with contributions of all partners in development interventions.

In Nigeria, these reforms currently indicated a move from a monopolistic, supply-driven, and a virtually non-accountable public extension service to a competitive pluralistic, accountable, and demand-driven extension service system. Since the mid-2000s, extension has regained interest by public policies which lead to its strengthening. Perhaps, the most recent and most authoritative agricultural extension policy pronouncement by government was enunciated in the 2008 through the National Food Security Program document which provided for the establishment of 'One-stop' Agricultural Extension Services (FMARD, 2002). Therefore, as a matter of policy, these projects provide for both pluralistic extension service delivery and multi-plural funding by development partners.

Today, the paradigm of advisory work (persuasive and participatory) can be seen where government organizations or private consulting companies respond to farmers' enquiries with technical prescriptions. It also takes the form of projects managed by donor agencies and NGOs that use participatory approaches to promote pre-determined packages of technology: for instance, SG2000 and Ecclesial Church of West Africa (ECWA) projects in Nigeria. In this period, the focus on farming remains as well as the valorization of participatory processes and pluralism of extension providers. Interestingly, the protective approach to rural extension and development processes has generated increasing interest. This approach highlights the importance of the multi-actor and inter-institutional nature of development processes, thus aiming at supporting the articulation of institutions and actors at a local level in the context of development platforms.

Another paradigm shift is in the facilitation for empowerment (educational and participatory). This paradigm involves methods such as experiential learning and farmer-to-farmer exchanges. Knowledge is gained through interactive processes and the participants are encouraged to make their own decisions. In terms of agricultural extension and advisory service providers, farmer-to-farmer extension has become an important mechanism for most of Africa's smallholder farmers, especially in the face of dwindling funding to extension service for recruitment and capacity building. The National Program for Food Security and Food and Agriculture Organization (NPFS/FAO) initiated the implementation of the Farmer Field School (FFS) extension approach in 2007 with significant changes in extension delivery and impact in the participating states. Unfortunately, however, the successes could not be sustained and scaled up due to the lack of funding. The effectiveness and efficiency of the approach for farmer empowerment and productivity improvement were clearly proven (Arokoyo, undated).

Nonetheless, the FFS extension system is a unique participatory extension approach that provides an alternative to traditional extension approaches, while the traditional approaches and the World Bank-promoted Training and Visit (T&V)

system view farmers as passive recipients of information and provide supply-driven extension services. The FFS is a learning process where farmers are gradually presented with new technologies, ideas, situations, and ways of responding to problems. The knowledge acquired during the learning process builds on existing knowledge, enabling farmers to adapt existing technologies to become more productive, more profitable, and more responsive to changing conditions or to adopt new technologies.

In retrospect, the NPFS and the Fadama Project (II and III) provide interesting platforms in agricultural research and extension service delivery in Nigeria. Anandajayasekeram et al., cited in Zhou (2008), suggested the need for a pluralistic extension system globally and locally, which includes potential provision of extension services from the public sector, the private non-profit sector, and the private for-profit sector. They pointed that 'the key issue of creating a pluralistic service is a need to find an appropriate "mix" of public and private funding and delivery mechanisms for extension which will achieve differing agricultural goals and serve diverse target populations'.

However, because of the different actors in the value chains approach, agricultural extension is expected to provide for multiple extension delivery systems/approaches (an appropriate mix of both supply-and demand-driven participatory approaches) to meet the varied needs of all actors. The process includes facilitating a wide range of activities such as access to inputs, strengthening the delivery of business and financial services, enabling the flow of information, and facilitating improved linkages between actors in the higher-value markets. All these activities are potential sources of income generation and employment creation for both skilled and unskilled labor.

5 Nigeria's Vulnerability to Climate Change

Nigeria is vulnerable to climate change impacts due to its geography, climate, vegetation, soils, economic structure, population, settlement, energy demands, and agricultural activities. The location, size, and the characteristic relief in Nigeria gave rise to a variety of climates ranging from tropical maritime climate characterized by the rainforest along the coastal and southern section to the tropical hinterland climate associated with the Sahel in the north-eastern part of the country. The paradox in climate change problem is that developing countries who contribute the least to global carbon emissions are the most vulnerable to its impacts. They are the most vulnerable because they are least endowed with resources and technology to adapt to these impacts and their economies are based largely on natural resource-dependent sectors that are climate-sensitive. Nigeria happens to be one of those countries.

There are various reasons why Nigeria is vulnerable to climate change apart from being a developing country. For instance, it is established that every country is affected by climate change but, the degree to which it is impacted to produce damage differs, depending on geographical circumstances, the capacity with which to withstand the impact and the nature of the economy. Nigeria's vulnerability to climate change comes both from being located in the tropics and from various socio-economic, demographic, and policy leanings limiting its adaptive capacity.

Nigerian Environmental Study/Action Team (NEST) (2003) indicated that constant loss of forest cover and biodiversity in Nigeria is linked to global warming. Nigeria is a country already being plagued by diverse ecological problems which have been directly linked to a changing climate. Thus, the effects will be more pronounced due to existing low level of coping capabilities in Nigeria and other parts of Africa.

6 Extension in the Context of Climate Change

The nexus between agriculture and climate is distinct and often complex. In both positive and negative ways, crops and livestock are sensitive to climate change. Therefore, agricultural systems are most sensitive to extreme climatic events such as droughts, floods, hailstorms, seasonal variability, and changing rainfall patterns. As such, Kajfez-Bogataj (2005) observed that farmers' adaptation is influenced by many factors, including agricultural policy, prices, technology, research and development, and agricultural extension services. Then, the poor often bear a disproportionate burden of direct damage from catastrophes and climate change as concluded by most studies in developing countries (IPCC, 2001). Also, the role of inadequate institutional support is frequently cited in literature as a hindrance to adaptation; for example, Adger and Kelly (1999) and Huq et al. (1999) showed how institutional constraints and deficiencies affect managerial capacities to cope with anticipated natural events.

In recent times, many observers of rural development have commented on the frequent manifestations of unsatisfactory extension performance (Rivera et al., 2001). Feder et al. (2001) have suggested interrelated characteristics of extension systems in the developing world that jointly result in a deficient performance namely: low staff morale, reduced efficiency, and financial stress. One more key factor is the number of clients and the vast spectrum of information/services needed to be covered by extension systems: for instance, the current situation on the ground, 1 extension agent/2,500–10,000 farm family ratio in Nigeria depending on the state.

However, policy makers in developing countries have reacted to this with the deployment of more extension personnel, which has continued the emphasis on more centralized, hierarchical, and top-down management systems. Hitherto, the requirement for combining a bottom-up approach with the conventional extension process is not fulfilled and the limitations on the extension process to influence issues such as credit availability, input supplies, market linkages, and logistics facilitation continued without change. In effect, there has been no visible impact due to such changes within the extension system in many parts of the developing world.

Nevertheless, Sulaiman and Hall (2006) described a range of extension initiatives from public and private sectors that explain the way extension agenda is expanding as embodied in the concept of 'extension plus', and have pleaded for new experiments in extension. Pluralistic institutional arrangements are emerging and are finding wider acceptance; this is mainly because developing countries have realized the need for extension to engage in a wider range of issues beyond merely disseminating production-oriented technologies. Therefore, extension pluralism is at the core of farmer's adaptation strategies and can offer new advantages in enabling reliable and rapid access to expert information support, which is much needed in the realization of adaptation strategies on a large scale.

7 Information Needs Under Climate Change

7.1 Information Need Assessment

Building Nigeria's Response to Climate Change (BNRCC) (2011) observed that the level of public awareness on issues related to climate change in Nigeria is considered to be low. For instance, results of a household survey indicated that the level of awareness of local communities of climate change impacts is low with close to 60% of respondents knowing little or nothing about climate change and its impacts (Nzeadibe, 2011). Also, studies show that the Nigerian media did not give adequate attention to climate change issues. Ajaero and Anorue (2018) attributed this to inadequate funding and the perception of climate change stories as 'hard sell'. They further pointed that there is a need for news editors to look for ways to make climate change reports more interesting perhaps, by framing stories more to reflect the human angle.

In another study Ozor et al. (2012) revealed that the high level of awareness among farmers of climate change (84%) can be linked to extensive awareness creation made through the print and electronic media and through other social and religious networks. Hitherto, in Nigeria, identification of the need to educate farmers and rural communities through agricultural extension services has not been matched with corresponding capacity building. That is, improved knowledge/skill or information and responsive attitude for extension professionals. Extension professionals in such situations are likely to Dimelu et al. (2014):

- Build awareness,
- Facilitate generation,
- Transfer of knowledge and technologies, and
- Implement actions for effective management of climate change risks.

As such, extensionists need to be competent in terms of technical knowledge of climate change issues, responses, and information management skills. They require training to act as educators and information/service providers. All these necessitate comprehensive capacity building of extension professionals and systems in Nigeria and elsewhere (Dimelu et al., 2014). Therefore, the government has a role to play in building the capacity of agricultural extension systems to achieve these various functions in relation to climate change (Ayanlade et al., 2017).

7.2 *Availability, Access, and Use of Climate and Weather Information*

The level of information available influences the level of awareness on climate change issues (Duru & Emetumah, 2016). As such, lack of adequate information is considered to be one of the key constraints faced by farmers in adapting to climate change in Africa (Otitoju & Enete, 2016). Consequently, there is a pressing need to improve information dissemination, access, public awareness, and understanding of the impacts of climate change. BNRCC (2011) suggested that these include access to information regarding historical climate, projections of future climate change, potential impacts, causes of vulnerability, technologies and measures for managing climate risks and the know-how for implementing these technologies.

In a study on knowledge of climate change in Nigeria, mass media (radio, television, and newspapers) was the largest source of information followed by contacts with friends. Other sources include the internet, researchers, extension workers, and farmers' co-operative (Nzeadibe, 2011). National organizations charged with managing environmental issues such as the National Emergency Management Agency have had some success at sensitizing people and raising awareness through radio, newspapers, and television in addition, to on the ground contacts with vulnerable people (Ozor et al., 2012). But, among Nigerian students, radio was the most accessible information source followed by mobile phone, television, and newspapers as reported by Duru and Emetumah (2016).

However, in another study it was found that the internet is considered to be accessible by more than half of the respondents. Still, for it to be a vital source of information on climate change issues, the quality and the ability of the respondents to understand that information is subjective (Duru & Emetumah, 2016). The study suggested greater awareness creation on climate change issues through public participation and technological tools (Duru & Emetumah, 2016). Moreover, BNRCC (2011) recommended that policies and programs should be aimed at increasing access to information for and raising awareness among public policy makers, the organized private sector, civil society organizations, users of natural resources and those working in agriculture and managers of infrastructure.

Besides, on information provision and awareness-raising of climate change impacts and responses, Otitoju and Enete (2016) averred that access to specific weather information, early warning and forecast technologies can help farmers to develop and readjust coping or adaptation strategies. Further, poor climate change information and lack of access to weather forecasts have been observed to be a major barrier to informed decision-making, useful farming plans, and adaptation. This renders farmers increasingly vulnerable to the impacts of changes in the climate and weather. This has been the case among farming households in Southern Nigeria as suggested by Otitoju and Enete (2016).

Also, socio-economic factors affect access to information and level of awareness in relation to climate change. Ozor et al. (2012) in a study found that income is positively and significantly related to institutional adaptation. Only farmers with the resources to buy radios and television were found to be able to listen to weather forecasts. Additionally, men in general listen to the radio and television where weather forecasts are usually transmitted more than women. Therefore, information and knowledge on the causes of climate change, its impacts, mitigation, and adaptation strategies need to be available. This should be in simplified forms and translated into and conducted in vernacular. It includes all mediums whether through radio, television, newspapers, magazines, seminars, workshops, and manuals (Anabaraonye et al., 2019).

7.2.1 Agricultural Extension and Climate Change Adaptation

Akande et al. (2017) stated that Nigeria has a tropical climate with two precipitation regimes: low precipitation in the North (short grass and marginal savanna) and high precipitation in parts of the Southwest and Southeast (rainforest and mangrove). On one hand, this can lead to aridity, persistent drought, and desertification in the north. On the other, it can lead to erosion and large-scale flooding in the south. As such, while climate change is a national phenomenon, the challenges associated with the climate change vary across the country (Federal Ministry of Environment, 2014).

Moreover, Madu (2016) observed that vulnerability analysis based on sensitivity, exposure to climate change, and degree of adaptive capacity reveals variations across Nigeria. Sensitivity to climate change relates to how readily a particular system such as agriculture would react to changes related to climate (Federal Ministry of Environment, 2014). However, exposure refers to the extent of climate stress to which a particular unit or system is exposed (Federal Ministry of Environment, 2014). However, adaptive capacity depends on sufficient education, assets, information, and income. Regions with the least adaptive capacity are more vulnerable to climate change (Madu, 2016). Thus, an understanding of geographic variation and impacts is essential to efficient deployment of adaptation actors (Federal Ministry of Environment, 2014).

Therefore, in Nigeria, there is a critical need to educate farmers and fishermen in rural areas on climate change issues and to build their technical capacity to engage in mitigation and adaptation. Presently, many of them have little, inadequate understanding or knowledge of climate change and possible responses (Anabaraonye et al., 2019). A study conducted finds that there is a need for capacity building of farmers to cope with changing climate. It suggests that agricultural extension workers should develop climate smart agriculture training (Ayanlade et al., 2017).

Nevertheless, agricultural extension services involve the application of scientific research and knowledge to agricultural practices, delivery of information, and skills training to farmers. Education and training to communicate specialized climate change knowledge and skills is needed to equip farmers. This will improve the ability to understand and address climate change risks and opportunities, to engage in possible adaptive and mitigation measures feasible in their locality (Nkechi et al., 2016). In addition, farmers need to have access to adaptive technology and innovations—through extension and information systems (Dimelu et al., 2014). Similarly, farmers need to learn about and understand the consequences of some of their farming practices that can exacerbate ecological problems. For instance, monocropping, bush burning, indiscriminate use of synthetic agrochemicals, and felling of trees (Dimelu et al., 2014).

It is recognized that the greater the contact farmers have with agricultural extension personnel and services, the better their production, productivity, efficiency in the use of resources, and profitability (Otitoju & Enete, 2016). It was also, observed that farmers with better access to information and knowledge of the changing climate through extension services have a greater likelihood of adopting agricultural technologies and other adaptation measures (Solomon & Edet, 2018). The absence of adequate agricultural extension programs directed to meet climate change adaptation strategies in food crop production and/or poor agricultural extension delivery have been key constraints in the adoption of adaptation measures, such as among cocoa agroforestry households (Otitoju & Enete, 2016).

7.2.2 Agricultural Extension and Climate Change Mitigation

The two main responses to climate change are mitigation and adaptation. Ifeanyi-obi and Nnadi (2014) opined that mitigation refers to 'measures that may either reduce the increase in greenhouse emissions (abatement) or increase terrestrial storage of carbon (sequestration)'. Adaptation refers to 'all the responses to climate change that may be used to reduce vulnerability'. Climate change mitigation and adaptation initiatives should be integrated in development projects and programs. This is in order to reduce the vulnerability of people to the impact of climate change (Akeh & Mshelia, 2016).

Mitigation

Energy: Federal Ministry of Environment (2014) pointed that the energy sector in the country is the most important sector for climate change mitigation. In Nigeria, Yahaya and Nwabuogo (2016) reported that conventional energy (oil and gas) with gas flaring has the highest percentage of carbon dioxide (CO_2). It is imperative to control CO_2 emission and other associated greenhouse gases by moving towards renewable energy development and an efficiency mechanism (Yahaya & Nwabuogo, 2016). Achike and Onoja (2014) recommended that the government provides incentives for investment in renewable energy resources and policies that will tax businesses emitting greenhouse gases. Then, tax proceeds could be used for research and building the capacities of farmers to engage in climate change adaptation.

Carbon Footprint: Also, Federal Ministry of Environment (2014) averred that evaluating carbon footprint by gender can contribute to more targeted mitigation strategies aimed at behavioral change. It reported that in Nigeria, men are likely to account for a larger proportion of CO_2 emissions and air pollution through their energy use. Generally, women adopt less emission-intensive modes of transport than men with lower levels of car-ownership and usage, less travel by air and greater use of public transport facilities (Federal Ministry of Environment, 2014).

Lifestyle choices: Dioha and Emodi (2018) observed that the energy sector has been identified as the largest contributor to Nigeria's national greenhouse gas inventory. Therefore, efforts toward climate change mitigation should be distributed across various sectors such as agriculture. Besides, there is a need to support sustainable lifestyle choices among Nigerians. These include less meat consumption, phasing out of inefficient appliances such as incandescent bulbs, greater access to and use of public transportation and greater use of non-motorized modes of transport (Elias & Omojola, 2015).

Infrastructure: Also, for effective waste reduction, public infrastructure and services need to be encouraged. This includes private-sector partnerships in collection, disposal of domestic, industrial wastes, and for waste reduction strategies (Elias & Omojola, 2015).

Afforestation: In Nigeria, more than 70% of the people living in rural areas use fuel wood which has been a key contributor to deforestation. However, reforestation is only about 10% of the deforestation rate (Elum & Momodu, 2017). There is an urgent need for a more aggressive tree planting.

Adaptation

Mitigation efforts take time and may only take place to a limited extent and are thus, essential (Otitoju & Enete, 2016). However, adaptation focuses on actions that would help to lessen the sensitivity of systems in different ecological zones in Nigeria (Federal Ministry of Environment, 2014). Equally, Jellason et al. (2019) pointed that planned adaptation strategies include government intervention and public policy such as investment in infrastructure, subsidies, research, innovation, and tax regimes. As such, autonomous adaptation involves coping strategies by farmers and others affected in rural settings. These include diversification, irrigation, changes in planting date, crop and livestock insurance and using tolerant varieties of crops (Jellason et al., 2019).

However, the term agricultural extension means different things to different people (Purcell & Anderson, 1997). Van den Ban and Hawkins (1996) arrived at a concept of extension that seems to synthesize diverse perspectives into five goals:

- Transferring knowledge from researchers to farmers,
- Advising farmers in their decision-making,
- Educating farmers to be able to make similar decisions in future,
- Enabling farmers to clarify their own goals and possibilities and to realize them, stimulating desirable agricultural developments (rural guidance), and
- Stimulating desirable agricultural development.

Similarly, agricultural extension has been defined as a series of embedded communicative interventions which apparently help to resolve problematic situations (Leeuwis, 2006). This definition offers agricultural extension the mandate to accommodate the issue of climate change in its duties (Ozor, 2009). But, in order to achieve results, there is need for change in roles and capacity in the extension system so as to accommodate the new dimensions brought about by climate change. Therefore, with the emerging significant threat posed by climate change, the agricultural extension system will need to address the following issues in order to overcome the militating effects of climate in the country:

Creating awareness on issues of climate change: The agricultural extension system should create awareness on climate change, its effects, and adaptation options available to farmers. Studies have shown that the level of awareness on issues of climate change is still low (Nzeadibe, 2011). There is a need for massive awareness campaigns in the country on issues concerning climate change. This will help farmers to be prepared as well as equip themselves to combat issues of climate change. As such, this calls for a review of the communication system in the agricultural system in Nigeria,

Technologies and information management: Traditionally, extension has played a major role in providing information and promoting new technologies or new ways of managing crops and farms. It also links farmers to researchers and other actors in the national and global innovation system. Therefore, farmers, extension agents, and researchers must work together on farmers' fields to prioritize, test, and promote new crop varieties and management techniques (Ozor et al., 2012). Whereas extension must at present go further than such methods, there is still a need for simple technology transfer in order to increase resilience to climate change.

Presently, in Nigeria, farmers will need to be able to quickly respond to climate change and skillfully manage risk. In areas of knowledge and information systems this will be especially difficult for extension. Farmers need to have access to this kind of information—be it climatic information, forecasts, adaptive technology innovations, or markets—through extension and information systems. However, extension agents can introduce locally appropriate technologies and management techniques that will enable farmers to adapt to climate change. For example, dissemination of local cultivars of drought-resistant crop varieties with information about the crops. In addition, extension staff can share with farmers their knowledge of cropping and management systems that are resilient to changing climate conditions such as intercropping, sequential cropping, and no-till agriculture (Davis, 2009).

Some of these practices have the additional advantage of improved natural resource management. It is imperative to provide farmers with information about how the various options will potentially increase income and yields, protect household food security, improve soils, and enhance sustainability. In general, this will help to alleviate the effects of climate change.

Therefore, the major challenge facing extension in the country is to move from providing 'packages' of technological and management advice to supporting farmers. This support is with the skills they need to choose the best option to deal with the climate uncertainty and variability. Some farmers will also, need access to new technologies and management options in those areas where climate change leaves their current farming systems attractive.

Capacity Building and Co-learning: Over time, one of the extension's major activities has been adult and informal education. This role continues today and is currently, even more important in light of the climate change scenario. Also, extension is accountable in providing information using techniques ranging from flyers and radio messages to field demonstrations (Anyadike, 2009). Recent innovative extension activities include the adult education and experiential learning approaches utilized in farmer field schools, an extension and education approach already working with farmers on issues of climate change.

Usually, climate change causes extreme events like sudden onset of disasters and new vectors of human and livestock diseases. Evidence is emerging that the biggest impacts will be in the form of small droughts, floods, and other events that cause severe hardship but do not attract the attention of the international community. IPCC (2007) observed that the capacity of farmers to cope with such different forms of risk will become ever more crucial and extension efforts must pay special attention to educating farmers about their options to enhance resilience and response capacity.

Therefore, there is a need for capacities to engage new sets of actors including humanitarian agencies. Thus, education must move beyond technical training to enhance farmers' abilities for planning, problem-solving, critical thinking and prioritizing, negotiating, building consensus and leadership skills, working with multiple stakeholders, and finally being practical. Within extension capacity development is all the more important as well. Extension agents have traditionally been trained only in technical expertise and often lack 'soft' skills such as communication, development of farmer groups, systems thinking, knowledge management, and networking (Davis, 2009).

Therefore, in order to realize better and positive outcomes out of rural development initiatives, farmers and extension agents need new skills that will require sound agricultural education and an updated extension curricula. It is imperative to attach greater value and need to better understand the knowledge and experiences of rural people and co-learning (that is, farmers and extension agents learning together rather than extension agents train farmers in a one-way information transfer). There are different ways and means to inform and educate farmers about adaptation options.

Indigenous technical knowledge: Presently, there is a growing awareness of the importance of indigenous knowledge and its value for environmental management and sustainable development. To improve outcomes in sustainable rural development and climate change adaptation and mitigation, indigenous knowledge should be documented. This could help to counter the gradual disappearance of valuable indigenous knowledge that remains solely in the memory of elderly people living in remote rural areas (Dimelu et al., 2014).

African indigenous religious values and practices can also serve as a response to climate change. Nkechi et al. (2016) observed that African traditional religion is very close to nature with structures, practices, and sanctions that can effectively protect the environment. For example, sacred forests have special place and significance. Customarily, farming and hunting is prohibited there. Such beliefs and practices can serve as alternate tools in mitigating climate change in Nigeria such as countering deforestation (Nkechi et al., 2016). Thus, agricultural education and extension curricula should incorporate the knowledge and experiences of rural people and co-learning. This would help farmers to develop skills to address the effects of climate change (Dimelu et al., 2014). Also, policy makers should consider integrating indigenous values and practice into policy guidelines to address climate change issues. This would allow for more comprehensive measures (Nkechi et al., 2016).

Therefore, assessing local knowledge is important. For example, a study of community urban risk assessments revealed that respondents had valuable knowledge about flooding in the area, causes, and local community adaptation choices (Odemerho, 2014). The contribution of indigenous knowledge can promote greater community acceptance and involvement in the implementation of flood management solutions occurring from it (Odemerho, 2014). It can also be relied upon to develop a well-formulated grassroots adaptation strategy for urban flood risk management for resiliency. This can become the base for a successful national policy for action (Odemerho, 2014). Similarly, extension staff can play an important role in transferring indigenous technical knowledge to help farmers worldwide.

Policies and Programs: Another role of extension which will be critical for climate change adaptation is that of bringing together different actors within the rural sector. Traditionally, this has meant linking farmers to transport agents, markets, and input suppliers, among others (Davis, 2009). With climate change, it will be increasingly important for the extension system to link farmers and other people in rural communities directly with private and public institutions that disseminate adaptation technologies and funding programs for adaptation investments. Increased access to meteorological information will be very important.

Besides, extension has a huge challenge in bringing farmers' concerns together and those of other actors as they address both climatic and market uncertainties together. Extension has the chance to make a significant contribution in overcoming this gap through enhanced farmer decision-making. Also, extension agents may play a role in assisting farmers in implementing policies and programs that deal with climate change adaptation. For instance, extension agents could be employed to

- Educate farmers in their area,
- Assist in forming community groups,
- Link farmers to governmental, non-governmental, and private service organizations at the national and international levels, and
- Assist with proposal preparation or negotiations with other stakeholders in climate change issues.

7.2.3 Potential Synergies Between Climate Change Adaptation and Mitigation

Ng'ang'a et al. (2016) defined adaptation as the long-term measures that provide vulnerable populations with the potential to deal with the risks and shocks posed by climate change. Climate change adaptation is critical to households and communities to secure their livelihoods and build their resilience to hazards such as floods and droughts. According to Field et al. (2014), adaptation is a process of adjusting to the actual or expected climate and its effects with the intent to lessen the negative impacts and/or exploit the valuable opportunities.

However, Food and Agriculture Organization of the United Nations (FAO) (FAO, 2018) explained three types of adaptation that could occur at different scales. These are autonomous or incremental adaptation, planned or systematic adaptation and transformational adaptation. FAO defines autonomous adaptation as the changes in management that farmers undertake within their existing systems. For instance, changes in planting dates, reallocation of land and other resources among different crops and livestock systems. Also, reallocation of time and labor requirements among farm and non-farm activities falls into this category. Ciscar et al. (2010) presented other examples of autonomous adaptation innovations to include switching to drought-tolerant crops, adopting salinity-tolerant crops, changing farming methods, and altering crop-livestock rotations among others.

On one hand, planned or systematic adaptation mainly focuses on increased spending on research and development of new varieties, diversification strategies such as intercropping or rotational cropping systems, and risk management options such as index-based insurance (FAO, 2018). On the other hand, transformational adaptation strategies are those options that require substantial changes in the production systems in terms of institutional arrangements, priorities for investments, and changes in norms and behaviors (FAO, 2018). Despite emphasis being placed on the importance of the adaptation strategies in increasing productivity (Rigolot et al., 2015), building resilience (Keenan, 2015) and reducing vulnerability (Descheemaeker et al., 2018) in the face of climate change, the acceptance and scaling-up are still low.

8 Conclusion

Current dynamics of agricultural extension in Nigeria's rural vulnerable contexts under a changing climate were discussed. With reference to the Nigeria context, the chapter highlighted: vulnerability as a concept, key actors in the agricultural extension practice, agricultural extension dynamics, the country's vulnerability to climate change, extension and information needs in the context of climate change, availability, access and use of climate and weather information, agricultural extension and climate change adaptation and mitigation, and potential synergies between adaptation and mitigation. The agricultural extension dynamics showed shifts in extension paradigms as of colonial era to date from advisory to facilitation or empowerment. The vulnerability of agriculture and rural life in developing countries is compelling extension to think beyond technology transfer and perform better under pressure.

Climate change has only recently emerged as a critical issue and so most agricultural extension professionals would have received little or no specific training in relation to climate change in their formal training. In addition to this, most training organizations have little current capacity in this area. Accessing and using knowledge and information in general have certainly been an issue for most public agricultural extension organizations, which have often tended to be passive recipients of information (usually from public-sector research). This is starting to change, but many agricultural extension actors have limited capacity to actively seek and use new knowledge and information. This is a critical issue to address if agricultural extension organizations are to support successful responses to climate change.

In this context, a paradigm shift is required from 'technology' to 'knowledge'; 'research centric' to 'farmer centric'; and 'top-down' to 'bottom-up' solutions. In order to revitalize agriculture in developing countries, it is time to start engaging in evolving extension pluralism. By knowledge empowerment of key players in extension and rural communities, many of the agrarian challenges can be efficiently addressed. In the context of climate change, information needs assessment was discussed by looking at the complexities of climate change scenarios. It was concluded that agricultural extension has a role to play in informing farmers about the changing sets of incentives for different choices in agricultural management. Therefore, agricultural extension organizations have an important role to play in advising farmers and other actors on this issue.

Finally, the following recommendations should be put in place for agricultural extension to perform effectively and efficiently in Nigeria:

- Climate change related scientific research should have a real impact on farm productivity and livelihoods and new methodologies for dissemination of information have to be developed or adapted;
- Reform in agricultural extension should be oriented toward a learning rather than a teaching paradigm;
- Roles and responsibilities of researchers, extensionists, and local people to be re-defined and shared;
- Socio-economic analysis for each methodology should be performed before adapting any method, approach, or strategy;
- An integrated approach (comprising different strategies) is recommended in diverse socio-cultural, economic, and political situations in order to achieve desired goals; and
- A sound agricultural extension policy is indispensable to achieve success in transferring knowledge to farmers.

References

Achike, A. I., Onoja. A. O. (2014). Greenhouse gas emission determinants in Nigeria: Implications for trade, climate change mitigation and adaptation policies. *British Journal of Environment and Climate Change, 4*(1), 83–94. Retrieved August 05, 2021, from www.sciencedomain.org

Adger, W. N., & Kelly, P. M. (1999). Social vulnerability to climate change and the architecture of entitlements. *Mitigation and Adaptation Strategies for Global Change, 4*(3–4), 253–266.

Adger, W. N. (2006). Vulnerability. *Global Environmental Change, 16*, 268–281.

Akande, A., Costa, A. C., Mateu, J., & Henriques, R. (2017). Geospatial analysis of extreme weather events in Nigeria (1985–2015) using self-organizing maps. *Advances in Meteorology*. Retrieved August 05, 2021. https://doi.org/10.1155/2017/8576150

Akeh, G. I., & Mshelia, A. D. (2016). Climate change and urban flooding: Implications for Nigeria's built environment. Retrieved August 05, 2021. https://www.preventionweb.net/publications/view/50865

Ajaero, I., & Anorue, L. (2018). Newspaper framing and climate change mitigation in Nigeria and Ghana. *African Population Studies, 32*(2), 4228–4238. Retrieved August 05, 2021. https://doi.org/10.11564/32-2-1195

Anabaraonye, B., Okafor, C. J., & Ikuelogbon, O. J. (2019). Educating farmers and fishermen in rural areas in Nigeria on climate change mitigation and adaptation for global sustainability. *International Journal of Scientific and Engineering Research, 10*(4), 1391–1398. Retrieved August 07, 2021, from https://www.ijser.org/researchpaper/Educating-Farmers-and-Fishermen-in-Rural-Areas-in-Nigeria-on-Climate-Change-Mitigation-and-Adaptation-for-Global-Sustainability.pdf

Anyadike, R. N. C. (2009). Climate change and sustainable development in Nigeria: Conceptual and empirical issues. Enugu Forum Policy Paper 10 (pp. 13–18). African Institute for Applied Economics (AIAE); Enugu, Nigeria.

Ayanlade, A., Radenyb, M., & Mortonc, J. F. (2017). Comparing smallholder farmers' perception of climate change with meteorological data: A case study from southwestern Nigeria. *Weather and Climate Extremes, 15*, 24–33. Retrieved August 07, 2021. https://doi.org/10.1016/j.wace.2016.12.001

Arokoyo, T. (undated). Final report of: The agricultural extension transformation agenda. Abuja, Nigeria.

Bonilla, G. A., & Gruat, J. V. (2003, November). *Social protection: A life cycle continuum investment for social justice, poverty reduction and development, Geneva* (pp. 2)

Building Nigeria's Response to Climate Change (BNRCC). (2011). *National adaptation strategy and plan of action on climate change for Nigeria* (NASPA-CCN). Prepared for the Federal Ministry of Environment Special Climate Change Unit. Retrieved August 07, 2021, from http://csdevnet.org/wp-content/uploads/NATIONAL-ADAPTATION-STRATEGY-AND-PLAN-OF-ACTION.pdf

Ciscar, J. C., Paci, D., & Vergano, L. (2010). *Issues on the economics of adaptation to climate change* (pp. 29–49). Edward Elgar Publishing.

Dasgupta, P. J. F., Morton, D., Dodman, B., Karapinar, F., Meza, M. G., Rivera-Ferre, A., Toure, S., & Vincent, K. E. (2014). Rural areas: In climate change 2014: Impacts, adaptation, and vulnerability. Part A: Global and sectoral aspects. contribution of working group II to the fifth assessment report of the Intergovernmental Panel on Climate Change [Field, C. B., Barros, V. R., Dokken, D. J., Mach, K. J., Mastrandrea, M. D., Bilir, T. E., Chatterjee, M., Ebi, K. L., Estrada, Y. O., Genova, R. C., Girma, B., Kissel, E. S., Levy, A. N., MacCracken, S., Mastrandrea, P. R., & White, L. L. (Eds.)] (pp. 613–657). Cambridge University Press.

Davis, K. E. (2009). *Agriculture and climate change: An agenda for negotiation on Copenhagen "The Important Role of Extension Systems"*. International Food Policy Research Institute. Washington, D.C., USA, Focus 16, Brief 11.

Descheemaeker, K., Oosting, S. J., Tui, S. H.-K., Masikati, P., Falconnier, G. N., & Ken, E. G. (2018). Effects of climate change and adaptation on the livestock component of mixed farming systems: a modeling study from semi-arid Zimbabwe. *Agricultural Systems, 159*, 282–295. Retrieved August 07, 2021.

Dimelu, M. U., Edoka, M. H., & Emodi, A. I. (2014). Challenges in building climate change mitigation and adaptation capacity of extension professionals in Nigeria. *PAT June, 10*(1), 110–122. Retrieved August 09, 2021, from https://pdfs.semanticscholar.org/82c2/4dfe13159cd47be90c418d2c67b087f69991.pdf

Dioha, M. O., & Emodi, N. V. (2018). Energy-climate dilemma in Nigeria: Options for the future. IAEE Energy Forum. Retrieved 09, 2021, from https://www.google.com/utions%2Fnewsletter CXox9HBg4d-I

Duru, N. P., & Emetumah, C. F. (2016). Evaluating the effects of information literacy on climate change awareness among students in Imo State University. *Archives of Current Research International*, *4*(3), 1–10. Retrieved August 09, 2021, from https://www.researchgate.net/profile/Faisal_Emetumah/publication/305177025_Evaluating_the_Effects_of_Information_Literacy_on_Climate_Change_Awareness_among_Students_in_Imo_State_University/links/5784051008aee45b8442ef66/Evaluating-the-Effects-ofInformation-Literacy-on-Climate-Change-Awareness-among-Students-in-Imo-StateUniversity.pdf

Elias, P., Omojola, A. (2015). Case study: The challenges of climate change for Lagos, Nigeria. *Current Opinion in Environmental Sustainability*, *13*, 74–78. Retrieved August 10, 2021, from https://reader.elsevier.com/reader/sd/pii/S1877343515000214?token=157FC47FCE33B92E5012D43084678311AECF686653F471D31E9469543A83864E8ABB5EDF3453A32B1ECA8EFDAFEB9F59

Elum, Z. A., & Momodu, A. S. (2017). Climate change mitigation and renewable energy for sustainable development in Nigeria: A discourse approach. *Renewable and Sustainable Energy Reviews*, *76*, 72–80. Retrieved August 10, 2021.https://doi.org/10.1016/j.rser.2017.03.040

FMARD. (2002). Agriculture in Nigeria: Policy (Before and Now) analysis of the existing (1988) agricultural policy and the revised policy. Federal Ministry of Agriculture and Rural development, Abuja.

Federal Ministry of Environment. (2014). United nations climate change Nigeria. National Communication (NC). NC 2. 2014. Retrieved August 10, 2021, from https://unfccc.int/sites/default/files/resource/nganc2.pdf

Feder, G., Willet, A., & Zijp, W. (2001). Agricultural extension: Generic challenges and the ingredients for solutions. In S. Wolf & D. Zimmerman (Eds.), *Knowledge generation and technical change: Institutional innovation in agriculture* (pp. 313–356). Kluwer.

Field, C. B., Barros, V. R., Dokken, D. J., Mach, K. J., Mastrandrea, M. D., Bilir, T. E., Chatterjee, M., Ebi, K. L., Estrada, Y. O., Genova, R. C., Girma, B., Kissel, E. S., Levy, A. N., MacCracken, S., Mastrandrea, P. R., & White, L. L. (2014). *Contribution of working group II to the fifth assessment report of the intergovernmental panel on climate change*. Cambridge University Press, Cambridge, United Kingdom and New York, NY.

FAO. (2018). In: L. Lipper, N. McCarthy, D. Zilberman, et al. (Eds.). *Climate smart agriculture: Building resilience to climate change*. Springer International Publishing, Switzerland. Retrieved August 10, 2021.

Huq, S., Karim, Z., Asaduzzaman, M., & Mahtab, F. (Eds.). (1999). *Vulnerability and adaptation to climate change in Bangladesh* (p. 147p). J. Kluwer Academic Publishers.

Ifeanyi-obi, C. C., & Nnadi, F. N. (2014). Climate change adaptation measures used by farmers in South-South Nigeria. *Journal of Environmental Science, Toxicology and Food Technology*, *8*(4). Retrieved August 11, 2021, from https://pdfs.semanticscholar.org/a63d/22bf8b8ebde892a7c9d761ee1653e0e11df6.pdf

International Labor Office (ILO). (2015). Report V, ILC, 105th Session. Employment and decent work for peace and resilience, revision of the employment (Transition from war to peace) Recommendation, 1944 (No. 71).

IPCC. (2001). *The regional impacts of climate change: An assessment of vulnerability*. Cambridge University Press.

IPCC. (2007). *Climate change—impacts, adaptation and vulnerability: Contribution of working group JP II to the fourth assessment report of the intergovernmental panel on climate change* (p. 976). Cambridge University Press.

Jellason, N.P., Richard, B.N., Conway, John S., and Ogbaga, Chukwuma. C. (2019).Climate change perceptions and attitudes to smallholder adaptation in Northwestern Nigerian Drylands. *Social Science*, *8*, 31. Retrieved August 11, 2021, from https://app.dimensions.ai/details/publication/pub.1111649369?and_facet_journal=jour

Jones, G. E. (1997). The history, development and the future of agricultural extension in improving agricultural extension: A reference manual by Burton E Swanson et al., FAO, Rome.

Kajfez-Bogataj, L. (2005). Climate change and agriculture vulnerability. *Acta Agriculturae Slovenica, 85*(1), 25–40.

Keenan, R. J. (2015). Climate change impacts and adaptation in forest management: A review. *Annals for Science, 72*, 145.

Leeuwis, C. (2006). *Communication for rural innovation: Re-thinking agricultural extension* (3rd ed., pp. 27). The Netherlands, Publishing.

Madu, I. A. (2012). Spatial vulnerability of rural households to climate change in Nigeria: Implications for internal security. *International Journal of Climate Change; Impacts and Responses, 3*(4), 79–97.

Madu, I. A. (2016). Rurality and climate change vulnerability in Nigeria: Assessment towards evidence based even rural development policy. In *Paper presented at the 2016 Berlin conference on global environmental change, 23–24 May 2016 at Freie Universität Berlin*. Retrieved August 11, 2021, from https://pdfs.semanticscholar.org/508b/94cab07b84a703b44eca1089326cc98d7495.pdf?_ga=.154518008.112403230.1572433568-162569160.1557482164

Madukwe, M. (2008). *Nigeria's agricultural extension service: Practice without policy*. University of Nigeria, Nsukka.

Ng'ang'a, S. K., Bulte, E. H., & Giller, K. E., et al. (2016). Livestock wealth and social capital as insurance against climate risk: A case study of Samburu County in Kenya. *Agricultural Systems, 146*, 44–54.

Nigerian Environmental Study/Action Team (NEST). (2003). Climate change in Nigeria: A communication guide for reporters and educators (Ibadan: NEST publishers, 2003).

Nkechi, O., Alphonsus, N. A., & Ekenedilichukwu, E. (2016). Mitigating climate change in Nigeria: African traditional religious values in focus. *Mediterranean Journal of Social Sciences, 7*(6), 299–308. Retrieved August 11, 2021, from https://www.mcser.org/journal/index.php/mjss/article/view/9612

Nzeadibe, T. C. (2011). *Climate change awareness and adaptation in the Niger Delta Region of Nigeria*. African Technology Policy Studies Network. Working paper, no. 57. Nairobi: Published by the African Technology Policy Studies Network. Retrieved August 11, 2021, from https://atpsnet.org/wp-content/uploads/2017/05/wps57.pdf

Odemerho, F. O. (2014). Building climate change resilience through bottom-up adaptation to flood risk in Warri, Nigeria. *Environment and Urbanisation, 27*(1), 139–160. Retrieved August 11, 2021. https://doi.org/10.1177/0956247814558194

Otitoju, M. A., & Enete, A. A. (2016). Climate change adaptation: Uncovering constraints to the use of adaptation strategies among food crop farmers in South-west, Nigeria using principal component analysis (PCA). *Cogent Food and Agriculture*, 2. Retrieved August 12, 2021. https://doi.org/10.1080/23311932.2016.1178692

Ozor, N. (2009). Understanding climate change: Implications for Nigerian agriculture, policy and extension. In A *paper presented at the national conference on "Climate change and the Nigerian Environment": Held at the University of Nigeria*, Nsukka, 29 June–2 July.

Ozor, N. M. C., Enete, A. A., Amaechina, E. C., Onokala, P., Eboh, E. C., Ujah, O., & Garforth, C. (2012). A framework for agricultural adaption to climate change in Southern Nigeria. *International Journal of Agriculture Sciences, 4*(5), 243–251. Retrieved 12 August, 2021, from http://centaur.reading.ac.uk/28884/1/13425102734_5_4_IJAS.pdf

Population Reference Bureau. (2013). *World population data sheet*, New York PRB.

Purcell, D. L., & Anderson, J. R. (1997). *Agricultural extension and research: Achievements and problems in national systems. World Bank operations evaluation study*. World Bank.

Ranger, N., & Garbett-Shiels, S. (2012). Accounting for a changing and uncertain climate in planning and policymaking today: Lessons for developing countries. *Climate and Development, 4*(4), 288–300.

Rishi, P., Omprakash, M.D. and Mudaliar, R. (2010). Behavioral mapping of Indian urban settlements towards changing climate: An empirical study. A paper presented at International Conference on "Human Security and Climate Change," organized by The Royal Norwegian Society of Science and Letters, Trondheim, Norway, 21–24 June, 2010.

Rivera, W. M., Qamar, M. K., & Crowder, L. V. (2001). *Agricultural and rural extension worldwide: Options for institutional reform in developing countries*. FAO.

Rigolot, C., Douxchamps, P. S., Prestwidge, D., van Wijk, M., Thornton, P., Henderson, B., Medina, H. D., Rodriguez, D., & Herrero, M. (2015). Additive impacts of climate-smart agriculture practices in mixed crop-livestock systems in Burkina Faso. Climate Smart Agriculture, Montpellier, France.hal-01195408. Retrieved 10 August, 2021, from https://hal.archives-ouvertes.fr/hal-01195408

Sulaiman, R., & Hall, A. (2006). Extension policy analysis in Asian nations: In A. Van den Ban & R. K. Samanta, *Changing roles of agricultural extension in Asian nations* (pp. 23–54). B.R. Publishing, Delhi.

Solomon, E., & Edet, O. G. (2018). Determinants of climate change adaptation strategies among farm households in Delta State, Nigeria. *Current Investigations in Agriculture and Current Research, 5*(3). Retrieved August 10, 2021, from https://lupinepublishers.com/agriculture-journal/fulltext/determinants-of-climate-change-adaptation-strategies-among-farm-households-in-delta-state-nigeria.ID.000213.php

Unique Forestry and Land Use. (2013). Rural development and adaptation to climate change: What do we know? UNIQUE Forestry and Land use FREIBURG Germany.

United Nations Environment Program (UNDP). (2013). Research priorities on vulnerability, impacts and adaptation: Responding to the climate change challenge, UNEP, Nairobi.

U.S. Agency for International Development (USAID). (2007). Adapting to climate variability and change: A guidance manual for development planning, Washington, DC USAID.

Van den Ban, A. W., & Hawkins, H. S. (1996). *Agricultural extension* (2nd ed.). Blackwell Science.

World Bank. (2010). *World development report 2010: Development and climate change*. The World Bank.

Yahaya, O., & Nwabuogo, O. E. (2016). Renewable energy deployment as climate change mitigation in Nigeria. *Global Journal of Human-Social Science, 16*(4). Retrieved August 12, 2021, from https://globaljournals.org/item/6195-renewable-energy-deployment-as-climate-change-mitigationin-nigeria

Zhou, Y. (2008). *Reinventing agricultural extension to smallholders*. Syngenta Foundation.

Climate-Smart Water Management Technologies in Pakistan

Muhammad Ashraf, Faizan ul Hasan, Bareerah Fatima, and Mufeezah Ahsan

Abstract Pakistan is among the most vulnerable countries due to climate change and variability. The most important impact of climate change is on the ecosystem, particularly on water resulting into frequent droughts and floods, which have devastating impacts on infrastructure, agriculture, livestock, and humans. Currently, urban flooding has become a common phenomenon. Moreover, climate change has a great impact on crops, on sowing and harvesting time, in addition to an increase in crop water requirements. This has not only varying impacts on the upper, middle, and lower riparian but also has a greater impact on drylands as compared to irrigated areas. Climate change has now become a reality. This chapter analyzes the impacts of climate change on the ecosystem particularly on the water, the underlying reasons and proposes simple and smart water management techniques to cope with the climate change for the upper Indus Basin, the plain irrigated area, the lower Indus Basin, and the rainfed areas. The main purpose of this chapter is to present the potential of already tested climate-smart water management technologies for adaptation at larger scale.

Keywords Climate smart · Water management · Climate change · Pakistan

M. Ashraf · F. Hasan (✉) · B. Fatima · M. Ahsan
Pakistan Council of Research in Water Resources (PCRWR), Islamabad, Pakistan
e-mail: faizan.ulhasan@canberra.edu.au

M. Ashraf
e-mail: muhammad_ashraf63@yahoo.com

B. Fatima
e-mail: breerahftm@gmail.com

M. Ahsan
e-mail: mufeezahahsan@live.com

F. Hasan
University of Canberra, Canberra, Australia

M. Behnassi et al. (eds.), *Climate-Smart and Resilient Food Systems and Security*,
https://doi.org/10.1007/978-3-031-65968-3_17

1 Situation Analysis

Climate change was not an important subject until 1970 when the world started experiencing environmental hazards of anthropogenic activities, especially the depletion of the ozone layer. For the first time during the Earth Summit in 1972, the attention of global leaders was drawn toward a changing climate (Jackson, 2007). However, climate change could not take much priority on the global agenda until 1992 and global leaders were more interested in the protection of natural resources for their socio-economic benefits. It took more than 20 years to realize that the climate is changing. During the COP-26 in 2021, it was realized that it might lead to the extinction of 'Homosapien'. Particularly, people in developing countries are facing this issue, and Pakistan is no exception. The country is reported to be among the top ten countries hit by climate change. In fact, it is facing both climate change and climate variability issues.

Extreme weather events in recent years, like cyclones and heavy rains in Karachi and Baluchistan and snowstorm in Murree, remind the remarks of UN Secretary General, Antonio Guterres, who said "*climate change is indeed running faster than we are, and we have the risk to see irreversible damage that will not be possible to recover if we don't act very, very quickly*" (UN, 2018). Therefore, climate change awareness among the common masses is of utmost importance to combat its impacts.

Pakistan is not a high contributor to climate change; however, it is among the most vulnerable countries in the world. Nevertheless, the knowledge about climate change and its impact is still limited, even at the highest professional levels (Shahid & Piracha, 2010). According to a study on climate change, about half of the highly qualified respondents were not even aware of climate change, and those who were aware, did not take it seriously (Jan et al., 2020).

Figure 1 explains climate change and its impacts. Pakistan has traveled fast from climate variability to climate change, resulting into increased unpredictable extreme events. Changing climate is impacting the whole ecosystem. Consequently, socio-economic sectors of the society are also being affected.

Pakistan Meteorological Department (PMD) analyzed long-term data and found that the mean annual temperature of Pakistan had increased by 0.47 °C from 1960 and 0.57 °C 1991 to 2000 (PMD, 2020). Due to variability in climate and mass urbanization, temperature increase is not linear. Resultantly, the summer season has prolonged and increased the number of heat days with temperatures exceeding 50 °C. The increased number of sunny days also affected the coastal city of Karachi where the temperature rose up to 43 °C in 2015, and approximately 1,200 people died because of hot and humid weather.

Another important aspect of climate change is the intensity of rain. The increase in mean rainfall is not significant; however, the intensity and changes in precipitation patterns are of great concern. For example, the coastal areas of Karachi are experiencing heavy rains in recent years, resulting in urban flooding. At the same time, a large area of Balochistan and some parts of Sindh have experienced drought-like conditions due to high temperatures. The rise in temperature is associated with

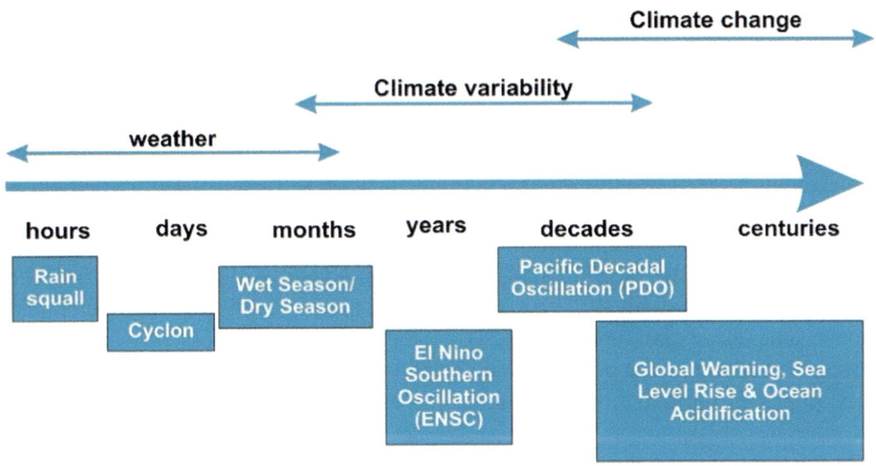

Fig. 1 Weather and climate scales. *Source* Mishra (2017)

increased salinity in the lower Indus due to high evaporation. Moreover, groundwater is also depleting in many parts of the Indus Basin (Iqbal et al., 2020).

Pakistan's agriculture sector is highly susceptible to climate change. Hasan et al. (2021) concluded that climate variability could not be ignored as it was posing several crop managements challenges. Local farmers, though not fully aware of climate change, yet are observing variability in weather and unpredictable seasonal variation of summer and winter seasons. They are reporting a decline in agricultural yield owing to seasonal unpredictability. Siddiqui et al. (2012) estimated significant negative impacts on widely grown staple crops, namely rice and wheat.

Adaptation to climate change in agriculture is considered important for reducing the negative effects on agricultural productivity. Moreover, traditional practices also depict the lack of information dissemination to farmers (Abid et al., 2015). It is also a fact that the literature on adaptation to climate change in agriculture in Pakistan is limited, and only a few studies are available in this area (Abid et al., 2015; Esham & Garforth, 2012). At local levels, farmers traditionally change sowing and harvesting dates as well as change cropping patterns and grow alternate crops. Studies have shown that farmers are only interested in those adaptation practices, which have either increased their crop productivity or the productivity in other farmers' land (Khuda & Kamran, 2019).

2 Climate Change Policy Instruments

Pakistani parliament was the first to adopt Sustainable Development Goals (SDGs) in 2016, including Goal 13 related to climate change. The National Climate Change Policy, 2012 provides guidelines at the federal level to develop climate-resilient plans for socio-economic development. Pakistan is also a signatory to the 2015 Paris Agreement, which emphasizes the reduction of greenhouse gas emissions up to 20%. In 2021, the National Climate Change Policy has been updated. The National Water Policy, 2018 also emphasizes on climate change impacts. All these policy frameworks underline the need for water conservation and reduction of irrigation system losses, and provide incentives for adoption of more efficient irrigation techniques and climate-smart water management technologies in all water-use sectors.

After the 18th Amendment to the Constitution of Pakistan in 2010, the primary responsibility for the development and management of water resources has been shifted to provinces. This shift provides Pakistan's provinces with greater potential to modify adaptation plans and actions, so that they can better address the water sector's different challenges and priorities. However, sufficient resources and capacity need to be in place to utilize this potential.

3 Responding to Climate Change-Climate Smart Technologies

Pakistan has an agro-based economy as most livelihood originates from agriculture. This is practiced in every region of the country with diverse climatic, soil, and water use patterns. In other words, the farming community is at the front line of climatic changes. Therefore, there is a need to develop, introduce, and promote climate-smart technologies for farming communities.

Being a research organization, the Pakistan Council of Research in Water Resources (PCRWR) has investigated and piloted some climate-smart technologies in different parts of the country to build resilience of farming communities. A brief description of these technologies is provided below.

3.1 Technologies for Northern Areas

The northern areas are the mother well of Pakistan's water resources. These areas mostly consist of marginal lands with low or no vegetation and, therefore, are prone to soil erosion. In these areas, it is always difficult to lift water due to the non-availability of energy. Climate variability and disasters, especially Glacial Lake Outburst Flood (GLOF), are threatening the livelihood options of communities residing there. Therefore, community resilience to climate change can be developed by introducing simple and smart technologies such as: (i) solar-powered pumping coupled with drip irrigation system; and (ii) hydraulic ram pump.

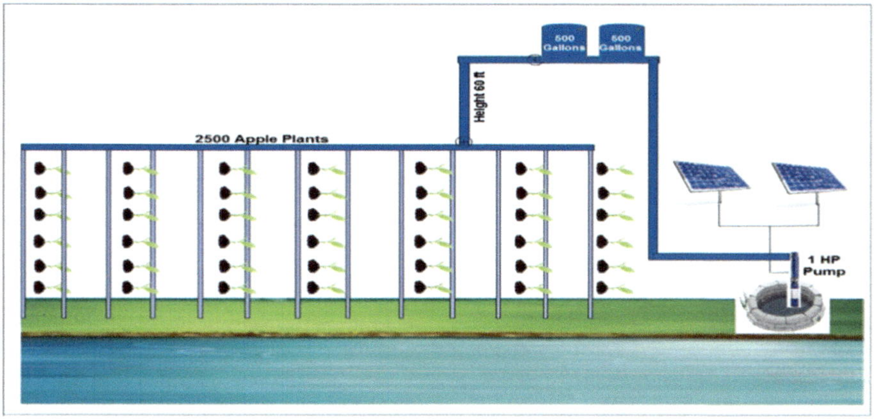

Fig. 2 Schematic diagram of solar-powered drip irrigation system. *Source* PCRWR (2018)

3.1.1 Solar-Powered Pumping Coupled with Drip Irrigation System

Solar pumping system can be used to pump water for domestic and small-scale agriculture. The system consists of a solar-powered pumping unit capable to deliver 6 gallons/minute through a DC submersible pump (400 watts) lifting water to 30 m height with four poly-crystalline solar panels. The water was lifted through a solar pump placed with an outer filter to avoid sediment entry into the pump. A UPVC pipe was wrapped with a finely meshed green net, and was tightly bound together. The filter was placed in the river, transverse to the direction of flow. The river flow forced the sediment water to enter into the outer filter through green net and fine openings. Initially a sump within the riverbed was dug for water pumping but it failed as the hydraulic conductivity of the parent material even within the riverbed was almost negligible. The water was collected into two PVC tanks each having 1.9 m^3 capacity. A low-pressure drip irrigation system was installed to irrigate an apple orchard planted on 2.4 ha barren patch. Figure 2 shows the schematic diagram of the system installed.

3.1.2 Hydraulic Ram Pump

Hydraulic ram pump works on the principle of 'water hammering', where a large amount of quickly moving water is pushed through a small opening to create pressure. As pressure builds within the system, it reaches a critical point that lifts a fraction of the water flow (Fig. 3). Generally, it lifts water with a ratio of 1:10, meaning that one meter of inlet water head can pump water to 10 m. First, such a system was installed in Khyber village which is located on the main Karakoram Highway between Passu and Sust. Though two water sources were available, both were located below the command area. Therefore, Hydra Ram Pump was installed for water lifting from the

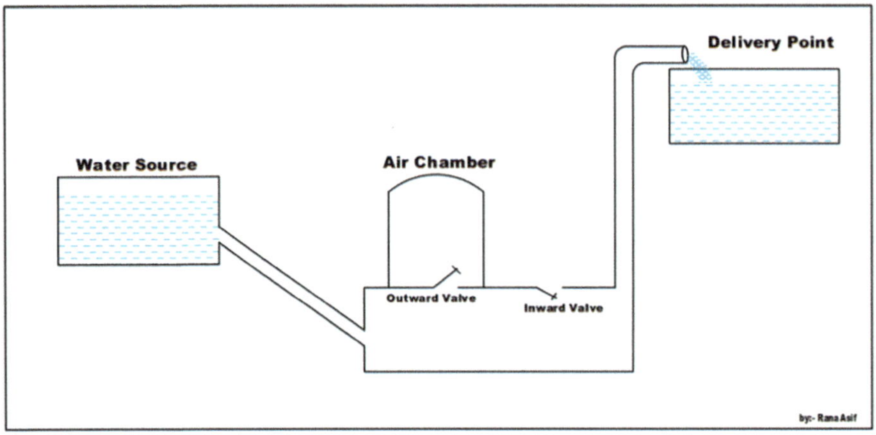

Fig. 3 Schematic Diagram of a Hydraulic Ram Pump

river. Through GI and HDPE pipeline, the pump is supplying water into 4 m^3 tanks placed at a head of 55 m. This pump can operate for 24 h without any operational cost. On the contrary, solar-powered pump can only operate during sunshine hours. These pumps are further supplying water through low head drip lines for one hectare apple orchard, integrated drip lines for alley cropping and the drip lines of a tunnel. Now, this technology is being up-scaled by other government and development organizations.

3.2 Technology for Irrigated Areas: Irrigation Advisory Service for Climate Resilience

Pakistan has one of the largest irrigation systems in the world where more than 93% of the total available water is used in agriculture. The overall irrigation efficiency is less than 40%. This is mainly due to low conveyance and application efficiencies. The major reasons for low application at the farm level are the conventional irrigation methods and the lack of knowledge about irrigation scheduling (when and how much to apply water to a particular crop). Due to lack of knowledge on irrigation scheduling, farmers keep on applying water even when the crop does not need water (Ashraf & Saeed, 2006), resulting in huge water and nutrient losses. Irrigation scheduling is either based on information on soil moisture in the root-zone or on the weather parameters involving empirical equations. Both these approaches have been out of the reach of common farmers.

The Pakistan Council of Research in Water Resources (PCRWR), with the help of University of Washington, launched a satellite-based irrigation advisory service (IAS) in 2016 to improve irrigation efficiency at the farm level. This IAS service is based on remotely sensed observation of potential evaporation (ETo), combined with crop coefficient (Kc) values from lysimetric stations of PCRWR. The information thus produced is broadcasted on a weekly basis using a Short Message Service (SMS). This advisory SMS, written in Urdu language, contains two sets of information: water consumed by the crops during the previous week, rainfall and wind speed expectancy and crop water requirement during the next week. Presently, PCRWR is sending this advisory SMS messages to 20,000 farmers in 43 districts of Pakistan covering all major crop zones. The farmer feedback regarding the use of advisory services shows that farmers are using the information provided through IAS. Farmers' associate high value to the timely forecast of rain and windstorms as it prevents them from unnecessary irrigation, pesticide, and fertilizer application before an upcoming rainfall event. In the scenario of a changing climate, farmers in canal irrigated zones need to be more prepared. This not only helps them in building resilience, but also enables water saving on large scale and relieving pressure on the dwindling groundwater resources.

3.3 Technologies for Dryland

Out of 30 million hectares culturable land, dryland agriculture is practiced on about 12 Mha, which is 40% of the total culturable land (Hasan et al., 2015). This is the area where livelihood is totally dependent on rainfall and where the level of poverty is very high. In addition, this area has been neglected in the past as most Pakistan's investment has been on the irrigated agriculture. Livestock is the only livelihood option for the inhabitants. Due to overgrazing, land is being degraded with time. Soil erosion is also a common feature with low productivity and high temperature, wind, and evaporation. Drought-like conditions are common. Rainfall is low to medium and erratic in nature. The groundwater either do not exist or is highly saline. This is the area that is most vulnerable to climate change.

3.3.1 Rainwater Harvesting and Desertification Control

Rainfall is the sole source of water for humans and livestock. The rainwater is stored in natural depressions, locally called *tobas* and in underground dug wells called *kunds*. Initially, the population was small in these areas, therefore, these *tobas* were sufficient to cater to the needs of the local community and their livestock. Now, the population and their livestock have increased substantially. Moreover, most of these *tobas* have silted up and the rainwater is available only for a small part of the year.

Once the water in the local *tobas* disappears, local communities (mostly nomads) have to migrate in search of water for themselves and for their livestock as their livelihood depends on livestock. For example, in the Cholistan desert there are about 276,674 human population whereas animal heads are around 1,334,037. Similarly, in the Thar Desert, the human population is 1.5 million whereas their animal heads are about 6.0 million. They also migrate if the rangelands are overgrazed in search of fodder for their livestock. During migration, infants suffer a lot mainly due to less

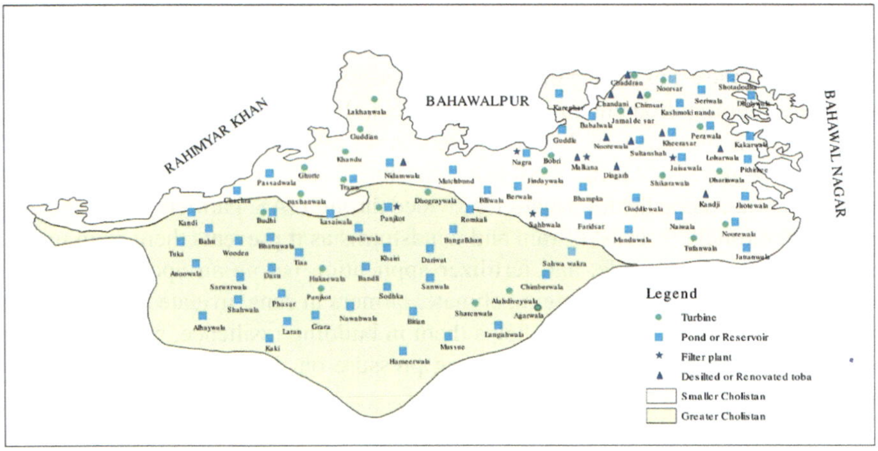

Fig. 4 Network of rainwater harvesting reservoirs in Cholistan desert. *Source* Ashraf and Hasan (2018)

immunity and malnutrition and number of casualties take place (Ashraf & Hasan, 2018). Therefore, there is a need to develop water resources (rainwater ponds) close to their hamlets in addition to rangelands for their livestock.

PCRWR established a network of rainwater harvesting system in the Cholistan desert (92 reservoirs with 15,000 m^3 capacity each) (Fig. 4). The same model is adopted and replicated by the Cholistan Development Authority and migration has stopped in the Cholistan desert. These reservoirs have not only changed the local ecology, but also act as a buffer against drought and climate change (Figs. 5 and 6).

3.3.2 Climate-Smart Land and Water Management Practices at Dingarh, Cholistan

In order to utilize collected rainwater, a solar-based pumping system (0.5 hp DC pump with 500 watts solar panel) can pump water from rainwater harvesting reservoir to elevated storage tanks (Fig. 7). This water may be linked with a low-pressure drip irrigation system. Presently, such system is being used to irrigate various plants such as *Salvadora persica* (pilu), *Cactaceae* (cactus), *Plantago ovata* (ispaghol), and *artiplex* (saltbush) (Fig. 8).

3.3.3 Saline Agriculture in Tharparkar

Water scarcity is the key problem in the Thar Desert. Rainfall is scanty and groundwater is saline. The masses depend on rainfed agriculture and livestock for their

Fig. 5 A herd of animal drinking water from rainwater harvesting reservoir. *Courtesy* Photographed by Muhammad Imran, PCRWR, Bahawalpur

Fig. 6 Improved ecosystem in Cholistan. *Courtesy* Photographed by Muhammad Imran, PCRWR, Bahawalpur

livelihoods. The consistent droughts in Thar have hit a severe blow to both agriculture and livestock. This has led to the seasonal migration of local communities along with livestock to the irrigated areas. The saline groundwater can be used for saline aquaculture and saline agriculture. Keeping in view above, saline-tolerant wheat, canola, and barley crops can be grown in different levels of saline water. A few glimpses of such interventions are given in Fig. 9.

Fig. 7 Solar-based water pumping from rainwater harvesting reservoir. *Courtesy* Photographed by FU Hasan (author)

Fig. 8 Growing *Cactaceae* (cactus) and *Plantago ovata* (ispaghol) in desert environment. *Courtesy* Photographed by FU Hasan (author)

Fig. 9 Cultivation of salt tolerant wheat and barley at Mithi, Tharparkar. *Courtesy* Photographed by FU Hasan (author)

Fig. 10 Pitcher irrigation on saline water at Mithi, Tharparkar. Courtesy: Photographed by FU Hasan (author)

3.3.4 Pitcher Irrigation System in Tharparkar

Sandy soils cannot retain soil moisture for a longer period. Moreover, saline water develops osmotic pressure in the root zone. Despite the availability of water, the plant's roots remain unable to extract the water from the root zone. However, pitcher irrigation system offers several advantages such as filtering the salts from the saline water; eliminating the chances of evaporation from the soil as water is supplied directly to the root zone; and no or negligible deep percolation losses. In this technology, a pitcher made of clay is laid below the ground surface. After placing the pitcher, it is covered with soil except the top hole with a cover. Water is stored inside the pitcher from where it is supplied to the plant's roots through pores. Presently, this technology is being used to grow Neem, *Spodilla* (Chiku), Lemon, *Mulberries* (toot), *Tamarind* (Imli), *Pongamia Pinata* (Sukh Chain) and Moringa (Fig. 10). The saline water (EC $= 9000$ μS/cm) is being used for filling the pitchers.

3.3.5 Use of Groundwater for Livestock and Domestic Purpose

Population of Thar is managing their livelihood primarily on livestock by raising goats, sheep, cattle, and camels. The high population growth has increased demand for water. After every five years, Thar experiences drought conditions. The community migrates from Thar to the lower parts of the Sindh. Keeping in view these issues, the groundwater abstracted through tubewell (having highly saline water having EC of 8000 uS/cm) and stored in small vertical tanks, locally called 'Tankas', is used for livestock and community on a daily basis (Fig. 11). Normally, the groundwater salinity ranges from 7500 to 35,000 uS/cm.

Fig. 11 Provision of water to livestock at Mithi, Tharparkar. *Courtesy* Photographed by B. Fatima (author)

3.4 Technologies for Uplands Balochistan

Balochistan is an arid mountainous region, and its diverse climatic zones have contributed to a rich animal and plant biodiversity while building a cultural heritage that allows survival in this rugged and harsh landscape. However, the province has suffered from the inefficient use of its resources—water, biodiversity, rangelands, forest, and soils. It is the most affected province by droughts, which resulted in human migration along with loss of livestock, orchards, and cash crops. The main reason is the shortage of water as the water resources are extremely limited as compared to other parts of the country. Surface water resources are mostly non-perennial. Groundwater is the only dependable source of water for domestic, agricultural, and industrial uses. However, this resource has been heavily over-exploited, resulting in a continuous decline in the water table, and drying of karezes and springs (Ashraf & Hasan, 2020).

Sustainable groundwater management requires a balanced approach of water abstraction and recharge. Since the groundwater abstraction is more than the recharge, therefore, efforts are needed to recharge the aquifer artificially and to control the groundwater abstraction by management techniques. The delay action dams were introduced to recharge the depleting aquifers. However, due to high silt load and impervious dam body, these dams were silted up within a short period of time. Subsequently, there is little or no recharge to the groundwater. Most of such dams work as evaporation ponds, and can not serve the purpose for which they were built (Ashraf & Ashfaq, 2017).

PCRWR introduced the concept of leaky dams. These structures are built in cascade approach wherein the purpose is to reduce velocity of water rather than storage of water. Unlike the delay action dams, these structures slowly release water. The flowing water has more surface area to recharge the aquifers. Inverted wells are also constructed downstream of these structures to enhance the recharge. The monitoring of groundwater in the study areas provided encouraging results in terms of more opportunity for rainwater to infiltrate, less silt deposition, and an increase in

Fig. 12 Leaky dam constructed in recharging zone of a Karez near Quetta. *Courtesy* Photographed by FU Hasan (author)

groundwater levels in the downstream areas (dug wells, karezes). A glimpse of such techniques is provided in Fig. 12.

3.5 Technologies for the Indus Delta

The Indus River System is the lifeline of Pakistan, originating from Gilgit-Baltistan and draining into the Arabian Sea. In its upper reaches, it is fed by snow and glacier melt. Despite this, the river system has brought huge sediments deposits along the coast and developed the Indus Delta. In this way, it connects the mountain and coastal ecosystems. Climate change, associated with recurrent floods, droughts, and high variability in monsoon rains, is affecting the overall water, food, and energy nexus in Pakistan. Such effects are more pronounced in the upstream mountainous and coastal areas, especially in the Indus Delta such as sea-level rise, soil erosion, seawater intrusion, and frequent cyclones. Moreover, sharp fluctuations in freshwater flows downstream Kotri followed by probable saltwater intrusion, altogether negatively impacting the Indus Delta. Consequently, a large productive land in the area has become unsuitable for agriculture, drinking water sources have become scarce or vanished, mangroves are depleting, and fish catch is declining.

Halophytes are the plant species known to have a high tolerance for soil salinity. They are often regarded as 'Salt Loving' plants. Halophytes are not just plants that may grow and survive in saline environment, rather they offer multiple benefits to local communities and may serve as an important livelihood opportunity. Halophytes may also be used as a source of food, forage, fuel, medicine, animal feed, ornamental plants, timber, chemicals, and other raw materials. More than 400 halophyte species exist in Pakistan (Khan & Qaiser, 2006). These species are acceptable to saline and

Ripened Saliconia Virignica bush *Atriplex (Saltbush)*

Fig. 13 Halophytes species in Pakistan

waterlogged environment having specialized and distinguished environmental characteristics (Fig. 13). The species range from grasses and shrubs to trees occurring in diversified environment, such as coastal swamps, marshes, saline and waterlogged areas. Some halophytes possess special characteristics and habitats. An adult tree is reported to utilize 235–2300 mm of water annually, and thus acts like a pump and controls water table naturally. Some herbs, shrubs, and trees extract salt from soil to reclaim the land and produce important bio-mass (Rashid & Chang, 2002).

Eucalyptus Canaldulensis (Suphaida) is known to be the most tolerant of all species toward waterlogging and salinity with high growth rate. Scientists suggest that it should not be planted on normal soil and fresh groundwater areas. This plant is valuable for its timber, whereas leaves and seeds have high medicinal value. Salix Babilonica (Baid) is also tolerant to waterlogging and salinity and serves as an expensive raw material in furniture industry. *Atriplex* (Saltbush) is very tolerant to salinity and drought and easily propagated. Its leaves are palatable for animals and humans. *Salicornia Virginca* is another bush species having enormous potential in the Indus Delta. This plant has multiple benefits. It is highly tolerant to waterlogged and saline soil. Its pods can be used as a food source and fiber for grazing animals. This plant is highly effective in Phytoremediation and Phyto-desalination. Its seeds have high medicinal value and can also be used for biofuel. Glasswort is another species of Salicornia, its residue is used for making soda ash for the glass and soap industry. *Salvadora persica* (chota Pilu) or toothbrush tree-shrub has high medicinal value and is tolerant to waterlogging and salinity. Its roots are used for the oral/dental hygiene instrument locally called *Miswak*. This plant is native to local environment of the Indus Coastal Delta.

4 The Way Forward

There is awareness and realization at the high political level regarding climate change and its impacts. The local population is also dealing and coping with the adverse climate impacts. Moreover, some policies and plans are under implementation at national and sub-national levels. Appropriate knowledge of climate-smart practices is also available and piloted. However, the important question is how to upscale and outscale the promising climate-smart interventions. Exotic solutions may not work sustainably; only nature-based solutions are sustainable solutions to cope with climate change impacts. Capacity building efforts need to be doubled at all levels for organizations and communities.

References

Abid, M., Scheffran, J., Schneider, U. A., & Ashfaq, M. (2015). Farmers' perceptions of and adaptation strategies to climate change and their determinants: The case of Punjab province, Pakistan. *Earth System Dynamics, 6*, 225–243. https://doi.org/10.5194/esd-6-225-2015

Ashraf, M., & Ashfaq, A. A. (2017). Sustainable groundwater management in Balochistan. Pakistan Council of Research in Water Resources (PCRWR) (pp. 34).

Ashraf, M., & Hasan, F. U. (2018). Pakistan's Deserts: Land and Water Resources Development. Monograph, Higher Education Commission of Pakistan, Islamabad (pp. 38).

Ashraf, M., & Hasan, F. U. (2020). Groundwater Management in Balochistan, Pakistan: A Case Study of Karez Rehabilitation (Managing Groundwater for Drought Resilience in South Asia. SAWI MGW4DR Case Study 07. International Water Management Institute (IWMI) (pp. 15), World Bank, Washington DC.

Ashraf, M., & Saeed, M. M. (2006). Effect of improved cultural practices on crop yield and soil salinity under relatively saline groundwater applications. *Irrigation and Drainage Systems., 20*, 111–124.

Esham, M., & Garforth, C. (2012). Agricultural adaptation to climate change: Insights from a farming community in Sri Lanka. *Mitigation and Adaptation Strategies for Global Change* 1–15.https://doi.org/10.1007/s11027-012-9374-6

Hasan, F. U., Ashraf, M., & Farooque, M. (2015). Guidelines and Training Module for the Development and Implementation of Rainwater Harvesting Systems, PCRWR, 2015 (pp. 32).

Hasan, F. U., Bareerah, F., & Ashraf, M. (2021). Challenges to irrigated crop zones of Punjab and Sindh province in the wake of climate variability. *International Journal of Environmental Sciences & Natural Resources*; *27*(2), 556207 (2021). https://doi.org/10.19080/IJESNR.2021.27.556207

Iqbal, N., Ashraf, M., Imran, M., Salam, H. A., Hasan, F. U., & Khan, A. D. (2020). Groundwater investigations and mapping in the Lower Indus Plain. Pakistan Council of Research in Water Resources (PCRWR), Islamabad (pp. 70).

Jackson, P. (2007). From Stockholm to Kyoto: A brief history of climate change. UN Chronicle 44.

Jan, A., Khan, T. A., & Mahsud, M. I. (2020). The climate change awareness and literacy in Pakistan: Role of media and social actors. *Liberal Arts and Social Sciences International Journal*, 4, 256–266. https://doi.org/10.47264/idea.lassij/4.2.20

Khan, M. A., & Qaiser, M. (2006). Halophytes of Pakistan: Distribution, ecology, and economic importance. In M. A. Khan, H. Barth. G. C. Kust, & B. Boer (Eds.), Sabkha ecosystems: Volume II: The south and central Asian countries (pp. 129–154). Kluwer Academic Press, Netherlands.

Khuda, B., & Kamran, M. (2019). Adaptation to climate change in rain-fed farming system in Punjab, Pakistan. *International Journal of the Commons, 13*, 833–847. https://doi.org/10.5334/ijc.887

Mishra, S. (2017). Impact of climate variability on soil-microbe interaction. In: S. Mishra, C. Srinivas, S. Singh, & D. Savant (Eds.), *Current perspectives in sustainable environment management* (pp.135–145, 1st ed.). SIES Indian Institute of Environment Management.

Pakistan Meteorological Department (PMD). (2020). State of Pakistan's Climate in 2020, Pakistan Meteorological Department http://www.pmd.gov.pk/cdpc/Pakistan_Climate_2020.pdf

Pakistan Council of Research in Water Resources (PCRWR). (2018). Agricultural water, energy and hazard management in the upper Indus Basin for improved livelihood and building resilience. Final Technical report (unpublished).

Rashid, A., & Chang, M. H. (2002, July–December). Salinity control and environmental protection through halophytes. *Journal of Drainage and Water Management*, *6*(2).

Siddiqui, N. U., Li, X., Luo, H., Karaiskakis, A., Hou, H., Kislinger, T., Westwood, J. T., Morris, Q., & Lipshitz, H. D. (2012). Genome-wide analysis of the maternal-to-zygotic transition in Drosophila primordial germ cells. *Genome Biology, 13*(2), R11. https://doi.org/10.1186/gb-2012-13-2-r11

Shahid, Z., & Piracha, A. (2010). Climate change impacts in Pakistan: Awareness and adaptation. *International Journal of Climate Change: Impacts & Responses*, *2*(1), 119–130.

United Nations (UN). (2018). Remarks at high-level event on climate change. Retrieved from https://www.un.org/sg/en/content/sg/speeches/2018-09-26/remarks-high-level-event-climate-change

Applications of Smart Agriculture in Irrigation Water Management in Developing Countries: The Way Forward

Narmilan Amarasingam, Muneeb Muhamed Musthafa, Mohamed Mujithaba Mohamed Najim, and Mirza Barjees Baig

Abstract Agriculture is the backbone of the world's population, being the primary source of food production. Climate change affects agriculture in various ways, including changes in typical temperatures, precipitation, and atmospheric conditions. Advanced automation of agronomic activities for land preparation, irrigation water management, pesticide application, fertilizer application and harvesting has substantially improved agricultural productivity. In many agricultural regions, determining the exact amount of water to provide for many crops is an unsolved challenge due to the impact of climate change. One of the best solutions for overcoming this difficulty could be to use precision farming technologies to increase agricultural productivity, reduce pollution, and lower production costs while overcoming climate change issues. Smart agriculture plays an important role in agriculture to maintain the sustainability of agricultural systems and improve the country's economic growth. It means that smart farming supports agronomic practices such as estimating crop area, monitoring crop growth, estimating soil moisture, evaluating soil fertility, detecting crop stress, detecting diseases and pests, yield estimation, and weather forecasting. The smart irrigation system is being used in agriculture to address water scarcity due to climate change. This review covers the importance of smart farming applications in irrigation water management to increase farm productivity by overcoming climate

N. Amarasingam · M. M. Musthafa (✉)
Department of Biosystems Technology, Faculty of Technology, South Eastern University of Sri Lanka, Oluvil, Sri Lanka
e-mail: muneeb@seu.ac.lk

N. Amarasingam
e-mail: narmilan@seu.ac.lk

M. M. M. Najim
Faculty of Agriculture, Sultan Sharif Ali Islamic University, Sinaut Campus, Km 33, Jln Tutong, Kampong Sinaut, Tutong TB1741, Brunei
e-mail: najim.mujithaba@unissa.edu.bn

M. B. Baig
Prince Sultan Institute for Environmental, Water & Desert Research, King Saud University, Riyadh, Kingdom of Saudi Arabia
e-mail: mbbaig@ksu.edu.sa

M. Behnassi et al. (eds.), *Climate-Smart and Resilient Food Systems and Security*,
https://doi.org/10.1007/978-3-031-65968-3_18

change, using less human energy, increasing efficiency, and improving management techniques through precision farming technologies.

Keywords Climate change · Crop water requirement · Internet of Things · Precision farming · Remote sensing · Smart irrigation

1 Introduction

Climate change is already influencing agriculture, with effects distributed unevenly over the globe (Barapatre & Patel, 2019). The necessity for good and sustainable irrigation technologies is in high demand due to extreme climatic changes and a lack of water. At such climatic conditions, the crops must be irrigated according to their water requirements (Ranjith et al., 2017). The agriculture sector has been transformed to greater heights by new technological developments such as smart farming, precision agriculture (PA), and Climate-Smart Agriculture (CSA), which will enable this primary sector to move to a higher level of farm productivity and profitability while maintaining sustainability (Saiz-Rubio & Rovira-Más, 2020). The latest advances have been applied to boost farm productivity, including the Internet of Things (IoT), artificial intelligence, wireless connectivity, and remote sensing technology. The primary goal of CSA is to strengthen agricultural systems' ability to support food security strategies. CSA can be used to manage agricultural field factors such as meteorological variables (air temperature, relative humidity, wind, light intensity, and so on), soil quality, agricultural system biodiversity, and post-harvest procedures, among others (Fig. 1). Furthermore, CSA involves incorporating information and communication technology into agricultural machines, sensors, actuators, and network equipment (Pivoto et al., 2018). CSA is linked to several technologies, including sensors, robots, IoT, mapping, decision-making, and statistical processes, which are heavily involved in improving agricultural systems (Pivoto et al., 2018).

Embedded sensors, processes, actuators, cloud computing, transceivers, security protocols and architectures, web services, the Internet, and search engines are all used to support IoT. Furthermore, sensors and actuators provide precise values for crop status, atmospheric fluctuations, soil nutrient levels, and irrigation control (Villa-Henriksen et al., 2020). Regarding production and long-term sustainability, irrigation water management is critical (Chartzoulakis & Bertaki, 2015), where IoT has resulted in a higher crop output at a lower cost (Saiz-Rubio and Rovira-Más 2020). Higher agricultural productivity has been demonstrated using a smart irrigation system combined with remote sensing techniques and cloud technologies (García et al., 2020).

This study evaluates the literature on the applications and usefulness of remote sensing for irrigation water management and crop water requirements. The chapter is systematized into four main sections: (1) challenges in irrigation water management; (2) smart farming applications in precision farming; (3) precision agricultural system in irrigation and water management; and (4) challenges and possible solutions of

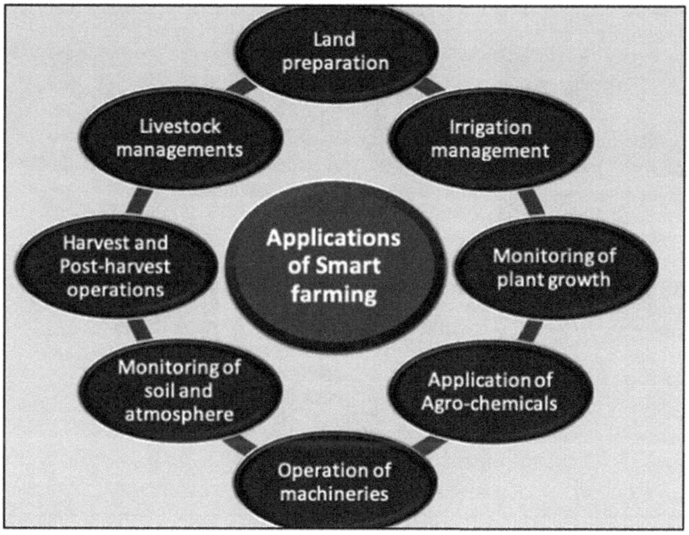

Fig. 1 Application of Precision Technologies in agriculture

smart application in irrigation water management, which summarize the approaches and the outcomes of existing remote sensing studies, creating outcomes considering the state-of-the-art technology.

Source Developed by the authors.

1.1 Internet of Things (IoT) in Precision Agriculture

IoT is a promising technology for driving the agricultural industry, which is the backbone for sustainable development, particularly in developing countries, where rapid population growth, stressed natural resources, reduced agricultural productivity due to climate change, and massive food waste are all issues (Kuaban et al., 2019). The Internet of Things (IoT) has transformed everything into smart and intelligent, revolutionizing every area of human life. The IoT will be the axis for the next generation of smart computing, with 500 billion devices expected to be connected by 2030 (Zikria et al., 2021). Almost all services and products used by the public on a daily basis, including agriculture, have been redefined by the IoTs (Zikria et al., 2021). In agriculture, IoT applications aim to empower farmers by providing decision tools and automation technologies that connect the knowledge, products, and services for enhanced productivity, quality improvement, and profit while retaining sustainability (Elijah et al., 2018) (Fig. 2). IoT devices can capture huge amounts of data on the environment, soil, and crop performance, providing time series data that can be used to anticipate and make recommendations, as well as transmit vital information to farmers in real-time (Adamides et al., 2020). These technologies provide

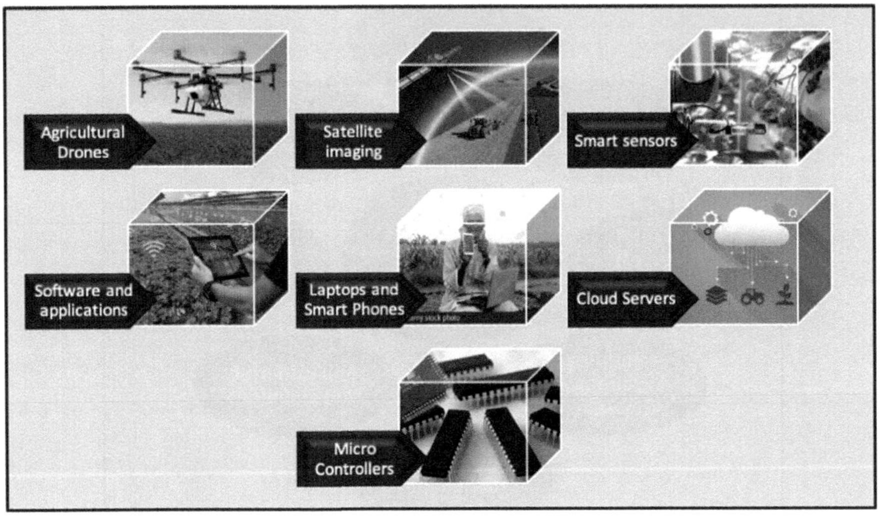

Fig. 2 Components of the IoT-based smart farming. *Source* Developed by the authors

the collection of spatial data and the monitoring of real-time occurrences, and they are the driving force behind the agricultural sector's current viability (Saiz-Rubio & Rovira-Más, 2020; Villa-Henriksen et al., 2020; Walter et al., 2017). Agriculture has faced a number of issues in recent years, including global warming, water scarcity, and crop pests and diseases, etc. (Raza et al., 2019; Skendžić et al., 2021; Turral et al., 2011).

Precision agricultural technology, which can closely monitor environmental elements such as temperature and relative humidity of the atmosphere to improve productivity while minimizing cost and pollution, can be utilized to tackle the issues that farm industries face today. PA carries the prospect of applying the correct cultivation practice at the right point with high precision, which is correctly stated by Zhang et al. (2002) as what is 'right'. The number of features of PA has been established with fitting technologies to determine the environmental effects and economic outcomes. The scientific community has extensively studied the environmental and economic benefits of IoT-based precision technology (Akhtar et al., 2021; Saiz-Rubio & Rovira-Más, 2020). PA is considered one of the most effective methods for improving plantation management practices and gathering accurate data for decision-making (Sofonia et al., 2019). PA's major goal is to apply the right input quantity at the right time and place to create greater results (Singhal et al., 2018). Data collection and mapping of agricultural land variability, data analysis, farming management decisions based on the study's findings, and controlled application of pesticides and fertilizers are all part of precision farming (Singhal et al., 2018).

The PA strives to increase productivity with low inputs and minimal environmental damage (Mondal & Basu, 2009). Furthermore, PA will not only assist in lowering the cost of agriculture but will also provide significant environmental benefits. PA ensures that the precise number of agrochemicals are applied at the right time, lowering costs and reducing the water required by the crops. PA is a cutting-edge crop management system that integrates modern agricultural fields and industry technologies to provide precise amounts and types of inputs to varied farm areas. In this sense, drones have lately been adopted in agriculture for large-area surveillance and smart, targeted irrigation and fertilization (Gago et al., 2015). The ability to detect places where heavy irrigation is required or where a leaf disease is spreading using a drone and an infrared camera can assist agronomists in saving time, water, and agrochemicals. At the same time, innovative agricultural practices have the potential to boost crop output and quality.

1.2 Remote Sensing in Precision Agriculture

Recent advancements in remote sensing (RS) application for precision agricultural practices have been critical in enhancing crop health and management. Scientific advances in RS, IoT, AI, machine learning, and robotics have benefited the development of agricultural technology, which must be hastened to keep up with global population increase and environmental deterioration. Unmanned Aerial Vehicles (UAVs) are mostly used for remote sensing purposes and can capture images and videos for post-processing (Eisenbeiss, 2004; Merlino et al., 2020). The UAV-derived images are used for planning, navigation, and georeferencing in a range of applications, including agriculture, forestry, environmental management, landscape management, weather forecasting, biodiversity management, cultural heritage conservation, and urban area monitoring (García et al., 2020; Reza Ghafarian Malamiri et al., 2021). The rapid development of UAVs and Wireless Sensor Networks (WSNs) can enable useful and cost-effective applications such as monitoring crop status and soil moisture status for irrigation practices and applying agrochemicals by agricultural UAVs (Radoglou-Grammatikis et al., 2020).

1.3 ICT Tools for Climate Change and Risk Management

There are several techniques at the intersection of climate change and agriculture because of the large number of crops and the difficulty of duplicating the same conditions across diverse places (Narmilan, 2017). Every tool provides for the analysis of many agricultural processes, ranging from local crop modelling under climate change conditions to the management of economic implications of climate change on agriculture (soil value fluctuations, demand and supply, production, and so on), among other things. Given a large number of tools available, it is more fascinating to concentrate on their common characteristics rather than their unique features (Singh et al., 2015). Before we can use a tool to prevent climate change, we must first understand how much they contribute to it. Otherwise, we will end ourselves circling a desert. According to the European Commission (2006), ICTs are far from blameless in contributing to climate change, with the expansion of user devices, all of which require power and emit heat, accounting for the majority of ICTs' impact on climate change. When it comes to mitigating climate change, information and communication technologies play a critical role, and these technologies can be used in both developed and developing countries, even though developing countries lack the necessary ICT infrastructure.

Barapatre and Patel (2019) stated that the impact of climate change on irrigation water could be overcome by using IoT technology to make decisions and meet the crops' water needs. Therefore, a smart irrigation system is one of the approaches to overcome this impact (Sudharshan et al., 2019). Smart management systems can perform water quality and quantity control, effective drainage, demand and flow management, habitat conservation, flood and drought management, stormwater and sanitation management, and future planning (Tadokoro et al., 2011). Water supply, distribution, and management of difficulties are becoming more prominent in industrialized and developing countries due to climate change (Tadokoro et al., 2011). Smart irrigation systems have a variety of advantages over conventional irrigation systems. Smart irrigation systems considering factors such as soil moisture and weather forecasts can help optimize water levels (Tadokoro et al., 2011). Smart water management is designed to reduce the amount of human labor required and boost the productivity of irrigation systems (Tadokoro et al., 2011). Moreover, smart irrigation has several advantages, including reduced water use, time savings, water conservation, cost savings, fewer weeds, disease control, and soil health, according to Kamil and Ojoawo (2019). Furthermore, because global warming is a severe concern that poses a global threat to the environment, low CO_2 emission procedures should be encouraged. The agricultural sector must seek more sustainable and economical options to overcome these difficulties.

1.4 Importance of Advanced Techniques in Irrigation Water Management

Digital, real-time ICT technologies such as sensors, screens, GIS, and satellite imagery, as well as other data exchange platforms, are used in smart water management (Tadokoro et al., 2011). Smart water management will address existing water challenges in both developed and developing countries by providing comprehensive water management approaches at all levels and in various scenarios (Tadokoro et al., 2011). Using crop evapotranspiration data and moisture sensors to properly predict when and how much to water has been termed as modern irrigation scheduling. Water management strategies are used in irrigation planning to avoid excess application of water and to limit performance loss due to water constraints or drought. As the major goal of irrigation is to keep plants hydrated, it seems rational that irrigation should be scheduled based on particular plant water status indicators (Ertek et al., 2006).

Quantitatively, irrigation planning methods are based on three approaches, namely, crop monitoring, soil monitoring, and the water balance technique (George et al., 2000). In agriculture, smart water management for PA is critical for boosting crop output and lowering expenses while also contributing to environmental sustainability (Kamienski et al., 2019). The smart irrigation system monitors the weather conditions, soil type and moisture level, evaporation, and water usage of the crops, which automatically modifies the watering schedule. It aids farmers in maximizing water utilization, and improving crop growth quality and yield in their fields (Anitha et al., 2019). Therefore, it is necessary to use smart tools to determine the accurate amount of water needed for crops.

1.5 Implications and Potential for Irrigated Agriculture

Irrigated agriculture is the main user of freshwater resources (Vuolo et al., 2015). The agricultural sector is one of the world's largest users of freshwater, accounting for more than 70% of all freshwaters diverted from rivers and groundwater for agricultural purposes (Barker & Molle, 2004) and irrigation represents 40% of world agricultural production and 60% of cereal products (Ahmad et al., 2009). In truth, irrigation has aided global and regional food security; yet, with the rapid growth of the world's population, a corresponding rise in agricultural productivity would be required to meet the growing demand for food. According to recent studies, many significant aquifers in agricultural areas are under stress due to intensive exploitation paired with a reduced recharge rate (Richey et al., 2015). In addition, our understanding of the overall amount of water stored and the systems' resistance is frequently poor (Richey et al., 2015). As a result, proper irrigation management is critical, especially in the face of growing issues such as increased groundwater consumption and deteriorated water quality. This should be accomplished either by placing more land under irrigation or by raising the yields of current farmed areas while utilizing similar or even lower water resources, with irrigation area growing by around 6% by 2050 and irrigation water withdrawals increasing by about 10% by 2050 (Ahmad et al., 2009).

2 Challenges in Irrigation Water Management

Water scarcity has a strong pursuit in agriculture and non-agriculture sectors, where irrigated agriculture requires precise mechanisms to counter this issue (Mojid & Mainuddin, 2021). Agriculture production needs to be increased to feed hunger, necessitating an increase in the irrigated area regardless of available water resources. As a result, irrigation methods necessitate the efficient and effective use of water using various water conservation techniques (Mojid & Mainuddin, 2021). Many technologies are utilized in agriculture, even though smart irrigation systems are the greatest way to address climate change challenges in the agricultural sector. As a result, irrigation management will benefit from IoT, remote sensing, and artificial intelligence.

2.1 Global Climate Change Impact on Agricultural Production and Irrigation Water Management

Climate change will likely worsen the frequency and length of floods and draughts, which must be predicted and managed (Tadokoro et al., 2011). Water safety and security limit the habitability of many localities in developing countries, while leak detection, sewage, and stormwater control impact metropolitan areas in developed regions (Tadokoro et al., 2011). The depletion of the water table, combined with a decrease in rainfall due to climate change, has prompted communities to seek new ways to conserve water (Kalezhi et al., 2019). Crop cultivation is extremely sensitive to both short-term and long-term climatic change. Temperature increases shorten the life of most cereal crops by hastening their phenological development (Wheeler et al., 2000).

The principal crops of Sri Lankan agriculture, such as rice, are influenced by rising CO_2 concentrations in the atmosphere, rising temperatures and precipitation patterns, and evapotranspiration regimes (Wijeratne et al., 2007). Short-term high-temperature variability could have a major negative impact on flowering and crop yield formation. Hence the consequences of short-term climate fluctuations and severe weather events should be carefully evaluated (Prasad et al., 2006). There are inconsistencies in the reports on whether greater CO_2 emissions will offset rising temperatures' negative effects on crop harvest (Leakey et al., 2009). Water management is the most significant difficulty facing irrigation, particularly in locations where the consequences of climate change are visible. Climate change is currently one of the most pressing environmental issues. Continuous or unexpected flooding and drought and storm damage are some of the key threats. They are likely to become more frequent and intense. Climate variability and change are anticipated to have a direct impact on land and water management, which are the key drivers of change in the agricultural industry. In sensitive locations, the agricultural output can be lost due to poor agricultural yields (Rosenzweig et al., 2002).

2.2 Poor Technological Adaptation in Developing Countries Compared with Developed Countries

Precision farming, such as smart water and irrigation technologies, has amplified the interest due to its potential to increase agricultural efficiency. However, many farmers in poor countries are hesitant to accept the technique. In underdeveloped countries, technological literacy, the amount of data to analyze, and the expense of establishing and maintaining the application are all issues. As a result, it is critical to creating user interfaces for technical gadgets that are easy, cost-effective, and efficient. As a result, there has been a lot of research into smart irrigation and water distribution (Byishimo & Garba, 2016). In fact, in underdeveloped countries, where 75% of the world's irrigated land is concentrated, high yields are unusual. It is not uncommon to find water productivity that is less than half of its maximal potential value. The main cause of these inappropriate results is frequently the lack of capacity among farmers and irrigation authorities in charge of water distribution. Because of the unfavorable institutional environment, improving the issue is typically difficult. It is worth noting that traditional irrigation techniques are more efficient than some modern approaches, because local communities adapt to the former much better.

2.3 Other Minor Challenges in Irrigation Water Management

The lack of sufficient water supply is one of the major issues in meeting domestic requirements, economic growth, and environmental needs in many regions due to several issues and challenges (Cosgrove & Loucks, 2015). The current performance of irrigation water management systems is environmentally unsustainable according to external variables (Nazari et al., 2018). Water levels in many lakes have changed as a result of changing climatic circumstances (Ayenew, 2007). However, contemporary human pressure has resulted in conflicting lake level shifts, such as expansion and decline. Due to the presence of huge amounts of soluble salts in irrigation water, insufficient precipitation, and excessive evaporation, soil quality has been considerably degraded, affecting crop output sustainability (Nazari et al., 2018). Another issue is salinization, which has been exacerbated and accelerated by high temperatures that increase evaporation (Ayenew, 2007). Furthermore, higher levels of fertilizers and salinity have degraded the quality of the surface and groundwater resources (Nazari et al., 2018). Eutrophication is the outcome of increased nutrient loading. This event, in turn, can result in fish fatalities due to low oxygen levels, algal blooms, and other interconnected repercussions (Vijverberg et al., 2012). As a result, farmers are unable to use saline water for irrigation. Other anthropogenic factors on lake catchments also directly and indirectly impact lakes.

Forest clearing, animal grazing lands, and another vegetation decrease in watershed areas have increased dramatically in recent years. Sediment and nutrient loads in water increased as a result of such dynamics. Forest clearing is currently a severe concern due to the current amount of deforestation, which is caused by the use of wood for fuel and construction, as well as the population-driven demand to develop cultivable lands (Ayenew, 2007). Most of the conditions that increase soil erosion are relatively widespread, and the repercussions of the connected processes of plant cover reduction, erosion, and sedimentation have been experienced by some of the larger lakes. Floods, glacial lake outburst floods (GLOFs), erosion, and sedimentation have all impacted irrigation systems and have diminished their capacities (Dhakal et al.,

2021). When erosion occurs, torrents of soil particles, which are usually high in nutrients, may end up in lakes (Ayenew, 2007). Understanding variations in evapotranspiration is critical for irrigation engineers responsible for water resource planning and management where crop water demand often exceeds precipitation, necessitating irrigation from surface and groundwater resources to make the difference (Gowda et al., 2008). The water utilized for irrigation is lost to runoff, deep percolation, and evaporation from crop or soil surfaces (Fereres et al., 2003). Therefore, smart water conservation aims to improve irrigation efficiency by switching from conventional irrigation methods.

3 Smart Irrigation Water Management

Smart irrigation systems with IoT nodes dispersed across broad areas are one of the solutions for climate change in the agricultural sector (Froiz-Míguez et al., 2020). Precision watering, one of the other significant advantages of a smart irrigation system, also deals with efficiency in water distribution. Smart irrigation systems have been proposed as a viable solution to irrigation-related problems worldwide (García et al., 2020). Fourati et al. (2015) developed an Internet-based decision support system for irrigation scheduling in olive fields, which used a wireless sensor network to assess environmental data such as relative humidity, sunlight radiation, ambient temperature, and precipitation.

Another review by Hashim et al. (2015) suggested using an Arduino device to monitor and control temperature and soil moisture using an Android-based smartphone application, which has various advantages, including low cost, user-friendliness, and a wide range of functions. Goap et al. (2018) created open-source technology based on a smart system to estimate the crop water requirements of an agricultural area. The development of new sensing technologies could be implemented in real-world irrigation application tools to improve irrigation scheduling and maximize irrigation water's value (El-Naggar, 2020). The fundamental soil and crop characteristics employed by the soil water balance, such as soil water content, soil water status, crop height, and crop evapotranspiration, are the focus of these sensing technologies (Vanino et al., 2018).

There are several studies on smart irrigation system conducted all over the world (Table 1). A low-cost smart irrigation system for paddy fields was presented by Bamurigire et al. (2020), which would automatically control irrigation based on seasonal and daily irrigation needs. In Taiwan, Li et al. (2020) proposed another promising water-saving technology based on Smart Field Cultivation Server (SFCS) for paddy field irrigation using information and communication technologies. Barkunan et al. (2019) suggested a drip irrigation automation system in which the smartphone collects a soil image, assesses its wetness level, and occasionally transmits the data to the microcontroller through the GSM module. The microcontroller controls the irrigation and sends the field's status to the farmer's phone. The technology is tested on a paddy field for three months. Compared to traditional flood and drip irrigation methods, it saves roughly 41.5% and 13% of water, respectively, according to the experimental setting. The three-step process of applying remote sensing for irrigation water management is described in Fig. 3 by Papadavid et al. (2011). To optimize water use for agricultural purposes, Nisha and Megala (2015) suggested an automated irrigation system based on wireless sensors. A network of distributed wireless soil temperature and humidity sensors mounted in the crop field to make up the system. So, utilizing the above-mentioned revolutionary concept, such as a smart irrigation system, farmers may irrigate the exact amount of water in the field to control water pollution owing to surplus water, which leads to contaminated surface water bodies and groundwater.

Adamides et al. (2020) investigated the potential benefits of implementing smart farming options in the developing world to reduce the impacts of climate change. This chapter further elucidates the opportunities for innovation targeting and climate adaptation that could help farmers in developing countries reduce their environmental impact. Karar et al. (2021) demonstrated a new mobile-based application using the IoT, embedded systems, and a UAV in agriculture. Arduino microcontroller boards, Wi-Fi modules, water pumps, and electronic environmental sensors, such as temperature, humidity, and soil moisture, were used to create and implement the suggested agricultural monitoring system. Sciences et al. (2013) developed an intelligent irrigation under sprinkler irrigation system to irrigate wheat crops. The findings of this system were then compared to those of an irrigation control system (ICS) based on an automatic weather station. Overall, this technology is suggested for effective automated irrigation systems, and the IIS technique could be a useful tool for water conservation planning and irrigation scheduling for wheat and other crops. Corbari et al. (2021) demonstrated the regional application and need to improve actual irrigation practices across different Italian irrigation consortia for evaluating the effects of optimized irrigation management on the minimization of percolation losses and maximization of crop yield for different soil types within the same region.

4 Challenges and Possible Solutions of Smart Farming Application in Irrigation Water Management by Farmers

4.1 Challenges

Despite the growing importance of modern and commercial farms, in most developing countries, agricultural production (particularly food crop production) is still dominated by small, often poorly trained farmers who employ a wide range of agricultural approaches (mixed cultivation, sowing, and continuous harvesting in small irregularly shaped plots, etc.). When farmers do not record or utilize conventional units of measurement, it is difficult to measure some of the most basic variables of

Table 1 Different types of smart irrigation system

Name of system	Location	Objective	Protocols/Design	Features	Tools/Components	Source
An Architectural Framework for IoT Driven Agriculture	Africa	To propose a three-layer architecture for IoT-driven agriculture for smart irrigation	The first layer: IoT devices and smart irrigation The second layer: gateways and local processing servers The third layer: Cloud layer	Energy efficiency Internet connectivity Security Cost savings	The low-power LoRaWAN network connects the IoT devices to the gateways	(Kuaban et al., 2019)
Smart energy for smart irrigation	United States & Greece	To examine the feasibility of using smart energy systems to reduce irrigation costs	The pump system is powered by a solar panel, which pumps and stores the water in the reservoir	Energy efficiency Cost reduction Renewable energy sources are being used	Energy Producing and Consuming devices Auxiliary Modules Power Distribution System	(Nasiakou et al., 2016)
Smart irrigation in greenhouse	Spain	To determine how to adjust irrigation frequency in greenhouses to maximize water use efficiency	The smart system is activated or deactivated by measuring solar radiation, irrigation volume, and the EC and volume of the DS	The quality of the fruit, specifically the TSS, was improved when the temperature was raised or when low-frequency watering was used	'HortiControl Expert' software was used Air T° and RH sensor, Solar radiation sensor, EC sensor, Drainage volume sensor and PC were used to monitor and control the system	(Rodriguez-Ortega et al., 2017)

(continued)

Table 1 (continued)

Name of system	Location	Objective	Protocols/Design	Features	Tools/Components	Source
Smart Photovoltaic Irrigation Manager (SPIM)	Spain	To provide daily smart irrigation plans for crops at plot scale using solar energy pump	The strategy is based on using solar energy to feed irrigation water directly to the network, without the need for intermediate storage devices (water tanks and batteries)	Reduction of cost Supply energy in isolated areas A low GHG emissions Increasing the sustainability and profitability	MATLAB software was used to develop the model A 13-kW submersible pump, drip irrigation network, photovoltaic system, and solar irradiance sensor were used	(García et al., 2018)
Renewable Energy Based Smart Irrigation System	India	To compare the growth of plant with automated irrigation and normal irrigation	The solar panels are wired to an Arduino, which provides power to the sensors and solenoid valve	Increase the efficiency Eco-friendly Cost-effective Easy installation More versatility More extendibility Allows the user to add more sensors with minor programming effort	Temperature, humidity, and soil moisture sensors are used in the system, and the solenoid valve is controlled by fuzzy logic The Solar Panel, Servo Motor, Two Light Sensors, and Arduino comprise this system	(Sudharshan et al., 2019)

(continued)

Table 1 (continued)

Name of system	Location	Objective	Protocols/Design	Features	Tools/Components	Source
Smart Irrigation System using IoT and Fuzzy Logic Controller	UAE	To propose an IoT based irrigation system that uses fuzzy logic to reduce irrigation frequency while enhancing production rate	The system comprises a fuzzy controller that uses specialized sensors to collect environmental identifiers to control water flow from the water pump and provide irrigation at the proper time and frequency	Better water conservation Maintaining the soil moisture above a pre-set value Easy to implement Economically justifiable	The fuzzy controller is designed and implemented using MATLAB. Light intensity, LM 393 soil moisture, DHT11, and flow meter sensors are used	(Alomar & Alazzam, 2019)
A DC Microgrid Smart-Irrigation System using IoT	Africa	To propose a sustainable and smart DC microgrid irrigation system using multi-agent systems along with IoT technologies	The sensors capture the monitored data and send it to a wireless network and an IoT gateway. The sensor data is analyzed and compared to recommended values at the IoT gateway which communicates with the actuator node to manage irrigation actuators based on the comparison	This can be utilized in areas when water is scarce Rural farmers living in off-grid areas will benefit from the system	Temperature sensor, Humidity sensor, Rain gauge sensor, and Leaf wetness sensor were used PV Solar panel, Storage Battery, Water Tank, Wireless Communication and Actuator nodes were used in this system	(Kalezhi et al., 2019)

(continued)

Table 1 (continued)

Name of system	Location	Objective	Protocols/Design	Features	Tools/Components	Source
Smart & Green: Smart Irrigation	Brazil	To propose the Smart & Green framework to provide smart irrigation services such as data monitoring, preprocessing, fusion, synchronization, storage, and irrigation control enhanced by soil moisture prediction	The application layer includes features for automating irrigation control and predicting soil moisture The service layer saves data provided by Smart & Green users and synchronizes crop, weather station, and soil data when any irrigation management module requires it. The communication layer is a piece of software that uses network protocols to collect meteorological and soil data	Water saving technology Soil moisture prediction using machine learning	Python 3 was used to implement the Smart & Green framework These methods determine how supervised machine learning techniques are used to create prediction models	(Campos et al., 2020)
Smart Water Management Platform	Italy, Spain, and Brazil	To develop an IoT-based smart water management platform for precision irrigation in agriculture with a hands-on approach based on four pilots in Brazil and Europe	Layer 1: Device & Communication Layer 2: Data Acquisition, Security & Management: Layer 3: Data Management Layer 4: Water Irrigation & Distribution Models Layer 5: Water Application Services	Adaptability Flexible enough to Optimization of water distribution	Sensor Simulating Environment (SenSE), Mosquitto MQTT Broker: Eclipse Mosquitto is an open source MQTT message broker, MongoDB	(Kamienski et al., 2019)

(continued)

Table 1 (continued)

Name of system	Location	Objective	Protocols/Design	Features	Tools/Components	Source
Cloud Based Automated Irrigation	India	To propose a smart irrigation system that can automatically control irrigation using an Android mobile application	The embedded system uses data from a soil moisture sensor and a temperature sensor to autonomously irrigate the plants. The motor pump, which is used to irrigate the plants using a sprinkler, is controlled by a relay switch. The user can operate the embedded system in automatic or manual mode using the mobile application, which is connected to the embedded system through the cloud	Simple and cost effective User can also interact and view the current, readings by android application	An embedded system (micro controller (ATmegas16), a soil moisture sensor, a temperature sensor, a relay switch, a comparator, and a Wi-Fi module) An android application Cloud services	(Ranjith et al., 2017)

(continued)

Table 1 (continued)

Name of system	Location	Objective	Protocols/Design	Features	Tools/Components	Source
IoT based smart plant irrigation system with enhanced learning	Turkey	To propose a smart irrigation model based IoT system that gradually learns the watering nature of a plant without any pre-prepared data initially given to it	The first component is the sensors and actuators section, which collects sensor data from the outside world. The second component uses a cloud server to run the proposed irrigation decision algorithm and returns a decision result. The third component is the end-user component, which allows a mobile device to transmit manual irrigation requests and see monitored irrigation statistics	The results show that the suggested system can successfully learn the watering habits of various plants	Four different supervised machine learning algorithms and adapted the Gradient Boosting Regression Trees (GBRT) method in our IoT solution	(Cagri Serdaroglu et al., 2020)
IoT Based Water Irrigation System	Saudi Arabia & Pakistan	To present an architecture for an IoT-based water irrigation system that utilizes a wireless sensor network to address farmer issues	To provide information on water entry and outflow, an ultrasonic sensor is chosen. The primary components of the current system that monitors and manages the entire system are sensors, cloud servers, and android applications	Communication with a real-time environment Saving 95% water, labor and time issues	Software component is an android-based application connected to a cloud server. Sensors detect physical parameters in the surroundings and transmit them to the Raspberry Pi using the ZigBee wireless communication protocol	(Hamdi et al., 2021)

(continued)

Table 1 (continued)

Name of system	Location	Objective	Protocols/Design	Features	Tools/Components	Source
An IoT System for Smart Energy Consumption and Smart Irrigation	Pakistan	To propose an intelligent solution for efficient irrigation that includes a database of daily water demands for various plant types and determines the quantity of water needed for each plant type	This method uses the IoT and a series of sensors to efficiently record plant data and watering requirements, and it is combined with a mobile phone application interface to continuously monitor and control the efficient watering system	This approach not only saves water by efficient utilization, but also supports smart consumption of energy The results of this study are easy to reproduce as the sensors used are cheap and easy to access	The proposed system with fuzzy logic approach has a device and perception layer that includes all the hardware devices such as sensors, actuators, microcontrollers and Wi-Fi devices	(Munir et al., 2018)
Smart Irrigation System	India	To propose the use of soil moisture sensors and solenoid valves to automate irrigation systems	The framework for an automated irrigation system consists of three phases: 1. Moisture detection and monitoring in the soil 2. Operation of water pumps 3. The action of solenoid valves	It reduces human effort, saves time, and improves the framework's precision The system has the ability to send out alerts and cautions	Using soil moisture sensors and solenoid valves, as well as Arduino Uno and Esp8266 Node MCU microcontrollers The CROPWAT 8.0 is used to calculate water requirements	(Barapatre & Patel, 2019)

(continued)

Table 1 (continued)

Name of system	Location	Objective	Protocols/Design	Features	Tools/Components	Source
A solar powered smart irrigation system (SPSIS)	Africa	To design a smart irrigation management system driven by solar energy, offer a way for implementing wireless sensor networks as an IoT device	Solar power is employed for two reasons in the proposed system: the first is to pump water from the source to a storage tank for irrigation The wireless sensor network is fed for the second purpose	Reduce water usage, accuracy, flexibility, low cost, and ease of deployment Overcome the problem of energy scarcity	Pump, controller, and solar panels with following sensors; Soil moisture, Soil temperature, Environmental temperature, RH, Intensity of illumination CO_2, Rain flow, and Wind speed /Direction	(Difallah et al., 2018)
IoT Smart Irrigation System	Spain	To present a unique communications architecture for a smart irrigation system based on LoRa/ LoRaWANIoT nodes and gateways	The IoT node layer is made up of smart irrigation IoT nodes that exchange data with local gateways, the Fog computing layer is made up of local gateways scattered across multiple places, and the Remote service layer collects data from the smart irrigation system's deployed components	In comparable huge scenarios, the proposed tool gives important suggestions for future developers on the design of smart irrigation systems	Irrigation network with IoT nodes, local gateways, and cloud database	(Froiz-Míguez et al., 2020)

(continued)

Table 1 (continued)

Name of system	Location	Objective	Protocols/Design	Features	Tools/Components	Source
A Hybrid Method for Smart Irrigation System	India	To propose a hybrid irrigation framework for smart irrigation systems	Depending on the weather and climate, either a sprinkler or a dripper system can be used. It's known as the hybrid irrigation approach. Through User Interfaces such as a cell phone or a laptop, this system may be managed from anywhere. Through the border router, the sensor data received by various sensors is stored in the cloud, similar to ThingSpeak	Conservation of water, Planning the irrigation scheduling, automatically control the dripper or sprinkler method, Minimize the human effort, and Farmers can keep an eye on and control the irrigation system from any location	Node MCU, Soil moisture sensor, temperature sensor, humidity sensor, 5v Relay, Sprinkler, Dripper, Solenoid valve and Water tank are used in this system	(Anitha et al., 2019)
A Smart Irrigation Tool	Africa	To determine how the ENSO phenomena influences tomato crop water requirements in Mozambique	The irrigation tool in AgroClimate Mozambique was created with input from potential users and based on other commonly used tools. The modules and material are displayed when the user enters the irrigation tool The planting date is in the first module, followed by the ENSO phase in the second, crop type in the third, and growth season length in the fourth	In Mozambique, researchers are working to better understand climate variability and enhance tomato irrigation schedule It displays a user-friendly interface	AgroClimate Mozambique irrigation tool is used with irrigation tools	(Gelcer et al., 2018)

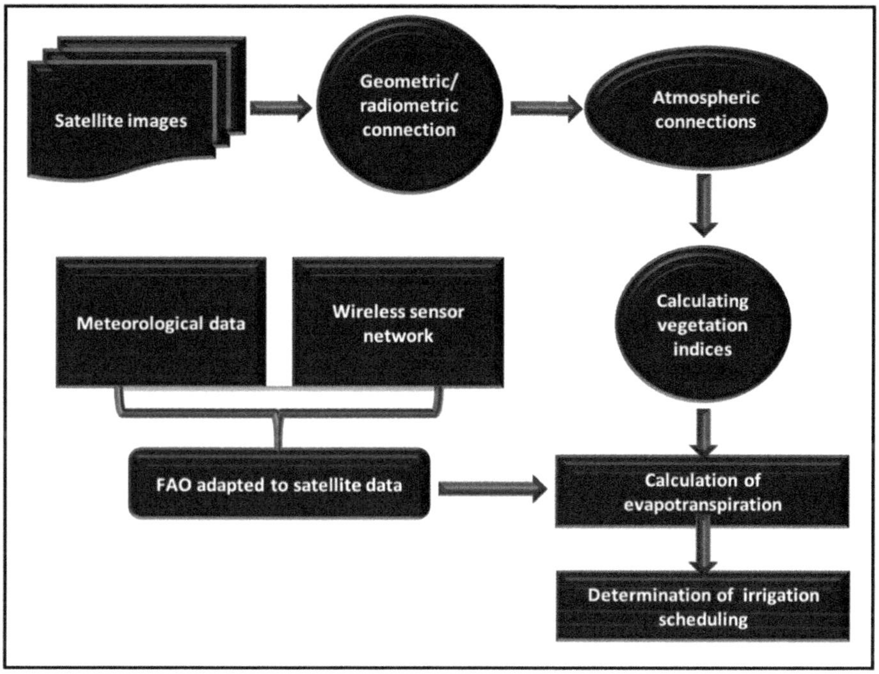

Fig. 3 Methodology of the application of remote sensing for irrigation water management. *Source* Papadavid et al. (2011)

agriculture, such as cultivated area, yield, and production (especially output for self-consumption), with an acceptable degree of precision. Challenges of smart farming application in irrigation water management by farmers can be presented as follows:

- The expenditure on tools (IoT devices, components of micro irrigation systems and other remote sensing devices) and employee training are the most evident limitations of remoteness. In every instance, the initial cost of technology must be offset by a large demand for it;
- Many governmental and non-governmental organizations lack the manpower, training, and technology resources needed to successfully exploit the new technologies;
- Technological transfer is difficult where there is no proper electricity and Internet connectivity, especially in remote areas;
- The usefulness of smart technologies is frequently limited by complicated political, scientific, and data consumption dynamics;
- Despite technical advances, the adoption of smart methodologies for environmental assessment policies remains low.

4.2 Possible Remedial Measures and Suggestions

To develop smart farming technologies, such an effort is also extremely desirable in developing countries. Some of the challenges highlighted above could be overcome by.

- Developing a system to ensure faster access and better distribution of smart farming data at an affordable price (for example, by creating a data infrastructure);
- Increasing public and private sector awareness and mastery of smart farming technologies through improved/appropriate training;
- Increasing public–private partnerships in implementation;
- Promoting scientific research in the field to ensure that developing smart farming technologies are effectively applied for the long-term development and management of critical groundwater resources;
- Implementing smart farming curriculum as well as suitable irrigation water management courses at the university and college level to promote greater education and training in these specialist areas, which would increase effective enforcement;
- Fostering collaboration between remote sensing scientists, GIS analysts, politicians, and indigenous peoples could help solve training and resource issues and improve indigenous participation in mapping projects.

4.3 Reflections and Future Research

PA, especially on small farms, is particularly valuable for evaluating various crops' biotic and abiotic stresses and diagnosing and managing various crop issues. Governments can use remote sensing data to make important policy decisions or solve national agricultural problems, and demonstration projects should be developed with the participation of potential end users to demonstrate the power of surveys from space, allowing IoT, AI, and remote sensing technologies to become more than just a specialized research tool. Clearly, these and other novel methodologies will emphasize the importance of remote sensing in agricultural output analyses in the future.

5 Conclusion

Agriculture is particularly vital for global food security and plays a prominent role in the economies of both developed and developing countries. Remote sensing and smart IoT technology have altered all aspects of human life, making everything precise and smart. Due to the automation of agricultural processes, this sort of solution appeals to farmers, thereby boosting production performance. Smart farming is a useful tool in irrigation water management since it gives data that can be used to plan and evaluate irrigation system performance. The remote sensing station is well positioned since irrigation is a critical component of agricultural productivity. Agriculture's development of smart farming practices will alter daily agricultural production by enhancing it and making it more profitable, decreasing waste, and reducing pollution. Smart agriculture is currently mostly used as a research tool and is rarely used to help with agricultural system management or other aspects of water allocation in watershed systems. Smart technology, in this context, might pave the way for PA based on more sustainable and logical use of resources (water and land) and chemical goods (fertilizers and pesticides) to boost crop yields while protecting ecosystems. As a result, education and training should strive to explain specialists' perceptions of the utility of such technologies so that PA can be properly taught in universities and rural regions. Expert continuing education courses should be planned, a network of experts, teachers, and technicians should be formed, and expert internship programs should be developed and managed. Finally, authorities and governments can use the findings of this study to develop strategic plans for disseminating this technology.

References

Adamides, G., Kalatzis, N., Stylianou, A., Marianos, N., Chatzipapadopoulos, F., Giannakopoulou, M., Papadavid, G., Vassiliou, V., & Neocleous, D. (2020). Smart farming techniques for climate change adaptation in Cyprus. *Atmosphere, 11*(6), 1–17. https://doi.org/10.3390/ATMOS11060557

Ahmad, M. D., Turral, H., & Nazeer, A. (2009). Diagnosing irrigation performance and water productivity through satellite remote sensing and secondary data in a large irrigation system of Pakistan. *Agricultural Water Management, 96*(4), 551–564. https://doi.org/10.1016/j.agwat.2008.09.017

Akhtar, M. N., Shaikh, A. J., Khan, A., Awais, H., Bakar, E. A., & Othman, A. R. (2021). Smart sensing with edge computing in precision agriculture for soil assessment and heavy metal monitoring: A review. *Agriculture (switzerland), 11*(6), 1–37. https://doi.org/10.3390/agriculture11060475

Alomar, B., & Alazzam, A. (2019). A Smart Irrigation System Using IoT and Fuzzy Logic Controller. In *ITT 2018 - Information Technology Trends: Emerging Technologies for Artificial Intelligence* (pp. 175–179). https://doi.org/10.1109/CTIT.2018.8649531

Anitha, A. A., Stephen, A., & Arockaim, D. L. A hybrid method for smart irrigation system. *International Journal of Recent Technology and Engineering (IJRTE)*, ISSN, 2277–3878

Ayenew, T. (2007). Water management problems in the Ethiopian rift: Challenges for development. *Journal of African Earth Sciences, 48*(2–3), 222–236. https://doi.org/10.1016/j.jafrearsci.2006.05.010

Bamurigire, P., Vodacek, A., Valko, A., & Ngoga, S. R. (2020). Simulation of Internet of Things water management for efficient rice irrigation in Rwanda. *Agriculture (switzerland), 10*(10), 1–12. https://doi.org/10.3390/agriculture10100431

Barapatre, P., & Patel, J. N. (2019). Development of Internet of Things (IoT) based smart irrigation system for sugarcane crop. *International Journal of Innovative Technology and Exploring Engineering, 8*(9 Special Issue), 650–654. https://doi.org/10.35940/ijitee.I1104.0789S19.

Barker, R., & Molle, F. (2004). Evolution of irrigation in South and Southeast Asia. In *Research Report 5*. http://www.iwmi.cgiar.org/assessment/files/pdf/publications/ResearchReports/CARR5.pdf.

Barkunan, S. R., Bhanumathi, V., & Sethuram, J. (2019). Smart sensor for automatic drip irrigation system for paddy cultivation. *Computers and Electrical Engineering, 73*, 180–193. https://doi.org/10.1016/j.compeleceng.2018.11.013

Byishimo, A., & Garba, A. A. (2016). Designing a farmer interface for smart irrigation in developing countries. In *Proceedings of the 7th Annual Symposium on Computing for Development, ACM DEV-7 2016*, (pp. 10–12). https://doi.org/10.1145/3001913.3006639.

Cagri Serdaroglu, K., Onel, C., & Baydere, S. (2020). IoT based smart plant irrigation system with enhanced learning. In *2020 IEEE Computing, Communications and IoT Applications, ComComAp 2020*. https://doi.org/10.1109/ComComAp51192.2020.9398892.

Campos, N. G. S., Rocha, A. R., Gondim, R., da Silva, T. L. C., & Gomes, D. G. (2020). Smart & green: An internet-of-things framework for smart irrigation. *Sensors (switzerland), 20*(1), 1–25. https://doi.org/10.3390/s20010190

Chartzoulakis, K., & Bertaki, M. (2015). Sustainable Water Management in Agriculture under Climate Change. *Agriculture and Agricultural Science Procedia, 4*, 88–98. https://doi.org/10.1016/j.aaspro.2015.03.011

Corbari, C., Ben Charfi, I., & Mancini, M. (2021). Optimizing irrigation water use efficiency for tomato and maize fields across italy combining remote sensing data and the aquacrop model. *Hydrology, 8*(1), 1–23. https://doi.org/10.3390/HYDROLOGY8010039

Cosgrove, W. J., & Loucks, D. P. (2015). Water management: Current and future challenges and research directions. *Water Resources Research, 51*, 4823–4839. https://doi.org/10.1002/2014WR016869

Dhakal, M. P., Ali, A., Khan, M. Z., Wagle, N., Shah, G. M., Maqsood, M. M., & Ali, A. (2021). Agricultural water management challenges in the Hunza River Basin: Is a solar water pump an alternative option? In *Irrigation and Drainage, December 2020*, 1–15. https://doi.org/10.1002/ird.2563.

Difallah, W., Benahmed, K., Draoui, B., & Bounaama, F. (2018). Design of a solar powered smart irrigation system (SPSIS) using WSN as an IoT device. *ACM International Conference Proceeding Series, 124–128*,. https://doi.org/10.1145/3178461.3178482

Eisenbeiss, H. (2004). A mini unmanned aerial vehicle (UAV): system overview and image acquisition. *International Archives of Photogrammetry. Remote Sensing and Spatial Information Sciences, 36*(5/W1), 1–7

El-Naggar, A. (2020). *New sensing methods for scheduling variable rate irrigation to improve water use efficiency and reduce the environmental footprint*. Massey University. http://hdl.handle.net/10179/15751.

Elijah, O., Rahman, T. A., Orikumhi, I., Leow, C. Y., & Hindia, M. N. (2018). An Overview of Internet of Things (IoT) and Data Analytics in Agriculture: Benefits and Challenges. *IEEE Internet of Things Journal, 5*(5), 3758–3773. https://doi.org/10.1109/JIOT.2018.2844296

Ertek, A., Şensoy, S., Gedik, I., & Küçükyumuk, C. (2006). Irrigation scheduling based on pan evaporation values for cucumber (Cucumis sativus L.) grown under field conditions. *Agricultural Water Management, 81*(1–2), 159–172. https://doi.org/10.1016/j.agwat.2005.03.008.

Fereres, E., Goldhamer, D. A., & Parsons, L. R. (2003). Irrigation water management of horticultural crops. *HortScience, 38*(5), 1036–1042. https://doi.org/10.21273/hortsci.38.5.1036

Fourati, M. A., Chebbi, W., & Kamoun, A. (2015). Development of a web-based weather station for irrigation scheduling. *Colloquium in Information Science and Technology, CIST, 2015-Janua*(January), 37–42. https://doi.org/10.1109/CIST.2014.7016591

Froiz-Míguez, I., Lopez-Iturri, P., Fraga-Lamas, P., Celaya-Echarri, M., Blanco-Novoa, Ó., Azpilicueta, L., Falcone, F., & Fernández-Caramés, T. M. (2020). Design, implementation, and empirical validation of an IoT smart irrigation system for fog computing applications based on Lora and Lorawan sensor nodes. *Sensors (switzerland), 20*(23), 1–33. https://doi.org/10.3390/s20236865

Gago, J., Douthe, C., Coopman, R. E., Gallego, P. P., Ribas-Carbo, M., Flexas, J., Escalona, J., & Medrano, H. (2015). UAVs challenge to assess water stress for sustainable agriculture. *Agricultural Water Management, 153*, 9–19. https://doi.org/10.1016/j.agwat.2015.01.020

García, L., Parra, L., Jimenez, J. M., Lloret, J., & Lorenz, P. (2020). IoT-based smart irrigation systems: An overview on the recent trends on sensors and iot systems for irrigation in precision agriculture. *Sensors (Switzerland), 20*(4). https://doi.org/10.3390/s20041042.

Gelcer, E., Fraisse, C. W., Zotarelli, L., Perondi, D., Malia, H. A., Ecole, C. C., & Migliaccio, K. W. (2018). A smart irrigation tool to determine the effects of ENSO on water requirements for tomato production in Mozambique. *Water (Switzerland), 10*(12). https://doi.org/10.3390/w10121820.

George, B. A., Shende, S. A., & Raghuwanshi, N. S. (2000). Development and testing of an irrigation scheduling model. *Agricultural Water Management, 46*(2), 121–136. https://doi.org/10.1016/S0378-3774(00)00083-4

Goap, A., Sharma, D., Shukla, A. K., & Rama Krishna, C. (2018). An IoT based smart irrigation management system using Machine learning and open source technologies. *Computers and Electronics in Agriculture, 155*(October), 41–49. https://doi.org/10.1016/j.compag.2018.09.040

Gowda, P. H., Chavez, J. L., Colaizzi, P. D., Evett, S. R., Howell, T. A., & Tolk, J. A. (2008). ET mapping for agricultural water management: Present status and challenges. *Irrigation Science, 26*(3), 223–237. https://doi.org/10.1007/s00271-007-0088-6

Hamdi, M., Rehman, A., Alghamdi, A., Nizamani, M. A., Missen, M. M. S., & Memon, M. A. (2021). Internet of Things (IoT) Based Water Irrigation System. *International Journal of Online and Biomedical Engineering, 17*(5), 69–80. https://doi.org/10.3991/ijoe.v17i05.22081

Hashim, N. M. Z., Mazlan, S. R., Abd Aziz, M. Z. A., Salleh, A., Ja'Afar, A. S., & Mohamad, N. R. (2015). Agriculture monitoring system: A study. *Jurnal Teknologi, 77*(1), 53–59. https://doi.org/10.11113/jt.v77.4099.

Kalezhi, J., Rwegasira, D., Dhaou, I. B., & Tenhunen, H. (2019). A DC Microgrid Smart-Irrigation System Using Internet of Things Technology. *IEEE PES/IAS PowerAfrica Conference: Power Economics and Energy Innovation in Africa, PowerAfrica, 2019*, 318–322. https://doi.org/10.1109/PowerAfrica.2019.8928795

Kamienski, C., Soininen, J. P., Taumberger, M., Dantas, R., Toscano, A., Cinotti, T. S., Maia, R. F., & Neto, A. T. (2019). Smart water management platform: IoT-based precision irrigation for agriculture. *Sensors (Switzerland), 19*(2). https://doi.org/10.3390/s19020276.

Kamil, I. A., & Ojoawo, O. J. (2019). Smart Irrigation: An ICT Application in Agriculture. *The International Journal of Engineering and Science (IJES), 8*(3 series), 53–58. https://doi.org/10.9790/1813-0803015358.

Karar, M. E., Alotaibi, F., Al Rasheed, A., & Reyad, O. (2021). A pilot study of smart agricultural irrigation using unmanned aerial vehicles and IoT-based cloud system. *Information Sciences Letters, 10*(1), 131–140. https://doi.org/10.18576/isl/100115.

Kuaban, G. S., Czekalski, P., Molua, E. L., & Grochla, K. (2019). An Architectural Framework Proposal for IoT Driven Agriculture. In *Communications in Computer and Information Science* (Vol. 1039). Springer International Publishing. https://doi.org/10.1007/978-3-030-21952-9_2.

Leakey, A. D. B., Ainsworth, E. A., Bernacchi, C. J., Rogers, A., Long, S. P., & Ort, D. R. (2009). Elevated CO_2 effects on plant carbon, nitrogen, and water relations: Six important lessons from FACE. *Journal of Experimental Botany, 60*(10), 2859–2876. https://doi.org/10.1093/jxb/erp096

Li, W., Awais, M., Ru, W., Shi, W., Ajmal, M., Uddin, S., & Liu, C. (2020). Review of Sensor Network-Based Irrigation Systems Using IoT and Remote Sensing. *Advances in Meteorology, 2020*,. https://doi.org/10.1155/2020/8396164

Merlino, S., Paterni, M., Berton, A., & Massetti, L. (2020). Unmanned aerial vehicles for debris survey in coastal areas: Long-term monitoring program to study spatial and temporal accumulation of the dynamics of beached marine litter. *Remote Sensing, 12*(8), 1260.

Mojid, M. A., & Mainuddin, M. (2021). Water-saving agricultural technologies: Regional hydrology outcomes and knowledge gaps in the eastern gangetic plains-a review. *Water (Switzerland), 13*(5). https://doi.org/10.3390/w13050636.

Mondal, P., & Basu, M. (2009). Adoption of precision agriculture technologies in India and in some developing countries: Scope, present status and strategies. *Progress in Natural Science, 19*(6), 659–666. https://doi.org/10.1016/j.pnsc.2008.07.020

Munir, M. S., Bajwa, I. S., Naeem, M. A., & Ramzan, B. (2018). Design and implementation of an IoT system for smart energy consumption and smart irrigation in tunnel farming. *Energies, 11*(12), 1–18. https://doi.org/10.3390/en11123427

Narmilan, A. (2017). E-Agricultural Concepts for Improving Productivity : A Review. *Scholars Journal of Engineering and Technology (SJET), 5*(March), 10–17. https://doi.org/10.21276/sjet.2017.5.1.3.

Nasiakou, A., Vavalis, M., & Zimeris, D. (2016). Smart energy for smart irrigation. *Computers and Electronics in Agriculture, 129*, 74–83. https://doi.org/10.1016/j.compag.2016.09.008

Nazari, B., Liaghat, A., Akbari, M. R., & Keshavarz, M. (2018). Irrigation water management in Iran: Implications for water use efficiency improvement. *Agricultural Water Management, 208*(July), 7–18. https://doi.org/10.1016/j.agwat.2018.06.003

Nisha, G., & Megala, J. (2015). Wireless sensor Network based automated irrigation and crop field monitoring system. In *6th International Conference on Advanced Computing, ICoAC 2014*, 189–194. https://doi.org/10.1109/ICoAC.2014.7229707.

Papadavid, G., Hadjimitsis, D., Fedra, K., & Michaelides, S. (2011). Smart management and irrigation demand monitoring in Cyprus, using remote sensing and water resources simulation and optimization. *Advances in Geosciences, 30*, 31–37. https://doi.org/10.5194/adgeo-30-31-2011

Pivoto, D., Waquil, P. D., Talamini, E., Finocchio, C. P. S., Dalla Corte, V. F., & de Vargas Mores, G. (2018). Scientific development of smart farming technologies and their application in Brazil. *Information Processing in Agriculture, 5*(1), 21–32. https://doi.org/10.1016/j.inpa.2017.12.002

Prasad, P. V. V., Boote, K. J., Allen, L. H., Jr., Sheehy, J. E., & Thomas, J. M. G. (2006). Species, ecotype and cultivar differences in spikelet fertility and harvest index of rice in response to high temperature stress. *Field Crops Research, 95*, 398–411.

Radoglou-Grammatikis, P., Sarigiannidis, P., Lagkas, T., & Moscholios, I. (2020). A compilation of UAV applications for precision agriculture. *Computer Networks, 172*(January), 107148. https://doi.org/10.1016/j.comnet.2020.107148

Ranjith, Anas, S., Badhusha, I., Ot, Z., Faseela, K., & Shelly, M. (2017). Cloud based automated irrigation and plant leaf disease detection system using an android application. In *Proceedings of the International Conference on Electronics, Communication and Aerospace Technology, ICECA 2017, 2017-Janua*, 211–214. https://doi.org/10.1109/ICECA.2017.8212798.

Raza, A., Razzaq, A., Mehmood, S. S., Zou, X., Zhang, X., Lv, Y., & Xu, J. (2019). Impact of climate change on crops adaptation and strategies to tackle its outcome: A review. *Plants, 8*(2). https://doi.org/10.3390/plants8020034.

Reza Ghafarian Malamiri, H., Arabi Aliabad, F., Shojaei, S., Morad, M., Band, S.S., 2021. A study on the use of UAV images to improve the separation accuracy of agricultural land areas. Comput. Electron. Agric. 184 (March), 106079 https://doi.org/10.1016/j.compag.2021.106079.

Richey, A. S., Thomas, B. F., Lo, M. H., Reager, J. T., Famiglietti, J. S., Voss, K., Swenson, S., & Rodell, M. (2015). Quantifying renewable groundwater stress with GRACE. *Water Resources Research, 51*(7), 5217–5237. https://doi.org/10.1002/2015WR017349

Rodriguez-Ortega, W. M., Martinez, V., Rivero, R. M., Camara-Zapata, J. M., Mestre, T., & Garcia-Sanchez, F. (2017). Use of a smart irrigation system to study the effects of irrigation management on the agronomic and physiological responses of tomato plants grown under different temperatures regimes. *Agricultural Water Management, 183*, 158–168. https://doi.org/10.1016/j.agwat.2016.07.014

Rosenzweig, C., Tubiello, F. N., Goldberg, R., Mills, E., & Bloomfield, J. (2002). Increased crop damage in the US from excess precipitation under climate change. *Global Environmental Change, 12*(3), 197–202. https://doi.org/10.1016/S0959-3780(02)00008-0

Saiz-Rubio, V., & Rovira-Más, F. (2020). From smart farming towards agriculture 5.0: A review on crop data management. *Agronomy, 10*(2). https://doi.org/10.3390/agronomy10020207.

Sciences, P., Mohammad, F. S., El Marazky, M. S. A., & Sciences, A. (2013). Effect Of Intelligent Irrigation On Water Use Efficiency Of Wheat Crop In Arid Region. *The Journal of Animal & Plant Sciences, 23*(6), 1691–1699.

Singh, K. M., Kumar, A., & Singh, R. K. P. (2015). Role of Information and Communication Technologies in Indian Agriculture: An Overview. *SSRN Electronic Journal, October 2017*, 1–11. https://doi.org/10.2139/ssrn.2570710.

Singhal, G., Bansod, B., & Mathew, L. (2018). Unmanned Aerial Vehicle Classification , Applications and Challenges : A Review. *Preprint, November*, 1–19. https://doi.org/10.20944/preprints 201811.0601.v1.

Skendžić, S., Zovko, M., Živković, I. P., Lešić, V., & Lemić, D. (2021). The impact of climate change on agricultural insect pests. *In Insects, 12*(5). https://doi.org/10.3390/insects12050440.

Sofonia, J., Shendryk, Y., Phinn, S., Roelfsema, C., Kendoul, F., & Skocaj, D. (2019). Monitoring sugarcane growth response to varying nitrogen application rates: A comparison of UAV SLAM LiDAR and photogrammetry. *International Journal of Applied Earth Observation and Geoinformation, 82*(December 2018), 101878. https://doi.org/10.1016/j.jag.2019.05.011.

Sudharshan, N., Karthik, A. V. S. K., Kiran, J. S. S., & Geetha, S. (2019). Renewable Energy Based Smart Irrigation System. *Procedia Computer Science, 165*(2019), 615–623. https://doi.org/10.1016/j.procs.2020.01.055

Tadokoro, H., Onishi, M., Kageyama, K., Kurisu, H., & Takahashi, S. (2011). Smart water management and usage systems for society and environment. *Hitachi Review, 60*(3), 164–171.

Turral, H., Burke, J., & Faurès, J.-M. (2011). *Climate change, water and food security.*

Vanino, S., Nino, P., De Michele, C., Falanga Bolognesi, S., D'Urso, G., Di Bene, C., Pennelli, B., Vuolo, F., Farina, R., Pulighe, G., & Napoli, R. (2018). Capability of Sentinel-2 data for estimating maximum evapotranspiration and irrigation requirements for tomato crop in Central Italy. *Remote Sensing of Environment, 215*(June), 452–470. https://doi.org/10.1016/j.rse.2018.06.035

Vijverberg, J., Dejen, E., Getahun, A., & Nagelkerke, L. A. J. (2012). The composition of fish communities of nine Ethiopian lakes along a north-south gradient: Threats and possible solutions. *Animal Biology, 62*(3), 315–335. https://doi.org/10.1163/157075611X618246

Villa-Henriksen, A., Edwards, G. T. C., Pesonen, L. A., Green, O., & Sørensen, C. A. G. (2020). Internet of Things in arable farming: Implementation, applications, challenges and potential. *Biosystems Engineering, 191*, 60–84. https://doi.org/10.1016/j.biosystemseng.2019.12.013

Vuolo, F., Essl, L., & Atzberger, C. (2015). Costs and benefits of satellite-based tools for irrigation management. *Frontiers in Environmental Science, 3*(JUL), 1–12. https://doi.org/10.3389/fenvs.2015.00052.

Walter, A., Finger, R., Huber, R., & Buchmann, N. (2017). Smart farming is key to developing sustainable agriculture. *Proceedings of the National Academy of Sciences of the United States of America, 114*(24), 6148–6150. https://doi.org/10.1073/pnas.1707462114

Wheeler, T. R., Craufurd, P. Q., Ellis, R. H., Porter, J. R., & Vara Prasad, P. V. (2000). Temperature variability and the yield of annual crops. *Agriculture, Ecosystems and Environment, 82*(1–3), 159–167. https://doi.org/10.1016/S0167-8809(00)00224-3

Wijeratne, M. A., Anandacoomaraswamy, A., Amarathunga, M. K. S. L. D., Ratnasiri, J., Basnayake, B. R. S. B., & Kalra, N. (2007). Assessment of impact of climate change on productivity of tea (Camellia sinensis L.) plantations in Sri Lanka. *Journal of the National Science Foundation of Sri Lanka, 35*(2), 119–126. https://doi.org/10.4038/jnsfsr.v35i2.3676.

Zhang, N., Wang, M., & Wang, N. (2002). Precision agriculture - A worldwide overview. *Computers and Electronics in Agriculture, 36*(2–3), 113–132. https://doi.org/10.1016/S0168-1699(02)00096-0

Zikria, Y. Bin, Ali, R., Afzal, M. K., & Kim, S. W. (2021). Next-Generation Internet of Things (IoT): Opportunities, Challenges, and Solutions. *Sensors, 21*(1174), 1–7. https://doi.org/10.3390/s21041174

Trending Smart Agricultural Techniques in Postharvest Handling to Mitigate Climate Change Impacts in Sri Lankan Context

M. B. F. Jemziya, Muneeb M. Musthafa, Mohamed Mujithaba Mohamed Najim, and Mirza Barjees Baig

Abstract Postharvest losses are very common and occur due to poor infrastructure and logistical arrangements, lack of technology, and insufficient knowledge, skills and management capacity of supply chain stakeholders. In Sri Lanka, the annual loss of fruits and vegetables during postharvest operation is around 30%–40% of the harvest and 20%–40% of the total harvest. High levels of postharvest losses may be the result of the lack of awareness and concern on proper postharvest practices, improper storage, harvesting methods, and packaging techniques. Most agricultural systems are mainly sensitive to climate change such as disparity in temperature and precipitation, which results in the high cost of agricultural management. Global warming, greenhouse gas emission, and increased atmospheric temperature not only threaten the crops but also the postharvest life of agricultural products. The postharvest losses are intensified by climate change, which leads to food insecurity and severely hampers smallholder farmers. Therefore, the food losses in the postharvest supply chain due to the differences in climate change must be reduced using advanced and innovative technologies. These include climate-smart processing technologies, solar energy-based techniques, thermal and non-thermal technologies, cutting-edge

M. B. F. Jemziya (✉) · M. M. Musthafa
Department of Biosystems Technology, Faculty of Technology, South Eastern University of Sri Lanka, Oluvil #32360, Sri Lanka
e-mail: jemziya@seu.ac.lk

Department of Biosystems Technology, Faculty of Technology, South Eastern University of Sri Lanka, Oluvil #32360, Sri Lanka

M. M. Musthafa
e-mail: muneeb@seu.ac.lk

M. M. M. Najim
Climate-Smart and Resilient Food Systems and Security, Sultan Sharif Ali Islamic University, Sinaut Campus, Sinaut, Brunei
e-mail: najim.mujithaba@unissa.edu.bn

M. B. Baig
Prince Sultan Institute for Environmental, Water & Desert Research, King Saud University, Riyadh, Kingdom of Saudi Arabia
e-mail: mbbaig@ksu.edu.sa

M. Behnassi et al. (eds.), *Climate-Smart and Resilient Food Systems and Security*,
https://doi.org/10.1007/978-3-031-65968-3_19

technology, ZigBee and sensor technologies, zero energy cooling chambers, novel, and intelligent packaging technologies, Internet of things (IoT), advanced plant breeding technologies, advanced food quality management strategies, automation, and advanced imaging technologies. Decisively, fostering awareness and education of farmers through extension services with the modern applications of postharvest operations will reduce the overall food losses and ensure food security by mitigating climate change.

Keywords Advanced technologies · Climate change · Food security · Mitigation · Postharvest technology

1 Introduction to Postharvest Technology

Postharvest technology is the discipline and system applied to agricultural products after harvest until it reaches the final consumer with the desired nutritional requirements. This system consists of product protection, processing, packaging, marketing, and delivery (NIC, 2021). The postharvest processing comprises many common operations implemented after harvesting of produce including drying, threshing, shelling, milling, storage, packaging, and marketing. The objective of using postharvest technology is to preserve and improve the quality of the products and make them readily saleable in order to cater ever-increasing food demand (NIC, 2021).

Postharvest technology should meet the food required by the expanding population with the elimination of losses and make available more nutritive food with appropriate processing methods and value addition techniques. The world's population has been forecasted to reach 9.7 billion in 2050 (The World Bank, 2016; Tian et al., 2016) and the food supplies for the entire population will be more demanding. Therefore, it is necessary to boost global food production while ensuring food security, maintaining food demands of the expanded population, and combating malnourishment; hence, most countries have concentrated on expanding agricultural productivity (Jayarajan et al., 2021).

2 Postharvest Losses

Postharvest losses are very common and occur spontaneously from harvesting to consumption, where postharvest processing loss of fruits and vegetables is up to 30% per year (FAO, 2011; Yahaya & Mardiyya, 2019). Around 25–30% of the worldwide food is wasted from on-farm food production to storage facility at a retail, because of inadequate food supply chain management and food spoilage (Shafiee-Jood & Cai, 2016). The farm-level postharvest loss is around 75% and the remaining is lost at the market level (Basavaraja et al., 2007).

Postharvest losses can be categorized into two: primary and secondary. The former is mainly caused by biological, chemical, physical, and mechanical factors. Postharvest pests, microorganisms, sterilant, fumigant, processing methods and conditions, climate and weather conditions can be listed as primary causes. Inadequate knowledge on processing operations and lack of logistic facilities and services such as transport, drying equipment, packaging materials, storage solutions, market access, receiving low prices are considered as secondary causes of postharvest losses. These factors lead to quantitative losses including reduction of physical weight, spillage, breakage, rodent and pest damage, and spoilage due to temperature changes, chemical changes, and changes in humidity content. Further, qualitative losses are measured by loss of nutrition value, edibility, caloric value, and acceptability of agricultural produce (Jayarajan et al., 2021).

Sri Lanka is primarily an agriculture-based nation, where postharvest losses are occurring at a wide range of influences by several factors and reasons. The annual loss of fruits and vegetables during postharvest operations represents about 210,000 MT (US$ 90 million) of fruits, which is about 30%–40% of the harvest and about 370,000 MT (US$ 110 million), which is about 20%–40% of the total harvest (Annual report, 2018). The total postharvest losses of crops in Sri Lanka are expressed in Table 1. Inappropriate postharvest practices and handling, issues in the incorporation of cold chain practices, and poor knowledge and awareness on postharvest practices at grass root farmer level are the primary reasons for elevating postharvest losses in Sri Lanka (Leelananda et al., 2021). However, environmental concerns including deforestation, soil depletion, degradation, and poor groundwater qualities due to agrochemical residues and climate change including greenhouse gas emissions, temperature changes, rainfall and precipitation patterns, flooding, global warming, dry spells, drought, and high winds, storms also exert great influence on postharvest losses. Invariable weather and climate conditions are sensitive and problematic concerns in Sri Lanka due to having limited cultivated land and water resources.

Table 1 Postharvest losses of crops in Sri Lanka

Crops	Total postharvest losses (%)
Paddy	15
Papaya	46
Pineapple	18
Avacado	40
Mango	41
Banana	30
Pumpkin	60

Source: Daily news (2016); Palipane (2000); Udara and Malsha (2019);Wijayaratne et al. (2009)

3 Impact of Climate Change in Postharvest Technology, Sri Lanka

Sri Lanka is extremely susceptible to the effects of climate change and ranked as the fourth most affected country in 2016 (UNICEF, 2018). Climate change mainly impacts tourism, commercial agriculture, and manufacturing sectors (USAID, 2018). Most of the natural disasters in Sri Lanka such as severe rainfall, extended drought, floods, landslides, and increasing sea levels impact agricultural productivity, increase the spread of infectious illnesses, and finally undermine living standards (Anuradha, 2019). Most of the agriculture systems are mainly sensitive to climate change such as disparity in temperature and precipitation. The production of rice and coconuts is mainly threatened by saltwater intrusion and extreme weather events, particularly drought (USAID, 2018). Further, climate change results in the high cost of grain management by the ways of damaging storage structures, the necessity for re-drying, need for repeated treatments, struggle in the forecasting of the storage period, scheduling of postharvest management, and related investments.

Global warming threatens the postharvest life of agricultural produces. The largest proportion of produces is spoiled by the outbreaking of pests and diseases because high temperature leads to shortening of the life cycles of insect pests and diseases. The high mean atmospheric temperature increases the drying rate of agricultural products; besides it causes the heat stress to laborious works and increases the performance of fungi. Further, high humidity and temperature result in the reduction of the effectiveness of active ingredients of commercial grain protectants. Unpredictable rain seasons also adversely impact the agricultural produces after harvest (Martin, 2018).

Greenhouse gas emission is a worldwide serious issue, and in Sri Lanka the greenhouse gas emissions have raised by 43% during the period 1990–2011. Based on the fact sheet, 45 million MT of CO_2 was emitted in 2011, of which the largest contributor of emission was energy sectors (40%) followed by waste sectors (28%), land-use change and forestry (15%), agriculture (14%), and industrial processes (3%) (USAID, 2016). In Sri Lanka, greenhouse gases are emitted frequently at high volumes by on-farm agriculture emissions such as livestock manure, on-farm energy use, and fertilizer emissions, electricity and heat used to manufacture and process food, energy used for transport, storage and cooking, landfill emissions and emissions for land-use change and deforestation.

The greenhouse gases (CO_2, O_3) have the capacity to absorb and release radiant energy within the thermal infrared range, which originates the greenhouse effect. CO_2 is the primary gas to create the greenhouse effect, which originates the problematic concern on tuber malformation, the occurrence of common scab, and changes in reducing sugars in agricultural produce. Similarly, O_3 also negatively impacts agricultural produce, where it reduces photosynthesis, growth, and biomass accumulation, increases vitamin C, beta carotenes, lutein, and lycopene content, and drops of volatile esters (Morettia et al., 2010).

Ultimately, climate change yields elevated atmospheric temperature in Sri Lanka. It negatively impacts on photosynthesis, which leads to alteration of sugar, organic acids, flavonoid content, antioxidant activity, and firmness of crops. The high temperature causes extreme drought and leads to high demand for water. On the other hand, high atmospheric temperature accelerates the evapotranspiration leading to lowering the water reservoir in soils. The high-water stress in crops results in a low level of plant growth, production, and ripening (Morettia et al., 2010).

4 Need of Technology

The postharvest losses are aggravated by climate change while inducing food insecurity and ultimately diminishing the smallholder farmers' income. Climate change raises postharvest food losses via elevated temperatures, higher pest and disease incidents, and storage issues. As a result of inadequate farming techniques or inappropriate farming techniques used, the productivity of land and water is not achieved up to the maximum or optimum potential. This substantial loss is due to poor food grain management and insufficient facilities (Bradford et al., 2018).

Most of the postharvest operations are mingled with climatic conditions. In Sri Lanka, on-field drying of cereal grains, bulk storage and handling of high moisture content fruits, transportation of exotic fruits and vegetables, non-scientific packing, irrigation scheduling, and crop planning are the major contributors for total postharvest losses. These factors expose the fresh produces to the extreme conditions and fail to secure their quality and quantity. Ultimately, the produces lose their distinct features and properties and fall into the column of wastages.

Around 25% reduction in food loss and waste may help reduce the food gap by 12%, the land gap by 27%, and the greenhouse gas mitigation gap by 15% (Ayaz et al., 2019). Food losses in the postharvest supply chain equate to substantial losses in packing, shipping, and storage activities (Jayarajan et al., 2021; Nicastro & Carillo, 2021). The amount of fresh food available for consumption would rise if these losses were reduced (Jayarajan et al., 2021). However, food wastage, because of postharvest losses, has become a challenge in achieving food security which is further aggravated as a result of postharvest losses, necessitating special vigilance. The bulk of losses of agricultural produce occur between harvest and distribution to the customers, particularly in developing countries (Jayarajan et al., 2021). The agriculture industry is an important source of income in developing nations and considerable amounts of loss of their produce occurred at the farm level. It is a serious concern that requires serious attention from all stakeholders (FAO, 2002). Therefore, the food losses in the postharvest supply chain due to the disparity of climate change must be reduced using advanced and innovative technologies.

5 Technologies

Local farmers are unable to purchase expensive food protection systems. Therefore, strategies to decrease postharvest losses with low investment and maximize the return to farmers are required. Every stage of loss cannot be totally controlled, but it may be minimized. One-third of the crop is lost owing to improper harvest management. Therefore, the food should be carefully kept and monitored until they are used (Kaushik & Singhai, 2019; Onibonoje et al., 2019).

Climate-smart processing technologies, including thermal and non-thermal energy for deactivation of microorganisms, solar energy-based techniques (greenhouse technology and solar inverters), thermal (ohmic and dielectric heating), and non-thermal (high hydrostatic pressure processing, pulsed electric field, and ultrasound novel processing) technologies may ensure the food security and enhance farmers' living with response to climate change (Srinivasarao et al., 2020).

Temperature fluctuations along the food supply chain cause serious problems in food quality during the supply chain, where the air temperature is the most influential factor (Goedhals-Gerber & Khumalo, 2020) with other contributing factors such as poor packaging, improper monitoring technology, improper air velocity monitoring, and relative humidity variations in cold chain systems, extended shipment length, and the heterogeneity of fruits and vegetables are also notable contributors (Onwude et al., 2020).

Temperature of food may be monitored and controlled in a variety of ways. The manual technique of checking a thermometer and recording the temperature has several disadvantages. Implementing an automated system that employs wireless sensors to electronically detect and record temperatures, on the other hand, can significantly enhance food safety. This technology simply enables for a continuous temperature data stream—24 h a day, seven days a week. Notifications may be set up to provide real-time alerts if the temperature exceeds predefined limits, allowing rapid action to be taken to correct the problem.

The primary problem of food industry is transporting food to more people in a safe and efficient manner (Mandyck & Schultz, 2015). Loading of food is also crucial in alleviating hunger challenge. Therefore, proper food storage may be the better solution to make available sufficient food for the people. Postharvest loss is a complex issue that changes based on storage procedures, weather conditions, and economic situation of the country. Many countries rely on foods to meet the requirements of individual citizens. Farmers' profits are increased by proper food storage and care after harvesting. However, an unscientific storage building design may result in loss and cannot guarantee protection against pests, dampness, and other unforeseen events (Kumar & Kalita, 2017). The majority of farmers store their food in open areas, in baskets, or polypropylene bags at household level. These ineffective storage methods result in food loss and insecurity. Therefore, regulating the humidity of non-perishable foods is crucial for preserving their quality throughout storage. To avoid postharvest loss of food grains, an effective infrastructure must be established using cutting-edge technology. In order to protect food quality, it is critical to keep track of factors associated with quality characteristics on a regular basis (Kitinoja et al., 2011).

Storage monitoring is a crucial activity that necessitates technological intervention due to its critical relevance. To maintain safety and to increase farmer revenue, a comprehensive monitoring system with various measures and methods must be implemented. A secure storage management system is required to prevent food spoiling. Implementation of Internet of Things (IoT) based system that monitors environmental characteristics such as temperature, humidity, and CO_2 levels in storage areas, as well as designing a notification system to dispatch goods in the proper order with proper monitoring process is required (Jayarajan et al., 2021). Zhang et al. (2009) recommended ZigBee and sensor technologies in large-scale grain barns to assess and monitor the grain temperature and to reduce the radio frequencies. They were unable to develop an optimization technique to extend the lifespan of the network (Zhang et al., 2009). In addition to that, cold chain management with zero energy cooling chambers in storage and transport is an efficient way to warrant food security. Further, the backbone of precision agriculture is communication. It ensures to collect and report the data periodically to protect and monitor the status of stored produces. A robust, trustworthy, and secure link among numerous participating objects is required to achieve the definite goal. Inventory tracking must be done flawlessly even if there is no direct interaction with the goods, according to warehouse inventory management.

Packaging of food is a key element and contributes main proportion to minimize the quality losses and extend the shelf life of products by protecting and making them convenient during the postharvest operations. Plastics is one of the main packaging materials and used versatile as reusable plastic crates. However, due to the concern on the usage of polymers, modern and novel technologies such as developing nano packaging materials are introduced to minimize the harmful loading on earth. Intelligent packaging including data carriers (barcode labels, radio frequency identification (RFID) tags) (Yam et al., 2005; Kozak & Cierpiszewski, 2010; Realini & Marcos, 2014; Fang et al., 2017; İçöz and Eker 2018), and package indicators (time–temperature indicators, gas indicators, biosensors) (Barska & Joanna, 2017; Day, 2001; Müller & Schmid, 2019) have a magnificent potential in the evolution of food packaging. The implementation of new packaging technologies can considerably improve the quality, freshness, and safety of food to ensure a longer shelf life in preventing food losses (Drago et al., 2020; Han et al., 2018).

Smart retailing, which is the use of smart technology such as the IoT in conjunction with traditional retailing, will be of benefit to obtain better results. Localization is the process of determining the location of a product, and it is crucial for smart retail. Location, data filtering, data collection, and data mining are the four localization components. First-expired-initially-out (FEFO), which was first suggested at the end of the 1980s, is one proposal for a better supply chain management solution. The primary concept is to rotate stock such that each item remaining its shelf life is best suited to the remaining transit length possibilities, reducing product waste during transportation and ensuring product uniformity at the retailing (Jedermann et al., 2014).

The Global Positioning System (GPS) and Radio Frequency Identification (RFID) play an important role in delivering products to clients. The data is collected in the cloud using tagging so that the system can follow the item and notify the person in charge of its state. During online buying, retailers can readily determine where the goods have been mistreated. As a result, IoT devices are linked to mobile, desktop, and social media platforms in order to capture all responses of clients. Thereby, the information between customers and retailers are shared in between so the postharvest loss can be reduced. Wireless technologies such as Wi-Fi, Bluetooth, Infrared, or Cellular Networks, as well as conventional technologies such as Ethernet, USB, and Serial connections, are all options for IoT devices. To monitor and automate, all of these distinct communication mechanisms must be chosen depending on environmental conditions (Jat et al., 2019).

Advanced plant breeding technologies are focused to overcome the climatic challenges. The plant breeding technology implies on developing hybrid seeds, the emerging plant which is capable to resist heat stress and adverse humidity status, and tolerant varieties which withstand climate change and postharvest pest attack. Precise development of improved climate-smart crops is another milestone to be comfortable in the worst climatic conditions.

Agricultural crops are either seasonal or year-round yielders. When the crops are exposed to suitable growing conditions such as temperature, humidity, and irrigation, as a result, the highest yield can be obtained throughout the system. Over-production with insufficient facilities and infrastructures leads to high losses of produce in terms of quantity and quality due to climatic changes. Therefore, based on the prediction of climatic facts, cropping has to be scheduled to limit the production. In addition to that, the safeness of crops is to be ensured at every growing stage to overcome the devastations by heavy rains and hailstorms. At the end, the food security and effective earnings throughout the systems can be achieved.

There are several ways to emit greenhouse gas into the atmospheric gas composition. The results revealed that the home cooking accounts for as much as 61% of total emissions associated with specific foods. The wise choice of alternative and readily available cooking practices assists to substantially reduce greenhouse gas emissions (Frankowska et al., 2020). Additionally, the public should understand the application of basic food safety principles during the food preparation and cooking practices, which should diminish the harmful emissions that contribute to climate change.

Drying is one of the most energy-intensive processing in postharvest technology. This unit operation is applied to reduce the moisture content of harvest (Mujumdar & Law, 2010). Optimum drying is a necessity of extended postharvest life of agricultural commodities. The conventional drying is mainly depending on the weather and climatic condition, mainly air temperature, air velocity, humidity, free mist, and free clouds. Alteration of optimum conditions in the drying process impacts on the postharvest life of produce. Therefore, adequate and protected drying can be obtained by established and green solar driers with zero energy.

Several advanced food quality management strategies have been used in postharvest supply chains to monitor, manage, and anticipate the quality development of diverse fruits and vegetables in recent years. At present, advanced technologies such as visualization techniques, InfraRed Spectroscopy, multi-sensors methods, electronic noses (E-noses), acoustic impulse response (AIR), radio frequency identification (RFID), printed sensors (PTS), and mathematical modeling are some of the approaches used (García-Esteban et al., 2018; Luning & Marcelis, 2020; Onwude et al., 2020). The IoT uses wired or wireless connections to link numerous devices. The IoT is a sensor-enabled, cloud-based technology that uses agriculture in all elements of technology to improve crop development and make agricultural systems more commercially viable. The IoT is a new technology that provides farmers with creative strategies for increasing profits while reducing hazards (Yin & He, 2021). IoT items share information and give intelligent services to users. Sensing, collecting, and analyzing data are all done by IoT devices. The suggested system uses an IoT-based approach to safeguard food from postharvest loss while also providing real-time storage information (Jayarajan et al., 2021). Therefore, smart agriculture is believed to be a modern solution to alleviate the hunger of the human population in the near future as well. Some of the key technologies available for this purpose are Compliance mate, Laird's sentries, CCP Smart Tag (RC4), and Tempreporter (Ayaz et al., 2019).

Automation, from an industrial standpoint, refers to the use of electrical/electronic, mechanical, and computer-based systems to operate and control manufacturing processes. As a result, robotics is classified as part of the industrial automation category. Robots are utilized in agriculture and the food sector for a variety of tasks, including sowing, watering, harvesting, and cutting, processing, and packing foods. Furthermore, several High Definition (HD) cameras are used for flow detection and inspection through robot learning. Therefore, precision and accuracy of processing are increased with automation, thereby a considerable amount of food wastage is reduced (Yaghoubi et al. 2013).

The food and agro-allied sectors use advanced imaging technologies to detect changes in food quality. Computer vision (CV), hyperspectral imaging (HSI), multispectral imaging (MSI), thermography, and X-ray imaging are all examples of this technology. External quality parameters may be detected and evaluated using image technology. Imaging technology is a cutting-edge approach for monitoring changes in food quality in the food and agro-allied industries. Image technology is very beneficial for identifying and analyzing exterior quality characteristics (Du et al., 2020). It helps sort the agricultural commodities based on their quality at the initial point rather than rejecting them at the final stage.

6 Conclusions

The postharvest losses are heightened by climate change, which directly affects smallholder farmers in developing countries like Sri Lanka. The postharvest losses can be reduced by farmer training and awareness programs through the extension services on modern applications of postharvest operations including processing, packaging, and transport. Ultimately, the postharvest losses can be minimized by technology adoption in the ever-changing modern world.

References

Annual report. (2018). *Central Bank, Sri Lanka*. Available at: https://www.cbsl.gov.lk/en%20/publications/economic-and-financial-reports/annual-reports/annual-report-2018 (Accessed on 18. 06. 2021)

Anuradha, W. (2019). *'Here's how Climate Change will impact you', UNDP in Sri Lanka*, Sri Lanka. Available at: https://www.lk.undp.org/content/srilanka/en/home/blog/2019/03/22032019.html (Accessed on 20.06.2021)

Ayaz, M., Ammad-Uddin, M., Sharif, Z., Mansour, A., & Aggoune, E. H. M. (2019). Internet-of-Things (IoT)-based smart agriculture: Toward making the fields talk. *IEEE Access, 7*, 129551–129583.

Barska, A. & Joanna, W. (2017). Innovations in the food packaging market – intelligent packaging – a review. *Czech Journal of Food Sciences*, 35(1), 1–6. https://doi.org/10.17221/268/2016-cjfs

Basavaraja, H., Mahajanashetti, S.B., Udagatti, and Naveen, C. (2007). Economic Analysis of Post-harvest Losses in Food Grains in India: A Case Study of Karnataka. *Agricultural Economics Research Review*, Agricultural Economics Research Association (India), 20(1).

Bradford, K., Dahal, P., Van, A. J., Kunusoth, K., Bello, P., Thompson, J., & Wu, F. (2018). The dry chain: Reducing postharvest losses and improving food safety in humid climates. *Trends in Food Science and Technology., 71*, 84–93.

Daily news. (2016). *Minimize post-harvest losses of fruits to ensure food security*. Available at: https://www.dailynews.lk/2016/11/10/business/98641 (Accessed on 10.11.2016).

Day, B. P. F. (2001). Active packaging – a fresh approach. *The Journal of Brand Technology, 1*(1), 32–41.

Drago, E., Campardelli, R., Pettinato, M., & Perego, P. (2020). Innovations in Smart Packaging Concepts for Food: An Extensive Review. *Foods, 9*, 1628. https://doi.org/10.3390/foods9111628

Du, Z., Zeng, X., Li, X., Ding, X., Cao, J., & Jiang, W. (2020). Recent advances in imaging techniques for bruise detection in fruits and vegetables. *Trends in Food Science and Technology, 99*, 133–141.

Fang, Z., Zhao, Y., Warner, R. D., & Johnson, S. K. (2017). Active and Intelligent Packaging in Meat Industry. *Trends in Food Science and Technology, 61*, 60–71.

FAO. (2002). *The Role of Agriculture in the Development of Least-Developed Countries and their Integration into the World Economy*. Italy.

FAO. (2011). *Global Food Losses and Food Waste. Extent, Causes and Prevention; Food and Agriculture Organization of the United Nations: Rome, Italy*. Available at: http://www.fao.org/3/a-i2697e.pdf (Accessed on 07.05.2021).

Frankowska, A., Rivera, X.S., Bridle, S. et al. (2020). Impacts of home cooking methods and appliances on the GHG emissions of food. *Nature Food*, 1, 787–791.https://doi.org/10.1038/s43016-020-00200-w.

García-Esteban, J.A., Curto, B., Moreno, V., González-Martín, I., Revilla, I. and Vivar-Quintana, A. (2018). A digitalization strategy for quality control in food industry based on Artificial Intelligence techniques. In: *IEEE 16th International Conference on Industrial Informatics (INDIN)*, pp. 221–226. https://doi.org/10.1109/INDIN.2018.8471994.

Goedhals-Gerber, L. L., & Khumalo, G. (2020). Identifying temperature breaks in the export cold chain of navel oranges: A Western Cape case. *Food Control, 110*,. https://doi.org/10.1016/j.foodcont.2019.107013

Han, J. W., Ruiz-Garcia, L., Qian, J. P., & Yang, X. T. (2018). Food Packaging: A Comprehensive Review and Future Trends. *Comprehensive Reviews in Food Science and Food Safety, 17*, 860–877.

İcoz, A. and Eker, B. (2018). Role of active, smart packaging in reduction of food loss. In: *3rd International conference on Quality of Life, Center for Quality, Faculty of Engineering.* University of Kragujevac. pp. 13–22.

Jat, D.S., Limbo, A.S., and Singh, C. (2019). Internet of Things for Automation in Smart Agriculture: A Technical Review. In R. Poonia, X. Gao, L. Raja, S. Sharma, and S. Vyas (Ed.), *Smart Farming Technologies for Sustainable Agricultural Development*, IGI Global. pp. 93–105. http://10.4018/978-1-5225-5909-2.ch005

Jayarajan, K., Poongothai, T. and Patra, P.S.K. (2021). Smart IoT System for the Reduction of Post-harvest Spoilage in Food Grains. *Implication of IoT in public projects*, Academic province, 1(1), 77–86. ISBN: 978–81–948555–1–4.

Jedermann, R., Nicometo, M., Uysal, I. and Lang, W. (2014). Reducing food losses by intelligent food logistics. *Philosophical transactions. Series A, Mathematical, physical, and engineering sciences,* 13; 372(2017): 20130302. https://doi.org/10.1098/rsta.2013.0302

Kaushik, R. and Singhai, J. (2019). An Approach for the Development of a Sensing System to Monitor Contamination in Stored Grain. In: *6th International Conference on Signal Processing and Integrated Networks (SPIN), Noida, India*, pp. 880–884.

Kitinoja, L., Saran, S., Roy, S. K., & Kadar, A. A. (2011). Postharvest technology for developing countries: Challenges and opportunities in research, outreach and advocacy. *Journal of the Science of Food and Agriculture, 91*(4), 597–603. https://doi.org/10.1002/jsfa.4295

Kozak, W., & Cierpiszewski, R. (2010). *Active Packaging. Przemysł Spożywczy, 64*(3), 36–38.

Kumar, D., & Kalita, P. (2017). Reducing postharvest losses during storage of crops to strengthen food security in developing countries. *Foods, 6*(1), 8. https://doi.org/10.3990/foods6010008

Leelananda, R., Champathi, G., Pathirana, S. M., & Fernando, T. N. (2021). Reducing post-harvest losses in fruits and vegetables for ensuring food security – Case of Sri Lanka. *MOJ Food Processing and Technology., 9*(1), 7–16. https://doi.org/10.15406/mojfpt.2021.09.00255

Luning, P. A., & Marcelis, W. J. (2020). Food quality management: Technological and managerial principles and practices. *Wageningen Academic Publishers.* https://doi.org/10.3920/978-90-8686-899-5

Mandyck, J. and Schultz, E.B. (2015). *Food foolish: waste, hunger and climate change.* Green Biz. Available at: http://www.greenbiz.com/article/food-foolish-waste-hunger-and-climate-change. (Accessed on 15.09.2021).

Martin, J.C. (2018). Climate change and post-harvest agriculture. In: *Agricultural Adaption to Climate Change in Africa.* Publisher: Routledge, Ed. Cyndi Spindell Berck, Peter Berck, Salvatore Di Falco.

Morettia, C. L., Mattosa, L. M., Calbob, A. G., & Sargentc, S. A. (2010). Climate changes and potential impacts on postharvest quality of fruit and vegetable crops: A review. *Food Research International, 43*(7), 1824–1832. https://doi.org/10.1016/j.foodres.2009.10.013

Mujumdar, A. S., & Law, C. L. (2010). Drying Technology: Trends and Applications in Postharvest Processing. *Food and Bioprocess Technology, 3*, 843–852. https://doi.org/10.1007/s11947-010-0353-1

Müller, P. and Schmid, M. (2019). Intelligent Packaging in the Food Sector: A Brief Overview. *Foods*, 8(1). https://doi.org/10.3390/foods8010016.

Networking of Agricultural Market Intelligence Centers (NIC). (2021). 'Post-Harvest Technology'. Kerala Agriculture University, Thrissur. Available at: https://www.amickau.nic.in/index.php?option=com_content&task=view&id=53&Itemid=87 (Accessed on 18.06.2021).

Nicastro, R., & Carillo, P. (2021). Food Loss and Waste Prevention Strategies from Farm to Fork. *Sustainability, 13*, 5443. https://doi.org/10.3390/su13105443

Onibonoje, M.O., Nwulu, N and Bokoro, P. (2019). A wireless sensor network system for monitoring environmental factors affecting bulk grains storability. *Journal of food process engineering*, 42(4). https://doi.org/10.1111/jfpe.13256.

Onwude, D. I., Chen, G., Eke-emezie, N., Kabutey, A., Khaled, A. Y., & Sturm, B. (2020). Recent advances in reducing food losses in the supply chain of fresh agricultural produce. *Advances in Postharvest Process Systems, 8*(11), 1431. https://doi.org/10.3390/pr8111431

Palipane, K.B. (2000). *Milling and quality improvement in rice. Proceeding of rice symposium.* Gannoruwa, Sri-Lanka, 14–15.

Realini, C. E., & Marcos, B. (2014). Active and intelligent packaging systems for a modern society. *Meat Science, 98*(3), 404–419. https://doi.org/10.1016/j.meatsci.2014.06.031

Shafiee-Jood, M., & Cai, X. (2016). Reducing food loss and waste to enhance food security and environmental sustainability. *Environmental Science and Technology, 50*(16), 8432–8443. https://doi.org/10.1021/acs.est.6b01993

Srinivasarao, C. H., Prasad, J. V. N. S., Choudhari, S. K., & Singh, A. K. (2020). Mainstreaming climate resilient villages in national programs towards sustainability of agriculture and environment in India. *Climate Change and Environmental Sustainability, 8*(2), 116–133. https://doi.org/10.5958/2320-642X.2020.00013.7

The World Bank. (2016). *Population, Total [Internet].* The World Bank. Available at: http://search.worldbank.org/all?qterm=world+population&title=&filetype=.(Accessed on 01.07.2021)

Tian, J., Bryksa, B. C., & Yada, R. Y. (2016). Feeding the world into the future – food and nutrition security: The role of food science and technology. *Frontiers in Life Science, 9*(3), 155–166. https://doi.org/10.1080/21553769.2016.1174958

UNICEF. (2018). *Sri Lanka needs stronger disaster preparedness and response to combat climate change.* Available at: https://www.unicef.org/srilanka/press-releases/sri-lanka-needs-stronger-disaster-preparedness-and-response-combat-climate-change (Accessed on 20.06.2021).

Udara, A., & Malsha, D. (2019). Development of Extruded Snacks Using Pumpkin Flour. *International Journal of Scientific and Technology Research, 08*(12), 1564–1566.

USAID. (2016). *Greenhouse Gas Emissions Factsheet: Sri Lanka. Climate links.* Available at: https://www.climatelinks.org/resources/greenhouse-gas-emissions-factsheet-sri-lanka (Accessed on 20.06.2021)

USAID. (2018). *Climate Risk Profile: Sri Lanka. Climate links.* Available at: https://www.climatelinks.org/resources/climate-risk-profile-sri-lanka (Accessed on 20.06.2021)

Wijayaratne, L. K. W., Fernando, M. D., & Palipane, K. B. (2009). Control of insect pests under ware-house conditions using smoke generated from partial combustion of rice (paddy) husk. *Journal of National Science Foundation of Sri Lanka, 37*, 125–134.

Yahaya, S.M. and Mardiyya, A.Y. (2019). Review of postharvest losses of fruits and vegetables. *Journal of scientific and technical research*, 13(4). https://doi.org/10.26717/BJSTR.2019.13.002448

Yam, K.L., Takhistov, P.T. and Miltz, J. (2005). Intelligent Packaging: Concepts and Applications. *Journal of Food Science, 70.* Available at: http://lib3.dss.go.th/fulltext/Journal/Journal%20of%20food%20science/2005%20v.70/no.1/23279jfsv70n1pR0001-0010ms20040382%5B1%5D.pdf

Yin, W. and He, C. (2021). Raman perception monitoring application of IoT technology in agricultural planting. In: *New developments of IT, IoT and ICT applied to agriculture. Smart innovation, systems and technologies*, vol. 183. https://doi.org/10.1007/978-981-15-5073-7-17

Zhang, F., Zhou, H., Zhou, X. and Liao, O. (2009). Simulation of temperature measurement system for grain storage based on ZigBee technology. In: *2nd IEEE International Conference on Computer Science and Information Technology, Beijing, China.* https://doi.org/10.1109/ICCSIT.2009.5234864.

Mitigation of the Effects of Climate Change on Agriculture Through the Adoption of Precision Agriculture Technologies

Muharrem Keskin and Yunus Emre Sekerli

Abstract Agriculture is one of the main causes of greenhouse gas (GHG) emissions while climate change affects agriculture with mainly higher water demand and reduced crop yields due to drought. New technologies offer enhanced precision management practices from tillage to harvest. Precision Agriculture (PA) aims to manage crops in a more efficient manner. PA technologies have the potential to mitigate the impacts of climate change by providing agricultural input reductions (water, fertilizer, pesticide, etc.), fuel savings, lower GHG emissions, less environmental pollution, higher yield, lower costs, increased profitability, and sustained natural resources. The adoption of PA technologies has been increasing especially in developed countries with bigger farm sizes, while they are also being adopted in developing countries. The most widely adopted technology is automatic steering reaching up to 80% in some regions of various countries. Aerial application of fertilizers and pesticides using drones has the potential to reduce water consumption. Yet, there are concerns regarding the threats to PA technologies such as data privacy, theft of data, signal disruption, and cyber-attacks on farming systems. This chapter presents some basic information about PA technologies and highlights their benefits, adoption rate, and future trends from the perspective of climate change, sustainability, and food security.

Keywords Climate change · Global warming · Precision agriculture · Sustainability · Food security

---

The original version of this chapter has been revised: Figures 1 and 2 have been replaced. A correction to this chapter can be found at
https://doi.org/10.1007/978-3-031-65968-3_24

M. Keskin (✉) · Y. E. Sekerli
Department of Biosystems Engineering, Hatay Mustafa Kemal University, Hatay, Türkiye
e-mail: keskin@mku.edu.tr

Y. E. Sekerli
e-mail: yesekerli@mku.edu.tr

M. Behnassi et al. (eds.), *Climate-Smart and Resilient Food Systems and Security*,
https://doi.org/10.1007/978-3-031-65968-3_20

1 Introduction

Numerous causal factors, including population growth, urbanization, climate change, environmental pollution, labor shortage, need for sustainability, traceability, food safety, cost reduction, yield increase, higher profitability demand, and some advancements in electronics, computing, and decision-making tools result in new concepts, methods, and techniques in agriculture such as organic farming, sustainable farming, no-till farming, conservation agriculture, climate-smart agriculture, vertical farming, and so on. Precision agriculture (PA) is among these new methods and technologies.

Climate change has been one of the crucial challenges to the planet Earth and its living organisms in the twenty-first century. It is the changes in weather patterns caused by the global temperature rise due to the increased concentration of heat-trapping greenhouse gases (GHGs)—mainly CH_4, CO_2, N_2O—in the atmosphere caused primarily by fossil-fuel emissions (Venkatramanan et al., 2020). The negative impacts of climate change comprise recurrent and severe storms, floods, and droughts threatening agricultural production and food security (Venkatramanan et al., 2020). In addition to these impacts, problems such as spreading pests and diseases, loss of biodiversity, modified ecosystems, and water shortage will intensify (FAO, 2019). Regarding the causes of the GHGs, the top three factors are reported to be industry (32%), agriculture and deforestation (27%), and power-heat production (18%) (McKinsey, 2020). Hence, agriculture is considered as one of the main causes of GHG emissions, greenhouse effect, and climate change.

2 Interactions Between Agriculture and Climate Change

There is a bidirectional relationship between agriculture and climate change (Table 1). While the agricultural sector triggers climate change due to GHG emissions, climate change threatens the environmental conditions needed for the sustainability of agricultural production. Some of the agricultural factors, such as internal combustion engines working with fossil fuels, result in a great amount of GHG emissions contributing to global warming and climate change. Conversely, climate change has a major impact on agriculture, such as yield reductions through variations in temperature and precipitation.

Agriculture has a significant share in climate change due to the activities accounting for approximately 13.5% of the total global anthropogenic GHG emissions (Soto et al., 2019). It was reported that about 12% of the European Union's net GHG emissions originated from agriculture (European Commission, 2022). The major GHGs produced through agricultural operations are CH_4, CO_2, and N_2O (Table 1). CO_2 is mainly emitted by the combustion of fossil fuels utilized in farm operations, transportation, crop processing, pumping irrigation water, tillage, harvesting, and the production of fertilizers. N_2O is sourced from microbial conversion of nitrogen in manure and soil supplied by ruminant livestock as urines and

Table 1 Bidirectional relationship between agriculture and climate change

Relationship	The mechanism of the relationship
Impact of agriculture on climate change	Agricultural activities result in GHG emissions mainly in three ways: – CO_2 from the combustion of fossil fuels in engines and irrigation pumps. Land-use change causes the release of CO_2 locked in plant biomass and soil by tillage – N_2O from manures applied to soil and excess use of synthetic nitrogen fertilizers – CH_4 is released from the enteric fermentation of livestock and the manures applied to soil and the cultivation of wet paddy rice
Impact of climate change on agriculture	– Changes in temperature and precipitation (droughts and floods) cause yield reduction – Depleted water resources increase water demand, restricting crop efficiency – Elevated atmospheric CO_2 gas improves photosynthetic activity (carbon fertilization) and reduces respiration that may compensate yield reductions – Raise in atmospheric temperature may extend the length of the growing season in northern temperate parts, but decrease it almost everywhere else

Source Developed by the authors

dungs or as a result of excessive use of synthetic nitrogen fertilizers, while CH_4 is released from the enteric fermentation of livestock and from the manures applied to soil and the cultivation of wet paddy rice (Balafoutis et al., 2017; FAO, 2011; Kozicka et al., 2021; Lal, 2022). N_2O poses a relatively lower warming effect but has a longer atmospheric existence, and thus higher influence than CO_2 (about 300 times higher) (NSAC, 2019). Apart from factors such as cultivation, excessive use of fertilizers, fossil fuels, and management of livestock wastes, GHG emissions also occur due to land-use change. To make a way for agriculture, destroying forests and tilling uncultivated land contribute to the release of CO_2 locked in plant biomass and soil (IIED, 2012). As a result of conversion with agricultural purposes of native grassland and forests that can hold 20–50 tons or more soil organic carbon (SOC) in the top 39 inches of soil profile, 30–50% of the SOC can oxidize into atmospheric CO_2 over a period of 50 years. This CO_2 oxidizing rate may increase to 75% within 25 years when tropical forests are cleared (NSAC, 2019).

Climate change has also crucial impacts on agricultural production (Table 1). Crop and livestock production in most regions are susceptible to climate change, particularly due to the variations in temperature and precipitation, especially in developing countries (IIED, 2012), although some northern temperate zone countries may actually benefit from these changing conditions (OECD, 2016). Elevated evaporation caused by higher temperatures along with drought results in plant stress, reduced crop growth, lower yields, depleted water resources as well as frequent and intense wildfires (NSAC, 2019). Climate change and reduced water availability increase water demand restricting crop efficiency, especially in regions where irrigation is mostly desired (FAO, 2011). Predictions of water needed to meet future demand for agriculture under climate change range from 40 to 100% extra water needed under no global warming conditions (FAO, 2011). An example is the dry summer experienced in 2018 in northern Europe in which only Sweden faced a 50% drop in cereal crops and a decline in livestock due to a lack of fodder and feed (Grusson et al., 2021). On the other hand, elevated atmospheric CO_2 quantities improve the photosynthetic activity of C3 crops and reduce respiration, which may compensate for the loss of production due to higher temperature (FAO, 2011).

3 Improved Technologies in Agricultural Production

Technological progress offers better management practices in many production sectors including agriculture (Fig. 1). Novel technologies provide more precision in farming tasks from tillage to harvest to lower inputs, increase profits, and protect the environment by reducing chemical fertilizers and pesticides and mitigating GHGs (Ess & Morgan, 2003; Ozguven, 2018; Rains & Thomas, 2009; Topakci et al., 2010). In fact, electronic, computer, and information technologies can be utilized in all types of crop and animal productions, including crop farming, horticulture, viticulture, forestry, aquaculture, turfgrass management, and livestock farming (Fig. 1) to get more efficient production and reduce their adverse effects on environment and climate change. Precision Agriculture (PA) or Precision Farming (PF) can be described as the management of the spatial and temporal variations of the fields and crops in a more efficient way by using advanced data collection, analysis, and application technologies to reduce agricultural inputs, improve crop performance, and sustain environmental quality. PA can also be named with various terms including Agriculture 4.0, smart agriculture, digital agriculture, and site-specific farm management.

4 Historical Development of Farming and Precision Agriculture

Historical developments in agriculture can be referred to as Agriculture 1.0 through Agriculture 5.0 in tandem with the developments in industry (Fig. 2) (Zambon et al., 2019; Dayioglu and Turker, 2021). Humans used animal power and simple implements to grow crops and raise animals until the 1850s (Agriculture 1.0). Until the 1950s, steam power, internal combustion engines, first models of tractors, and farm machinery were employed (Agriculture 2.0) (Fig. 2). The transistor was invented in 1947, which led to the fast development of computing technologies. Satellite systems were developed for remote sensing and geographical positioning and navigation

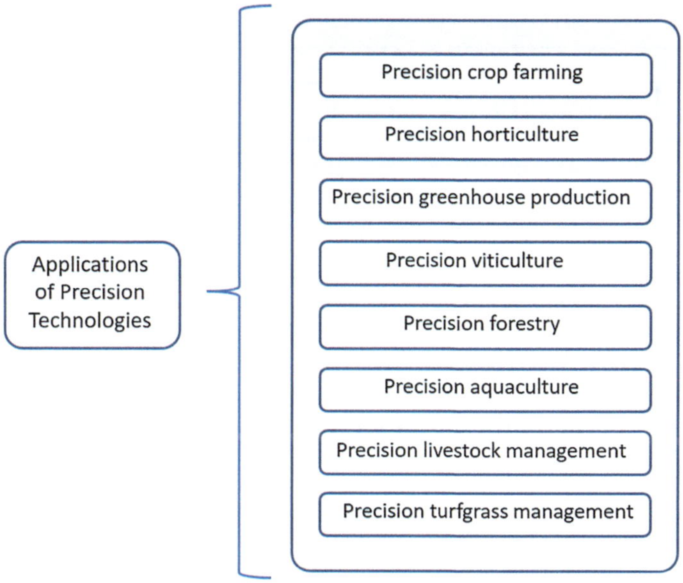

Fig. 1 Application of precision technologies to various production systems. *Source* Developed by the authors

during the 1960s and 1970s. A significant increase was accomplished in crop yield from 1950 to 1990 by utilizing better seeds, chemical fertilizers, and pesticides with less animal and human power and more mechanization, which was called 'green revolution' (Agriculture 3.0). However, excessive usage of chemical synthetic fertilizers and pesticides during this time period resulted in environmental pollution and negative impacts on the well-being of humans, other plants, and animals leading to better methods for the more careful usage of these inputs including integrated pest management (IPM). The first application of precision agriculture can be seen as agricultural remote sensing, satellite-based positioning systems (GNSS), yield monitoring and yield mapping systems, and automatic steering and guidance systems during the 1980s and 1990s (Agriculture 4.0). Further innovations such as machine vision, artificial intelligence, big data, cloud technology, Internet of Things, and blockchain are being employed within various technologies in agriculture. Autonomous electric-powered and hybrid tractors, and field and aerial robots are still being developed for farming (Agriculture 5.0) (Fig. 2). Electric-powered tractors and robots have the potential to reduce the use of fossil-based fuels, and therefore GHGs that have the potential to mitigate the impact of climate change.

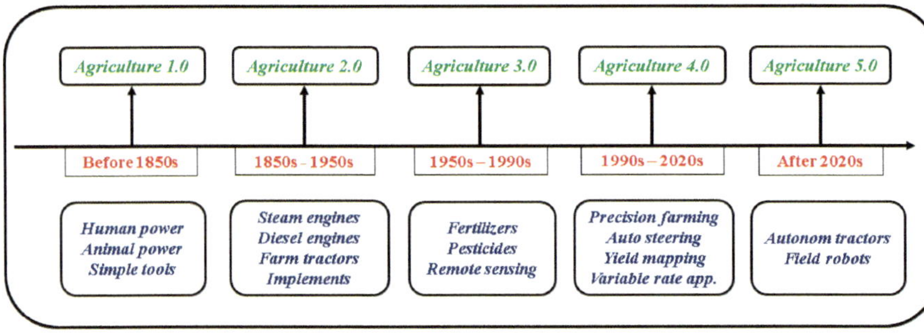

Fig. 2 Historical development stages of agricultural technologies. *Source* Developed by the authors

5 Precision Agriculture (PA) Technologies

PA encompasses various technologies that can be classified into three groups such as data collection, data processing—decision-making, and application technologies (Dodd et al., 1999; Keskin & Gorucu Keskin, 2012; Keskin et al., 1999) (Fig. 3).

Within the data collection technologies, positioning systems are utilized to determine the coordinates of a location in the field or of farm machinery (Fig. 3). Satellite-based positioning (GNSS) systems along with various augmentation services provide high precision positioning up to 2 cm accuracy. Yield monitoring systems are mounted on harvesting machinery to quantify and record the yield along with location data on the go, and then these data are used to create yield maps that show spatial yield variability to determine lower yielding and higher yielding management zones

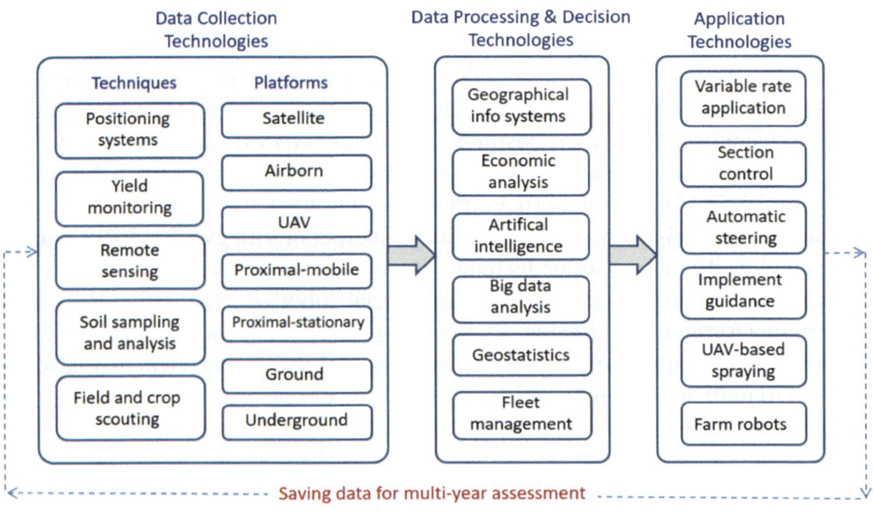

Fig. 3 Classification of the basic PA technologies. *Source* Developed by the authors

within a field. Remote sensing offers data collection from soil and crops using sensors mounted on satellites, aircraft, and unmanned aerial vehicles (UAV) or drones. Soil analysis and mapping technology is used to quantify various properties of soil such as organic matter, electrical conductivity, moisture content, pH, texture, salinity, and nutrient content. Field and crop scouting provides an opportunity to check the status of the field and the crops against any problems, including biotic and abiotic stress factors such as nutrient deficiency, water stress, drainage, pests, and diseases by utilizing ground or aerial platforms. These techniques are used on various platforms (Fig. 3).

Data processing and decision-making technologies are employed to convert raw data into useful information (Fig. 3). Among them, geographical information systems (GIS) are used to process the spatial (geo-referenced) data to create various maps, including yield maps and variable rate prescription maps by employing various data analyzing techniques such as geostatistic methods. For large fields, it is common to use more than one farm tractor and machinery concurrently to finish the work in a planned time period; in this case, fleet management methods are used to organize and plan the multi-machinery operations.

As for the application technologies (Fig. 3), variable rate application (VRA) deals with the application of farm inputs including tillage, lime, seeds, water, fertilizer, and pesticides in variable amounts based on the spatial variability of the soil and crop. These inputs are applied in two ways: (a) based on the data obtained from an application map (map-based VRA) with the aid of a GNSS receiver or (b) based on the data collected from a sensor (sensor-based VRA) without a GNSS receiver. VRA offers reducing GHG emissions through savings of water, energy, fertilizers, pesticides, and other inputs. Section control technique allows turning on or off a section of machinery especially in fertilizer and pesticide application controlling individual sections or nozzles of a sprayer to prevent unnecessary application of chemicals, and this could be done automatically using location data from GNSS receivers.

By using automatic steering/auto-guidance (AS/AG), it is possible to steer a farm tractor or self-propelled machinery such as a combine harvester based on the geo-location data from a GNSS receiver or imaging system. Farm implements such as sowing machinery can also be guided using active or passive implement guidance, especially in turns and in sloped fields. GNSS-assisted guidance of tractors allows to reduce the overlapping of passes during chemical fertilizer and pesticide application, which results in saving of fuel, water, chemicals, labor, and time. Aerial and field robots are being developed to be used indoors (in greenhouses and animal barns) or in open fields. Unmanned drones are currently being utilized in the application of foliar pesticides and fertilizers providing many benefits such as savings of water, fossil fuels, and energy and reducing GHG emissions.

6 Benefits and Potential of PA for the Mitigation of Climate Change

PA technologies have the potential to provide some important benefits (Table 2), such as reduced agricultural inputs (seed, lime, water, energy, fertilizer, pesticide, etc.), fuel savings, less GHG emissions and environmental pollution, increased yield, lower production costs, increased profitability, better managed farms, sustained natural resources (water, soil, etc.), improved work efficiency (easiness, convenience, working at night, etc.), and less labor-dependence. For example, automatic steering of farm tractors and self-propelled machinery provide many benefits such as fossil-fuel savings, less emissions, and reduction of fertilizers, water, and pesticides (Table 2). Similarly, drone-based aerial application of fertilizers and pesticides offers many benefits including savings in water, fertilizer, and pesticides, reduction in fossil-fuel use and emissions, reduction in soil compaction due to machinery traffic, etc. (Aslan et al., 2022). Amounts of fertilizers, pesticides, and irrigation water can also be reduced through the use of VRA technologies. Overall, PA technologies have a very big potential to reduce the causes and effects of climate change as summarized in Table 2.

Results of some previous studies are presented revealing the benefits that can be obtained from the auto-guidance (AG) systems (Table 3), variable rate (VR) irrigation systems (Table 4), variable rate (VR) pesticide application systems (Table 5), and variable rate (VR) fertilizer application systems (Table 6). As confirmed by these studies, PA technologies have a significant potential for savings in fuel consumption resulting in less GHG emissions and less global warming and climate change effects. In addition, significant amounts of energy, water, fertilizer, and pesticide savings were also reported in these studies.

Many studies were conducted to reveal the advantages of automatic steering in agriculture against climate change compared to manual steering (Table 3). These studies were carried out in the form of monitoring agricultural processes such as tillage, planting, spraying, fertilizing, and harvesting for intensively cultivated field crops such as wheat, oats, canola, corn, legumes, and peanuts. When the results of these studies were examined, the main positive feature of the automatic steering for climate change was observed as the lower GHG emission in agricultural production through fuel savings. In addition, fertilizer, pesticide, seed, labor, and time savings were achieved due to the reductions in overlaps and spaces with better accuracy in parallel passes of the machinery (Keskin et al., 2018) (Table 3).

PA technologies offer an alternative to uniform applications of agricultural inputs in many farm operations, including variable-level irrigation. According to the results of the studies performed with field crops and vineyards, variable rate irrigation provided less irrigation water and pumping energy consumption compared to uniform applications (Table 4). Especially, for field crops, significant savings were obtained by irrigating only the needed place in the required amounts with the center pivot systems with section or zone control features based on soil maps. The most remarkable benefit of the variable rate irrigation was savings of water and energy with no

Table 2 Mitigation potential and benefits of PA related to climate change

PA technology	Mitigation impact level	Benefits and mitigation impact
Variable depth tillage	Moderate	– Fuel and energy savings – Less GHG emissions due to fuel savings – Less CO_2 release from soil
Variable rate sowing	Low	– Optimized crop growth and yield – Less fertilizer and pesticide may be possible
Variable rate irrigation	High	– Less irrigation water – Better crop growth due to optimized use of water – Optimized soil health and lower cost
Variable rate fertilizer application	High	– Fertilizer savings and less N_2O release – Better crop growth and yield due to optimized fertilizer use – Energy savings and lower cost
Variable rate chemical herbicide and pesticide application and section control	High	– Savings in chemical herbicides and other pesticides – Water savings and lower cost – Better crop growth and yield due to reduced toxicity – Less herbicide drift and environmental pollution – Less chemical residue and healthier agricultural products
Automatic steering/implement guidance	High	– Reduced overlaps in pesticide applications – Savings of fuel, energy, fertilizer, pesticide, and water – Savings of labor and time – Less GHG emissions due to fuel savings – Ergonomic working conditions, operation at night
Unmanned aerial vehicles (UAVs) for foliar spraying of fertilizers and pesticides	High	– Fuel and energy savings – Less GHG emissions due to fuel savings – Savings of water, labor, and time – Timely application in muddy fields after rain or irrigation – Less soil compaction due to less machinery traffic

(continued)

Table 2 (continued)

PA technology	Mitigation impact level	Benefits and mitigation impact
Farm robots (electric-powered machinery)	Moderate	– Fuel savings – Less GHG emissions due to electrical energy source
Precision livestock farming, PLF	Moderate	– Reduced GHG emissions caused by farm animals – Lower GHG emissions from animal urines and dung

Source Developed by the authors

decrease in yield or quality of the crop. In some studies, an increase in yield was observed along with water and energy savings (Table 4).

Variable rate pesticide application is the most common variable rate technology, especially in orchards offering the opportunity to spray effectively by detecting and targeting only the fruit trees by means of infrared, machine vision, laser, LIDAR, photoelectric, and ultrasound sensors (Table 5). These eco-friendly technologies provide advantages in terms of sustainability by minimizing the amount of water and chemical fertilizer or pesticide inputs making sustainable agricultural production possible. In some studies, it was observed that up to 86% decrease in spray volume rate were reported (Table 5). Another important technology is the field robot-based selective herbicide application. This type of robot determines the weeds using machine vision and then burns the weeds by using high-powered LASER beam energy without using any water and pesticide resulting in a significant savings of water and chemical herbicides (Delbery, 2021).

Map-based or sensor-based variable rate fertilizer applications have noteworthy potential to provide significant fertilizer savings, especially in the cultivation of field crops by managing the spatial variability of plants or soil (Table 6). In addition, utilizing these methods in the use of nitrogenous fertilizers, which causes N_2O emissions as a result of evaporation in excessive use, can make a significant difference in terms of global climate change. In addition to its positive impact on climate change due to lower fertilizer evaporation, the use of variable rate fertilization has the potential to prevent agricultural sustainability threats such as soil salinity problems or groundwater and surface water pollution by minimizing excessive use of chemical fertilizers.

Table 3 Results of some previous research studies showing the benefits of auto-guidance (AG) systems

AG system	Plant/application	Result	Source
Semi AG (SAG) versus manual guidance	Wheat/sowing	2.4% less sowing cost 6.3% lower fuel consumption per ha 2.7% lower seed usage with SAG	Scarfone et al. (2021)
Manual versus LASER assisted AG	Wheat/harvest	Less fuel consumption (2 L ha^{-1})	Juostas and Jotautiene (2021)
Manual versus RTK-assisted AG	Canola, beans, oats, peas, wheat/harvest	22.0% less fuel consumption 4.8% less global warming emissions	Jotautiene et al. (2021)
Manual versus AG	Cotton, peanut, corn/spraying	Lower overlaps and spacing with AG (4–8%) compared to manual guidance	Topcueri and Keskin (2019)
Manual versus AG	Field crops/tillage, sowing, fertilizing	Reduction in overlaps 15% savings in fuel, seeds, fertilizer	CECE (2018)
Manual versus AG (survey study)	Field crops/all applications	AG offers benefits: flexible working hours, time-savings, fuel savings	Keskin et al. (2018)
Route planning with AG	Grasses/tillage, sowing, mowing	8% reduction in fuel consumption	Rodias et al. (2017)
Manual versus RTK-assisted AG	Weeds/mechanical control, herbicide application	7–30 USD saving in herbicide per acre	Griffin and Lowenberg-DeBoer (2017)
Manual versus RTK-assisted AG	Field crops/seedbed preparation, tillage, seeding	1–6% overlap reduction of the machine working width	Kviz et al. (2014)
Manual versus RTK-assisted AG	Peanut/harvesting by digger	Higher net return of 94–695$ ha^{-1} (includes fuel saving)	Ortiz et al. (2012)
Manual versus AG	Field crops/tillage, planting, spraying, harvesting	6.3% fuel saving 1647–1866 L fuel saving per farm	Bora et al. (2012)
Manual versus AG	Soybean, corn/tillage, planting, spraying, harvesting	0.9% (3.3$ per acre) increase in net return (fuel saving and less overlaps)	Shockley et al. (2011)

Table 4 Results of some previous studies showing the benefits of variable rate (VR) irrigation systems

Irrigation method	Plant/application	Result	Source
Drip irrigation	Vineyard/VR irrigation	18% water saving without loss in yield and product quality	Ortuani et al. (2019)
Center pivot, lateral-move	Field crop/VR irrigation	Reduction in irrigation 27–55 mm ha^{-1}	Hedley et al. (2017)
Center pivot	Soybean, corn/VR irrigation	25% water saving, 0.8–2.8% yield increase	Sui and Yan (2017)
Drip irrigation	Vineyard/VR irrigation	20% water saving, 17% yield increase	Nadav and Schweitzer (2017)
Center pivot	Field crop/VR irrigation	Reduction in irrigation up to 18 mm	Miller et al. (2017)
Center pivot	Field crop/VR irrigation	0.3% or 1.3% (51 or 25 mm) less annual irrigation pumping	Lo et al. (2016)
Center pivot	Field crop/VR irrigation	19% better irrigation accuracy, 73–93% of the field had optimal irrigation	LaRue and Evans (2012)
Reel hose irrigation	Potato/VR irrigation	474 m^3 ha^{-1} water saving 250 kWh ha^{-1} electric power saving	Jobbágy et al. (2011)
Center pivot	Pasture, maize/VR irrigation	22–26% annual water saving	Hedley and Yule (2009)
Center pivot	Sugar beet/VR irrigation	13% water saving	Al-Kufaishi et al. (2006)
Center pivot	Potato, corn/VR irrigation	13% yield increase, 15% water saving for potato, 9–21% water saving for corn	Sadler et al. (2005)

7 Adoptions of PA Technologies

7.1 Current Status of PA Adoption

PA technologies have significant potential for savings in the consumption of fuel, energy, water, fertilizer, and pesticides resulting in less GHG emissions, and less global warming and climate change effects as confirmed by previous studies depicted in Tables 3, 4, 5 and 6. In order to maximize these effects, these technologies should

Table 5 Results of some previous studies showing the benefits of variable rate (VR) pesticide application

Application system	Plant/application	Result	Source
Orchard sprayer	Pre-emergence VR herbicide application	Up to 34% herbicide (and water) saving	Gundy and Dille (2022)
Field sprayer	VR pesticide application	Up to 66% pesticide (and water) saving	Rajmis et al. (2022)
Orchard sprayer	Apple, cherry trees/VR pesticide application	22–51% pesticide (and water) saving	Dou et al. (2021)
Orchard sprayer	VR pesticide application	54% pesticide and water saving	Asaei et al. (2019)
Orchard sprayer	Pistachio trees/VR pesticide application	34% pesticide and water saving	Maghsoudi et al. (2015)
Orchard sprayer	Apple trees/VR pesticide application	47–73% spray mixture saving	Chen et al. (2013)
Field sprayer	Wheat/VR insecticide application	13% insecticide (and water) saving	Dammer and Adamek (2012)
Orchard sprayer	Apple trees/VR pesticide application	50–75% pesticide (and water) saving	Xiongkui et al. (2011)
Orchard sprayer	Nursery tree/VR pesticide application	71–86% spray volume saving	Jeon et al. (2011)
Orchard sprayer	Vineyard/VR pesticide application	47–67% spray volume saving	Llorens et al. (2010)
Orchard sprayer	Olive, pear, apple trees/VR pesticide applications	28–70% pesticide (and water) saving	Solanelles et al. (2006)
Field sprayer	Cereals, maize, sugar beet/VR herbicide application	54% herbicide (and water) saving	Timmermann et al. (2003)

be used to a higher extent. However, the adoption rate of PA technologies is influenced by numerous parameters, including the cost and profitability of the equipment, features of the farms, properties of the technologies such as compatibility and complexity, personality of the farmer, legal frameworks, and the availability of firms and institutions offering technical support for these technologies (Table 7).

The adoption rate is variable in various countries and also in different regions within a country. Developed countries have an increasing trend in the adoption of PA, mainly in the US. Significant growth is also visible in some other developed countries, particularly in those having bigger farm sizes such as Canada and Australia since big farm sizes and higher income have a significant encouraging effect toward the acceptance of the PA systems. In addition, farmers in some developing countries including Türkiye started using PA equipment, particularly auto-steering systems and drone-based aerial pesticide applications in the last decade (Akdemir, 2016; Keskin et al., 2018). Among the PA technologies, auto-steering and guidance systems are

Table 6 Results of some previous studies showing the benefits of variable rate (VR) fertilizer application

Application system	Plant/application	Result	Source
Fertilizer spreader	Wheat/VR N fertilizer application	Up to 33.4 kg N ha^{-1} saving without significant change in yield	Santaga et al. (2021)
Fertilizer spreader	Maize/VR N fertilizer application	Up to 17 kg N ha^{-1} saving without significant change in yield	Schillaci et al. (2021)
Fertilizer spreader	Canola/VR N fertilizer application	7–19 kg N ha^{-1} saving	Khakbazan et al. (2021)
Fertilizer spreader	Barley, oilseed rape, wheat, bean/VR NPK fertilizer application	25% saving in NPK fertilizer 3463.1 MJ ha^{-1} energy saving 341 kg CO_{2eq} reduction per ha	Kazlauskas et al. (2021)
Pneumatic fertilizer applicator	Coffee/VR NPK fertilizer application	17–24% P_2O_5, 17–21% K_2O fertilizer saving per season	Angnes et al. (2021)
Applied by hand	Apple trees/VR NPK fertilizer application	59–63% fertilizer saving	Liakos et al. (2020)
Fertilizer spreader	Rice/VR N fertilizer application	Up to 11.9% N saving per ha	Ghazali et al. (2020)
Applied by hand	Barley/sensor-based VR N fertilizer application	75 kg ha^{-1} N fertilizer saving	Fabbri et al. (2020)
Fertilizer spreader	Wheat/satellite-based VR N fertilizer application	25% N fertilizer saving per ha	Vizzari et al. (2019)
Fertilizer spreader	Turfgrass/camera-based VR N fertilizer application	16–23% N fertilizer saving per ha	Kabir et al. (2019)
Center pivot fertigation	Cotton/sensor-based VR N fertilizer application	66–69% reduction in N requirement, no significant change in yield	Williams (2018)
Fertilizer spreader	Corn/UAV-based VR N fertilizer application	Up to 28 kg N saving per ha	Thompson (2018)
Liquid fertilizer applicator (RAMP)	Corn/sensor-based VR N fertilizer application	21–34% reduction in N application rate, no significant change in yield	Khalilian et al. (2017)
Fertilizer applicator	Wheat/VR fertilizer application	4–37% fertilizer savings	Tekin (2010)

Table 7 Factors affecting the adoption level of PA technologies

Factor	Source
Lack of familiarity and awareness about the PA technologies	Keskin and Sekerli (2016), György et al. (2018), Bucci et al. (2019), Vecchio et al. (2020)
Sufficiency of classical methods	Paudel et al. (2011)
Personality of the farmer, age, education, computer usage	Daberkow and McBride (2003), Edwards-Jones (2006), Lowenberg-DeBoer and Griffin (2006), Barnes et al. (2019)
Influence of other farmers	Barnes et al. (2019)
Communication, social media	Veltheim et al. (2021), Colussi et al. (2022)
Confidence of the farmer on agronomic recommendations	Erickson et al. (2017)
Psychological factors of the farmer, risk aversion, farmer perception	Foguesatto et al. (2020)
Family structure of the farmer	Edwards-Jones (2006)
Social factors, social interactions	Edwards-Jones (2006), Kutter et al. (2011), Gotor et al. (2020)
Farm size, income level, and the other features of the farm	Isgin et al. (2008), Erickson et al. (2017), György et al. (2018), Kernecker et al. (2020)
Legal issues on environmental protection, sustainability	Edwards-Jones (2006)
Confidentiality, data privacy	PPAEP (2018)
Economic factors, cost of equipment, return on investment time, profitability	Barnes et al. (2019), György et al. (2018), Kernecker et al. (2020), Sylvère and D'amour (2020), Spati et al. (2021)
Features of the system, complexity, technical support, compatibility	Paudel et al. (2011), Kutter et al. (2011), Pathak et al. (2019), Sylvère and D'amour (2020), Masi et al. (2023)
Difficulty of the data processing and decision support systems	Fountas et al. (2005)
Supporting firms, institutions, dealers	Edwards-Jones (2006), Fountas et al. (2005)
Technical service	Whipker and Akridge (2009), Kutter et al. (2011)
Advertisement, fairs, workshops, demonstration farms, field days	Kutter et al. (2011)

the most widely adopted technology in the world (up to 80–90% in some countries) (McFadden et al., 2023; Topcueri & Keskin, 2019).

7.2 Future Trends for the Adoption of PA Technologies

PA market is expected to continue growing steadily in the near future. A report predicts that the PA market is projected to grow from USD7.0 billion to 12.8 billion in 2020–2025 (Markets & Markets, 2020). The future trends for the PA technologies can be summarized as below.

Currently, auto-steering can be considered the very extensively adopted PA technology. It is estimated that it will be used extensively in the future while light bar guidance technology has been on a decreasing trend (Erickson et al., 2017; Miller et al., 2019). Farmers can get many benefits from automatic guidance technology, including reduced operator fatigue, reduced overlapping, reduced skips, and less agricultural inputs (fuel, water, seeds, fertilizer, pesticides, etc.) (Grisso et al., 2009; Keskin et al., 2018). In the upcoming years, auto-guidance is expected to be a standard technology for new farm tractors. Moreover, the first prototypes of driverless (autonomous) farm tractors and electric-powered tractors are currently being introduced in various countries, particularly in the US.

The farm robot market is also estimated to grow widely from USD2.7 billion in 2016 to USD12.8 billion by 2022 (GPS World, 2017). Also, it is thought that field swarm robots will be used in the future and they have the ability to obtain minimum production costs in relatively small farm sizes (Lowenberg-DeBoer et al., 2019). In a survey study in the Bavaria region of Germany, 22.2% of the surveyed farmers ($n = 174$) intended to invest in field crop robots while larger farms prefer large autonomous tractors for financial benefits; however, small-scale or organic farms favor small robots due to their environmental benefits (Spykman et al., 2021).

Drones or unmanned aerial vehicles (UAVs) are anticipated to be another technology to be widely adopted in agriculture. The US commercial drone market was about USD 0.6 billion in 2017 and is expected to grow to about USD1.0 billion in 2022 and agriculture composed of about one-fourth of this amount (Grandviewresearch, 2016). The global market of agricultural drones is expected to expand from USD0.3 billion in 2016 to about USD3.7 billion in 2024 (Researchnester, 2021). UAVs are expected to be used not only for remote sensing purposes but also for the aerial application of pesticides and fertilizers. Also, satellites that can obtain data more frequently (nearly daily) will be expected to gain wider adoption for remote sensing purposes since the UAVs have some disadvantages such as covering limited area with a full battery, strict flight regulations, and difficulties in their operation (Grassi, 2019).

The use of Internet of Things (IoT) is expected to increase as well. According to a report, over 75 million IoT devices on farms were expected to be used by 2020 and an average farm is anticipated to yield 4.1 million data points (big data) daily in 2050 up from 0.2 million in 2014 (De Clercq et al., 2018). However, based on the extensive data produced by the IoT sensors, big data and cloud computing technologies are expected to grow rapidly. In addition, variable rate application (VRA) technology is expected to be expansively adopted. A report states that 330 000 VRA systems will be sold in 2023, up from about 40 000 in 2015 (Kanicki, 2017).

7.3 Threats for the Broader Adoption of PA

Some factors could be threats to the wider adoption of PA technologies that can be classified under three titles (PPAEP, 2018):

- First, *threats to confidentiality*: data privacy, theft of data, publishing of confidential information from the PA industry, foreign access to drones, and sale of confidential data.
- Second, *threats to integrity*: falsification of data and insufficiently modeled algorithms.
- Last, *threats to equipment availability*: timing of availability, GNSS signal disruption or loss, interruption of RTK augmentation signals, disruption in rural communication systems, foreign access to equipment and machinery, and cyber-attacks on farming systems such as animal heating ventilation and air conditioning (HVAC), and feeding systems.

In a report covering a PA dealer survey results, the three most important barriers to the expansion of PA technologies were found to be the difficulty of finding PA service employees (62%); the quick changes of the PA equipment which increase costs (50%); and incompatibilities across types of PA equipment (different data formats, inability to share information) (41%) (Erickson et al., 2017). In the same report, for the producers, the three most important barriers were farm income level (65%); cost of PA services (34%); and farmers' lack of confidence in the site-specific agronomic recommendations (24%).

The use of Internet-connected devices may also pose vulnerabilities within the PA system (Sontowski et al., 2020; Faguy, 2021). Cyber-attackers or hackers can exploit these vulnerabilities and disrupt or control data transfer security among the field sensors and autonomous tractors, combines, and drones causing significant technical and economic problems (Sontowski et al. 2020). Cyber-attacks demanding ransom from farms and agriculture-related companies are on an increasing trend in recent years in the US and Canada (Griffiths, 2022).

Data privacy can also be considered as an important threat. In an online survey with 561 grain farmers, only 36% reported their readiness to share their agricultural data with a big data platform. Farmers want to share their data with university researchers to a higher extent; however, fairly small financial and non-financial benefits considerably improved participation (Turland & Slade, 2019).

Dealers involved in PA services report that the difficulty of finding PA service employees is one of the biggest problems in the PA service offerings (Erickson et al., 2017). Thus, an increase in the number of training and education programs on PA is eminent.

Compatibility between different models and brands of farm tractors and PA equipment has also been considered as an important issue and barrier to the adoption of PA technologies (Kutter et al., 2011; Pathak et al., 2019). ISOBUS technology can be a solution to this problem and the adoption of ISOBUS-compliant tractor and PA equipment is on an increasing trend. In a survey in the US, 66% of the respondent farmers reported that they use ISOBUS-compliant implements, while 75% of them used one or two tractors that are ISOBUS-compliant (NoTill Farmer, 2019).

Denying the right to repair (R2R) newer technology products such as farm tractors and PA equipment can also be considered as another barrier to the adoption of PA technologies. Farmers want to have an option to repair their equipment wherever they want to reduce repair costs. Some senators submitted bills to provide farmers the right to repair in the US (Fatka, 2022).

8 Conclusions and Recommendations

This chapter presented some basic information about PA technologies and highlighted the benefits that can be obtained from them from the perspective of climate change, food security, and sustainability. PA technologies offer benefits and have the potential to mitigate the impacts of climate change by means of offering reductions in agricultural inputs (seed, lime, water, fertilizer, pesticide, etc.), fuel savings, lower GHG emissions and environmental pollution, optimized crop growth, increased yield, lower production costs, increased profitability, better managed farms, sustained natural resources (water, soil, etc.), improved work efficiency (easiness, convenience, working at night, etc.), and less human labor dependence.

The adoption of PA systems is on a rising trend, especially in various developed countries with bigger farm sizes and higher income levels. In recent years, these systems have been adopted in some developing countries as well. The most widely adopted technology is automatic steering reaching an adoption rate of up to 80–90% in some regions of some countries. Aerial application of fertilizer and pesticides using drones or UAVs has a big potential to be utilized to a significant extent in the near future due to significant advantages including savings of fuel, GHG emissions, water, fertilizers, and pesticides contributing to mitigate the effects of climate change. The most important PA technologies with a significant influence on reducing the effect of climatic changes are thought as variable rate fertilizer application, variable rate pesticide application, variable rate irrigation, automatic steering, and drone-based aerial application of pesticides and fertilizers.

On the other hand, there are some concerns regarding the threats to these PA technologies such as confidentiality, data privacy, theft of data, foreign access to drone systems, sale of confidential data, falsification of data, equipment availability, timing of availability, GNSS signal disruption or loss, interruption of GNSS augmentation signals, disruption in rural communication systems, foreign access to equipment and machinery, and cyber-attacks with ransom demand on farming systems.

References

Akdemir, B. (2016, October). Evaluation of precision farming research and applications in Turkey. In *7th International scientific agriculture symposium, Jahorina, Bosnia.*

Al-Kufaishi, S. A., Blackmore, B. S., & Sourell, H. (2006). The feasibility of using variable rate water application under a central pivot irrigation system. *Irrigation and Drainage Systems, 20,* 317–327.

Angnes, G., Martello, M., Faulin, G. D., Molin, J. P., & Romanelli, T. L. (2021). Energy efficiency of variable rate fertilizer application in coffee production in Brazil. *AgriEngineering, 3*, 815–826.

Asaei, H., Jafari, A., & Loghavi, M. (2019). Site-specific orchard sprayer equipped with machine vision for chemical usage management. *Computers and Electronics in Agriculture, 162*, 431–439.

Aslan, M. F., Durdu, A., Sabanci, K., Ropelewska, E., & Gultekin, S. S. (2022). A Comprehensive survey of the recent studies with UAV for precision agriculture in open fields and greenhouses. *Applied Science, 12*, 1047. https://doi.org/10.3390/app12031047

Balafoutis, A., Beck, B., Fountas, S., Vangeyte, J., Wal, T. V., Soto, I., Gómez-Barbero, M., Barnes, A. P., & Eory, V. (2017). Precision Agriculture technologies positively contributing to GHG emissions mitigation, farm productivity and economics. *Sustainability, 9*, 1339.

Barnes, A. P., Soto, I., Eory, V., Beck, B., Balafoutis, A., Sanchez, B., Vangeyte, J., Fountas, S., van der Wal, T., & Gomez-Barbero, M. (2019). Exploring the adoption of precision agricultural technologies: A cross regional study of EU farmers. *Land Use Policy, 80*, 163–174.

Bora, G. C., Nowatzki, J., & Roberts, D. C. (2012). Energy savings by adopting precision agriculture in rural USA. *Energy, Sustainability and Society, 2*, 22.

Bucci, G., Bentivoglio, D., Finco, A, & Belletti, M. (2019). Exploring the impact of innovation adoption in agriculture: How and where Precision Agriculture technologies can be suitable for the Italian farm system? In *IOP conference series: Earth and environmental science* (Vol. 275, p. 012004). https://doi.org/10.1088/1755-1315/275/1/012004

CECE. (2018). CECE & CEMA Optimizing our industry to reduce emissions. Retrieved November 20, 2021, from www.cece.eu/news/new-brochure-cece-and-cema-optimising-our-industry-2-reduce-emissions

Chen, Y., Ozkan, H. E., Zhu, H., Derksen, R. C., & Krause, C. R. (2013). Spray deposition inside tree canopies from a newly developed variable-rate air-assisted sprayer. *Transactions of the ASABE, 56*(6), 1263–1272.

Colussi, J., Morgan, E. L., Schnitkey, G. D., & Padula, A. D. (2022). How Communication affects the adoption of digital technologies in soybean production: A survey in Brazil. *Agriculture, 12*, 611. https://doi.org/10.3390/agriculture12050611

Daberkow, S., & McBride, W. (2003). Farm and operator characteristics affecting the awareness and adoption of precision agriculture technologies in the US. *Precision Agriculture, 4*, 163–177.

Dammer, K., & Adamek, R. (2012). Sensor-based insecticide spraying to control cereal aphids and preserve lady beetles. *Agronomy Journal, 104*(6), 1694–1701.

Dayioglu, M. A., & Turker, U. (2021). Digital transformation for sustainable future—Agriculture 4.0: A review. *Journal of Agricultural Sciences, 27*(4), 373–399. https://doi.org/10.15832/ankutbd.986431

De Clercq, M., Vats, A., & Biel, A. (2018). *Agriculture 4.0: The future of farming technology.* World Government Summit 25pp. Retrieved September 22, 2021, from www.worldgovernmentsummit.org/api/publications/document?id=95df8ac4-e97c-6578-b2f8-ff0000a7ddb6

Delbery, C. (2021). This badass robot uses lasers to slay 100,000 weeds per hour. Retrieved February 12, 2022, from www.popularmechanics.com/technology/robots/a36331690/autonomous-farming-robot-kills-weeds

Dodd, R. B., Han, Y. J., Khalilian, A., & Keskin, M. (1999, April). Farm mechanization in USA for environment-friendly agriculture - environmentally friendly agriculture: Is it possible? In *International symposium on farm mechanization for environment—friendly agriculture*, Seoul, South Korea.

Dou, H., Zhai, C., Chen, L., Wang, X., & Zou, W. (2021). Comparison of orchard target-oriented spraying systems using photoelectric or ultrasonic sensors. *Agriculture, 11*, 753.

Edwards-Jones, G. (2006). Modeling farmer decision-making: Concepts, progress and challenges. *Animal Science, 82*, 783–790.

Erickson, B., Lowenberg-DeBoer, J., & Bradford, J. (2017). *2017 precision agriculture dealership survey* (27pp.). Purdue University. Retrieved December 10, 2021, from www.agribusiness.purdue.edu/wp-content/uploads/2019/07/croplife-purdue-2017-precision-dealer-survey-report.pdf

Ess, D., & Morgan, M. (2003). *The precision-farming guide for agriculturists* (138 pp.). Deere & Company, Moline, Illinois.

European Commission. (2022). Common agricultural policy for 2023–2027: 28 CAP strategic plans at a glance. Retrieved July 16, 2023, from https://agriculture.ec.europa.eu/system/files/2022-12/csp-at-a-glance-eu-countries_en.pdf

Fabbri, C., Napoli, M., Verdi, L., Mancini, M., Orlandini, S., & Dalla Marta, A. (2020). A sustainability assessment of the Greenseeker management tool: A lysimetric experiment on barley. *Sustainability, 12*, 7303.

Faguy D. (2021). Farms are great targets for hackers. Retrieved February 7, 2022, from www.farms.com/ag-industry-news/farms-are-great-targets-for-hackers-562.aspx

FAO. (2011). Climate change, water and food security. Retrieved September 2, 2021, from www.fao.org/3/i2096e/i2096e.pdf

FAO. (2019). FAO's work on climate change. Retrieved September 2, 2021, from www.fao.org/3/ca7126en/ca7126en.pdf

Fatka, J. (2022, February 2). Senator Tester's bill gives farmers 'right to repair'. Retrieved May 30, 2022, from www.farmprogress.com

Foguesatto, C. R., Borges, J. A. R., & Machado, J. A. D. (2020). A review and some reflections on farmers' adoption of sustainable agricultural practices worldwide. *Science of the Total Environment, 729*, 138831. https://doi.org/10.1016/j.scitotenv.2020.138831

Fountas, S., Pedersen, S.M., & Blackmore, S., (2005). ICT in precision agriculture—diffusion of technology. In E. Gelb & A. Offer (Eds.), *ICT in agriculture: Perspective of technological innovation.* Retrieved February 10, 2022, from www.departments.agri.huji.ac.il/economics/gelb-main.html

Ghazali, M. S., Bookeri, M. A., Bakar, B. A., Bujang, A. S., Mohd., N. N., Chuang, T. C., Jusoh, M. S. M., Ahmad, M. T., Khadzir, M. K., Isa, M. F. M., Ahmad, M. H., Omar, M. F. Z., Sharu, E. H., & Samsuri, N. S. (2020). Implementation of fertilizer variable rate technology (VRT) of rice precision farming in FELCRA Seberang Perak. *International Journal of Agriculture, Forestry and Plantation, 10*, 349–355.

Gotor, A. A., Marraccini, E., Leclercq, C., & Scheurer, O. (2020). Precision farming uses typology in arable crop oriented farms in northern France. *Precision Agriculture, 21*, 131–146. https://doi.org/10.1007/s11119-019-09660-y

GPS World. (2017). Agricultural robots market worth $12.8B by 2022. *GPS World staff.* Retrieved January 5, 2022, from www.gpsworld.com/agricultural-robots-market-worth-12-8b-by-2022/

Grandviewresearch. (2016). Commercial drone market analysis by product, by application and segment forecasts to 2022. Retrieved February 10, 2022, from www.grandviewresearch.com/industry-analysis/commercial-uav-market

Grassi, M. J. (2019). 2019 state of the industry aerial imagery market evolving embracing efficiencies. Retrieved February 11, 2022, from www.precisionag.com/in-field-technologies/imagery/2019-state-of-the-industry-aerial-imagery-market-evolving-embracing-efficiencies/

Griffin, T. W., & Lowenberg-DeBoer, J. (2017). Impact of automated guidance for mechanical control of herbicide resistant weeds in corn. *Journal of Applied Farm Economics, 1*(2), 62–74.

Griffiths, C. (2022). Protect your farm: Don't let cyberattacks hold your data for ransom. Retrieved February 16, 2022, from www.agweb.com/news/business/technology/protect-your-farm-dont-let-cyberattacks-hold-your-data-ransom

Grisso, R. D., Alley, M. M., & Groover, G. E. (2009). *Precision farming tools: GPS navigation.* Virginia cooperation extension, Publication no: 442-501. Retrieved February 11, 2022, from https://vtechworks.lib.vt.edu/bitstream/handle/10919/51374/442-501.pdf

Grusson, Y., Wesström, I., & Joel, A. (2021). Impact of climate change on Swedish agriculture: Growing season rain deficit and irrigation need. *Agricultural Water Management, 251*, 106858.

Gundy, J. G., & Dille, J. A. (2022). Implementing variable-rate herbicide applications based on soil physical properties in grain sorghum. *Precision Agriculture, 23*, 768–790. https://doi.org/10.1007/s11119-021-09860-5

György, K., Lamfalusi, I., Molnar, A., Sulyok, D., Gaal, M., Kemenyne, H. Z., Doman, C., Illes, I., Kiss, A., Peter, K., & Kemeny, G. (2018). Precision agriculture in Hungary: Assessment of perceptions and accounting records of FADN arable farms. *Studies in Agricultural Economics, 120*, 47–54.

Hedley, C. B., & Yule, I. (2009). Soil water status mapping and two variable-rate irrigation scenarios. *Precision Agriculture, 10*, 342–355.

Hedley, C., Ekanayake, J., & McCarthy, A. (2017, September). Precision irrigation: Trials to assess impacts on crop yield. In *18th Australian society of agronomy conference, Ballarat, Australia*. www.agronomyaustraliaproceedings.org/images/sampledata/2017/27_ASA2017_Hedley_Carolyn_Final.pdf

IIED. (2012). Climate change and agriculture. Retrieved February 11, 2022, from www.pubs.iied.org/sites/default/files/pdfs/migrate/16512IIED.pdf

Isgin, T., Bilgic, A., Forster, L., & Batte, M. T. (2008). Using count data models to determine the factors affecting farmers' quantity decisions of precision farming technology adoption. *Computers and Electronics in Agriculture, 6*, 231–242.

Jeon, H. Y., Zhu, H., Derksen, R. C., Ozkan, H. E., Krause, C. R., & Fox, R. D. (2011). Performance evaluation of a newly developed variable-rate sprayer for nursery liner applications. *Transactions of the ASABE, 54*(6), 1997–2007.

Jobbágy, J., Šimoník, J., & Findura, P. (2011). Evaluation of efficiency of precision irrigation for potatoes. *Research in Agricultural Engineering, 57*, 14–23.

Jotautiene, E., Juostas, A., & Venslauskas, K. (2021, May). Evaluation of harvesting driving modes from environmental point of view. In *IECAG 2021: 1st international electronic conference on agronomy*. Retrieved February 10, 2022, from https://sciforum.net/manuscripts/10178/manuscript.pdf

Juostas, A., & Jotautiene, E. (2021, May). Evaluation of combine harvester parameters using manual and auto-steering modes. In *20th international scientific conference engineering for rural development, Jelgava-Latvia*. Retrieved February 10, 2022, from www.tf.llu.lv/conference/proceedings2021/Papers/TF072.pdf

Kabir, M. S., Chung, S., Jang, B. E., Kim, Y., Lee, G., Lee, K., Okayasu, T., & Inoue, E. (2019). Variable fertilizer recommendation for grass production by image–based growth status. *Journal of the Faculty of Agriculture Kyushu University, 64*(1), 145–155.

Kanicki, D. (2017). Will Ag lead the way in autonomous vehicles? Retrieved February 11, 2022, from www.farm-equipment.com

Kazlauskas, M., Bruciene, I., Jasinskas, A., & Šarauskis, E. (2021). Comparative analysis of energy and GHG emissions using fixed and variable fertilization rates. *Agronomy, 11*, 138.

Kernecker, M., Knierim, A., Wurbs, A., Kraus, T., & Borges, F. (2020). Experience versus expectation: Farmers' perceptions of smart farming technologies for cropping systems across Europe. *Precision Agriculture, 21*, 34–50. https://doi.org/10.1007/s11119-019-09651-z

Keskin, M., Han, Y. J., & Dodd, R. B. (1999, May). A review of yield monitoring instrumentation applied to the combine harvesters for precision agriculture purposes. In *7th international congress on agricultural mechanization and energy, Adana, Turkiye*.

Keskin, M., & Gorucu Keskin, S. (2012). *Precision agriculture technologies (Hassas Tarım Teknolojileri)*. Mustafa Kemal University, Turkiye.

Keskin, M., & Sekerli, Y. E. (2016). Awareness and adoption of precision agriculture in the Cukurova region of Turkey. *Agronomy Research, 14*(4), 1307–1320.

Keskin, M., Sekerli, Y. E., Say, S. M., & Topcueri, M. (2018). Farmers' experiences with GNSS-based tractor auto guidance in Adana province of Turkiye. *Journal of Agricultural Faculty of Gaziosmanpasa University, 35*(2), 172–181.

Khakbazan, M., Moulin, A. P., & Huang, J. (2021). Economic evaluation of variable rate nitrogen management of canola for zones based on historical yield maps and soil test recommendations. *Scientific Reports, 11*, 4439.

Khalilian, A. T., Rogers, N. G., Williams, P. B., Han, Y. J., Nafchi, A. M., Maja, J. M., Marshall, M. W., & Payero, J. O. (2017). Sensor-based algorithm for mid-season nitrogen application in corn. *Open Journal of Soil Science, 7*, 278–287.

Kozicka, K., Gozdowski, D., & Wójcik-Gront, E. (2021). Spatial-temporal changes of methane content in the atmosphere for selected countries and regions with high methane emission from rice cultivation. *Atmosphere, 12*, 1382. https://doi.org/10.3390/atmos12111382

Kutter, T., Tiemann, S., Siebert, R., & Fountas, S. (2011). The role of communication and co-operation in the adoption of precision farming. *Precision Agriculture, 12*, 2–17.

Kviz, Z., Kroulik, M., & Chyba, J. (2014). Machinery guidance systems analysis concerning pass-to-pass accuracy as a tool for efficient plant production in fields and for soil damage reduction. *Plant Soil and Environment, 60*, 36–42.

Lal, R. (2022). Reducing carbon footprints of agriculture and food systems. *Carbon Footprints, 2022*, 1–3. https://doi.org/10.20517/cf.2021.05

LaRue, J., & Evans, R. (2012, February). Considerations for variable rate irrigation. In *24th annual central plains irrigation conference, Colby, Kansas, USA*. Retrieved February 11, 2022, from www.ksre.k-state.edu/irrigate/oow/p12/Larue12.pdf

Liakos, V., Smith, E. D., Fountas, S., Nanos, G., Kalfountzos, D., & Gemtos, T. A. (2020). On-farm evaluation of variable rate fertilizer applications using yield-based mathematical formulae in a Greek Apple orchard. *International Journal of Fruit Science, 20*(2), 48–65.

Llorens, J., Gil, E., Llop, J., & Escolà, A. (2010). Variable rate dosing in precision viticulture: Use of electronic devices to improve application efficiency. *Crop Protection, 29*, 239–248.

Lo, T. H., Heeren, D. M., Martin, D. L., Mateos, L., Luck, J. D., & Eisenhauer, D. E. (2016). Pumpage reduction by using variable-rate irrigation to mine undepleted soil water. *Transactions of the ASABE, 59*(5), 1285–1298.

Lowenberg-DeBoer, J., & Griffin, T. W. (2006). *Potential for precision agriculture adoption in Brazil*. Site Specific Management Center Newsletter. Purdue University. Retrieved September 12, 2021, from www.agriculture.purdue.edu/ssmc/frames/SSMC_newsletter6_06.pdf

Lowenberg-DeBoer, J., Behrendt, K., Godwin, R., & Franklin, K. (2019, April). The impact of swarm robotics on arable farm size and structure in the UK. In *93rd annual conference of the agricultural economics society*. University of Warwick, England.

Maghsoudi, H., Minaei, S., Ghobadian, B., & Masoudi, H. (2015). Ultrasonic sensing of pistachio canopy for low-volume precision spraying. *Computers and Electronics in Agriculture, 112*, 149–160.

Markets and Markets. (2020). Precision Farming Market by Technology, Application, Offering and Geography: Global Forecast to 2025. Retrieved April 7, 2020, from www.marketsandmarkets.com

Masi, M., Di Pasquale, J., Vecchio, Y., & Capitanio, F. (2023). Precision farming: Barriers of variable rate technology adoption in Italy. *Land, 12*, 1084. https://doi.org/10.3390/land12051084

McFadden, J., Njuki E., & Griffin, T. (2023). *Precision agriculture in the digital era: Recent adoption on US farms*. Retrieved July 16, 2023, from www.ers.usda.gov/webdocs/publications/105894/eib-248.pdf

McKinsey. (2020). Agriculture and climate change. Retrieved September 2, 2021, from www.mckinsey.com/~/media/mckinsey/industries/agriculture/our%20insights/reducing%20agriculture%20emissions%20through%20improved%20farming%20practices/agriculture-and-climate-change.pdf

Miller, K. A., Luck, J. D., Heeren, D. M., Lo, T. H., Martin, D. L., & Barker, J. B. (2017). A geospatial variable rate irrigation control scenario evaluation methodology based on mining root zone available water capacity. *Precision Agriculture, 19*, 666–683.

Miller, N. J., Griffin, T. W., Ciampitti, I. A., & Sharda, A. (2019). Farm adoption of embodied knowledge and information intensive precision agriculture technology bundles. *Precision Agriculture, 20*, 348–361. https://doi.org/10.1007/s11119-018-9611-4

Nadav, I., & Schweitzer, A. (2017). VRDI: Variable rate drip irrigation in vineyards. *Advances in Animal Biosciences, 8*(2), 569–573.

NoTill Farmer. (2019). *Tracking farm technology adoption: Trends, transitions & takeaways* (9 pages). Lessiter Media. Retrieved February 11, 2022, from www.no-tillfarmer.com/articles/9173

NSAC. (2019). Agriculture and climate change: Policy imperatives and opportunities to help producers meet the challenge. Retrieved September 3, 2021, from www.sustainableagriculture.net/wp-content/uploads/2019/11/NSAC-Climate-Change-Policy-Position_paper-112019_WEB.pdf

OECD. (2016). Agriculture and climate change: Towards sustainable, productive and climate-friendly agricultural systems. OECD Meeting of Agricultural Ministers. Retrieved September 3, 2021, from www.oecd.org/agriculture/ministerial

Ortiz, B. V., Balkcom, K., Duzy, L. M., Santen, E., & Hartzog, D. L. (2012). Evaluation of agronomic and economic benefits of using RTK-GPS-based auto-steer guidance systems for peanut digging operations. *Precision Agriculture, 14*, 357–375.

Ortuani, F., Facchi, A., Mayer, A., Bianchi, D., Bianchi, A., & Brancadoro, L. (2019). Assessing the effectiveness of variable-rate drip irrigation on water use efficiency in a vineyard in Northern Italy. *Water, 11*, 1964.

Ozguven, M. M. (2018). The newest agricultural technologies. *Current Investigations in Agriculture and Current Research, 5*(1), 573–579. https://doi.org/10.32474/CIACR.2018.05.000201

Pathak, H. S., Brown, P., & Best, T. (2019). A systematic literature review of the factors affecting the precision agriculture adoption process. *Precision Agriculture, 20*, 1292–1316. https://doi.org/10.1007/s11119-019-09653-x

Paudel, K., Pandit, M., Mishra, A., & Segarra, E. (2011, July). Why don't farmers adopt precision farming technologies in cotton production? In *2011 AAEA & NAREA joint annual meeting, Pittsburgh, PA, USA*.

PPAEP. (2018). *Threats to precision agriculture*. Public-Private Analytic Exchange Program 24pp. Retrieved September 12, 2021, from www.dhs.gov/sites/default/files/publications/2018%20AEP_Threats_to_Precision_Agriculture.pdf

Rains, C. R., & Thomas, D. L. (2009, March). *Precision farming: An introduction* (12 pp.). The University of Georgia, Bulletin 1186. Rev.

Rajmis, S., Karpinski, I., Pohl, J. P., Herrmann, M., & Kehlenbeck, H. (2022). Economic potential of site-specific pesticide application scenarios with direct injection and automatic application assistant in northern Germany. *Precision Agriculture, 23*, 2063–2088. https://doi.org/10.1007/s11119-022-09888-1

Researchnester. (2021). Agricultural drones market overview. Retrieved February 12, 2021, from www.researchnester.com/reports/agricultural-drones-market/462

Rodias, E., Berruto, R., Busato, P., Bochtis, D. D., Sørensen, C. A., & Zhou, K. (2017). Energy savings from optimized in-field route planning for agricultural machinery. *Sustainability, 9*, 1–13.

Sadler, E., Evans, R. G., Stone, K. C., & Camp, C. R. (2005). Opportunities for conservation with precision irrigation. *Journal of Soil and Water Conservation, 60*, 371–378.

Santaga, F., Benincasa, P., Toscano, P., Antognelli, S., Ranieri, E., & Vizzari, M. (2021). Simplified and advanced Sentinel-2-based precision nitrogen management of wheat. *Agronomy, 11*, 1156.

Scarfone, A., Picchio, R., del Giudice, A., Latterini, F., Mattei, P., Santangelo, E., & Assirelli, A. (2021). Semi-automatic guidance vs. manual guidance in agriculture: A comparison of work performance in wheat sowing. *Electronics, 10*(7), 825.

Schillaci, C., Tadiello, T., Acutis, M., & Perego, A. (2021). Reducing topdressing N fertilization with variable rates does not reduce maize yield. *Sustainability, 13*, 8059. https://doi.org/10.3390/su13148059

Shockley, J., Dillon, C. R., & Stombaugh, T. S. (2011). A whole farm analysis of the influence of auto-steer navigation on net returns, risk and production practices. *Journal of Agricultural and Applied Economics, 43*, 57–75.

Solanelles, F., Escolà, A., Planas, S., Rosell, J. R., Camp, F., & Gracia, F. (2006). An electronic control system for pesticide application proportional to the canopy width of tree crops. *Biosystems Engineering, 95*(4), 473–481.

Sontowski, S., Gupta, M., Chukkapalli, S. S. L., Abdelsalam, M., Mittal, S., Joshik, A., & Sandhu, R. (2020). Cyber-attacks on smart farming infrastructure. Retrieved February 11, 2021, from https://ebiquity.umbc.edu/_file_directory_/papers/1031.pdf

Soto, I., Barnes, A., Balafoutis, A., Beck, B., Sánchez, B., Vangeyte, J., Fountas, S., Van der Wal, T., Eory, V. & Gómez-Barbero, M. (2019). *The contribution of precision agriculture technologies to farm productivity and the mitigation of greenhouse gas emissions in the EU*. Publications

Office of the European Union, Luxembourg, ISBN 978-92-79-92834-5. https://doi.org/10.2760/016263,JRC112505

Spati, K., Huber, R., & Finger, R. (2021). Benefits of increasing information accuracy in variable rate technologies. *Ecological Economics, 185*(2021), 107047. https://doi.org/10.1016/j.ecolecon.2021.107047

Spykman, O., Gabriel, A., Ptacek, M., & Gandorfer, M. (2021). Farmers' perspectives on field crop robots: Evidence from Bavaria, Germany. *Computers and Electronics in Agriculture, 186*, 106176. https://doi.org/10.1016/j.compag.2021.106176

Sui, R., & Yan, H. (2017). Field study of variable rate irrigation management in humid climates. *Irrigation and Drainage, 66*, 327–339.

Sylvère, N., & D'amour, R. J. (2020). Updates on modern agricultural technologies adoption and its impacts on the improvement of agricultural activities in Rwanda: A review. *International Journal of Innovative Science and Research Technology, 5*(12), 222–229.

Tekin, A. B. (2010). Variable rate fertilizer application in Turkish wheat agriculture: Economic assessment. *African Journal of Agricultural Research, 5*(8), 647–652. https://doi.org/10.5897/AJAR09.562

Thompson, L. J. (2018, June). Using drone-based sensors to direct variable-rate, in-season, aerial nitrogen application on corn. In 14th I*nternational conference on precision agriculture, Montreal, Quebec, Canada*. Retrieved February 11, 2022, from www.ispag.org/proceedings/?action=download&item=4740

Timmermann, C., Gerhards, R., & Kühbauch, W. (2003). The economic impact of site-specific weed control. *Precision Agriculture, 4*, 249–260.

Topakci, M., Unal, I., Canakci, M., Celik, H. K., & Karayel, D. (2010). Design of a horizontal penetrometer for measuring on-the-go soil resistance. *Sensors, 10*, 9337–9348. https://doi.org/10.3390/s101009337

Topcueri, M., & Keskin, M. (2019). Effectiveness of GNSS-based tractor auto steering systems in crop spraying. *Mustafa Kemal University Journal of Agricultural Sciences, 24*, 78–90.

Turland, M., & Slade, P. (2019). Farmers' willingness to participate in a big data platform. *Agribusiness, 36*, 20–36.

Vecchio, Y., Agnusdei, G. P., Miglietta, P. P., & Capitanio, F. (2020). Adoption of precision farming tools: The case of Italian farmers. *International Journal of Environmental Research and Public Health, 17*, 869. https://doi.org/10.3390/ijerph17030869

Veltheim, F. R., Theuvsen, L., & Heise, H. (2021). German farmers' intention to use autonomous field robots: A PLS-analysis. *Precision Agriculture*. https://doi.org/10.1007/s11119-021-09854-3

Venkatramanan, V., Shah, S., & Prasad, R. (2020). *Global climate change: Resilient and smart agriculture*. Springer. https://doi.org/10.1007/978-981-32-9856-9

Vizzari, M., Santaga, F., & Benincasa, P. (2019). Sentinel 2-based nitrogen VRT fertilization in wheat: Comparison between traditional and simple precision practices. *Agronomy, 9*, 278.

Whipker, L. D., & Akridge, J. T. (2009). *2009 Precision agriculture services dealership survey results* (64 pp). Purdue University, West Lafayette, Indiana, USA. Retrieved February 11, 2022, from https://agribusiness.purdue.edu/2009-precision-agricultural-services-dealership-survey-results/

Williams, P. (2018). Development of a sensor-based, variable-rate fertigation technique for overhead irrigation systems. A Thesis Presented to the Graduate School of Clemson University. Retrieved February, 11, 2022, from https://tigerprints.clemson.edu/all_dissertations/2176

Xiongkui, H., Aijun, Z., Yajia, L., & Jianli, S. (2011). Precision orchard sprayer based on automatically infrared target detecting and electrostatic spraying techniques. *International Journal of Agricultural and Biological Engineering, 4*(1), 35–40.

Zambon, I., Cecchini, M., Egidi, G., Saporito, M. G., & Colantoni, A. (2019). Revolution 4.0: Industry vs. agriculture in a future development for SMEs. *Processes, 7*(1), 36. https://doi.org/10.3390/pr7010036

Effects of Climate Change and Food Security on Sustainable Agricultural Development and Potential Solutions: Evidence from China

Wang Fang, Chen Xiao-Dong, Shi Jun-Hui, Nawab Khan, and Ram L. Ray

Abstract Due to rapid changes in industrialization and urbanization, climate change has been an important issue that impedes sustainable development. Another unavoidable factor related to the national economy and people's livelihood is food security. It is a global issue, particularly in developing countries. Considering these adverse effects, we investigated the impacts of climate and food security on China's production efficiency, productivity growth, and sustainable agricultural development. Overall, we provided possible solutions addressing these dilemmas, which may significantly impact efficiency promotion and technological progress. To address these issues, we introduced the Slack-based approach to quantify the inefficiency of the overall agricultural production. Besides, we further analyzed the variable-specific decomposition. Results suggest that, with climate change and food security constraints, China's agricultural inefficiency is 0.61, mainly attributed to food security constraints (0.29). In this chapter, we tried to provide solutions for poverty alleviation, awareness creation, intelligent agricultural extension, and biotechnology generalization.

Keywords Data envelopment analysis (DEA) · Climate change · Food security · China · Agricultural sector · Sustainable agricultural development

W. Fang (✉) · C. Xiao-Dong · S. Jun-Hui · N. Khan
College of Management, Sichuan Agricultural University Chengdu Campus, Wenjiang, Sichuan 611100, China
e-mail: wangfangscnd@sicau.edu.cn

C. Xiao-Dong
e-mail: chenxiaodongtzxy@gmail.com

S. Jun-Hui
e-mail: 20147396@stu.sicau.edu.cn

N. Khan
e-mail: nawabkhan@stu.sicau.edu.cn

R. L. Ray
College of Agriculture, Food, and Natural Resources, Prairie View A&M University, Prairie View, TX 77446, USA
e-mail: raray@pvamu.edu

M. Behnassi et al. (eds.), *Climate-Smart and Resilient Food Systems and Security*,
https://doi.org/10.1007/978-3-031-65968-3_21

1 Introduction

Climate change, induced by intensive industrialization and economic development, has attracted the attention of the international community, both in developed and developing countries. However, for developing countries, such as China, a larger scale producer and emitter of pollutants, this issue is more serious. As the largest developing country, China has adopted various policies (measures) for reducing its greenhouse gas emissions (mainly carbon dioxide emissions) (Stephan & Stephan, 2020). For example, China achieved its carbon intensity reduction goal by 40–45% in 2020 compared with 2005 in the Medium and Long-Term Plan for National Economic and Social Development. The earlier carbon intensity reduction target was set for 2015 (Chen et al., 2020; Miao et al., 2019). In 2014, the National Development and Reform Commission of China promulgated the Interim Measures on the Management of Carbon Dioxide Emission Rights Trading. These Interim Measures help guide and promote the establishment and development of the national carbon market. Li et al. (2020) emphasized that achieving this target requires the unification of policies at the province level in China following the provincial targets for carbon intensity provided in the 12th Five-year Plan. However, it is a challenge to achieve this goal due to the uneven economic development of the provinces (e.g., the difference in growth among eastern and western provinces). The specific setting of the target values may render infeasible objectives. In order to avoid such issues, the principle of "common but differential" should be followed. Therefore, effective frameworks for target setting are needed. One of the effective measures for this issue is carbon emission permit markets, which were established in Beijing, Shanghai, Tianjin, Hubei, Guangdong, Shenzhen, and Chongqing in 2013.

Besides, 2020 was considered an exceptional year due to the outbreak of COVID-19, which raised food security concerns under the global pandemic. According to FSIN (2020), nearly 690 million people suffered from hunger in 2019, an increase of 10 million compared with 2018, and almost 60 million compared with that five years ago. High costs and low affordability mean that billions of people cannot eat healthy and nutritious food. Asia has the most significant number of hungry people, whereas Africa has the fastest growing number (FSIN 2020). According to the report, worldwide, the 2019 COVID-19 pandemic might have increased the number of long-term hungry people by more than 130 million by the end of 2020 (sudden severe hunger under the pandemic may further increase this number) (FSIN 2020). Under these circumstances, the investigation of food security has been an important issue.

According to the previous discussion, the effects of climate change and food security are significant, particularly in the agricultural sector. Hence, this chapter aimed to investigate the effects of productivity change on climate change and food security constraints. The main contributions of this chapter are as follows. First, we constructed an efficiency estimation framework that incorporates both climate change and food security. Second, we modified the Slack-based model, which provides a theoretical basis for static efficiency and dynamic productivity calculation.[1] Third, the corresponding policy implications were provided to alleviate the adverse effects of climate change and food security on sustainable agricultural development.

The remaining parts of the chapter are organized as follows: Sect. 2 reviews previous literature on these topics, as well as the rationality in incorporating these two factors (climate change and food security) into the framework; the applied framework is discussed in Sects. 3 and 4 includes details of the data on selected variables; the empirical analysis of China's agricultural sector is presented in Sect. 5; and conclusion and policy recommendations are summarized in Sects. 6 and 7, respectively.

2 Literature

In the past, many researchers investigated climate change and food security using various approaches, tools, techniques, and methods. For example, Chen et al. (2017) investigated the combined effects of CO_2 emissions and flight delays using a stochastic network DEA model to assess the efficiency of Chinese airlines. Mahdiloo et al. (2018) used the DEA model to investigate the static abatement efficiency of CO_2 emissions in terms of the electricity producers, involving the influence strategy of rewarding ecological efficiency and penalizing ecological inefficiency behaviors of electricity producers. Cecchini et al. (2018) performed the environmental efficiency analysis and estimation of CO_2 abatement costs in dairy cattle farms in Umbria (Italy), using the Slack-based model (SBM). Iftikhar et al. (2018) used the network DEA approach to globally estimate the energy and CO_2 emission efficiency of major economies. They found that 85% of energy consumption and 89% of CO2 emissions were just because of economic and distributive inefficiencies. Yang et al. (2020) improved the zero-sum gains of Data Envelopment Analysis (DEA) to assess and optimize the provincial CO_2 emission reduction scheme. They found that higher reduction targets are required for 12 provinces, whereas the other 18 provinces could have lower target values to achieve overall efficiency. Li et al. (2020) employed an integrated hybrid life cycle assessment (LCA) model for multiple regions of China to evaluate the life cycle CO_2 emission, energy consumption, and energy payback time of a 10-MW power tower CSP plant. Whitmarsh et al. (2020) conducted a large-scale survey globally to figure out the use of aviation by climate change scientists. They found that climate change researchers, particularly professors, fly more than other researchers. They would have taken steps to reduce or offset their air trips.

As CO_2 emission is a driving factor for climate change, food security is another essential factor that drew academic communities' attention due to its importance in maintaining a national livelihood. McKune et al. (2015) proposed a gendered

[1] This paper employs SBM for its additive nature, non-oriented and non-radial property. This can help us obtain the most objective result.

conceptual framework to understand the impact of climate change on food security among livestock holders, highlighting potential pathways of vulnerability and points of intervention to consider in global health strategies for improving household food security. Bizikova et al. (2020) explored how agricultural interventions enhance food security and sustainable development goals. Kansiime et al. (2021) examined the implications of the COVID-19 pandemic on household income and food security in two East African countries (i.e., Kenya and Uganda), using online survey data from 442 respondents. Hans et al. (2020) proposed a broader framework to explore the future of food and nutrition security, focusing on food availability, food access, and a reasonable proxy for food utilization. Badami and Ramankutty (2015) reviewed divergent perspectives on whether urban agriculture has greatly promoted food security and poverty alleviation. They assessed the potential for urban agriculture to contribute to urban food security in different regions, based on a low threshold of urban land required to grow the daily vegetable intake for the urban poor.

Climate change and food security are two main factors impeding sustainable agricultural development. However, previous studies did not pay enough attention to incorporate these two factors into one framework to support agriculture sustainability.

3 Material and Methods

3.1 Agricultural Production Technology

Constructing an analytical framework reasonably for undesirable outputs is the foundation for the research on total factor productivity. In recent years, with the increasing emphasis on environmental issues, pollutant emissions have been characterized as environmental constraints in various research frameworks. They have a significant impact on technical efficiency and total factor productivity indicators. Among these studies, Färe et al., (1989, 2007) first described the mechanism of Environmental Production Technology, laying a theoretical foundation for the study of environmental efficiency that includes undesirable outputs. Zhou et al. (2008) measured the performance of carbon dioxide emissions in some countries. Environmental Production Technology is briefly discussed as follows:

Environmental Production Technology analysis framework includes P input variables $x = \left(x_1, \ldots x_p\right) \in R_P^+$ and Q output variables $y = \left(y_1, \ldots y_q\right) \in R_Q^+$ for one decision-making unit (DMU) and R undesirable output variables $b = (b_1, \ldots b_r) \in R_R^+$. In period t, the input, desirable output, and undesirable output variables of the DMU i are $\left(x_i^t, y_i^t, b_i^t\right)$.

On the premise that all input and output variables satisfy substantial disability, the Environmental Production Technology can be defined as

$$P^t\left(x^t\right) = \left\{\left(y^t, b^t\right) : \lambda X \leq x_{ip}^t,\ \lambda Y \geq y_{iq}^t, \lambda B \leq b_{ir}{}^t \forall p, q, r, \lambda \geq 0\right\} \quad (1)$$

where λ is a weight vector greater than or equal to zero, and x, y, and b are variables of the input, desirable output, and undesirable output for constructing the production frontier, respectively. According to the different constraints on the value of λ, it can be characterized explicitly as variable returns to scale (VRS) and constant returns to scale (CRS).

3.2 Agricultural Production Inefficiency

The oriented and radical DEA method has been widely used in existing food security and climate change-related performance measurement studies. Still, the shortcomings of these studies are that they can only measure efficiency and productivity from the perspective of input or output and cannot consider the effects of all variables. To measure the performance of input and output variables more comprehensively, Fukuyama and Weber (2009), Wang et al. (2010), Sargani et al. (2020), and Tone (2001) proposed or improved the Slack-based Measure (SBM) method. This method completes the construction of the technological frontier by measuring the slack values of input and output variables. It can be applied to the non-radial measurement of all variables; thus, it is called the non-radial direction distance function (Zhou et al., 2008). When both input and output variables are redundant, the calculation results using the non-radial direction distance function are different from the traditional direction distance function. This chapter follows the distance function used by Miao et al. (2019). The specific form is as follows:

$$\begin{aligned}&\vec{S}^t\left(x_i^t, y_i^t, b_i^t; g^x, g^y, g^b\right) = \frac{1}{3}\max\left(\frac{1}{P}\sum_{p=1}^{P}\frac{S_p^x}{g_p^x} + \frac{1}{Q}\sum_{q=1}^{Q}\frac{S_q^y}{g_q^y} + \frac{1}{R}\sum_{r=1}^{R}\frac{S_r^b}{g_r^b}\right)\\&s.t.\lambda X + S_p^x = x_{ip}^t, \lambda Y - S_q^y = y_{iq}^t, \ \lambda B + S_r^b = b_{ir}^t;\\&\forall p, q, r, \lambda \geq 0; \ S_p^x, \ S_q^y, \ S_r^b \geq 0\end{aligned} \quad (2)$$

In Eq. (2), $\left(x_i^t, \ y_i^t \ b_i^t\right)$ represents the input and output variables of DMU i in period t, $\left(g^x, \ g^y \ g^b\right)$ represents the direction vector of reducing input, increasing desirable output, and reducing undesirable output, $\left(S_p^x, \ S_q^y \ S_r^b\right)$ represents slack variables for inputs, desirable outputs, and undesirable outputs, respectively. Using the additive structure to decompose all variables, we can further obtain the inefficiency value of each variable:

$$IE = \vec{S}^t = IE_x + IE_y + IE_b = \frac{1}{3P}\sum_{p=1}^{P}\frac{S_p^x}{g_p^x} + \frac{1}{3Q}\sum_{q=1}^{Q}\frac{S_q^y}{g_q^y} + \frac{1}{3R}\sum_{r=1}^{R}\frac{S_r^b}{g_r^b} \quad (3)$$

where $\frac{1}{3P}\sum_{p=1}^{P}\frac{S_p^x}{g_p^x}$ is the sum of inefficiencies of the input variables, $\frac{1}{3Q}\sum_{q=1}^{Q}\frac{S_q^y}{g_q^y}$ is the sum of inefficiencies of the desired output variables, and $\frac{1}{3R}\sum_{r=1}^{R}\frac{S_r^b}{g_r^b}$ is the sum of inefficiencies of the undesirable output variables.

4 Data

This chapter selected 30 provincial-level regions of China as the research area. The research in China, the largest developing country, may have more beneficial policy implications globally (Ye et al., 2015). Regarding the selection of input and output variables, the investment in fixed assets, seeded area, labor force, machinery, and fertilizer consumption of China's agricultural sector are considered inputs, whereas China's gross agricultural outputs are considered desirable outputs. When agricultural production activities cause CO_2 emissions, it is considered as a constraint of climate change. Agricultural disaster areas are taken as the constraints of food security in this book chapter. Using these assumptions, the production technology can be well-defined. Notably, all the data was collected from the National Bureau of Statistics (assessed on January 12, 2021, from *the National Bureau of Statistics*). Besides, the lack of data in the Tibet Autonomous Region for agricultural CO_2 emissions forced us to opt for the remaining 30 provinces as the research area. We selected the study duration from 2006 to 2015 to include China's 11th and 12th Five-Year Plan. Specifically, the descriptive data of input/output variables are presented in Table 1.

As implied in Table 1, on average, the constant gains across both these Five-Year Plans can be observed for investment in fixed assets, seeded areas, machinery, fertilizer consumption, CO_2 emissions, and value added in the agricultural sector. By contrast, on average, labor force and disaster areas witnessed a constant decline across this decade. Hence, the mitigation of agricultural CO_2 emissions should be a priority for the authorities of China.

5 Empirical Analysis

Using a previously defined Slack-based model and selected variables, we could then compute the inefficiency of China's agricultural sector of 30 province-level regions. Notably, we attributed the overall inefficiency scores to individual variables owing to the variable-specific decomposition. The detailed analysis is presented as follows.

Table 1 Summary data of agricultural input/output variables

Variable	Year	Maximum	Minimum	Mean	Standard deviation
Investment in fixed assets (I, 10^8 CNY)	2006	508.20	5.96	147.87	118.64
	2010	786.60	2.00	262.86	201.04
	2015	930.98	3.31	346.99	236.01
Labor force (L, 10^4 people)	2006	6505.00	206.00	2411.77	1660.99
	2010	5784.00	246.00	2200.37	1487.37
	2015	5039.00	269.00	1977.90	1308.97
Seeded areas (S, 10^3 hectare)	2006	13,922.73	234.96	5015.73	3539.60
	2010	14,320.79	218.86	5115.47	3785.93
	2015	14,879.73	172.14	5381.59	4076.24
Machinery (M, 10^4 Kilowatt)	2006	9199.33	96.46	2206.38	2305.30
	2010	11,628.97	104.06	2992.92	2886.79
	2015	13,353.02	119.01	3604.13	3296.88
Fertilizer consumption (F, 10^4 tons)	2006	518.14	4.21	153.75	128.53
	2010	655.15	4.74	179.41	146.84
	2015	716.09	6.03	194.28	153.98
CO_2 emissions (C, 10^4 tons)	2006	13.10	0.20	3.68	2.61
	2010	7.70	0.30	3.71	2.18
	2015	10.23	0.28	4.35	3.05
Value added (Y, 10^8 CNY)	2006	4058.62	71.40	1316.48	1009.81
	2010	6573.77	100.77	2185.91	1648.17
	2015	9283.92	149.46	3286.89	2333.91
Disaster areas (D, 10^8 CNY)	2006	1615.70	1.20	644.07	435.45
	2010	2136.50	0.00	598.00	510.57
	2015	1740.40	5.10	412.67	377.31

5.1 Variable-Specific Decomposition

As is shown in Table 2, the average value of inefficiency for 30 provinces in China during 2006–2015 is 0.61, whereas the inefficiency value related to the desirable output (GDP, Y) is 0.00. This indicates that possibilities for economic growth in the agricultural sector are relatively limited under the constraints of existing agricultural production technology as defined by the total agricultural mechanical power, agricultural CO_2 emissions, and agricultural disaster areas. Thus, the government should opt for an adjustment of agricultural structure to ensure agricultural productivity rather than increase the factor inputs.

Table 2 Input/output inefficiency of China's agricultural sector across 2006–2015

Province[1]	IE	I	L	S	M	F	Y	C	D
Beijing (BJ)	0.22	0.03	0.00	0.00	0.00	0.00	0.00	0.00	0.20
Tianjin (TJ)	0.60	0.07	0.00	0.00	0.06	0.05	0.00	0.00	0.42
Heibei (HB)	0.82	0.12	0.11	0.04	0.12	0.08	0.00	0.14	0.21
Shanxi (SHX)	0.50	0.12	0.06	0.00	0.12	0.04	0.00	0.00	0.15
Inner Mongolia (IN-MON)	0.08	0.00	0.00	0.00	0.00	0.00	0.00	0.00	0.07
Liaoning (LN)	0.58	0.12	0.02	0.00	0.09	0.06	0.00	0.00	0.29
Jilin (JL)	0.69	0.07	0.01	0.00	0.02	0.05	0.00	0.07	0.46
Heilongjiang (HLJ)	0.46	0.09	0.05	0.07	0.04	0.03	0.00	0.00	0.19
Shanghai (SH)	0.00	0.00	0.00	0.00	0.00	0.00	0.00	0.00	0.00
Jiangsu (JS)	0.85	0.08	0.07	0.01	0.03	0.05	0.00	0.00	0.61
Zhejiang (ZJ)	0.40	0.01	0.02	0.00	0.03	0.00	0.00	0.00	0.33
Anhui (AH)	0.88	0.09	0.07	0.04	0.07	0.07	0.00	0.04	0.48
Fujian (FJ)	0.68	0.06	0.04	0.00	0.04	0.06	0.00	0.00	0.48
Jiangxi (JX)	0.79	0.11	0.08	0.04	0.08	0.05	0.00	0.14	0.30
Shandong (SD)	0.70	0.09	0.08	0.04	0.08	0.07	0.00	0.00	0.34
Henan (HEN)	0.91	0.11	0.09	0.06	0.09	0.09	0.00	0.00	0.46
Hubei (HUB)	0.59	0.10	0.10	0.01	0.02	0.07	0.00	0.00	0.29
Hunan (HUN)	0.58	0.12	0.11	0.01	0.06	0.04	0.00	0.00	0.24
Guangdong (GD)	0.81	0.11	0.08	0.00	0.06	0.08	0.00	0.00	0.48
Guangxi (GX)	0.88	0.10	0.07	0.05	0.08	0.08	0.00	0.25	0.25
Hainan (HAN)	0.52	0.10	0.03	0.00	0.05	0.08	0.00	0.02	0.24
Chongqing (CQ)	0.44	0.02	0.01	0.00	0.01	0.01	0.00	0.01	0.37
Sichuan (SC)	0.80	0.12	0.11	0.06	0.06	0.07	0.00	0.01	0.37
Guizhou (GZ)	0.69	0.08	0.07	0.04	0.06	0.00	0.00	0.09	0.34
Yunnan (YN)	0.68	0.12	0.10	0.00	0.03	0.06	0.00	0.00	0.38
Shaanxi (SH'X)	0.81	0.11	0.08	0.05	0.09	0.10	0.00	0.19	0.19
Gansu (GS)	0.57	0.11	0.09	0.05	0.11	0.06	0.00	0.12	0.04
Qinghai (QH)	0.50	0.10	0.00	0.01	0.05	0.00	0.00	0.13	0.21
Ningxia (NX)	0.72	0.09	0.01	0.03	0.07	0.04	0.00	0.42	0.05
Xinjiang (XJ)	0.56	0.11	0.01	0.00	0.04	0.09	0.00	0.00	0.30
Average	0.61	0.09	0.05	0.02	0.06	0.05	0.00	0.05	0.29

[1]Note: we presented abbreviations of provinces in 1st column of Table 2, and in the following context we will mainly use the abbreviation for sake of brevity.

5.2 Tempo-Wise Decomposition

Table 3 presents the change in average annual inefficiency of China's agricultural sector across 2006–2015. As implied in Table 3, the average inefficiency score in 2006 is 0.47, whereas it turned to 0.64 in 2015. Hence, the decline of agricultural production efficiency reached 36.2%, indicating that structural adjustments are required with constraints of food security and climate change. Approximately, the inverted U-shape was observed across these 11th and 12th Five-Year Plans. Notably, Shanghai has been the only province-level region that is located in the frontier of agricultural production technology. Considering China's granary (i.e., Liaoning, Jilin, and Heilongjiang), Jilin's inefficiency distinguishes itself, which should be focused on further.

Moreover, it is important to analyze further the regional pattern of China's 30 provinces. For the sake of brevity, we clustered the 30 provinces into three groups (i.e., Western region, Inland region, and Eastern region) when analyzing the changing trend of inefficiency. Specifically, the Western region includes Beijing, Tianjin, Hebei, Liaoning, Shanghai, Jiangsu, Zhejiang, Fujian, Shandong, Guangdong, Guangxi, and Hainan; Inland region incorporates Shanxi, Inner Mongolia, Jilin, Heilongjiang, Anhui, Jiangxi, Henan, Hubei, and Hunan; Eastern region incorporates Sichuan, Chongqing, Guizhou, Yunnan, Shannxi, Gansu, Qinghai, Ningxia, and Xinjiang. The detailed results are presented in Fig. 1.

As is depicted in Fig. 1, the inefficiency scores for all Eastern, Inland, and Western regions range from 0.3 to 0.8. In general, the inefficiency in 2015 is higher than that of 2006, which can be attributed to all these three regions. A sharp increase in inefficiency scores for Eastern and Inland regions can be observed across 2007–2010, which may be associated with the global financial crisis. This suggests that food security and climate change in China's agricultural sector are fragile to the global environment. By contrast, the Western region is self-reliant. Notably, across 2014–2015, the inefficiency scores witnessed a decline for all these three regions.

6 Conclusion

This study introduced an agricultural production technology framework that modifies the previously used approach by incorporating variable-specific decomposition to evaluate the performance of the Chinese agricultural sector. Moreover, unlike the previous research, we also estimated the aggregated performance. Considering the data availability, we used panel data covering 2006–2015 of China's agricultural sector, which includes China's 11th and 12th Five-Year Plans. The non-parametric model allows the multi-input and multi-output analysis from the economic and environmental perspective, which incorporates food security and climate change. Notably, the overall inefficiency was broken down into variable-specific ones from the resource-specific perspective.

Table 3 Average annual inefficiency of China's agricultural sector across 2006–2015

Province[1]	2006	2007	2008	2009	2010	2011	2012	2013	2014	2015
Beijing (BJ)	0.00	0.00	0.09	0.69	0.92	0.47	0.05	0.00	0.00	0.00
Tianjin (TJ)	0.67	0.56	0.54	0.54	0.89	0.94	0.00	0.97	0.86	0.00
Heibei (HB)	0.89	0.87	0.86	0.81	0.83	0.85	0.74	0.83	0.77	0.74
Shanxi (SHX)	0.39	0.50	0.35	0.14	0.44	0.60	0.62	0.63	0.67	0.64
Inner Mongolia (IN-MON)	0.00	0.06	0.30	0.00	0.00	0.22	0.00	0.27	0.00	0.00
Liaoning (LN)	0.46	0.36	0.74	0.41	0.61	0.85	0.77	0.78	0.35	0.46
Jilin (JL)	0.47	0.00	0.85	0.57	0.74	0.88	0.88	0.88	0.84	0.80
Heilongjiang (HLJ)	0.30	0.00	0.58	0.00	0.67	0.74	0.64	0.38	0.75	0.57
Shanghai (SH)	0.00	0.00	0.00	0.00	0.00	0.00	0.00	0.00	0.00	0.00
Jiangsu (JS)	0.69	0.79	0.89	0.85	0.91	0.84	0.80	0.92	0.91	0.86
Zhejiang (ZJ)	0.59	0.20	0.26	0.57	0.84	0.63	0.36	0.00	0.52	0.00
Anhui (AH)	0.86	0.80	0.85	0.91	0.88	0.94	0.86	0.87	0.93	0.87
Fujian (FJ)	0.00	0.45	0.87	0.78	0.49	0.87	0.76	0.83	0.92	0.86
Jiangxi (JX)	0.74	0.74	0.71	0.78	0.74	0.83	0.85	0.77	0.85	0.85
Shandong (SD)	0.00	0.00	0.94	0.79	0.81	0.91	0.89	0.88	0.92	0.84
Henan (HEN)	0.90	0.86	0.89	0.86	0.91	0.93	0.94	0.94	0.87	0.98
Hubei (HUB)	0.45	0.54	0.00	0.75	0.66	0.66	0.67	0.62	0.82	0.76
Hunan (HUN)	0.58	0.53	0.00	0.75	0.58	0.66	0.74	0.48	0.71	0.79
Guangdong (GD)	0.65	0.83	0.66	0.89	0.90	0.93	0.89	0.78	0.80	0.76
Guangxi (GX)	0.84	0.86	0.87	0.90	0.86	0.86	0.90	0.93	0.85	0.90
Hainan (HAN)	0.77	0.56	0.43	0.60	0.50	0.00	0.58	0.49	0.33	0.88
Chongqing (CQ)	0.00	0.37	0.36	0.68	0.00	0.29	0.00	0.84	0.85	0.96
Sichuan (SC)	0.63	0.72	0.79	0.80	0.77	0.80	0.87	0.83	0.88	0.92
Guizhou (GZ)	0.46	0.74	0.51	0.74	0.55	0.61	0.87	0.65	0.81	0.93
Yunnan (YN)	0.60	0.65	0.63	0.69	0.41	0.71	0.78	0.80	0.81	0.77
Shaanxi (SH'X)	0.75	0.73	0.78	0.77	0.78	0.85	0.90	0.82	0.81	0.88
Gansu (GS)	0.00	0.48	0.62	0.67	0.63	0.62	0.70	0.66	0.64	0.68
Qinghai (QH)	0.00	0.51	0.57	0.58	0.68	0.73	0.63	0.69	0.66	0.00
Ningxia (NX)	0.83	0.00	0.78	0.78	0.85	0.78	0.80	0.78	0.82	0.83
Xinjiang (XJ)	0.51	0.50	0.23	0.46	0.56	0.75	0.64	0.75	0.50	0.66
Average	0.47	0.47	0.57	0.63	0.65	0.69	0.64	0.67	0.68	0.64

[1]Note that we presented abbreviations of provinces in Table 2, and in the following context we will mainly use the abbreviation for sake of brevity.

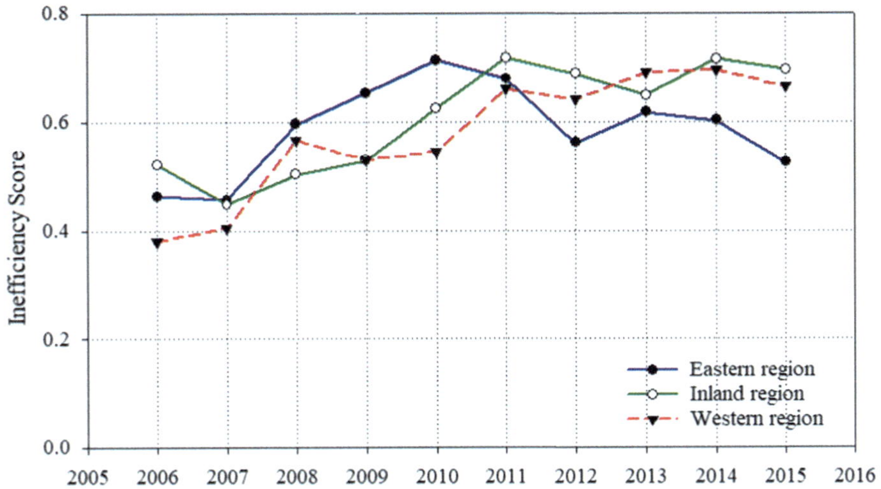

Fig. 1 The changing trend of inefficiency score in the Western region, Inland region, and Eastern region during the study period (2006–2015)

Results show that the overall inefficiency of China's agricultural sector is 0.61. The disaster areas account for 0.29 out of 0.61. Also, the analysis shows that CO_2 emissions, which account for 0.05, cannot be ignored. Besides, the comparatively exceeded inputs (represented by high inefficiency scores) and zero inefficiencies for gross agricultural output suggest the necessity of structural adjustments. Regionally, Shanghai has been the province-level region located in the frontier across this decade (2006–2015), which should be the best example for other provinces. Furthermore, regarding the granary of China (i.e., Liaoning, Jilin, and Heilongjiang), the inefficiency scores observed for these three provinces are 0.58, 0.69, and 0.46, respectively. The contribution of Jilin to the inefficiency score is much higher than others, which can be attributed to climate change and food security constraints. Hence, the local authorities ought to pay more attention to agricultural CO_2 mitigation and agricultural disaster prevention.

7 Policy Recommendations

In China, spring production is of great importance, which is significantly impacted by climate change and natural disasters. As the main part of summer grain, wheat plays an important role in ensuring food security. Concurrently, rape, an important oil crop, is also a spring crop. Therefore, spring production is directly related to the annual grain and oil supply, which is of great significance in promoting the sustainable supply and price of grain and other important agricultural products to increase farmers' income. Considering the normalization of epidemic prevention and control and the prominent La Nina phenomenon this year, the relevant departments should strengthen the understanding of the importance of early spring production, comprehensively promote the planting of winter crops, do everything possible to promote early spring production, increase food production, improve farmers' income, and ensure sustainable food production.

Based on our investigations, we recommend the following measures to mitigate the impact of climate change and food security for sustainable agricultural development.

(i) Fully understand the importance of agricultural production in spring

With population growth, the upgrading of consumption structure, and the slacking resource-carrying capacity, grain production, and demand should still maintain an appropriate balance. The unexpected new epidemic situation has impacted the trade chain and supply chain of grain and other bulk agricultural products, increased the uncertainty of supply in the international agricultural products market, and further highlighted the problem of "no immediate worries and long-term concerns" in the supply of China's grain and other major agricultural products. It is necessary to sustain domestic grain production to cope with the uncertainty brought by changes in the international situation. Among them, spring production is an important part, which directly determines whether the summer harvest is good or not. As the saying goes, "if you have a good harvest in summer, you will take the initiative all year round; if you supplement summer with autumn, you would be frightened". Therefore, ensuring to do a good job in spring production and a good harvest of summer grain will be the key direction of sustainable production and supply of grain and important agricultural products, providing substantial economic and social development support.

(ii) Take active precautions against possible meteorological disasters in winter and spring

In 2020, the temperature in most parts of the country was much lower than that of the same period of previous years. Hence, special attention should be paid to the deep winter and extreme weather to avoid a devastating impact on agricultural production. The National Climate Center predicts that China's cold air influence will be significantly more substantial around January next year. It may be an extensive range of low-temperature, rain, snow, and weather in the northern region, and periodic low-temperature, rain, snow, and freezing weather in the south, especially in the eastern part of Southwest China. In this regard, the northern region should be prepared to deal with the snow disaster and low temperature, and the southern region should guard against the adverse effects of low temperature, rain, snow, and freezing weather on transportation, electric power, and agricultural production. In the future, for example, in spring next year, the temperature in most areas of China is on the high side, and the precipitation is on the low side, so it is necessary to make preparations for drought prevention and disaster reduction in different areas in advance.

(iii) Strengthen the supervision of regional grain production to ensure the production area in spring

The land is the essential foundation of total production. Strictly abiding by the red line of 1.8 billion mu of arable land is an important prerequisite to ensuring grain and oil production capacity. All local authorities should.

- strengthen the supervision of grain production functional areas;
- consider the county as the unit;
- implement the action of farmland water conservancy construction;
- implement the "field leader system" management.

It will help put an end to the "abandonment" of cultivated land and curb the "non-grain" of cultivated land. Considering different land productivity conditions, we should consider the space–time distribution of grain productivity in the medium and long terms. We should also continue to maintain the strong production capacity of the land to meet our goal. The grain is stored in the ground. Simultaneously, the unreasonable use of farmland in the early stage should be avoided, and project-based land fertility restoration should be carried out systematically to ensure the improvement of soil fertility and production efficiency. Besides, it is necessary to increase reclamation work to ensure that the regional spring sowing surface is stable and rising.

(iv) Persisting in scientific and technological innovation to improve quality and efficiency of Spring production

We should:

- strengthen breeding of stress-resistant, high-quality, and high-yield varieties;
- guide the production of wheat varieties in autumn and winter each year;
- focus on "improving the sowing quality and cultivating strong seedlings before winter";
- select suitable machine harvest, and double low rapeseed varieties;
- focus on the base and park to carry out comprehensive integrated demonstration, to drive the development of surrounding areas.

We should also promote high-yield and efficient clean production technology, focus on key technology innovations such as annual coordination of small spring and big spring, variety updating, increase in plow layer thickness, soil organic matter fertilization, mechanized sowing of rice and wheat, rape machine harvesting, crop interplanting, weight loss and pesticide reduction, green prevention, control diseases and pests, straw resource utilization, and strengthening of core technologies such as using an unmanned aerial vehicle (UAV) for drug dosage and battery endurance breach.

Under a complex and changing climate, we should reserve a series of anti-adversity and disaster reduction technologies, such as anti-waterlogging, anti-low temperature, and cold damage, and efficiently increasing production and tapping potential in an all-round way, to realize "storing grain in technology".

(v) Adjusting the structure of grain subsidies and promoting the enthusiasm of small farmers in growing grain

The goal of food security is not to maximize the operating profit. The intensive cultivation and deep love for small farmers' land make them responsible for maintaining food security. For limited total subsidies, we need to consider the reverse tilt of the grain subsidy structure, in particular, farmers' decisions regarding whether returning to rural areas are affected by the new epidemic situation. Therefore, more subsidies should be given to small grain farmers to consolidate their enthusiasm for growing grain. Simultaneously, we should reduce the direct policy intervention to the scale operators and avoid cultivating the subsidy-dependent scale operators.

(vi) Increase investment in agricultural machinery and equipment and improve the socialized service system

Winter wheat is mainly distributed in the south of the Great Wall, and spring wheat is mainly distributed in the north of the Great Wall. The results show that more than 90% of farmers in the main wheat-producing areas, especially in the main winter wheat-producing areas, employ machinery in the production process. Therefore, the cross-regional agricultural machinery outsourcing service has increased significantly. Meanwhile, in Pidu District of Chengdu, by building the "Internet + agricultural machinery" cloud platform, using agricultural machinery with the Beidou navigation system, the cost of planting is reduced by 30%, and the efficiency is increased by 30%. Farmers enjoy "drip type" services, and the content of agricultural machinery cooperative services extends from the mechanical industry to information collection, production supervision, and so on. Therefore, it is imperative to strengthen the policy support and standardization construction of social outsourcing services in grain production. We should prioritize increasing investment in agricultural machinery and equipment, fill gaps in the development of 30 ~ 50 between China and abroad, easy access to cross-regional service green channels, and build an "Internet + agriculture" platform.

Acknowledgements We acknowledge the financial support from the National Social Science Foundation of China (19BGL152).

References

Badami, M., & Ramankutty, N. (2015). Urban agriculture and food security: A critique based on an assessment of urban land constraints. *Global Food Security, 4*, 8–15. https://doi.org/10.1016/j.gfs.2014.10.003

Cecchini, L., Venanzi, S., Pierri, A., & Chiorri, M. (2018). Environmental efficiency analysis and estimation of CO2 abatement costs in dairy cattle farms in Umbria (Italy): A SBM-DEA model with undesirable output. *Journal of Cleaner Production, 191*(1), 895–907. https://doi.org/10.1016/j.jclepro.2018.06.165

Chen, X., Miao, Z., Wang, K., & Sun, C. (2020). Assessing eco-performance of transport sector: Approach framework, static efficiency and dynamic evolution. *Transportation Research Part D: Transport and Environment, 85*, 102414. https://doi.org/10.1016/j.trd.2020.102414

Chen, Z., Wanke, P., Antunes, J., & Zhang, N. (2017). Chinese airline efficiency under CO_2 emissions and flight delays: A stochastic network DEA model. *Energy Economics, 68*, 89–108. https://doi.org/10.1016/j.eneco.2017.09.015

Färe, R., Grosskopf, S., Lovell, C. A. K., & Pasurka, C. (1989). Multilateral productivity comparisons when some outputs are undesirable: A nonparametric approach. *The Review of Economics and Statistics, 71*, 90–98. https://doi.org/10.2307/1928055

Färe, R., Grosskopf, S., & Pasurka, C. A. (2007). Environmental production functions and environmental directional distance functions. *Energy, 32*, 1055–1066. https://doi.org/10.1016/j.energy.2006.09.005

Food Security Information Network (FSIN) (2020). *Global report on food crises*. Technical report, pp. 240.

Fukuyama, H., Weber, W.L. (2009) A directional slacks-based measure of technical inefficiency. *Socio-Economic Planning Sciences, 43*(4):274–287. https://econpapers.repec.org/scripts/redir.pf?u=http%3A%2F%2Fwww.sciencedirect.com.

Iftikhar, Y., Wang, Z., Zhang, B., & Wang, B. (2018). Energy and CO_2 emissions efficiency of major economies: A network DEA approach. *Energy, 147*, 197–207. https://doi.org/10.1016/j.energy.2018.01.012

Kansiime, M., Tambo, J., Mugambi, I., Bundi, M., Kara, A., Owuor, C. (2021) COVID-19 implications on household income and food security in Kenya and Uganda: Findings from a rapid assessment. *World Development, 127*, 105199. https://doi.org/10.1016%2Fj.worlddev.2020.105199.

Li, J., Tian, Y., Deng, Y., Zhang, Y., & Xie, K. (2020). *Improving the estimation of greenhouse gas emissions from the Chinese coal-to-electricity chain by a bottom-up approach*. In Press. https://doi.org/10.1016/j.resconrec.2020.105237

Li, R., Zhang, H., Wang, H., Tu, Q., & Wang, X. (2019). Integrated hybrid life cycle assessment and contribution analysis for CO2 emission and energy consumption of a concentrated solar power plant in China. *Energy, 174*, 310–322. https://doi.org/10.1016/j.energy.2019.02.066

Livia Bizikova, Stefan Jungcurt, Kieran McDougal, Stephen Tyler. How can agricultural interventions enhance contribution to food security and SDG 2.1?, *Global Food Security*, 2020, 26, 100450. https://doi.org/10.1016/j.gfs.2020.100450.

Mahdiloo, M., Ngwenyama, O., Scheepers, R., & Tamaddoni, A. (2018). Managing emissions allowances of electricity producers to maximize CO_2 abatement: DEA models for analyzing emissions and allocating emissions allowances. *International Journal of Production Economics, 205*, 244–255. https://doi.org/10.1016/j.ijpe.2018.09.014

Mckune, S. L., Borresen, E. C., Young, A. G., Auria, T. D., Russo, S. L., Diao, A., et al. (2015). Climate change through a gendered lens: Examining livestock holder food security. *Glob. Food Secur., 6*, 1–8.

Miao., Z, Chen, X., Baležentis, T., Cun, C. (2019). Atmospheric environmental productivity across the provinces of China: Joint decomposition of range adjusted measure and Luenberger productivity indicator. *Energy Policy, 132*, 665–677. https://doi.org/10.1016/j.enpol.2019.06.019

Khan, N., Siddiqui, B. N., Khan, N., Ullah, N., Wali, A., Khan, I. U., & Ihtisham, M. (2020). Drastic impacts of COVID-19 on food, agriculture and economy. *Pure and Applied Biology (PAB), 10*(1), 62–68. https://doi.org/10.19045/bspab.2021.100008.

McKune, S. L., Borresen, E. C., Young, A. G., Auria, T. D., Ryley, S. L., Russo, A. D., Camara, M. C., & Ryan, E. P. (2015). Climate change through a gendered lens: Examining livestock holder food security. *Global Food Security., 6*, 1–8. https://doi.org/10.1016/j.gfs.2015.05.001

Sargani, G. R., Zhou, D., Raza, M. H., & Wei, Y. (2020). Sustainable entrepreneurship in the agriculture sector: The nexus of the triple bottom line measurement approach. *Sustainability, 12*(8), 3275. https://doi.org/10.3390/su12083275

Stephan, A., & Stephan, L. (2020). Achieving net zero life cycle primary energy and greenhouse gas emissions apartment buildings in a Mediterranean climate. *Applied Energy., 2020*(280), 115932. https://doi.org/10.1016/j.apenergy.2020.115932

Tone, K. (2001). A slacks-based measure of efficiency in data envelopment analysis. *European Journal of Operational Research, 130*, 498–509. https://doi.org/10.1016/S0377-2217(99)00407-5

Wang, B., Wu, Y., & Yan, P. (2010). Environmental efficiency and environmental total factor productivity growth in China's regional economies. *Economic Research Journal, 5*, 95–109.

Whitmarsh, L., Capstick, S., Moore, I., Köhler, J., & Quéré, C. (2020). Use of aviation by climate change researchers: Structural influences, personal attitudes, and information provision. *Global Environmental Change., 65*, 102184. https://doi.org/10.1016/j.gloenvcha.2020.102184

Yang, M., Hou, Y., Ji, Q., & Zhang, D. (2020). Assessment and optimization of provincial CO2 emission reduction scheme in China: An improved ZSG-DEA approach. *Energy Economics, 91*, 104931. https://doi.org/10.1016/j.eneco.2020.104931

Ye, L., Tang, H., Yang, G., & Ranst, E. (2015). Adopting Higher-yielding Varieties to Ensure Chinese Food Security under Climate Change in 2050. *Procedia Environmental Sciences., 2015*(29), 281. https://doi.org/10.1016/j.proenv.2015.07.211

Zhou, P., Ang, B.W., Poh, K.L. (2008) Measuring environmental performance under different environmental DEA technologies. *Energy Economics, 30*, 1–14. Handle: RePEc:eee:eneeco:v:30:y:2008:i:1:p:1–14.

Can Seasonal Climate Forecasts Improve Household Welfare? Insights from Rural Senegal

Thierno Malick Diallo and Malick Ndiaye

Abstract Despite the potential of climate forecasts to manage climate risks and build resilience, the extent to which they contribute to improving the welfare of rural farm households is still not well-known. We, therefore, ask how far the application of seasonal climate forecasts to production decisions is linked to better livelihood outcomes. In this study, we examine this question empirically by assessing if the uptake of seasonal climate forecasts is associated with, or even leads to, improved rural household welfare using the propensity score matching method. We find no evidence that the application of seasonal climate forecasts to farming activities contributes to enhancing rural household welfare in Senegal. Not only do the findings hold regardless of the welfare indicator chosen, but they are further confirmed when using different matching algorithms. The study suggests that co-production of climate services through a partnership between producers (scientists) and users (farmers) may be an effective way to boost the welfare effects of seasonal climate forecasts.

Keywords Seasonal climate forecasts · Propensity score matching · Rural households · Senegal

1 Introduction

Climate change is a major development issue in rural Senegal. In fact, rural households are highly sensitive to climate risks which, in turn, exacerbate poverty and food insecurity (Ilboudo Nébié et al., 2021). Senegal has a total population of over

T. M. Diallo (✉)
Laboratoire de recherche en économie de Saint-Louis (LARES), Gaston Berger University, Saint-Louis, Senegal
e-mail: malikidiallo@gmail.com

M. Ndiaye
Higher Institute of Agricultural and Rural Training, Alioune Diop University of Bambey, Bambey, Senegal
e-mail: malick1.ndiaye@uadb.edu.sn

M. Behnassi et al. (eds.), *Climate-Smart and Resilient Food Systems and Security*,
https://doi.org/10.1007/978-3-031-65968-3_22

15 million people of which 55% live in rural areas (ANSD, 2020). Agriculture is the main economic activity for 70% of the rural population (IFAD, 2019), but due to its rain-fed nature, most rural households are likely to be vulnerable to climate shocks (CIAT & BFS/USAID, 2016). Dry spells, erratic rainfall and rising temperatures are identified as the main climate shocks affecting farming activities (Lo and Dieng, 2015; Diouf et al., 2019). Future trends predict by 2035 an average rise in temperatures of 1.17–1.41 °C, a decrease in rainfall of 89 mm and higher frequency of extreme events, such as heat waves, drought and floods (CDN, 2020).

Rural households adopt many strategies to cope with climate shocks. These include the use of drought-resistance crops, soil moisture conservation techniques and changes in planting dates (Akinnagbe & Irohibe, 2014). Seasonal climate forecasting has also been introduced in Senegal to help rural households better manage climate shocks. In recent years, there have been important collaborations between scientists, the National Agency for Civil Aviation and Meteorology (ANACIM) and rural community radios to provide well-targeted and tailored climate information services (Zougmoré et al., 2018). In 2013, the number of people with access to climate information reached two million as a result of these policy efforts (CCAFS, 2015). This number is expected to increase over time following the new program "Accelerating the Impact of CGIAR Climate Research for Africa project (AICCRA)", which aims to improve the uptake of climate information services in many African countries, including Senegal (World Bank, 2020).

Several studies show that access and use of climate forecasts helped Senegalese farmers enhance their decision-making about the adoption of climate-smart agricultural practices (e.g. Roudier et al., 2014; Lo and Dieng, 2015; Ouedraogo et al., 2018; Diouf et al., 2019; Blundo-Canto et al., 2021). However, the question remains to what extent climate forecasts contribute to improving household welfare. In this study, we aim to investigate the impact of seasonal climate forecasts on rural household welfare in Senegal. We focus on rural households because they rely heavily on weather-sensitive activities such as agriculture (CIAT & BFS/USAID, 2016), and remain the country's poorest and most food-insecure populations (ANSD, 2021; Ilboudo Nébié et al., 2021). In this context, examining the determinants of climate forecasting use and its impact on rural household welfare is timely for policy design aimed at addressing poverty and food insecurity.

Our chapter adds to the literature on climate forecast adoption. While many studies have been conducted on this research area, they have mostly examined the factors influencing the use of climate forecast information (see Coulibaly et al., 2017; Diouf et al., 2019; Kirui et al., 2014; Mudombi & Nhamo, 2014; Muema et al., 2018; Ochieng et al., 2017; Oyekale, 2015; Yong, 2014), and how this affects farm management decisions (Coulibaly et al., 2017; Etwire et al., 2017; Ouedraogo et al., 2018; Roudier et al., 2014; Serra & McKune, 2016; Vaughan et al., 2019). There is still limited understanding of the impact of climate forecast use on farmers' livelihoods. We contribute to filling this gap by investigating the relationship between climate forecast use and the welfare of rural households using a nationally representative rural household survey. One of the advantages of our survey data is that it allows us to identify the main factors that influence farmers' use of climate forecasts and to link the use of climate forecasts with farmers' welfare. Specifically, we would like to assess the impact of seasonal climate forecast use on three welfare indicators, namely household income, consumption and food security.

The rest of the study is structured as follows: Sect. 2 reviews previous studies on the uptake of climate forecasts; Sect. 3 presents the empirical strategy; Sect. 4 describes the data and provides some descriptive statistics; Sect. 4 presents the results and discusses the findings; and Sect. 5 provides conclusions.

2 Literature Review

Although the agricultural sector plays an important role in poverty reduction in Africa (Christiansen & Martin, 2018), it is highly sensitive to changing climate, especially changes in precipitation and temperature (Dube & Sivakumar, 2015; Tall et al., 2012; Traore et al., 2013). Climate forecasting is one among many tools that can be used by farmers to reduce the risks posed by climate variability and change (Bruno Soares et al., 2018; Goddard, 2016; Muema et al., 2018; Ouédraogo et al., 2018; Roncoli, 2006). Specifically, climate forecasts provide information on the climatic conditions of the forthcoming seasons on two- to three-month' time scales, thus enhancing farmers' adaptation to climate risks (Clarkson et al., 2017; Hansen, 2002; Ziervogel & Calder, 2003). The forecasts usually include information such as onset and offset dates of the rainy season, daily precipitation, daily weather and risk of extreme events, among others (Coulibaly et al., 2017; Goddard et al., 2001).

The importance of seasonal climate forecasts to agriculture has been extensively documented. Many studies have established the central role of climate forecasts in improving both agricultural risk management (Bruno Soares et al., 2018; Meinke & Stone, 2005; Patt et al., 2005; Salack et al., 2011) and farmers' decision-making (Boulanger & Penalba, 2010; Diouf et al., 2019; Frisvold & Murugesan, 2012; Klopper et al., 2006; Ouédraogo et al., 2018; Singh et al., 2017). In the literature, evidence also exists that the use of climate forecasts leads to actual increases in yield (Parton et al., 2019; Roudier et al., 2014; Sultan et al., 2013).

Despite the potential of climate forecasts to manage climate risks and build resilience, their application to production decisions remains limited. Part of the problem is that climate information services are often not usable because of poor awareness and understanding of information by farmers (Dilling & Lemos, 2011; Hansen et al., 2011; Jones et al., 2015; Lemos et al., 2012; Roncoli, 2006; Roncoli et al., 2009). Lack of trust and reliability is also a major barrier to the effective uptake of climate forecasts (Mudombi & Nhamo, 2014; Muema et al., 2018; Ziervogel et al., 2005).

Recently, co-produced climate services have been proposed as a method of enhancing the accuracy and usability of climate forecasts in developing countries (e.g. Blundo-Canto et al., 2021; Carter et al., 2019; Harvey et al., 2021; Vaughan & Dessai, 2014; Vincent et al., 2018). This is consistent with findings from studies conducted in Senegal, which showed that participatory farmer workshops helped farmers better understand and use seasonal climate forecasts (Dayamba et al., 2018; Ouedraogo et al., 2018; Roudier et al., 2014).

In summary, most of the above studies have sought to determine the factors influencing the uptake of climate forecast information. However, none of these have investigated the impact of climate forecasts on farmers' livelihoods. Our study thus extends the literature by assessing the extent to which climate forecasts contribute to improving farmers' welfare. As such, Senegal can be a policy case study for other countries seeking to benchmark the impact of climate forecasts on rural development.

3 Empirical Strategy

3.1 Identifying the Treatment Variable

In this study, the use of seasonal climate forecasts is conceived as the main independent variable of interest, i.e. the treatment variable. We employ a dichotomous variable, in which 1 indicates that the farmer applied the forecasts to production decisions, and 0 otherwise.

3.2 Identifying the Outcome Variables

We use three welfare outcomes to compare treated and untreated rural households (i.e. users and non-users of climate forecasts). Specifically, we want to examine the effects of climate forecasts on per capita income, consumption and food security.

Rural household consumption may be better suited than income when investigating the impact of climate forecasts on welfare. This is because income is only received intermittently, while consumption is smoothed over time (O'Donnell et al., 2007). Unlike consumption, income is therefore an indicator that can be used to capture the welfare of rural households in the short run (Haughton & Khandker, 2009). In our study, we use both income and consumption as welfare indicators. This helped us test the robustness of our results because there might be measurement errors in the reported income and/or consumption.

There are several ways to measure food security (Ilboudo Nébié et al., 2021). Given the nature of the data available, we use the food consumption score (FCS) to capture the food security status of rural households. The FCS measures rural households' dietary diversity and nutrient intake. It is calculated by taking the weighted sum of all the food items consumed in a household during a 7-day reference period (World Food Program, 2008).

3.3 Analytical Model

We use impact evaluation techniques to determine the effect of seasonal climate forecasts on rural household welfare. Researchers typically choose either experimental or quasi-experimental methods to measure the impact of an intervention. Due to the cross-sectional nature of our survey data, the analysis is performed using a quasi-experimental approach, i.e. the propensity scores matching method (PSM).

The PSM is a non-parametric estimation technique that matches observations of users and non-users of climate forecasts based on the propensity score (Heckman et al., 1998; Rosenbaum & Rubin, 1983). The PSM model consists of two stages. In the first stage, a binary choice logistic regression is modelled taking the use of climate forecasts as the dependent variable. This is specified as follows:

$$T_i^* = \alpha x_i + u_i$$
$$\text{with } T_i = \begin{cases} 1 & if\ T_i^* > 1 \\ 0 & otherwise \end{cases} \tag{1}$$

where T_i^* is a latent variable, T_i is a binary dependent variable equal to 1 if the ith rural household uses climate forecasts and 0 otherwise, x_i is a vector of exogenous variables, α is a vector of parameters to be estimated and u_i is the error term. Equation (1) enables us to determine the factors influencing farmers' use of climate forecasts as well as the propensity scores, $P(x)$.

In the second stage, three matching methods, including nearest neighbour matching, radius matching and kernel matching, are used to estimate the average treatment effect.

Let Y_{i1} and Y_{i0} represent the welfare outcomes (household income, consumption, food security) of rural households using climate forecasts and those who do not, respectively. The average impact of treatment on the treated (ATT) is given by

$$\text{ATT} = E[Y_{i1}|T = 1, P(x)] - E[Y_{i0}|T = 1, P(x)] \tag{2}$$

As such, Eq. (2) cannot be estimated because the component $E[Y_{i0}|T = 1, P(x)]$ is unobserved. However, given the conditional independence assumption, $(Y_{1i}, Y_{0i}) \perp T|P(x)$, i.e. the potential outcomes are independent of the use of climate forecasts but conditional on the propensity score, one can write $E(Y_{0i}|T = 1), P(x)] = E(Y_{0i}|T = 0), P(x)]$.

Thus, one can compute the ATT as

$$\text{ATT} = E[Y_{i1}|T = 1, P(x)] - E[Y_{i0}|T = 0, P(x)] \tag{3}$$

That said, careful testing of the matching quality is required before estimating the average treatment effects (ATT) of using climate forecasts.

4 Data and Descriptive Statistics

In this study, we use data from a rural household survey (ERASAN) conducted in 2014 in Senegal by three government agencies under the Ministry of Agriculture and Rural Equipment. ERASAN uses a unified questionnaire that provides detailed information on the food and nutrition security status of rural households. The questionnaire is also designed to obtain information on farming activities of rural households, key climate-sensitive decisions and strategies adopted by rural households to reduce their vulnerability to climate risks. Special attention is also given in the questionnaire to the use of climate information services by rural households. A three-stage stratified sampling is applied for the selection of a sample and is described in ERASAN (2014). A total of 5270 rural farm households were interviewed. But given the purpose of the study, our sample consists of 1,050 households divided into two groups: households who use seasonal climate forecasts (430) and those who do not (626).

Table 1 shows descriptive statistics for the outcome variables and the set of socioeconomic variables that we use as household-level controls in the regression models. Results indicate a difference of around 68% in per capita income between rural households who use climate forecasts and those who do not. We also observe a difference of 4% in per capita consumption expenditures between rural households using and not using climate forecasts. For both groups, mean expenditures are below the national poverty line of 501.41 euros per capita and year (ANSD, 2021). Results furthermore reveal that rural households who use climate forecasts obtain lower food consumption scores, with a 2% difference on average. Although this difference in score is small, climate forecasts seem to have a limited impact on household food security.

Particularly noteworthy in Table 1 is no statistically significant differences are observed for any of the three outcome variables (column 3). This suggests that the two groups of rural households (i.e. users and non-users of climate forecasts) are not systematically different in terms of income, consumption expenditure and food security level. The two groups of rural households are also very similar in terms of most socioeconomic variables, including household size, household head's age and shock exposure.

5 Results and Discussion

5.1 *Propensity Score Estimation*

As mentioned above, we use a logistic regression model to estimate the propensity score (Table 2). We find that household size, number of female farmers in the household and shocks, such as increased input prices and poor rains, are the significant predictors of farmers' use of climate forecasts.

Table 1 Characteristics of rural households using and not using climate forecasts

	(1)	(2)	(3)
	Treated households[a]	Untreated households[b]	Mean difference
Outcome variables			
Income per capita (Euros)	229.62 [104.30]	136.38 [17.17]	93.23
Consumption expenditure per capita (Euros)	159.06 [21.82]	152.92 [17.20]	6.13
Food security (FCS)	57.11 [1.15]	58.42 [0.94]	−1.30
Household-level variables			
Age of head	52.11 [0.66]	52.12 [0.57]	−0.01
Household size	12.48 [0.17]	12.23 [0.15]	−0.25
Number of female farmers	2.93 [0.08]	3.21 [0.07]	−0.28**
Number of male farmers	4.31 [0.16]	4.29 [0.12]	0.02
Increase in price of inputs	0.74 [0.02]	0.72 [0.01]	0.01
Pests invasion	0.45 [0.02]	0.45 [0.02]	0.001
Poor rains	0.82 [0.01]	0.84 [0.01]	−0.021
Out-of-season rains	0.20 [0.01]	0.08 [0.01]	0.12***
Observations	430	626	1056

Notes [a,b] Treated and untreated households refer to households using and not using climate forecasts, respectively. Mean values are shown with standard deviations in [.]. ** $p < 0.05$. *** $p < 0.01$. The income and consumption are expressed in CFA in the database, but we converted the values to euros at the rate of 665. *Source* Own calculation based on survey data

In terms of household size, it appears that larger rural households have a better chance of using climate forecasts compared to the smaller ones. This contrasts with the work of Muema et al. (2018) who found that household size is significantly and negatively related to climate forecast use. That said, farmers' likelihood of using climate forecasts decreases with the number of female farmers, suggesting that rural women are less likely to integrate climate information into their farming activities. Our results also reveal that shocks such as increased input prices and poor rainfall are significantly and positively related to the use of climate forecasts. This indicates that rural households experiencing adverse shocks have a greater interest in

Table 2 Propensity score estimation: factors influencing the use of seasonal climate forecasts

Variables	Marginal effects	Robust standard errors	P-value
Female-headed household	−0.086	0.057	0.132
Age of head	−0.00018	0.00097	0.854
Household size	0.0077	0.0045	0.086*
Number of female farmers	−0.0304	0.0088	0.001***
Number of male farmers	0.0023	0.0051	0.647
Increase in price of inputs	0.124	0.036	0.001***
Pests invasion	−0.019	0.0298	0.524
Poor rains	0.166	0.045	0.000***
Out-of-season rains	−0.011	0.041	0.788
Region[a]	*Yes*	*Yes*	*Yes*
Number of observations	1044		
Prob > chi^2	0.0000		
Pseudo R^2	0.1769		
Wald $chi^2(20)$	171.33		

Notes ***, ** and * indicate statistical significance at the 0.1%, 1% and 5% levels. *Source* Own calculation based on survey data

applying climate forecasts to production decisions, probably because climate information services are crucial for risk management in agricultural practices (Diouf et al., 2019; Singh et al., 2017).

5.2 *Impact of Seasonal Climate Forecast Use on Income, Consumption and Food Security*

We apply three matching methods, i.e. nearest neighbour matching, radius matching and kernel matching, to estimate the impact of seasonal climate forecasts on rural household welfare.

Before presenting the results, it is important to carefully check the matching quality. This requires comparing the situations before and after the matching procedure. Figure 1 shows that the propensity score density in the "treated" and "untreated" groups is different before matching. This is indicative of an inadequate comparison of the sample characteristics in the "untreated" group. After matching, we find that the density distributions of the two groups become somewhat similar.

Finally, Fig. 2 shows the existence of a good overlap between the propensity scores of users (treated) and non-users (untreated) of climate forecasts, suggesting that the two groups have a large enough area of common support.

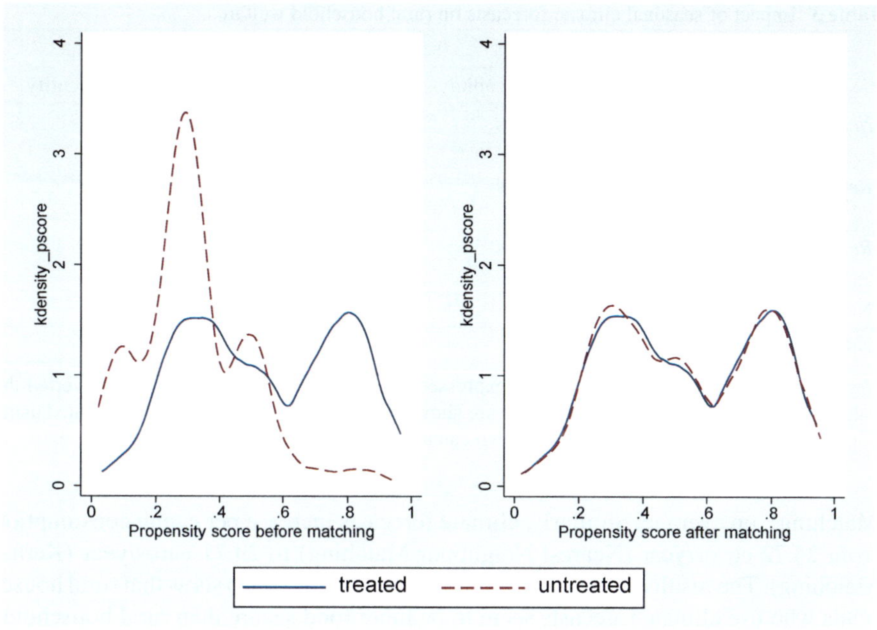

Fig. 1 Distribution of propensity scores before and after matching in treated and untreated groups. *Source* Own calculation based on survey data

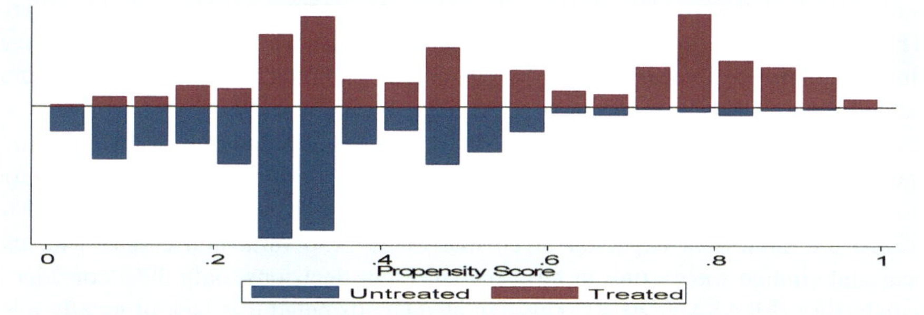

Fig. 2 Common support area. *Source* Own calculation based on survey data

Overall, these various tests indicate a well-matched and corrected selection bias for our model. It should therefore be possible to use the propensity scores to estimate treatment effects.

The estimated average effects of climate forecasts on the three outcome variables are summarized in Table 3. We used a bootstrap with 100 replications to calculate standard errors.

As shown, the impact of climate forecasts on per capita income is positive, ranging from 109.72 euros/year (Nearest Neighbour Matching) to 113.31 euros/year (Radius

Table 3 Impact of seasonal climate forecasts on rural household welfare

	(1)	(2)	(3)
	Income per capita	Consumption per capita	Food security
Nearest neighbour	109.72 euros [0.34]	24.72 euros [0.33]	2.68 [0.24]
Kernel matching	113.22 euros [0.319]	29.71 euros [0.21]	2.83 [0.19]
Radius matching	113.31 euros [0.41]	17.96 euros [0.46]	2.79 [0.25]
Number of observations	1,044	1,044	1,044
Number of replications	100	100	100

Notes The income and consumption are expressed in CFA in the database, but we converted the values to euros at the rate of 665. P-values are shown in [.] and standard errors are calculated using bootstrap with 100 replications. *Source* Own calculation based on survey data

Matching), on average. Similarly, climate forecasts increase per capita consumption from 24.72 euros/year (Nearest Neighbour Matching) to 29.71 euros/year (Kernel Matching). The results of the impact assessment in Table 3 also show that rural households who use climate forecasts seem to be more food secure than rural households who do not.

That said, none of the results are significant. In other words, there is no significant difference in income, consumption and food security between rural households who use climate forecasts and those who do not. The results are robust to the three matching algorithms and the three outcomes used to capture rural household welfare. The results are also consistent with the descriptive statistics in Table 1, which reveal that using climate forecasts does not translate into a better living standard for rural households.

Our study shows that the use of climate forecasting does not translate into a measurable change in income, consumption and food security, indicating that rural households using climate forecasts are not better off than those who do not. The results are somewhat expected given that of the 430 rural households who use seasonal climate forecasting in farm management decisions, only 39% consider it satisfactory (ERASAN, 2014). This can also be attributed to a lack of locally relevant climate information (Coulibaly et al., 2017; Jones et al., 2017; Muema et al., 2018; Singh et al., 2017), and poor awareness and understanding of information by farmers (Hansen et al., 2011; Mudombi & Nhamo, 2014; Roncoli, 2006; Roncoli et al., 2009). Our findings concur with the earlier study by Patt et al. (2005), as they reported that climate forecast application to agriculture led to small increases in yield and resulting low impact on the welfare of the farmers. However, the findings contrast with those of Roudier (2012) and Sultan et al. (2013), which showed that farmers who used climate forecasts were more likely to improve their livelihoods.

6 Conclusions

In this study, we use a unique dataset of 1056 rural households to empirically test the extent to which the application of seasonal climate forecasts in farm decisions can improve rural household welfare. Our findings show that climate forecasts have no effects on income, consumption and food security, suggesting that rural households who use climate forecasts are not significantly better off than those who do not.

For climate forecasts to improve rural household welfare, we argue that they should be accessible, accurate, timely and usable for climate risk management. To achieve this goal, there is a need to promote partnerships between climate forecast producers (scientists who generate the forecasts) and users (farmers receiving and applying the forecasts). In recent years, significant investments have been made to enable the Senegal Meteorological Department (ANACIM) to provide climate information services that are well-targeted and tailored. These efforts should be strengthened in order to make climate forecasts an effective policy instrument for improving the productivity and welfare of farmers.

While this study adds to the literature by investigating the impact of climate forecasts on rural household welfare, it nevertheless has some limitations. We accounted for potential self-selection bias using propensity score matching (PSM) analysis. Yet, PSM is a quasi-experimental method that only controls for observable factors. That said, our findings are robust to the three matching algorithms, but we are cautious in claiming any causal effect between climate forecasts and welfare outcomes. Hence, further research should be undertaken with longitudinal data in which rural households would be randomly assigned to groups.

References

Akinnagbe, O. M., & Irohibe, I. J. (2014). Agricultural adaptation strategies to climate change impacts in Africa: A review. *Bangladesh Journal of Agricultural Research, 39*, 407–418. https://doi.org/10.3329/bjar.v39i3.21984

ANSD. (2020). Situation Économique et Sociale du Sénégal 2017-2018, Juillet 2020, Ministère de l'Économie, du Plan et de la Coopération. République du Sénégal.

ANSD [Agence Nationale de la Statistique et de la Démographie]. (2021). Enquête harmonisée sur les Conditions de Vie des Ménages (EHCVM) au Sénégal. Rapport final.

Blundo-Canto, G., Andrieu, N., Soule Adam, N., Ndiaye, O., & Chiputwa, B. (2021). Scaling weather and climate services for agriculture in Senegal: Evaluating systemic but overlooked effects. *Climate Services, 22*, 100216. https://doi.org/10.1016/j.cliser.2021.100216

Boulanger, J.-P., & Penalba, O. (2010). Assessment of climate information needs in the Argentinean agro-business sector. *Climate Change, 98*, 551–563.

Bruno Soares, M., Alexander, M., & Dessai, S. (2018). Sectoral use of climate information in Europe: A synoptic overview. *Climate Services, 9*, 5–20. https://doi.org/10.1016/j.cliser.2017.06.001

Carter, S., Steynor, A., Vincent, K., Visman, E., Waagsaether, K. L., Araujo, J., et al. (2019). Co-production in Weather and Climate Services. Cape Town: Future Climate for Africa. Available online at: https://futureclimateafrica.org/coproduction-manual/

CCAFS. (2015). The impact of climate information services in Senegal. CCAFS Outcome Case No. 3. Copenhagen.
CDN [Contribution Déterminée au Niveau National du Sénégal]. (2020). Document approuvé décembre 2020. République du Sénégal.
Christiaensen, L., & Martin, W. (2018). Agriculture, structural transformation and poverty reduction: Eight new insights. *World Development, 109*, 413–416. https://doi.org/10.1016/j.worlddev.2018.05.02
CIAT; BFS/USAID. (2016). Climate-Smart Agriculture in Senegal. CSA Country Profiles for Africa Series. International Center for Tropical Agriculture (CIAT); Bureau for Food Security, United States Agency for International Development (BFS/USAID). Washington, D.C., p. 20.
Clarkson, G., Dorward, P., Kagabo, D. M., & Nsengiyumva, G. (2017). *Climate Services for Agriculture in Rwanda: Initial findings from PICSA monitoring and evaluation.* CCAFS Info Note. Wageningen, Netherlands: CGIAR Research Program on Climate Change, Agriculture and Food Security (CCAFS).
Coulibaly, J. Y., Birachi, E. A., Kagabo, D. M., & Mutua, M. (2017). *Climate services for agriculture in Rwanda: Baseline survey report* (CCAFS Working Paper No. 202). Copenhagen, Denmark: CGIAR Research Program on Climate Change, Agriculture and Food Security (CCAFS). Available online at.www.ccafs.cgiar.org
Dayamba, D. S., Ky-Dembele, C., Bayala, J., Dorward, P., Clarkson, G., et al. (2018). Assessment of the use of Participatory Integrated Climate Services for Agriculture (PICSA) approach by farmers to manage climate risk in Mali and Senegal. *Climate Services, 12*, 27–35.
Dilling, L., & Lemos, M. C. (2011). Creating usable science: Opportunities and constraints for climate knowledge use and their implications for science policy. *Global Environmental Change, 21*(2), 680–689. https://doi.org/10.1016/j.gloenvcha.2010.11.006
Diouf, N. S., Ouedraogo, I., Zougmoré, R. B., Ouedraogo, M., Partey, S. T., & Gumucio, T. (2019). Factors influencing gendered access to climate information services for farming in Senegal. *Gender, Technology and Development, 23*(2), 93–110. https://doi.org/10.1080/09718524.2019.1649790
Dube, O. P., & Sivakumar, M. (2015). Global environmental change and vulnerability of Least Developed Countries to extreme events: Editorial on the special issue. *Weather and Climate Extremes, 7*(2015), 2–7.
Enquête rurale sur l'Agriculture, la Sécurité alimentaire et la Nutrition (ERASAN). (2014). Dakar, Sénégal: SECNSA, DAPSA, and ANACIM.
Etwire, P. M., Buah, S., Ouedraogo, M., Zougmore, R., Partey, S. T., & Martey, E., et al. (2017). An assessment of mobile phone-based dissemination of weather and market information in the Upper West Region of Ghana. *Agriculture & Food Security, 6*, 8. https://doi.org/10.1186/s40066-016-0088-y
Frisvold, G. B., & Murugesan, A. (2012). Use of weather information for agricultural decision making. *Weather, Climate, and Society, 5*, 55–69.
Goddard, L. (2016). From science to service. *Science, 353*(6306), 1366–1367.
Goddard, L., Mason, S. J., Zebiak, S. E., Ropelewski, C. F., Basher, R., & Cane, M. A. (2001). Current approaches to seasonal-to-annual climate predictions. *International Journal of Climatology, 21*, 1111–1152.
Hansen, J. W. (2002). Realizing the potential benefits of climate prediction to agriculture: Issues, approaches. *Agricultural Systems, 74*(2002), 309–330.
Hansen, J. W., Mason, S. J., Sun, L., & Tall, A. (2011). Review of seasonal climate forecasting for agriculture in Sub-Saharan Africa. *Experimental Agriculture, 47*, 205–240. https://doi.org/10.1017/S0014479710000876
Harvey, B., Huang, Y.-S., Araujo, J., et al. (2021). Mobilizing climate information for decision-making in Africa: Contrasting user-centered and knowledge-centered approaches. *Frontiers in Climate, 2*, 589282. https://doi.org/10.3389/fclim.2020.589282
Haughton, J., & Khandker, S. R. (2009). *Handbook on Poverty and Inequality.* The World Bank.

Heckman, J., Ichimura, H., Smith, J., & Todd, P. (1998). Characterizing selection bias using experimental data. *Econometrica, 66*(5), 1017–1098.

IFAD [International Fund for Agricultural Development]. (2019). Country Strategic Opportunities Program 2019–2024. Republic of Senegal

Ilboudo Nébié, E. K., Ba, D., & Giannini, A. (2021). Food security and climate shocks in Senegal: Who and where are the most vulnerable households? *Global Food Security, 29*, 100513. https://doi.org/10.1016/j.gfs.2021.100513

Jones, L., Dougill, A., Jones, R., Steynor, A., Watkiss, P., Kane, C., Koelle, B., Moufouma-Okia, W., Padgham, J., Ranger, N., Roux, J.-P., Suarez, P., Tanner, T., & Vincent, K. (2015). Ensuring climate information supports long-term development objectives. *Nature Climate Change, 5*, 812–814.

Jones, L., Champalle, C., Chesterman, S., Cramer, L., & Crane, T. A. (2017). Constraining and enabling factors to using long-term climate information in decision-making. *Climate Policy, 17*, 551–572. https://doi.org/10.1080/14693062.2016.1191008

Kirui, V. C., Waiganjo, M., & Cheplogoi, S. (2014). Evaluating access and use of dissemination pathways for delivering climate information and services to women farmers in semi-arid Kenya. *International Journal of Advanced Research, 2*(9), 44–53.

Klopper, E., Vogel, C. H., & Landman, W. A. (2006). Seasonal climate forecasts—Potential agricultural-risk management tools? *Climatic Change, 76*, 73–90.

Lemos, M. C., Kirchhoff, C. J., & Ramprasad, V. (2012). Narrowing the climate information usability gap. *Nature Climate Change, 2*(11), 789–794. https://doi.org/10.1038/nclimate1614

Lo, H. M., & Dieng, M. (2015). *Impact assessment of communicating seasonal climate forecasts in Kaffrine, Diourbel, Louga, Thies and Fatick (Niakhar) regions in Senegal: Final report for CCAFS West Africa Regional Program (Report)*

Meinke, H., & Stone, R. C. (2005). Seasonal and inter-annual climate forecasting: The new tool for increasing preparedness to climate variability and change in agricultural planning and operations. *Climate Change, 70*(1–2), 221–253.

Mudombi, S., & Nhamo, G. (2014). Access to weather forecasting and early warning information by communal farmers in seke and murewa districts, Zimbabwe. *Journal of Human Ecology, 48*(3), 357–366.

Muema, E., Mburu, J., Coulibaly, J., & Mutune, J. (2018). Determinants of access and utilisation of seasonal climate information services among smallholder farmers in Makueni County, Kenya. *Heliyon, 4*(11), e00889. https://doi.org/10.1016/j.heliyon.2018.e00889

Ochieng, R., Recha, C., & Bebe, B. O. (2017). Enabling conditions for improved use of seasonal climate forecast in arid and semi-arid Baringo county—Kenya. *Open Access Library Journal, 4*, e3826. ISSN Online: 2333-9721 ISSN Print: 2333-9705.

Ouedraogo, I., Diouf, N. S., Ouedraogo, M., Ndiaye, O., & Zougmore, R. (2018). Closing the gap between climate information producers and users: Assessment of needs and uptake in Senegal. *Climate, 6*, 13. https://doi.org/10.3390/cli6010013

Ouédraogo, M., Barry, S., Zougmoré, R. B., Partey, S. T., Somé, L., & Baki, G. (2018). Farmers' willingness to pay for climate information services: Evidence from cowpea and sesame producers in Northern Burkina Faso. *Sustainability, 2018*(10), 611. https://doi.org/10.3390/su10030611

Oyekale, A. S. (2015). Factors explaining farm households'access to and utilization of extreme climate forecasts in sub-saharan Africa (SSA). *Environmental Economics, 6*(1), 91–103.

Parton, K. A., Crean, J., & Hayman, P. (2019). The value of seasonal climate forecasts for Australian agriculture. *Agricultural Systems, 174*, 1–10. https://doi.org/10.1016/j.agsy.2019.04.005

Patt, A., Suarez, P., & Gwata, C. (2005). Effects of seasonal climate forecasts and participatory workshops among subsistence farmers in Zimbabwe. *PNAS, 102*, 12623–12628.

Roncoli, C. (2006). Ethnographic and participatory approaches to research on farmers' responses to climate predictions. *Climate Research, 33*, 81–99.

Roncoli, C., Jost, C., Kirshen, P., Sanon, M., Ingram, K. T., Woodin, M., et al. (2009). From accessing to assessing forecasts: An end-to-end study of participatory climate forecast dissemination in

Burkina Faso (West Africa). *Climatic Change, 92*, 433–460. https://doi.org/10.1007/s10584-008-9445-6
Rosenbaum, P. R., & Rubin, D. B. (1983). The central role of the propensity score in observational studies for causal effects. *Biometrika, 70*(1), 41–55.
Roudier, P., Muller, B., d'Aquino, P., Roncoli, C., Soumaré, M. A., Batté, L., & Sultan, B. (2014). The role of climate forecasts in smallholder agriculture: Lessons from participatory research in two communities in Senegal. *Climate Risk Management, 2*, 42–55. https://doi.org/10.1016/j.crm.2014.02.001
Roudier, P. (2012). Climat et agriculture en Afrique de l'Ouest : Quantification de l'impact du changement climatique sur les rendements et évaluation de l'utilité des prévisions saisonnières, Paris: Ecole des Hautes Etudes en Sciences Sociales (EHESS).
Salack, S., Muller, B., & Gaye, A. (2011). Rain-based factors of high agricultural impacts over Senegal. Part I: Integration of local to sub-regional trends and variability. *Theoretical and Applied Climatology, 106*, 1–22.
Serra, R., & McKune, S. (2016). Climate Information Services and Behavioral Change: The Case of Senegal. Sahel Research Group Working Paper No. 010.
Singh, C., Daron, J., Bazaz, A., Ziervogel, G., Spear, D., Krishnaswamy, J., Zaroug, M., & Kituyi, E. (2017). The utility of weather and climate information for adaptation decision-making: Current uses and future prospects in Africa and India. Climate and Development 2017
Sultan, B., Roudier, P., & Quirion, P. (2013). Les bénéfices de la prévision saisonnière dans l'agriculture en Afrique de l'Ouest. *Sécheresse, 24*(4), 304–313.
Tall, A., Simon, J., Mason, M., van Aalst, P., Suarez, Y.-C., Diall, A. A., & Braman, L. (2012). Using seasonal climate forecasts to guide disaster management: The Red Cross experience during the 2008 West Africa floods. *International Journal of Geophysics, 2012*, 1–12.
Traore, B., Corbeels, M., van Wijk, M. T., Rufino, M. C., & Giller, K. E. (2013). Effects of climate variability and climate change on crop production in southern Mali. *European Journal of Agronomy, 49*, 115–125. https://doi.org/10.1016/j.eja.2013.04.004
van O'Donnell, E., Doorslaer, A. W., & Lindelow, M. (2007). *Analysing health equity using household survey data: A guide to techniques and their implementation.* The World Bank.
Vaughan, C., Hansen, J., Roudier, P., Watkiss, P., & Carr, E. R. (2019). Evaluating agricultural weather and climate services in Africa: Evidence, methods, and a learning agenda. *Wiley Interdisciplinary Reviews: Climate Change, 10*, e586. https://doi.org/10.1002/wcc.586
Vaughan, C., & Dessai, S. (2014). Climate services for society: origins, institutional arrangements, and design elements for an evaluation framework. *Wiley Interdisciplinary Reviews: Climate Change, 5*(5), 587–603. https://doi.org/10.1002/wcc.290
Vincent, K., Daly, M., Scannell, C., & Leathes, B. (2018). What can climate services learn from theory and practice of co-production? *Climate Services, 12*, 48–58. https://doi.org/10.1016/j.cliser.2018.11.001
World Bank. (2020). Advancing Research on Climate Change: World Bank Grants $60 Million to Help Strengthen the Resilience of the Agricultural Sector in Africa. Retrieved from World Bank: https://www.worldbank.org/en/news/press-release/2020/12/10/advancing-research-on-climate-change-world-bank-grants-60-million-to-help-strengthen-the-resilience-of-the-agricultural-sector-in-africa
World Food Program. (2008). Calculation and use of the food consumption score in food security analysis. Prepared by VAM unit HQ Rome.
Yong, D. N. (2014). Factors affecting the choice of adaptation measures to climate change: The case of famers in the sudano-sahelian area of Cameroon. *Tanzanian Economic Review, 4*(1 & 2), 56–72.
Ziervogel, G., & Calder, R. (2003). Climate variability and rural livelihoods: Assessing the impact of seasonal climate forecasts in Lesotho. *Area, 35*(4), 403–417.
Ziervogel, G., Bithell, M., Washington, R., & Downing, T. (2005). Agent-based social simulation: A method for assessing the impact of seasonal climate forecast applications among smallholder farmers. *Agricultural Systems, 83*, 1–26.

Zougmoré, R., Partey, S., Ouédraogo, M., Torquebiau, E., & Campbell, B. (2018). Facing climate variability in sub-Saharan Africa: Analysis of climate-smart agriculture opportunities to manage climate-related risks. *Cahiers Agriculture, 27*(3), 34001.

Contribution of Renewable Energy Resources for Food Production and Food Security

Abdul Munaf Mohamed Irfeey, Muneeb M. Musthafa, Mohamed Mujithaba Mohamed Najim, and Mirza Barjees Baig

Abstract The escalating demand for food, along with the fluctuating price of fossil fuels, has prompted Sri Lanka to investigate environmentally benign alternative energy sources. The feasibility of renewable energy sources in the agricultural production process will put the country on a secure footing in the face of an energy crisis and rising demand for fossil fuels. Solar energy is a more abundant resource in Sri Lanka and proper harvesting of solar photovoltaic and thermal energies will result in a contribution to energy needs for food production. Solar dryers, solar irrigation systems, and solar photovoltaics are only a few of the solar energy harvesting systems. Harvesting techniques that harness the kinetic energy of the wind for purposes such as water pumping, building heating, and product processing and storage would aid farmers in improving food production. Geothermal energy is an expensive resource to harness but used in some places in Sri Lanka. Bioenergy which includes bioresources, bioresidues, and biofuels can support a considerable portion of required energy. However, the use of renewable energy sources in food production in Sri Lanka is limited by economic, institutional, technological, and sociocultural factors. As a result, the responsible parties' efforts to promote all renewable energy sources among farmers and food production enterprises would alleviate the burden of national energy demand. Implementing policies to promote renewable energy usage

A. M. M. Irfeey · M. M. Musthafa (✉)
Department of Biosystems Technology, Faculty of Technology, South Eastern University of Sri Lanka, University Park, Oluvil 32360, Sri Lanka
e-mail: muneeb@seu.ac.lk

A. M. M. Irfeey
e-mail: irfeeymunaf@seu.ac.lk

M. M. M. Najim
Faculty of Agriculture, Sultan Sharif Ali Islamic University, Faculty of Agriculture, Sultan Sharif Ali Islamic University Sinaut Campus, Tutong TB1741, Brunei
e-mail: najim.mujithaba@unissa.edu.bn

M. B. Baig
Prince Sultan Institute for Environmental, Water & Desert Research, King Saud University, Riyadh, Kingdom of Saudi Arabia
e-mail: mbbaig@ksu.edu.sa

M. Behnassi et al. (eds.), *Climate-Smart and Resilient Food Systems and Security*,
https://doi.org/10.1007/978-3-031-65968-3_23

with better technologies to extract energy is essential for ensuring future food demand in Sri Lanka with the least burden on the country's economy and the environment.

Keywords Energy crisis · Food production · Fossil fuels · Renewable energy sources · Sri Lanka

1 Introduction

The challenge of global population growth and food security are some of the key issues that the globe has to deal with for the next few decades including Sri Lanka (Hussain & Lee, 2015; Premalatha et al., 2011). Currently, the island has a population of 21,497,310, and it showed a population growth rate of 0.39% in 2020 (Department of Economic and Social Affairs, United Nations, 2021). According to the report published in 2021 by the Food and Agriculture Organization of the United Nations, about 4.7 million people in Sri Lanka's population, or more than one-fourth of the population, lack access to sufficient food for a healthy life. Sri Lanka's population is predicted to grow by more than 600,000 people by 2030, resulting in a dramatic increase in food consumption. With the ever-increasing strain on the food supply, the need to address the food security problem is more pressing than ever (Thibbotuwawa, 2017). Food security refers to the state of having consistent access to an adequate supply of affordable, nutritious food and is highly reliant on agriculture. Thus, the emphasis must be on expanding national agriculture to improve the amount and diversity of nutritious food available to the population of the country and the globe.

Agriculture is the prime step of food production where food is cultivated and harvested. The introduction of agricultural practices with novel techniques is more beneficial to managing the issues faced by intensive food production. Precision technology and effective methods in the food production sector enable us to overcome and improve food production, rather than following traditional practices (Aroonsrimorakot & Laiphrakpam, 2019). However, certain basic difficulties have been identified in novel agricultural methods including raw material supply, labor availability, technological acceptance, and energy inputs for an increased output (Tisdell, 2014).

Among the issues faced by the agricultural sector, energy input plays a major role in each step of the agricultural practices such as agronomic and field operations, transportation and processing of the harvest, and handling of the products for consumption, livestock, and forestry (Bayrakci & Koçar, 2012; Chel & Kaushik, 2011; Mekhilef et al., 2013). The quantity of the energy requirements differs according to the type of work done in the food production process. Approximately 25% of the energy requirement in food production is consumed by agriculture from growing to harvesting. After harvesting, transportation takes place to store the production from the place of production (crop fields, livestock farms, and forestry lands) or supply from the store to retail sales (Hill, 2002). Based on the mode of transportation as roads, air, and sea or a combination of any modes together, the required energy types and quantity will vary. About 14% of the required energy for food production is spent on transportation. Some food products are ready for consumption after getting the harvest and some need processing to make them edible. In the latter step, electricity is utilized to operate machinery (Hill, 2002) for the production process. Therefore, food processing consumes 16% of the energy requirement in food production. Packaging and preservation are followed under food handling, and it consumes 50% of the energy from the total energy requirement in food production (Hill, 2002).

Fossil fuels provide the overwhelming majority of the energy required for food production (Bayrakci & Koçar, 2012; Chel & Kaushik, 2011; Mekhilef et al., 2013). Thus, relying only on fossil fuels in the future is a difficult undertaking in two ways: first, less fossil fuel deposits and increased use of fossil fuels are significant difficulties for farmers; second, fossil fuels generate greenhouse gases (GHGs) such as COx, NOx, SOx, and other trace gases (FAO, 2011). Global warming-related concerns are exacerbated by GHGs, and, as a result, global climate changes, or an imposed disturbance of the earth's energy balance with space, occur (Marks-Bielska et al., 2020). GHG emissions have increased by over 90% since 1970, and the fossil fuel burning and industrial processes are accounting for around 78% of the entire increase in greenhouse gas emissions (Intergovernmental Panel on Climate Change, 2014).

Sustainable agriculture is one of the best solutions for basic and applied issues related to increasing demand for food and food production processes in the future (Lal, 2008), and it reveals several aspects of nonconventional ways of agricultural practices such as organic inputs, low amounts of inputs, high-end technologies, regenerative and alternative resources for energy, and other inputs and also it has potential to reduce the negative environmental consequences (Laufer & Schäfer, 2011; Tisdell, 2014) reducing the use of scarce natural resources while maximizing crop yield and maintaining economic stability (Tisdell, 2014). Physical challenges, such as changes in soil characteristics (Lal, 2008) owing to long-term effects of agriculture and processes deemed crucial for crop production, and socioeconomic challenges, such as labor and energy resource management, will need to be addressed to achieve sustainable agriculture. In comparison to conventional fossil fuel-based technologies, renewable energy technologies are cost-effective for agriculture in any area, with the added benefit of collecting carbon credits. However, because of the high capital costs of these technologies, government support is required to deploy them in the agricultural sector. To achieve higher environmental sustainability in agriculture, it is important to encourage energy-efficient, low-carbon technology as well as a larger penetration of renewable energy.

2 Renewable Energy Technologies: Global Scenario

Renewable energy sources are derived from natural resources and replenished by natural processes at a faster rate than they are consumed (Chel & Kaushik, 2011). Investment in the renewable energy sector of any country will be a major asset for the future while combating climate change, energy, and food security (Sayigh, 2012). Sources such as solar energy (Hussain & Lee, 2015; Mekhilef et al., 2013; Sampaio & González, 2017; Sciences, 2010), hydro energy (The International Hydropower Association, 2021; Xu et al., 2015), wind energy (Ackermann & Soder, 2000; Acosta-Silva et al., 2019), biomass energy (Chel & Kaushik, 2011; National Geographic, 2021), geothermal energy (Dickson & Fanelli, 2020; Robins et al., 2021; U.S. Energy Information Administration, 2021), and ocean energy are included under renewable energy sources and more abundantly available in the world in varying quantities based on the geographic region (Mohtasham, 2015).

The renewable energy concept started to accelerate in the 1970s. In 2008, the contribution of renewable energy sources to global energy consumption was 10.2% (2.1% renewable electric energy, 1.1% solar thermal and geothermal heat energy, and 7% modern bioenergy). For the generation of electricity, 27.1% of the contribution was obtained from renewable energy sources, whereas, in the transportation sector, 3.4% of the power contribution was from renewable energy resources (Madeddu et al., 2020). According to a new report from the International Energy Agency, the world's renewable energy capacity has increased by 45% in 2020 and it is the highest annual growth rate since 1999 (International Energy Agency, 2021).

Among the renewable energy resources, demand for solar energy is increasing as it becomes the most cost-effective option for energy generation in an increasing number of sectors, including residential and commercial applications, as well as utility-scale projects, even before external costs associated with fossil fuels are included (Clack et al., 2019; Yan et al., 2019). In an increasing number of markets, this is also the case for solar-plus-storage. Estimated global production added at least 1 GW of new SPV capacity in 2020, up from 18 countries in 2019, and all continents played a substantial role in global expansion. At least 42 countries had a total capacity of 1 GW or higher by the end of 2020 (Jackson, 2020).

However, wind power also contributes a significant portion of global energy production, and more than 93 GW of wind energy generation capacity was constructed globally at the end of 2020, with more than 86.9 GW onshore and 6.1 GW offshore. It is 45% higher than the previous peak of capacity installed for wind power production compared with 2015 (63.8 GW), and marks a nearly 53% growth over 2019 installations. Total global wind power production capacity increased by 14% in 2019 to roughly 743 GW (including 707.4 GW onshore), and this was more than double the production capacity six years ago, at the end of 2014. This production growth assists the globe in the reduction of over 1.1 billion tonnes of CO_2 (Lee & Zhao, 2021). Similar to solar and wind energy, hydro energy is also one of the most ambitious initiatives for achieving a low-carbon, long-term energy source (Xu et al., 2015). The prospective development degree was determined by dividing the number of commercially accessible hydropower resources by the amount of energy produced by hydropower. In 2010, hydropower generated 3410 TW h globally, accounting for roughly 39% of the total 8721 TW h available. According to the World Atlas Industry Guide (2008), future hydropower installed capacity in Africa (the Nile and Congo rivers), Asia (such as China), and South America is anticipated to rise (The International Hydropower Association, 2021).

The utilization of biological materials for energy purposes is known as bioenergy. Agriculture and forestry leftovers, solid and liquid organic wastes [including municipal solid waste (MSW) and sewage], and crops produced specifically for energy production can all be utilized (International Energy Agency, 2021). These feedstocks may be converted into heat, power, and transportation fuels using a variety of methods (biofuels). Some of these processes are well established, while others are still in the development, demonstration, and commercialization stages. Other than solar, wind, hydro, and bioenergies, geothermal resources are nowadays very popular in the renewable energy sector. It is utilized for energy in two ways, either as a source of electricity or as a source of heat for different "direct-use" thermal applications, such as space heating and industrial heat input. In 2020, geothermal energy generation was estimated to be approximately 97 TWh, while direct usable thermal output was estimated to be over 128 TWh (462 PJ). In certain cases, geothermal facilities may generate both electricity and heat for thermal purposes (co-generation); however, this option is dependent on local thermal demand and geothermal resource availability (Robins et al., 2021).

Choosing a suitable renewable energy source has implications for greenhouse gas emissions, equipment manufacturing, accessibility, and water resource distribution (Mohtasham, 2015), and also effective regulations are required to convert local and international investments into renewable energy projects that will result in reduced carbon emissions and long-term economic growth. Renewable energy output grew from 16.6% to 18.1% of total energy consumption globally between 1990 and 2012. Furthermore, renewable energy is expected to expand at a pace of 0.17% each year, accounting for 21% of total final energy consumption by 2030 (Capuano, 2020).

3 Renewable Energy Technologies: Sri Lankan Scenario

Sri Lanka is regarded as one of the world's most susceptible countries in terms of climate change. Climate change has important consequences for the country's agriculture, tourism, society, health, and energy production, in addition to its environmental effects. Sri Lanka must focus on renewable energy investment in order to ensure its electricity supply (Wijayatunga & Kolantharaj, 2021). Sri Lanka, as a developing country, will continue to face escalating energy consumption. Fortunately, the country has a lot of potential for renewable energy. Transitioning to renewable energy provides a chance to cut fuel imports, which are putting a strain on government budgets, generate employment, particularly for women, and develop an economy based on clean, cheap energy. This shift will benefit the economy, the energy industry, and the environment in numerous ways.

The country's reliance on fossil fuel imports is being exacerbated by rising energy demand and poor progress on renewable energy initiatives. Climate change implications on hydropower output, which has grown increasingly unpredictable, have exacerbated the problem. In 2014, Sri Lanka's primary energy supply from fossil fuel (petroleum and coal) was 42.8% (Vithanage, 2019). But in 2017, it accounted for roughly 54% (Wijayatunga & Kolantharaj, 2021). The country's budgetary burden is exacerbated by high generating costs stemming from a growing percentage of fossil fuels in the power mix, as well as end-user tariff subsidies. Transitioning to renewables would diversify the country's energy mix and reduce the country's reliance on imported fossil fuels. Firewood, kerosene, LPG, and electricity are the primary sources of energy for Sri Lankan households (International Energy Agency, 2021). Residential energy use in Sri Lanka grew from 4.1% to insufficient capacity expansion, decreased efficiency in consumption and delivery, increased reliance on imported fuels, and rising fuel costs, meeting the rising demand for home energy has become a serious challenge (Rajmohan & Weerahewa, 2010).

The government of Sri Lanka wants the country to be energy self-sufficient by 2030. The goal is to raise the country's power generating capacity from 4,043 to 6,900 MW by 2025 (U.S. Department of Commerce, 2021), including a considerable increase in renewable energy. Sri Lanka has already achieved 98% grid connectivity, which is really high compared to South Asian standards (Ministry of Power & Energy, 2015). Sri Lanka generates electricity from three main sources: thermal power (coal and fuel oil), hydropower, and other non-conventional renewable energy sources (solar power and wind power) (U.S. Department of Commerce, 2021). Sri Lanka aims to increase its energy generating capacity by 842 MW of major hydro, 215 MW of mini-hydro, 1,389 MW of solar, 1,205 MW of wind, 85 MW of biomass, 425 MW of oil-based power, 1,500 MW of natural gas, and 2,700 MW of coal power between 2018 and 2037. The entire annual power usage is estimated to be around 14,150 GWh. Over the next 20 years, yearly power demand is anticipated to rise by 4.9%, a figure limited by high pricing (Ceylon Electricity Board STG, 2017; U.S. Department of Commerce, 2021).

Renewable energy for agricultural activities in Sri Lanka has been used for many centuries. From land preparation to post-harvest technologies, renewable energy sources such as solar, wind, hydro, biomass, and geothermal energies are considerably utilized in various ways. Solar energy has been used for drying crops, clothing, and other items since ancient times, and it has generally remained a non-commercial energy source (Sayanthan & Kannan, 2017). Due to constant cloud cover in such places, two-thirds of the country's low land area receive radiation of 4–5.5 kWh/m^2 per day, while the remaining area in the central highlands receives lesser radiation of 2–3.5 kWh/m^2 per day (Gunaratne, 1994).

Sri Lanka is situated in the Indian Ocean, with a large expanse of the unbroken ocean giving excellent wind energy potential. The South-Western monsoon (May–August) and the North-Eastern monsoon (September–November) are the two primary wind climates in the nation (October–December). During the era 300-200 B.C., archeologists discovered that iron-smelting using the South-Western monsoon without the use of a bellow to pump air to smelting furnaces was a commonly utilized method. Sri Lanka became the first country to use wind energy for productive purposes in the Balangoda region, making it the first country to do so (Public Utilities Commission of Sri Lanka, 2011).

4 Contribution of Renewable Energy Resources in Food Production

4.1 Solar Energy

Sun is the prime source of energy in the world in two forms as solar thermal energy and solar photovoltaic energy (Makdisie et al., 2018). Solar thermal energy is the heat energy of the radiation and is used in drying, and heating-related applications (El-Sebaii & Shalaby, 2012) of the food production and food preservation techniques. Solar photovoltaic energy generates electricity from the photon energy of solar radiation, and it is used in electrical equipment and machinery-related activities (Mekhilef et al., 2013).

From the ancient period onwards, the drying process of agricultural products such as grains and plant parts (Harijan, 2012) using solar radiation is practiced. It is a simple and less expensive method of drying which allows a natural way of drying in the field or outside of the field by spreading as a layer. Even though it has such advantages, birds, rodents, dust and dirt, and other environmental factors are the big challenge in this method (El-Sebaii et al., 2002). Technologies to prevent direct contact with the environment of animals will be the way to safe drying. While direct sun drying is still widely used in non-commercial applications at the rural household level in Sri Lanka such as drying meats, fruits, fish-like perishable foods, and parboiled rice, various attempts have been made to create improved food processing drying techniques (Kumarasiriwardhana et al., 2020).

Sun drying agricultural goods involves exposing them to direct sunlight under ambient conditions using a thin coating of product. During sun drying, heat is transmitted to the product by convection from the surrounding air caused by solar radiation absorption, both direct and diffuse. The procedure is repeated until the product's vapor pressure reaches that of the surrounding atmosphere. Solar drying is a technical alternative to the conventional process of sun drying. It entails increasing the vapor pressure of the drying air and reducing its relative humidity in the drying chamber, therefore improving the moisture-carrying capacity (Kumarasiriwardhana et al., 2020).

4.1.1 Solar Dryer

Removal of excess water from the food or agriculture stuff is called drying. Agricultural products usually have approximately 25–80% (El-Sebaii & Shalaby, 2012) moisture, and this much moisture content will cause the stuff to quickly degrade because the microorganisms such as bacteria and fungi prefer to grow in high moisture content. Metabolic activities of the microorganisms will spoil the foodstuff and reduce its nutrient contents. Reducing the water content of the foodstuffs will limit the growth of those microorganisms.

Sri Lanka is a producer of perishable goods as well as a consumer. The country has a bright future in a number of competitive global food markets, provided that appropriate technology is implemented to improve the quality and diversity of food products. Cost and quality are critical variables in food dehydration in Sri Lanka, as it is a totally energy-based process (Amunugoda et al., 2012). Instead of using fossil fuel, solar dryer technologies are cost-effective renewable energy sources and also free from environmental pollution and low carbon emission to the atmosphere. Except in the mountainous regions and particularly in the dry zone, there is considerable potential in Sri Lanka to utilize solar energy for fruit and vegetable drying. Despite this, solar dehydration technology has a relatively low penetration rate in these locations (Renné et al., 2014). As a result, the difficulties connected with solar-dried fruits and vegetables remain unresolved owing to a lack of effort in identifying and transferring appropriate technologies (Amunugoda et al., 2012).

Solar dryers are categorized into several categories based on their mode of heat or operation (El-Sebaii et al., 2002). In general, solar dryers are classified into three types as direct solar dryers, indirect solar dryers, and mixed-mode solar dryers (El-Sebaii & Shalaby, 2012; Patel et al., 2017). In another classification, it is categorized into two main types with the flow of air as natural convection and forced convection solar dryers (Harijan, 2012). Natural convection airflow is allowing airflow due to the buoyancy induced way and in the forced convection, airflow is provided by using a fan or any such mechanisms. Natural convection needs no energy requirements additionally, but the forced convection process needs additional energy input to operate the fan.

Solar dryers, in comparison to certain other solar technologies, continue to face resistance from commercial manufacturers of dried goods. The causes for this are numerous and complicated, and are dependent on a variety of circumstances (Bena & Fuller, 2002). However, several studies on solar dryers have been conducted in order to resolve practical difficulties that arise during their use.

4.1.2 Natural Convection Solar Dryer

Natural convection solar dryers require low cost with low maintenance and operational cost, but it is difficult to control or regulate the drying temperature and the rate of drying. It is an obvious choice for drying agricultural perishable products. These types of solar dryers are classified as (a) direct natural convection solar dryers (Fig. 1) (Harijan, 2012), (b) indirect natural convection solar dryers (Fig. 2) (Odhiambo, 2016), and (c) mixed-mode natural convection solar dryers (Fig. 3) (Bhagoria & Bhagoria, 2013; Prakash & Kumar, 2013).

In a direct natural convection solar dryer, the air heater contains the agricultural products inside and the solar radiation absorbed by the agricultural products which are coming through the transparent cover (Harijan, 2012). In indirect natural convection solar dryers, solar radiation does not have a direct hit, and the radiation is collected in a separate air heater and then heated air is passed through the agricultural products and gets heated. But, in the mixed-mode natural convection solar dryers, a direct hit

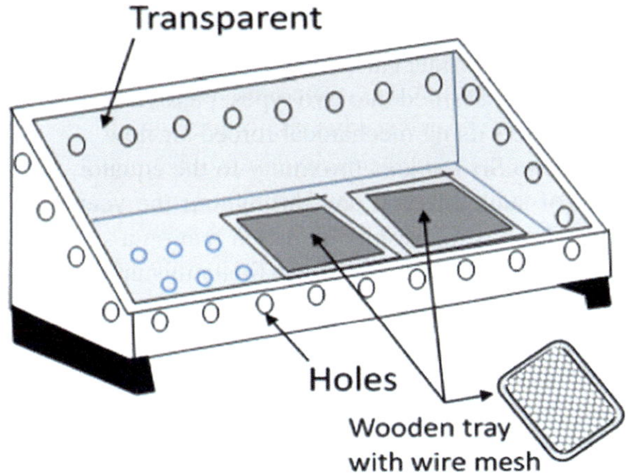

Fig. 1 Diagram of direct natural convection solar dryer. *Source* Harijan (2012)

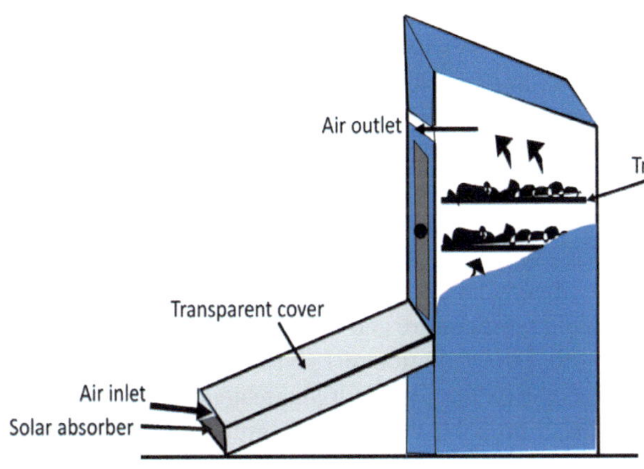

Fig. 2 Diagram of indirect natural convection solar dryer. *Source* Odhiambo (2016)

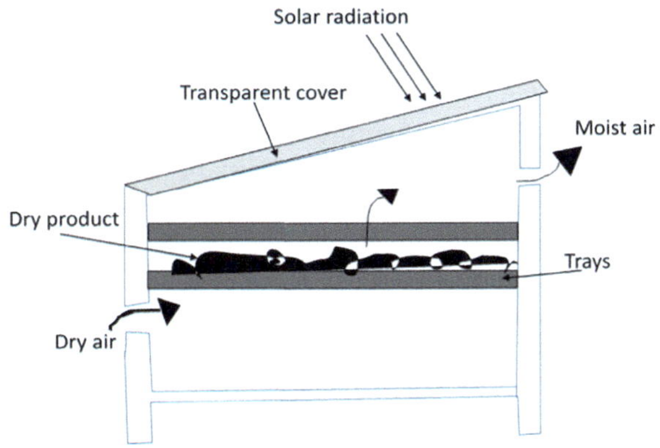

Fig. 3 Diagram of Mix mode natural solar dryer. *Source* Bhagoria and Bhagoria (2013)

of solar radiation to agricultural products and the heated airflow from a separate air heater to the product also happens (Harijan, 2012). According to the flow of air, solar dryers are classified into two types: passive as air allows to flow naturally, and active solar dryers using mechanical forced air flow.

Due to Sri Lanka's proximity to the equator, the country receives an average of 12 h of sunlight each day throughout the year, with the exception of a few rainy days. The testing results indicated that on a sunny day, perishable fruit flesh with a moisture level of 95% dries to a final moisture content of 13% after 12 sunlight hours. Cabinet-style driers are quite beneficial for drying fruits and vegetables in Sri Lanka. Historically, cabinet driers were widely used for fruit drying. A cabinet dryer with a wooden plenum that differed from the standard design was introduced, and this new alteration of plenum served as a guide for the dryer's intake of air as it entered. In addition, a long plywood chimney was installed to improve the flow of fresh air. This drier, compared to open-air solar drying, increased drying time by a factor of five (Phadke & Walke, 2015). According to the research studies conducted by Borah et al., 2015, the air temperature inside the direct sun dryer for drying was determined to be in the range 39–51 °C, compared to the ambient air temperature of 25–28°C. The damp content was lowered from 78.65 to 6.36% in solid foodstuffs and 5.50% in slice kinds after 12 h of efficient drying time. The overall thermal efficiency was estimated to be 55%.

Direct natural convection solar driers are excellent for drying tapioca (Diemuodeke & Momoh, 2011). According to them, the initial and ultimate moisture contents in this drier were determined to be 79% and 10% wet basis, respectively, by testing. The average ambient temperature and relative humidity were 32 °C and 74%, respectively, with daily global sun radiation of 13 MJ/m^2/day. A dryer prototype with a minimum collector area of 1.08 m^2 was built and tested under a variety of loading situations. A natural convection solar dryer with a multi-shelf design and intermediate heating has been created lately as a portable solar drier. It was comprised of four principal components: a multi-tray rack, moveable glazing, a shade plate, and trays. The multi-rack was angled according to the location's latitude. A moveable frame and a UV-stabilized plastic sheet comprise the movable glazing (Singh et al., 2004). The dryer was portable and economically viable due to its cheap cost. It may be utilized in cottage industries in Sri Lanka's rural areas. Increasing the drying air temperature and configuring the system to allow for mass flow of air are the primary variables affecting the effectiveness of solar dryers utilized in natural convection solar dryers (Gbaha et al., 2007). Numerous research projects have also been done to enhance such variables and to boost the dryer's efficiency.

Natural convection sun dryers are cost-effective for drying a variety of food items, even if adoption among the general public is still low or the technology is still gaining traction in the food industry. One important drawback of the solar dryer is that it is rarely utilized in conjunction with backup heating. This issue affects commercial farmers' capacity to process a harvest during periods of inclement weather. Additionally, it prolongs the drying period because drying may occur only during daylight hours, which limits output but may result in a substandard product. For commercial manufacturers, the capacity to process reliably and continuously is critical to meeting market demands (Bena & Fuller, 2002). The solar dryer, which employed a sawdust burner to generate heat during inclement weather and at night, was meant to address the problems raised. The burner was intended to supply the drying cabinet with 400 W m^{-2} of energy and used steam as the heat transfer medium. The sawdust burner was created independently of the drying cabinet, rather than being integrated into it.

4.1.3 Forced Convection Solar Dryer

Providing the required airflow rate much higher than the natural convection airflow rate by using a fan powered by any electricity source is done in a forced convection solar dryer (Fig. 4) (Gupta, 2016). Additional electricity sources needed for this are either electricity from the national grid or solar photovoltaic or fossil fuel used. For small-scale commercial production of quality dried items, such dryers can be used to dry fruits, vegetables, spices, and even fish.

It's difficult to keep dryers affordable to Sri Lankan farmers due to the high cost of imported materials, which is a barrier for many potential customers. Due to power being expensive and frequently absent in rural regions, dryers cannot utilize fans to circulate hot air, which adds yet another design limitation to rural areas. It is nevertheless possible to build practical solar crop dryers despite these obstacles. Solar hot-air collectors and drying chambers with a fan are designed regularly. Where natural circulation is used, however, low airflow rates have created problems in designs. There are no satisfactory design procedures for indirect natural-flow dryers because the fluid-flow and heat transfer processes involved in operating these dryers as a means of matching airflow rates and temperatures through the crops have not been actively investigated by researchers.

Forced convection solar dryers are classified as (a) solar tunnel dryers with polycarbonate cover (Fig. 5) (Janjai & Keawprasert, 2006), (b) solar tunnel dryers with plastic cover, (c) greenhouse solar dryers (Fig. 6) (Janjai et al., 2009), and (d) roof-mounted solar dryers (Fig. 7). Solar tunnel dryer consists of a heat absorber/collector,

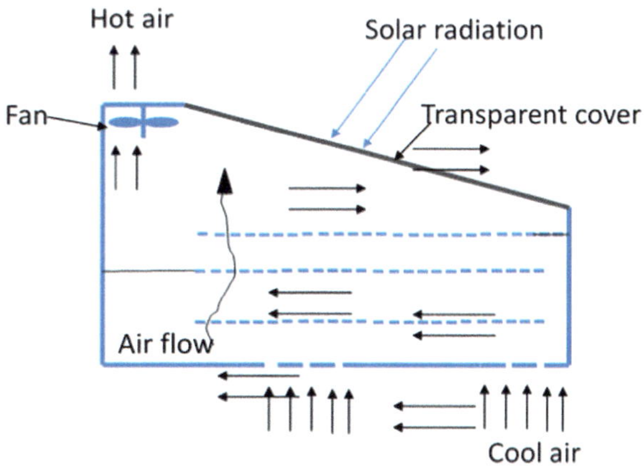

Fig. 4 Diagram of a direct forced convection solar dryer. *Source* Gupta (2016)

a tunnel unit, and a fan to provide the airflow inside the tunnel. The collector and drying chamber are constructed out of tiny portions of simple metal sheets and hardwood frames that are linked together in a sequence. Insulation material is used to reduce the heat loss from the dryer. Fruits, vegetables, cereals, legumes, oilseeds, spices, fish, and even meat may be dried correctly in the solar tunnel drier, according to numerous experiments conducted in various tropics and subtropics (Harijan, 2012; El-Sebaii & Shalaby, 2012).

A comparison study was carried out between the greenhouse dryer and the photovoltaic-thermal roof. The experiment was conducted in both natural and forced modes (Tiwari & Tiwari, 2016). It was discovered that increasing the PV module packing factor value lowered heat energy by 76.39% while increasing electrical

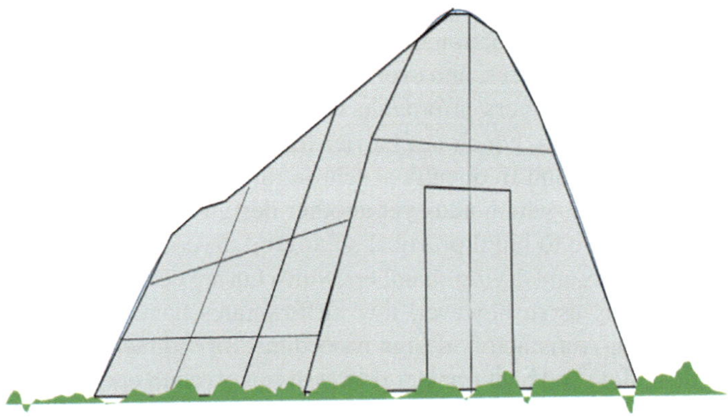

Fig. 5 Greenhouse-type solar dryer with polycarbonate sheet. *Source* Janjai et al. (2009)

Fig. 6 Schematic diagram of the solar tunnel dryer with polycarbonate cover. *Source* Janjai and Keawprasert (2006)

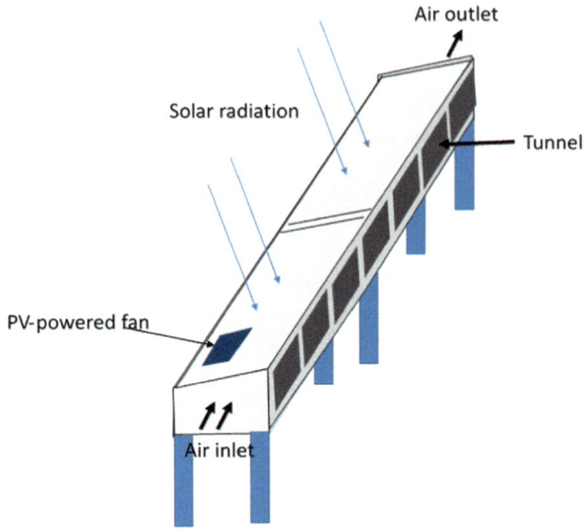

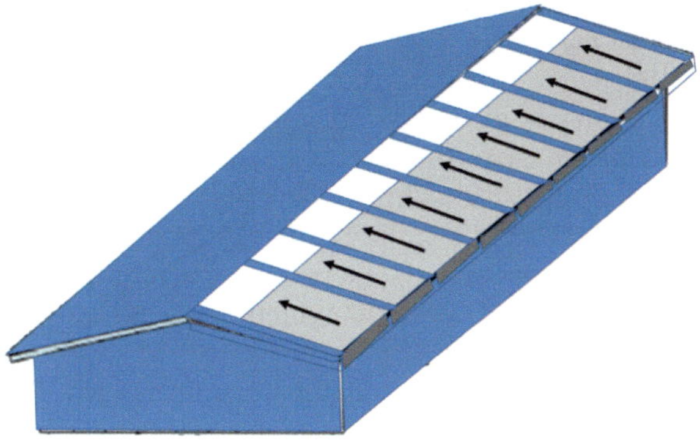

Fig. 7 Roof-integrated solar dryer. *Source* Harijan (2012)

energy by 88.73%. The increase in mass flow rate affected thermal energy by 89.44% in natural convection and 65.70% in forced convection mode. Red chili is dried in a greenhouse dryer and drying system at a low temperature. The wetness of chili was reduced by 0.02g/h on average during night drying and by 0.193g/h during day drying. Within 52 h, the drying procedures reduced the moisture level by 13.55% (Nimrotham et al., 2017).

4.1.4 Solar Heat Collectors for Agricultural Purposes

Solar heat collectors function as a type of heat exchange device, using solar radiation energy and converting it into internal energy of a transport medium. Solar energy is captured by absorbing sunlight and transferring the heat to a fluid such as water, air, and oil. The heat can be used directly or converted into electrical energy using turbines and generators (Karsli, 2007). Solar air collectors have a number of benefits, for example, as compared to wood fuel combustion. It is a straightforward and inexpensive technology; they are simple to run and maintain, do not require a highly skilled crew, and can be manufactured locally using locally accessible materials. Additionally, they consume no energy and are environmentally friendly. Solar air heaters, on the other hand, have limited thermal efficacy due to the air's low volumetric heat capacity, density, and heat conductivity (Karsli, 2007).

Flat plate collectors (FPCs) (Fig. 8) (Cruz-Peragon et al., 2012) are the most widely utilized solar thermal collectors for low-temperature applications, where temperatures typically do not exceed 100 °C. They can, however, reach 200 °C if specialized coatings for the absorption plate are used. Furthermore, FPCs are suitable for locations with a warm and bright environment (Kalogirou, 2004). They are relatively simple to operate due to their simple construction, have low capital costs, and require a minimal number of materials. Furthermore, FPCs may utilize both beam and diffuse solar energy, require very little maintenance, and do not require sun tracking. FPCs are most commonly used in the dehydration of industrial goods, drying of agricultural crops, and space heating (Labed et al., 2012).

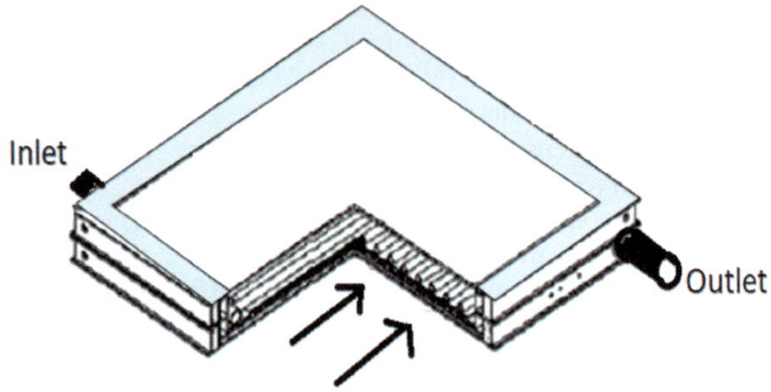

Fig. 8 Flat plate collector. *Source* Cruz-Peragon et al. (2012)

In livestock agriculture production such as milk, the cleaning process of milking equipment and the storage tank is mandatory in order to maintain low microbial counts (Jaisankar et al., 2011). Hot water of 60–75 °C is needed for this purpose. The required amount of hot water is calculated based on the frequency of milking and the size of the farm (Byrne et al., 2021). Solar water heating systems have been used to produce hot water by converting solar energy to thermal energy. The performance of the water heater depends on the efficiency of the absorber/collector plate's thermal absorptivity, the reflective index of the absorber material, the insulation of the system, environmental factors, etc. (Han et al., 2010). When solar radiation hits the absorber or solar collector (Fig. 8) surface of the solar heater, heat energy from the solar radiation is absorbed by the collector or absorber. Captured heat energy is then transferred to the inside fluid (typically a combination of water and glycol). Vacuum tubes are highly effective in capturing heat from the sun and transferring it to the fluid being heated. When the fluid is heated in the evacuated tube solar collector, it is pushed down into the heat exchanger in the sun tank, which heats the water in the solar tank (Harijan, 2012).

Sri Lanka is one of the world's largest producers and exporters of high-quality tea. The process of drying tea necessitates the use of thermal energy, which is now provided by the burning of fuel woods, on which the sector is heavy. With growing fuel wood costs in recent years, it has become increasingly important to find alternatives or complements to this source of energy. One potential option would be to use solar thermal energy in the tea production process by including solar collectors. Because Sri Lanka has perfect conditions for harvesting solar energy, solar applications would be an excellent fit in this setting (Brangstrup, 2014).

Solar process heat plants must be able to function reliably in all modes of operation. Solar thermal applications, unlike typical closed hot water or steam delivery systems, need unique technological solutions to deal with the issue of stagnation. Stagnation defines the condition of a solar thermal system in which the flow in the collector loop is halted and solar radiation is further absorbed by the solar thermal collector, heating the fluid in the collector to a temperature where the absorbed energy equals the losses (Mauthner & Fischer, 2015). In comparison to conventional heat supply technologies, this means that, in the event of technical defects, a power outage, or simply a lack of heat demand (i.e., temporarily no available heat sink), some solar thermal collector fields (depending on the type of collector used) cannot be simply shut down and heated up again. Solar thermal systems might overheat if suitable measures for overheating avoidance and stagnation management are not implemented. Depending on the solar thermal collector concept, various impacts must be avoided under normal operating conditions (Mauthner & Fischer, 2015).

4.1.5 Solar Irrigation System

About 66% of Sri Lanka's total geographical area is considered a dry zone, with annual rainfall of less than 1000 mm. According to the Sri Lanka Survey Department, 2019 land use map, the total agricultural area in Sri Lanka is estimated to be 2.86 million hectares, with roughly 632,000 ha irrigated (Aheeyar, 2012). Precipitation is primarily restricted to April and September, with drought conditions dominating the rest of the year. Food production in Sri Lanka's arid zone is comparably poor due to lower cropping intensity and non-cultivation of high-value crops due to unpredictable water availability. Farmers were forced to use high-cost water pumps to irrigate fields via water hoses for lengthy periods, resulting in increased fuel costs and wastage of scarce groundwater. Excessive water pumping also promotes the growth of weeds. Farmers that rely on existing systems confront a serious dilemma as fuel costs rise. The usage of gasoline pumps pollutes the environment due to fuel leaks and CO_2 emissions. Most farmers do not have access to grid energy for their farmlands in remote locations (Aheeyar, 2012).

When compared to traditional fuel, solar water pumping systems have several advantages. For example, in addition to no requirement for any fuel or maintenance, the system produces no noise or pollution, which is beneficial to the environment. Although high-capacity solar water pumps (up to 10 kW) (Saputra et al., 2018) are available, most solar water pumps utilized in rural regions are small-scale (usually less than 1500 W). Solar water pumping system is classified into two types: (a) Solar photovoltaic pumping system (Fig. 8) (Meah et al., 2008) and (b) Solar thermal water pumping system (Mekhilef et al., 2013).

Nonetheless, solar-powered irrigation systems have a somewhat high initial investment cost, and novel financing approaches are required to overcome this obstacle to adoption, particularly for small-scale farmers. To guarantee that the systems function and are maintained properly, some technical expertise or service infrastructure is required. Furthermore, if not properly controlled and regulated, solar-powered irrigation systems run the danger of encouraging unsustainable water usage, as cheap energy costs can lead to inefficient water consumption, over-abstraction of groundwater, and poor field application (Hartung & Pluschke, 2018).

Socioeconomic barriers to adopting a solar-powered irrigation system at the national level in Sri Lanka are very much boosting farmers' worries, such as current energy subsidies for fossil fuels and electricity, which distort the market; legislation and regulation of the energy and agricultural markets may impede the adoption and scaling of solar energy systems; tariffs on imported equipment may distort costs; high-risk investment, especially if solar-powered irrigation system rollout programs

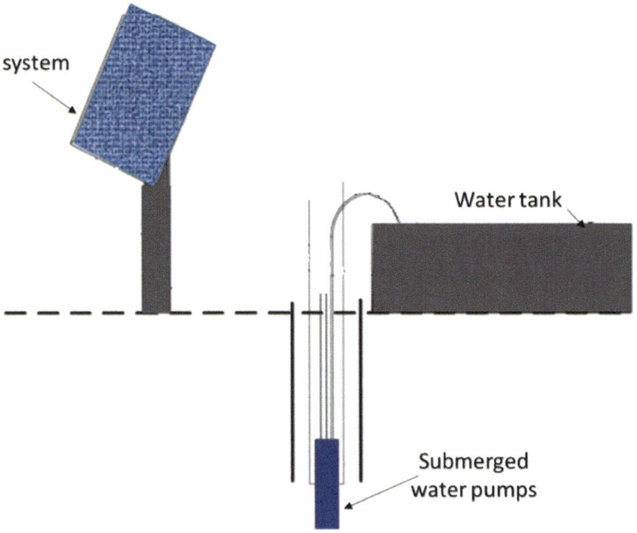

Fig. 9 Schematic diagram of photovoltaic water pumping system. *Source* Shinde & Wandre (2015)

do not adequately address onsite ownership; lack of groundwater management/ institutional framework for abstraction and codes and standards to ensure the quality of SPIS; lack of systematic training; thus, lack of skilled personnel (Hartung & Pluschke, 2018) (see Fig. 9).

4.1.6 Solar Photovoltaic for Electricity Generation

Electricity generation from solar photovoltaic energy by the solar photovoltaic system is a well-established and growing field in the renewable energy sector anywhere in the world (Green et al., 2019). When the solar radiation hits the surface of the semiconductor materials inside the solar cell, photon energy of the solar radiation is transferred to the semiconductor material and the electron from the valance band of the semiconductor material is excited to the conduction band of the semiconductor (Parida et al., 2011; Razykov et al., 2011; Sampaio & González, 2017). The separation of the hole and the electron creates the potential difference between the conduction band and the valance band and forms voltage differences. While connecting to a load, the system's flow of electrons creates electricity. Photovoltaic systems are a cost-effective way to provide energy to farms, ranches, orchards, and other agricultural enterprises in distant areas (Chel & Kaushik, 2011). In applications such as an electric fence, area or building lighting, and water pumping for animal watering or crop irrigation, PV systems can be significantly less expensive than installing power lines and step-down transformers (Tyagi et al., 2013).

In 2015, fossil fuels generated 52% of Sri Lanka's electric power (ADB and UNDP, 2017). Due to the scarcity of indigenous fossil fuel resources, the fossil fuel needed to generate power is imported, accounting for a sizable portion of Sri Lanka's import expenditure. Sri Lanka's need for energy will continue to grow in the future as a growing country. Sri Lanka must thus ensure its energy future by focusing on the development and deployment of indigenous, renewable energy sources to fulfill this rising demand and alleviate the economic burden associated with imports.

The decision occurred at a time when the development of renewable energy generation has been a key priority of the Sri Lankan government as well as private sector investors. To encourage individuals to produce their electricity, the government is presently undertaking "Soorya Bala Sangramaya" (Battle for Solar Energy), a solar power generating initiative. Under this program, any household or business with a valid energy account can export the power generated by a solar system to the national grid using one of three schemes: net metering, net accounting, or net plus (Public Utilities Commission of Sri Lanka, 2020).

According to the revised Sri Lanka Energy Act, no party is permitted to generate and sell electricity to the national grid without a license from the PUCSL (Public Utilities Commission of Sri Lanka, 2020). However, under the new judgment, any energy consumer might build a solar system at his or her residence/premises and generate and sell power in accordance with an agreement signed between the said consumer and the corresponding licensee (CEB or LECO). All such parties will be excluded from the need to get a generating license (Public Utilities Commission of Sri Lanka, 2020). This would provide more opportunities for Sri Lankan power users to obtain environmentally beneficial and renewable energy sources at a fair expenditure. It is believed that by encouraging solar-based generation among energy customers, it would be possible to fulfill daytime electricity demand through household-based solar plants, therefore replacing thermal plants that run during such times. This also makes it easier to manage water resources more efficiently and effectively.

Water pumping is, in fact, one of the most basic and practical applications for photovoltaics. Photovoltaic-powered pumping systems satisfy a wide range of water demands, from agricultural irrigation to animal watering to household usage (Saputra et al., 2018). Most of these systems also have the capability of storing water for usage when the sun isn't shining, obviating the need for batteries, simplifying the system, and lowering total expenses. Photovoltaics can also be used to power electric fans for air circulation. By rearing the animals in enclosed structures, modern pig and poultry farms may double or even triple their output (Chel & Kaushik, 2011). Photovoltaics are also useful for lighting agricultural structures and enclosures. Photovoltaic systems are potentially more cost-effective than traditional battery-powered lighting, flashlights, and fuel lamps (Chel & Kaushik, 2011). They also give more light of greater quality and do not create any smoke or odors. PV systems may also be used in the following ways on farms, ranches, and orchards: providing electric power to the feed/product grinding, electric-powered machinery, electric fencing for protection of livestock farms, battery charging, etc. (Chel & Kaushik, 2011).

Because the bulk of these farms are located in distant areas across the world, diesel is used to provide the majority of their power needs. Electric fences are an excellent option for farmers in distant locations, as well as being cost-effective in the long term, especially in nations with high labor expenses. As a result, PV-powered electric fences are the most practical pasture management alternative, particularly in off-grid locations. Solar energy may be used in a variety of agricultural applications, including veterinary clinics, livestock watering, egg incubators, crop spraying, and fruit preservation chilling (Bundschuh et al., 2014). The wireless sensor network (WSN) is an enabling technology for precision farming that can increase agricultural productivity and farming communities' standard of living. Among the most recent advances, the use of solar energy in WSN is more practical since it can save substantial amounts of energy. Its use in distant and challenging regions makes solar energy more cost-effective and practical as a source of electricity for this new technology (Bhullar & Lalwani, 2018).

4.2 Wind Energy as a Renewable Energy Source

Wind energy has been utilized since the early years of civilization. Until the early twentieth century, it was the only source for pumping water from the water source to agricultural-related works from the mechanical energy of wind (Ackermann & Soder, 2000). The fluctuation in the availability of wind power, and seeking a constant power source was the target at that time. Fossil fuel energy was the best substitute for constant power supply demand (Bitar et al., 2012). High demand for fossil fuel because of industrialization and urbanization, in the early 1970s, the fossil fuel crisis, and very low deposits triggered to rethink of renewable energy sources.

Electricity generation from wind turbines was coming into practice in the early twentieth century (Mathew, 2017). The kinetic energy of the wind source is used to rotate the wind turbines and the mechanical energy of the turbines is used to generate electricity. Step by step, the technology of wind-electricity generation was improved from the 1970s and then has been identified as the most sustainable energy source. Worldwide wind capacity increased every three years during the last decade of the twentieth century. Since the early 1980s, the cost of energy generated by wind power has decreased by roughly one-sixth, and the trend appears to be continuing. Vertical types of wind turbines were used in the earliest period of wind-electric power generation (Ackermann & Soder, 2000). These types of windmills were recorded in Afghan highlands to grind grains (Ackermann & Soder, 2000). Horizontal axis windmills were found in the Europe region, and it was used not only for grinding grains but also for pumping water for agricultural purposes. In the 1900s, North America started to use windmills for their agricultural activities including pumping water for livestock and this was very popular (Islam et al., 2013). Wind power is currently utilized for irrigation pumping, building heating, and product processing and storage.

The wind power sector in Sri Lanka first witnessed activity in 1988, when studies were done to develop a prototype wind project in the Southern Province (Asia Wind Energy Association, 2021). The state-owned 3 MW Hambantota Wind Farm was completed more than a decade later. The sector was idle until 2003, when the National Renewable Energy Laboratory conducted more wind power tests on the island, after which it was dormant for another seven years (Asia Wind Energy Association, 2021). Unlike many other sectors, the wind energy business in Sri Lanka saw rapid growth in 2010, with the commissioning of the Mampuri Wind Farms, the country's first private-sector wind project. Following several scandals involving major regulatory organizations such as the Sustainable Energy Authority and the Ceylon Electricity Board (CEB), as well as several top individuals, it then abruptly fell during the next four years (CEBSTG, 2017).

Sri Lanka has a large amount of accessible land with significant wind resource potential for development; nevertheless, the near-term prospective wind power capacity increase is constrained by the electrical transmission infrastructure. According to CEB, additional wind power of more than 7% of peak load, or about 100 MW, cannot be accommodated on the grid. Furthermore, CEB predicts that putting up more than 20 MW of wind capacity in any particular location may have a negative influence on local grid stability and electricity quality. Infrastructure upgrades, as well as grid-impact studies, should be carried out to further identify and overcome these constraints (Young & Vilhauer, 2003).

4.2.1 Wind Power Used for Water Pumping in Food Production

Irrigation consumes a significant amount of energy in producing agriculture, accounting for 13% of the energy consumed on farms (Saputra et al., 2018). Irrigation pumping is now the most energy-intensive stationary use in agriculture. As a result, replacing natural gas or electricity with wind power may significantly lower total fossil fuel consumption in agriculture as well as the farmer's operating costs. Pumping occurs on most irrigated fields for just a few months of the year, usually one and a half to three months (Saputra et al., 2018), drastically reducing the hours of useful usage of prospective wind-driven irrigation systems. This has the effect of lowering the economic value of wind electricity.

Wind power extraction has a greater potential in dry zone parts of Sri Lanka when yearly rainfall is less than 1200 mm. According to survey data, the location benefits from a strong wind stream with an average wind speed of 06 knots throughout the year during both monsoon seasons, and government research has estimated a staggering 500–1000 MW wind power potential in the Puttalam/Kalipitiya area (Sandagiri et al., 2003).

Agricultural load in rural regions increases dramatically during particular seasons due to concentrative irrigation arrangements, putting additional strain on the power grid and even leaving some residents without electricity (Belaud et al., 2020). Farmers utilize conventional waterwheels, which require a wind and a water source to be present. They can only cope with surface water and cannot transport significant amounts of water to far higher elevations (Zhu et al., 2008). Another method of capturing wind energy for irrigation is to use a tiny wind turbine, which is then turned into electric power by a generator with correct control. One or more pumps powered by electricity provided on-site by a generator or from hundreds of meters distant through a cable. Water is pumped into a pool at a higher elevation, which may have numerous water gates used to manage water storage or distribution through channels pointing in different directions. Each water channel might be a custom-built penstock or simply a hose (Zhu et al., 2008).

4.2.2 Wind Power for Building Heating in Agriculture

Heating is the second-largest on-farm energy usage in agriculture for stationary uses. Heating for the farmhouse accounts for the majority of this energy use, followed by heating for poultry brooding and baby pig brooding. Heating water for residential use and dairy sanitation consumes a significant amount of energy on farms, accounting for one-third of all-electric energy consumed in rural areas. Wind-driven heating systems for drying and building heating in agriculture are a highly controversial practice in Sri Lanka. However, the country of Sri Lanka's equatorial position is richer in wind sources, and the practicality of harvesting wind power for drying purposes will be more advantageous.

Propane and fuel oil are the most common energy sources used for heating in rural areas. Both of these energy sources' supply has become unreliable and expensive. Due to the rising expense of fossil fuels, an increasing number of rural users are opting for electric heating systems, causing the utility's peak power demand during the winter to increase in the upcountry of Sri Lanka. There is a link between heating requirements and wind speeds, especially in the sorts of structures utilized on farms. As a result, a wind-powered heating system will often supply more heat during times when it is most required; however, the extent to which this increases the value of wind-powered heating is unknown.

4.2.3 Product Processing and Storage

In certain types of farms, these processes may account for a significant portion of the overall energy consumed. If a way can be found to use wind energy to replace conventional energy in these operations, the farm operator may have a strong incentive to do so. Milk chilling on dairy farms and apple warehouse refrigeration on apple orchards are examples of such systems. Many of these enterprises may be more economically feasible since they are less seasonal than other farming operations. Dairy milk chilling, for example, is done every day, uses roughly the same amount of energy each day, and may store cooling capacity in an ice bank during windless periods.

4.3 Geothermal Power for Food Production

The utilization of waste heat as an energy source to operate a heat pump in a large-scale horticultural plant has been investigated through modeling studies (U.S. Energy

Information Administration, 2021). The study suggested that power plant waste heat be used as the primary source of heat with the best efficiency in horticulture. Depending on the cultured crop, heating expenses at a horticulture facility might range from 19 to 58% of the overall horticulture management cost (Kale, 2016).

The heating performance of heat pumps was examined using seawater versus air as a heat source and it was found that utilizing seawater as a heat source increased heating performance when compared to using air as a heat source. The end-user in this study was apartment buildings. Different uses of geothermal energy may be found in sustainable farming techniques. Organic agriculture makes use of high-temperature geothermal resources with temperatures over 150 °C. In greenhouse operations, geothermal energy is utilized to provide heat for growing vegetables, flowers, ornamentals, and tree seedlings. Aquaculture companies may use geothermal heat to raise many types of fish and other aquatic animals. Dehydration and drying of food and grains are both done with geothermal heat.

In Sri Lanka, geothermal energy has been expressed as ten low-enthalpy thermal springs (35–61 °C) in a short band roughly parallel to the lithological border between the Highland complex (HC) and the Vijayan complex (VC). These springs are not exploited economically, but some have been used recreationally (Kumara, 2014). Geothermometry investigations of Sri Lankan thermal springs reveal that the reservoir fluids have a low to moderate enthalpy, indicating that the resource is well-suited for direct use applications (Chandrajith et al., 2013).

Direct use of geothermal energy may encompass a broad range of end uses and is mostly based on current technology and straightforward engineering. When electricity is generated from geothermal energy with direct usage, numerous benefits accrue, including significantly increased energy efficiency, a shorter development period, and cheap capital investment. Direct use, on the other hand, is considerably more market-specific, as steam and hot water are rarely carried large distances from the geothermal site due to fast heat loss during transport (Kumara, 2014).

In Sri Lanka, the areas surrounding geothermal manifestations have a dry environment, and agriculture is the primary economic activity. Additionally, it is a famous tourist attraction due to its animals and coastline. These sectors can greatly benefit from geothermal energy's direct usage. Agricultural goods (cereals) drying, sugar processing, paper mill operations, cold storage, and tourist facilities such as warm springs, saunas, and steam rooms are all viable applications. Cold storage of fruits, vegetables, meat, and drinks is possible through the use of absorption chillers that run on hot water. The absorption cooling machine that is best suited for this activity employs a combination of water and ammonia as the working fluid. The cooling pump can be powered by geothermal energy. The application of this technology enables the storage of fresh agricultural goods produced in locations with thermal springs.

While geothermal energy is environmentally friendly, the primary downside is that it is location-specific. Sri Lanka is only accessible from a few places. Geothermal plants must be constructed in regions where the energy can be accessed, which implies that certain communities will be unable to utilize this resource. Geothermal energy is a costly resource to exploit, costing millions of dollars for a facility with a 1-megawatt capacity (Training Within Industry, 2018). Where upfront expenditures are considerable, however, they can be recouped over time as part of a long-term investment. While geothermal energy does not normally emit greenhouse gases, many of these gases are trapped beneath the Earth's surface and are released into the atmosphere during excavation (Rinkesh, 2021). While these gases are also naturally discharged into the atmosphere, they are produced at a higher rate around geothermal facilities. These gas emissions, however, are still far less than those connected with fossil fuels. Additionally, geothermal energy has the potential to cause earthquakes. This is owing to the fact that digging alters the Earth's structure. This issue is particularly frequent with improved geothermal power plants, which inject water into the Earth's crust in order to widen cracks and allow for more resource extraction. However, because the majority of geothermal facilities are located far from population centers, the consequences of these earthquakes are minimal (Training Within Industry, 2018).

4.4 Biomass Energy for Food Production

Sri Lanka has a significant potential to satisfy a significant percentage of its energy demands through bioenergy, including bioresources, bioresidues, and biofuels. Until recently, fuel wood was the most abundant bioresource, mostly used for domestic cooking, but its proportion is declining as petroleum is increasingly utilized for transportation and electrical generation (Jayasinghe, 2010). Sri Lanka has a biogas energy potential of around 4,000 MWh per day from agricultural, human, municipal, and animal waste (Asian Development Bank, 2019). Additionally, there is enormous potential for biofuels. Sri Lanka is predicted to be capable of generating approximately 24,000 GWh per year only from dendro power (both naturally occurring and through energy plantations). This is approximately four times the entire hydropower capacity of the country and is sufficient to fulfill the country's electrical energy requirement for several decades. In general, more study is needed to generate more current data on the energy potential of naturally occurring bioresources in order to support the need for energy crops (Musafer et al., 2010).

Even though electricity is available to 85% of houses in Sri Lanka, firewood is the primary source of cooking fuel. As a result, cooking consumes 81% of total biomass. This is mostly owing to electricity's high cost compared to wood's relative availability. Due to the fact that all alternative energy sources for cooking, such as Liquefied Petroleum Gases (LPG) and kerosene, are imported, wood provides a more economically and energy-secure choice (Musafer et al., 2010). At the household level, however, burning biomass without using a clean and efficient stove can result in local environmental and health problems, which have been addressed via a number of efforts in Sri Lanka over the last decades (Amarasekara et al., 2009).

Rice husks and rice straws (Marks-Bielska et al., 2019) are currently used as a source of energy in rice mills and tobacco barns, as well as in cement production (rice husks have an energy potential of 13–19 MJ/kg per kg). In some regions of Sri Lanka, electricity is now generated by the combustion of rice husks in thermal power plants (Fernando & Wimalarathna, 2021). Initiatives to use rice husk as a heating medium for brick manufacturing have had little impact thus far. Nonetheless, with 910,500 ha of rice cultivation and 3,246,000 metric tons of rice produced, rice husks may prove to be a useful supply (Central Bank of Sri Lanka, 2020).

Anaerobic digestion, which produces methane as biogas and digestive residue, is the most frequently used and sophisticated technique for treating organic waste (National Geographic, 2021). The residue includes the nutrients utilized in the crop's cultivation, and it may be used as an organic fertilizer to boost crop production and soil fertility while also closing the energy and nutrient cycles (Lal, 2008). The digestive by-product is the result of mineralization; it has a high nitrogen-to-carbon ratio, making it suitable as a fertilizer; it improves nutrient penetration into the soil, and it reduces the feedstock's odors. This high content is advantageous to crops, as they are primarily capable of utilizing ammonium nitrogen.

Anaerobic digestion of agricultural waste leftovers from the agricultural sector, agriculture industry, and grassland biomass is widely used in many nations to produce energy. Animal manure has also been used as a feedstock in anaerobic digestion, either alone or in combination with agricultural waste. Buffalo grass, for example, which is utilized as fodder for animals, has been found to boost biogas generation when combined with buffalo excrement.

The physical structure and chemical content of lignocellulosic materials that are ordinarily unsuitable for anaerobic digestion can be altered using different pre-treatment techniques. As a result, lignocellulosic waste may be pre-treated so that it can be anaerobically digested, and the methane concentration of the biogas can be increased. The best condition was reached at 100 °C with a two-hour retention time and a 1:1 ratio of co-digestion mixture with manure, with the best condition being obtained at 100 °C with a two-hour retention time and a 1:1 ratio of co-digestion mixture with manure (Kale, 2016).

Biomass may be utilized to generate energy in a variety of methods, including direct gasification or co-firing with coal at an existing power station. Cellulosic ethanol may also be made from biomass. Although maize is the most common energy crop, natural prairie grasses like switchgrass and fast-growing trees like poplar and willow can also be employed. Based on their energy content, biochar produced from torrefaction or pyrolysis of energy crops like switchgrass is a high-quality solid fuel. These perennial crops require less upkeep than annual row crops, making them more cost-effective and sustainable to grow. Agricultural biomass may be utilized to make a wide range of fuels, including liquid and gaseous fuels (Kale, 2016).

The difficulty in establishing biomass, biogas, and biodiesel facilities in Sri Lanka is getting high-quality biomass feedstock with low moisture content, whether it be fuel crops or solid waste. Additionally, much fine-tuning is required to improve the manufacturing process for these enhanced fuel sources. Its potential is enormous, given the cost of energy from agricultural and industrial waste facilities is just Rs. 9.43/kWh (0.046USD/kWh) (CEBSTG, 2017). As a result, it is the only renewable energy source that is cost equivalent to coal power generation, which cost the CEB an average Rs. 7.13/kWh in fuel expenses in 2017 (Ceylon Electricity Board, 2017).

This, along with its capacity to generate continuous baseload power, makes it an excellent answer for the island's power issue (Fernando & Wimalarathna, 2021).

5 Barriers to the Implication of Renewable Energy Technologies in Food Production

Even though renewable energy sources are more abundant and easily accessible everywhere, extracting energy from these sources remains one of the most significant challenges. Capital, competency, technical expertise, and operation and maintenance are important matters in the utilization of renewable energy sources. All these barriers are categorized under economic, institutional, technical, and socio-cultural stands and those negatively impact the way countries transition from fossil fuels to renewable energy sources (Seetharaman et al., 2019). The willingness of the public to renewable energy utilization and strong opposition were continuing from the early days of renewable practices to the present days due to a lack of understanding of the advantages of renewable energy, the disruption of the outlook, and the acquisition of property that could have been used for agriculture, tourism, and other purposes (Blechinger et al., 2015). The fulfillment of human aspirations through socially acknowledged technology systems, as well as appropriate policies and regulatory measures, is the source of sustainable development (Paravantis et al., 2014). The main issue with renewable energy power plants is the need for a lot of land (Collins, 2006; Bayrakci & Koçar, 2012). High initial capital, a lack of financial institutions, a lack of investors, competition from fossil fuels, and fewer subsidies than traditional fuels are all factors that influence economic and financial hurdles. These factors have hampered the use of renewable energy (Saulat, 2017).

Limited infrastructure availability, inefficient knowledge of operations and maintenance, insufficient research and development initiatives, and technical complexities such as energy storage and the lack of standards are all legitimate technological barriers to the widespread deployment of renewable energy (Zhao et al., 2016). Lack of national policies, bureaucratic and administrative barriers, insufficient incentives, unworkable government targets, and a lack of standards and certifications have all hampered the growth of renewable energy (Stokes, 2013). Biofuels for transportation fuel supply from ethanol, for example, are grown on land using corn and soybeans (Gulati et al., 2013). The global increase in food costs caused by the growth of biofuel crops on agricultural land has created serious concerns about the long-term trade-off between energy and food security (Gulati et al., 2013). Similarly, wind energy takes a large quantity of land to produce electricity. Capital cost for the implementation of power plants and the technological skills to operate plants such as solar, wind, and geothermal electric power plants are among serious matters and those influence on the decision-making of the people to shift to renewable energy utilization.

In addition to influencing the development of renewable energy and reducing pressure on agricultural land, organic agriculture's impact on community economic and social development was also discovered. The development of organic agriculture has great potential in the generation and use of renewable energy Furthermore, organic farming can help to minimize the use of costly and energy-intensive synthetic fertilizers. In poor nations, organic agriculture methods produce equivalent or even higher production than conventional agricultural operations. Traditional energy investments are projected to have a minimal societal impact in this nation, owing to high and prohibitive costs as well as restricted access to power. Old agricultural farming, harvesting, and processing methods and practices are the country's primary source of revenue and employment in many of the developing countries.

In Sri Lanka, grid integration barriers for the utility renewable energy power plants are a big issue. The highest demand for the Sri Lanka Power System is now at 2100 MW (Asian Development Bank, 2019). As a result, the scale of the power system itself is a barrier to absorbing Sri Lanka's renewable energy potential (mostly mini-hydro, solar PV, and wind) (Patabendi et al., 2020). Electricity cannot be kept in its pure form. As a result, power demand must be satisfied immediately by supply. Under some circumstances, wind and solar output may be necessary to be restricted to match demand (or held and released, which is still not economically viable). Moreover, during the monsoon season, wind and rain occur concurrently (JICA, 2018). The coming developers will see this as a possible danger because installed capacity is approaching that barrier and non-curtailment cannot be assured. It is necessary to implement a compensation system (Central Bank of Sri Lanka, 2020).

The process is made more complex by a lack of clearly assigned institutional responsibility in Sri Lanka (e.g., many distinct players participating in the execution of policies or the fulfillment of projects). Institutional disagreements can obstruct or hinder the rapid promotion of renewable energies if stakeholders cannot agree on aims and methods (Japan International Cooperation Agency (JICA, 2018). In Sri Lanka, inexperienced stakeholders and decision-makers are still ignorant of the economic, social, and environmental benefits of renewable energy. Renewable energy is frequently viewed as a costly investment by government stakeholders who fail to recognize its near- and long-term advantages. The benefits of entering into an energy swap are not well understood on a social and political level (Patabendi et al., 2020).

6 Conclusions

The energy input for production is the biggest challenge that the globe has been facing in the food production industry. From cultivation to reaching consumers, each sector needs a lot of energy in different forms. Fossil fuel contributes a major part of the energy requirement in each step of production. Hence, the limited resource availability and few deposits of fossil fuel, shifting from fossil fuel to an alternative resource for energy is a very important section in the research activities. Food security of the globe should be ensured by each and every country in the future with the rapidly increasing population. Sri Lanka is a developing country facing significant constraints in high production processes. Energy input will undoubtedly attract attention in the future due to the increasing demand and the rising cost of importing fossil fuels. There are lots of ways and opportunities in the utilization of the renewable energy sector. Even though further techno-economical improvements are needed in the implementation of the renewable energy sector in the application, common renewable energy technologies such as solar power, wind energy, biomass energy, and geothermal energy are easy to implement, and adopting them will not only act as an alternative to fossil fuels but also it will help to reduce production costs, strengthen the food production process, reduce the greenhouse gas emissions and help to produce green and clean products. Barriers should be addressed by all stakeholders, and finding the best ways to overcome those limitations will improve the diffusion of renewable energy utilization among the people and contribute to the safe and sustainable food production and supply chain of the country Sri Lanka in the future.

References

Ackermann, T., & Soder, L. (2000). Wind energy technology and current status : A review. *Renewable and Sustainable Energy Reviews, 4*, 315–374.

Acosta-Silva, Y.D.J., Torres-Pacheco, I., Matsumoto, Y., Toledano-Ayala, M., Soto-Zarazúa, G. M., Zelaya-Ángel, O., & Méndez-López, A. (2019). Applications of solar and wind renewable energy in agriculture: A review. *Science Progress, 102*(2), 127–140.https://doi.org/10.1177/0036850419832696

ADB & UNDP. (2017). *Assessment of Sri Lanka's power sector-100% electricity generation through renewable energy by 2050.* 122. https://www.undp.org/content/dam/LECB/docs/pubs-reports/UNDP-LECB-Assessment-Sri-Lanka-Power-Sector.pdf

Aheeyar, M. (2012). *Assessment of solar powered drip irrigation project implemented by ministry of agriculture—phase 1 hector Kobbekaduwa agrarian research and training institute* (Issue October).

Amarasekara, R. M., Bellanca, R., Konaris, T., & Musafer, N. (2009). scale-Up and commercialization of improved cookstoves in Sri Lanka: Lessons learnt from the Anagi experience. In *Design* (Issue December).

Amunugoda, P. N. R. J., Senanayake, N. S., Wijeratnam, R. S. W., & Kulatunga, K. D. G. (2012). Quality enhancement of dehydrated products through the modification of solar tunnel dryer for continuous operation in rural communities. In *Proceedings of the 2011 International Conference and Utility Exhibition on Power and Energy Systems: Issues and Prospects for Asia, ICUE 2011* (Vol. 3(1), pp. 7–14). https://doi.org/10.1109/ICUEPES.2011.6497747

Aroonsrimorakot, S., & Laiphrakpam, M. (2019). Application of solar energy technology in agricultural farming for sustainable development: A review article. *International Journal of Agricultural Technology, 15*(5), 685–692.

Asia Wind Energy Association. (2021). *Sri Lanka.* Sri Lanka—Asia Wind Energy Association 2021. https://www.asiawind.org/research-data/market-overview/sri-lanka/

Asian Development Bank. (2019). *Sri Lanka: Energy sector assessment, strategy, and road map* (Issue October). www. adb.orgwww. adb.org%0Awww. adb.org

Bayrakci, A. G., & Koçar, G. (2012). Utilization of renewable energies in Turkey's agriculture. *Renewable and Sustainable Energy Reviews, 16*(1), 618–633. https://doi.org/10.1016/j.rser.2011.08.027

Belaud, G., Mateos, L., Aliod, R., Buisson, M. C., Faci, E., Gendre, S., Ghinassi, G., Gonzales Perea, R., Lejars, C., Maruejols, F., & Zapata, N. (2020). Irrigation and energy: Issues and challenges. *Irrigation and Drainage, 69*(S1), 177–185. https://doi.org/10.1002/ird.2343

Bena, B., & Fuller, R. J. (2002). Natural convection solar dryer with biomass back-up heater. *Solar Energy, 72*(1), 75–83. https://doi.org/10.1016/S0038-092X(01)00095-0

Bhagoria, P., & Bhagoria, J. L. (2013). Development and performance evaluation of mixed-mode solar dryer with forced convection. *International Journal of Energy and Environmental Engineering, 23*(4), 1–8. https://doi.org/10.1186/2251-6832-4-23

Bhullar, S. S., & Lalwani, M. (2018). Performance analysis of 25 MW grid connected solar photovoltaic plant in Gujarat, India. In *3rd international conference on innovative applications of computational intelligence on power, energy and controls with their impact on humanity, CIPECH 2018* (pp. 20–25). https://doi.org/10.1109/CIPECH.2018.8724370

Bitar, E. Y., Rajagopal, R., Khargonekar, P. P., Poolla, K., & Varaiya, P. (2012). Bringing wind energy to market. *IEEE Transactions on Power Systems, 27*(3), 1225–1235. https://doi.org/10.1109/TPWRS.2012.2183395

Blechinger, P., Richter, K., & Renn, O. (2015, August). *Barriers and solutions to the development of renewable energy technologies in the Caribbean* (pp. 267–284). https://doi.org/10.1007/978-3-319-15964-5_24

Borah, A., Hazarika, K., & Khayer, S. M. (2015). Drying kinetics of whole and sliced turmeric rhizomes (Curcuma longa L.) in a solar conduction dryer. *Information Processing in Agriculture, 2*(2), 85–92. https://doi.org/10.1016/j.inpa.2015.06.002

Brangstrup, P. E. (2014). *Solar assisted air heating process—implementing solar collectors in Sri Lankan Tea industry.*

Bundschuh, J., Chen, G., & Mushtaq, S. (2014). Towards a sustainable energy technologies-based agriculture. *Sustainable Energy Solutions in Agriculture, 2012*, 3–15. https://doi.org/10.1201/b16643

Byrne, P. S., Carton, J. G., & Corcoran, B. (2021). Investigating the suitability of a heat pump water-heater as a method to reduce agricultural emissions in dairy farms. *Sustainability (Switzerland), 13*(10). https://doi.org/10.3390/su13105736

Capuano, L. (2020). International Energy Outlook 2020 (IEO 2020). *US Energy Information Administration, 2020*, 1–7. https://www.eia.gov/outlooks/ieo/pdf/ieo2020.pdf

CEBSTG. (2017). Statistical Digest 2017. In *Statistical digest*.

Central bank of Sri Lanka. (2020). *Economic and social statistics of Sri Lanka* (Vol. 42).

Ceylon Electricity Board. (2017). *Sales and generation data book 2017*.

Chandrajith, R., Barth, J. A. C., Subasinghe, N. D., Merten, D., & Dissanayake, C. B. (2013). Geochemical and isotope characterization of geothermal spring waters in Sri Lanka: Evidence for steeper than expected geothermal gradients. *Journal of Hydrology, 476*, 360–369. https://doi.org/10.1016/j.jhydrol.2012.11.004

Chel, A., & Kaushik, G. (2011). Renewable energy for sustainable agriculture. *Agronomy for Sustainable Development, 31*(1), 91–118. https://doi.org/10.1051/agro/2010029

Clack, C. T. M., Mckee, S., Clean, V., & March, E. (2019). The coal cost crossover: Economic viability of existing coal compared to new local wind and solar resources. *Energy Innovation*, September 2018, 1–24.

Collins, J. A. D. (2006). Evolution of renewable energy policy. In *Agricultural & applied economics association* (Vol. 21, Issue 1).

Cruz-Peragon, F., Palomar, J. M., Casanova, P. J., Dorado, M. P., & Manzano-Agugliaro, F. (2012). Characterization of solar flat plate collectors. *Renewable and Sustainable Energy Reviews, 16*(3), 1709–1720. https://doi.org/10.1016/j.rser.2011.11.025

Department of Economic and Social Affairs United Nations. (2021). *Population dynamics*. World Population Prospects 2019. https://population.un.org/wpp/Download/Standard/Population/

Dickson, M. H., & Fanelli, M. (2020). What is geothermal energy? *Renewable Energy*, 302–328. https://doi.org/10.4324/9781315793245-25

Diemuodeke, E. O., & Momoh, O. L. Y. (2011). Design and fabrication of a direct natural convection solar dryer for tapioca. *Leonardo Electronic Journal of Practices and Technologies, 10*(18), 95–104.

El-Sebaii, A. A., Aboul-Enein, S., Ramadan, M. R. I., & El-Gohary, H. G. (2002). Experimental investigation of an indirect type natural convection solar dryer. *Energy Conversion and Management, 43*(16), 2251–2266. https://doi.org/10.1016/S0196-8904(01)00152-2

El-Sebaii, A. A., & Shalaby, S. M. (2012). Solar drying of agricultural products: A review. *Renewable and Sustainable Energy Reviews, 16*(1), 37–43. https://doi.org/10.1016/j.rser.2011.07.134

FAO. (2011). Energy-smart food for people and climate. Issue Paper. *Food and Agriculture Organization of the United Nations*, 66.

Fernando, S., & Wimalarathna, R. (2021). *Renewable Energy P2—Biomass Energy.* Inside Access. https://www.insideaccess.lk/detail/view/143/industry-insights/renewable-energy-p2-biomass-energy

Food and Agriculture Organization of the United Nations. (2021). *The state of food security and nutrition in the world.*

Gbaha, P., Yobouet Andoh, H., Kouassi Saraka, J., Kaménan Koua, B., & Touré, S. (2007). Experimental investigation of a solar dryer with natural convective heat flow. *Renewable Energy, 32*(11), 1817–1829. https://doi.org/10.1016/j.renene.2006.10.011

Green, M. A., Dunlop, E. D., Levi, D. H., Hohl-Ebinger, J., Yoshita, M., & Ho-Baillie, A. W. Y. (2019). Solar cell efficiency tables (version 54). *Progress in Photovoltaics: Research and Applications, 27*(7), 565–575. https://doi.org/10.1002/pip.3171

Gulati, M., Jacobs, I., Jooste, A., Naidoo, D., & Fakir, S. (2013). The water–energy–food security nexus: Challenges and opportunities for food security in South Africa. *Aquatic Procedia, 1*, 150–164. https://doi.org/10.1016/j.aqpro.2013.07.013

Gunaratne, L. (1994). Solar photovoltaics in Sri Lanka: A short history. *Progress in Photovoltaics: Research and Applications, 2*(4), 307–316. https://doi.org/10.1002/pip.4670020406

Gupta, P. K. (2016). Air flow investigations in direct type solar food dryer using computational fluid dynamics. *Carbon—Science and Technology, 8*(2), 9–16.

Han, J., Mol, A. P. J., & Lu, Y. (2010). Solar water heaters in China: A new day dawning. *Energy Policy, 38*(1), 383–391. https://doi.org/10.1016/j.enpol.2009.09.029

Harijan, M. A. U. (2012). Solar drying technology: Potentials and developments. *Energy, Environment and Sustainable Development*, 69–98. https://doi.org/10.1007/978-3-7091-0109-4

Hartung, H., & Pluschke, L. (2018). The benefits and risks of solar powered irrigation—a global overview. In *Food and agriculture organization of the United Nations and Deutsche Gesellschaft für Internationale Zusammenarbeit.*

Hill, D. S. (2002). *Stages in food production.* Springer, Dordrecht. https://doi.org/10.1007/0-306-48131-6_2

Hussain, M. I., & Lee, G. H. (2015). Utilization of solar energy in agricultural machinery engineering: A review. *Journal of Biosystems Engineering, 40*(3), 186–192. https://doi.org/10.5307/jbe.2015.40.3.186

Intergovernmental Panel on Climate Change. (2014). Climate change 2014 mitigation of climate change. In *Climate change 2014 mitigation of climate change.* https://doi.org/10.1017/cbo9781107415416

International Energy Agency. (2021). Global Energy Review 2021. In *Global energy review 2020.* https://iea.blob.core.windows.net/assets/d0031107-401d-4a2f-a48b-9eed19457335/GlobalEnergyReview2021.pdf

Islam, M. R., Mekhilef, S., & Saidur, R. (2013). Progress and recent trends of wind energy technology. *Renewable and Sustainable Energy Reviews, 21*, 456–468. https://doi.org/10.1016/j.rser.2013.01.007

Jackson, F. (2020). *Solar soars as emerging markets renewables investment hits.* Forbes.

Jaisankar, S., Ananth, J., Thulasi, S., Jayasuthakar, S. T., & Sheeba, K. N. (2011). A comprehensive review on solar water heaters. *Renewable and Sustainable Energy Reviews, 15*(6), 3045–3050. https://doi.org/10.1016/j.rser.2011.03.009

Janjai, S., & Keawprasert, T. (2006). Design and performance of a solar tunnel dryer with a polycarbonate cover. *International Energy Journal, 7*(3), 187–194.

Janjai, S., Lamlert, N., Intawee, P., Mahayothee, B., Bala, B. K., Nagle, M., & Müller, J. (2009). Experimental and simulated performance of a PV-ventilated solar greenhouse dryer for drying of peeled longan and banana. *Solar Energy, 83*(9), 1550–1565. https://doi.org/10.1016/j.solener.2009.05.003

Japan International Cooperation Agency (JICA). (2018). *Project on electricity sector master plan study in democratic socialist Republic of Sri Lanka.*

Jayasinghe, P. (2010). *The biomass energy sector in Sri Lanka successes and constraints.*

Kale, S. A. (2016). Renewable energy systems. In *Renewable energy systems.*https://doi.org/10.1201/9780429354632-12

Kalogirou, S. A. (2004). Solar thermal collectors and applications. *Progress in Energy and Combustion Science, 30*(3), 231–295. https://doi.org/10.1016/j.pecs.2004.02.001

Karsli, S. (2007). Performance analysis of new-design solar air collectors for drying applications. *Renewable Energy, 32*(10), 1645–1660. https://doi.org/10.1016/j.renene.2006.08.005

Kumara, S. M. P. G. S. (2014). Potential for direct utilization of geothermal energy in Sri Lanka exploration of geothermal energy, Sri Lanka View project. *Sciscitator, 01*(March), 33–35. https://www.researchgate.net/publication/273132948

Kumarasiriwardhana, C. A., Kahandage, P. D., & Weerasooriya, G. V. T. V. (2020). Design, fabrication and evaluation of solar and biomass hybrid dryer with trackable solar collector. *Sri Lankan Journal of Agriculture and Ecosystems, 2*(2), 164. https://doi.org/10.4038/sljae.v2i2.45

Labed, A., Moummi, N., & Benchabane, A. (2012). Experimental investigation of various designs of solar flat plate collectors: Application for the drying of green chili. *Journal of Renewable and Sustainable Energy, 4*(4). https://doi.org/10.1063/1.4742337

Lal, R. (2008). Soils and sustainable agriculture. A review. *Agronomy for Sustainable Development, 28*(1), 57–64. https://doi.org/10.1051/agro:2007025

Laufer, D., & Schäfer, M. (2011). The implementation of Solar Home Systems as a poverty reduction strategy-a case study in Sri Lanka. *Energy for Sustainable Development, 15*(3), 330–336. https://doi.org/10.1016/j.esd.2011.07.002

Lee, J., & Zhao, F. (2021). Gobal Wind Report 2021. In *Global wind energy council.* http://www.gwec.net/global-figures/wind-energy-global-status/

Madeddu, S., Ueckerdt, F., Pehl, M., Peterseim, J., Lord, M., & Kumar, K. A. (2020). The CO_2 reduction potential for the European industry via direct electrification of heat supply OPEN ACCESS The CO_2 reduction potential for the European industry via direct electrification of heat supply (power-to-heat). *Environmental Research Letters, 15*(1), 1–14.

Makdisie, C., Haidar, B., & Alhelou, H. H. (2018). An optimal photovoltaic conversion system for future smart grids. *Handbook of Research on Power and Energy System Optimization*, 601–657. https://doi.org/10.4018/978-1-5225-3935-3.ch018

Marks-Bielska, R., Bielski, S., Novikova, A., & Romaneckas, K. (2019). Straw stocks as a source of renewable energy. A case study of a district in Poland. *Sustainability (Switzerland), 11*(17), 1–18. https://doi.org/10.3390/su11174714

Marks-Bielska, R., Bielski, S., Pik, K., & Kurowska, K. (2020). The importance of renewable energy sources in Poland's energy mix. *Energies, 13*(18), 1–23. https://doi.org/10.3390/en13184624

Mathew, S. (2017). *Wind energy.* In *13th German wind energy conference.* https://doi.org/10.1016/B978-0-12-804448-3/00004-9

Mauthner, F., & Fischer, S. (2015). Overheating prevention and stagnation handling in solar process heat applications. In *IEA SHC task 49 solar process heat for production and advanced applications.* https://doi.org/10.1016/j.rser.2009.07.014

Meah, K., Fletcher, S., & Ula, S. (2008). Solar photovoltaic water pumping for remote locations. *Renewable and Sustainable Energy Reviews, 12*(2), 472–487. https://doi.org/10.1016/j.rser.2006.10.008

Mekhilef, S., Faramarzi, S. Z., Saidur, R., & Salam, Z. (2013). The application of solar technologies for sustainable development of agricultural sector. *Renewable and Sustainable Energy Reviews, 18*, 583–594. https://doi.org/10.1016/j.rser.2012.10.049

Ministry of Power and Energy. (2015). Energy Empowered Nation. In *Sri Lanka energy sector development plan for a knowledge-based economy 2015–2025* (Vol. 1542).

Mohtasham, J. (2015). Review article-renewable energies. *Energy Procedia, 74*, 1289–1297. https://doi.org/10.1016/j.egypro.2015.07.774

Musafer, N., Pritchard, C., Hunt, E. S., Bloomfield, E., & Bellanca, R. (2010). *Bioenergy in Sri Lanka: Resources, applications and initiatives* (Issue May).

National Geographic. (2021). Biomass energy. In *Resource library encyclopedic entry* (p. 15). https://www.nationalgeographic.org/encyclopedia/biomass-energy/

Nimrotham, C., Songprakorp, R., Thepa, S., & Monyakul, V. (2017). Experimental research of drying red chili by two methods: Solar drying and low-temperature system drying. *Energy Procedia, 138*, 512–517. https://doi.org/10.1016/j.egypro.2017.10.237

Odhiambo, O. (2016). Development of improved orange flesh sweet potato dryers in Western Kenya. In *Technical report* (Issue November). https://doi.org/10.13140/RG.2.1.1142.6643

Paravantis, J. A., Stigka, E. K., & Mihalakakou, G. K. (2014). An analysis of public attitudes towards renewable energy in Western Greece. In *IISA 2014—5th international conference on information, intelligence, systems and applications* (Vol. 32, pp. 300–305). https://doi.org/10.1109/IISA.2014.6878776

Parida, B., Iniyan, S., & Goic, R. (2011). A review of solar photovoltaic technologies. *Renewable and Sustainable Energy Reviews, 15*(3), 1625–1636. https://doi.org/10.1016/j.rser.2010.11.032

Patabendi, K., Darshika, S., & Silva, D. (2020). Opportunities and challenges of solar energy application in energy sector of Sri Lanka. *Bulletin of the Korea Photovoltaic Society, 6*(1), 45–55.

Patel, A. H., Shah, S. A., & Bhargav, H. (2017). Review on solar dryer for grains, vegetables and fruits. *International Journal of Engineering Research & Technology (IJERT), 2*(January 2013), 1–7.

Phadke, P., & Walke, P. (2015). Direct type natural convection solar dryer: A review. *International Journal of Advance Research in Science and Engineering, 4*(2), 256–262.

Prakash, O., & Kumar, A. (2013). Historical review and recent trends in solar drying systems. *International Journal of Green Energy, 10*(7), 690–738. https://doi.org/10.1080/15435075.2012.727113

Premalatha, M., Abbasi, T., Abbasi, T., & Abbasi, S. A. (2011). Energy-efficient food production to reduce global warming and ecodegradation: The use of edible insects. *Renewable and Sustainable Energy Reviews, 15*(9), 4357–4360. https://doi.org/10.1016/j.rser.2011.07.115

Public Utilities Commission of Sri Lanka. (2011). *Achievements of renewable energy targets in Sri Lanka.*

Public Utilities Commission of Sri Lanka. (2020). Draft final report on development of merit order dispatch procedures. In *Public utilities commission of Sri Lanka* (Issue August). http://www.mautc.psu.edu/docs/WVU-2000-05.pdf

Rajmohan, K., & Weerahewa, J. (2010). Household energy consumption patterns in Sri Lanka. *Sri Lankan Journal of Agricultural Economics, 9*, 55. https://doi.org/10.4038/sjae.v9i0.1833

Razykov, T. M., Ferekides, C. S., Morel, D., Stefanakos, E., Ullal, H. S., & Upadhyaya, H. M. (2011). Solar photovoltaic electricity: Current status and future prospects. *Solar Energy, 85*(8), 1580–1608. https://doi.org/10.1016/j.solener.2010.12.002

Renné, D., George, R., Marion, B., Nrel, D. H., Gueymard, C., & Consulting, S. (2014). Solar resource assessment for Sri Lanka and Maldives. In *National renewable energy laboratory, U.S. Department of Energy NREL/TP-710-34645,* (Issue March 2014). http://www.osti.gov/bridge

Rinkesh. (2021). *Geothermal energy pros and cons.* Conserve Energy Future. https://www.conserve-energy-future.com/pros-and-cons-of-geothermal-energy.php

Robins, J. C., Kolker, A., Flores-Espino, F., Pettitt, W., Rising, G., Schmidt, B., Rising, G., Augustine, C., Blair, N., Cook, J., Cruce, J., Johnston, H., Kurtz, J., Levine, A., Porro, G., Rhodes, G.,

Rubin, R., Taverna, N., Turchi, C., & Warren, I. (2021). *2021 U.S. geothermal power production and district heating market report.*
Sampaio, P. G. V., & González, M. O. A. (2017). Photovoltaic solar energy: Conceptual framework. *Renewable and Sustainable Energy Reviews, 74*(December 2016), 590–601. https://doi.org/10.1016/j.rser.2017.02.081
Sandagiri, C. A., Peiris, L. R., Perera, L. A. A. N., Siriwardena, K. A. P., & Soysa, W. H. I. D. (2003). *Selection of suitable wind turbine for wind farm candidate site Narakkalliya, Kalpitiya Peninsula, Sri Lanka.* Department of Electrical Engineering, University of Moratuwa, Sri Lanka.
Saputra, M., Syuhada, A., & Sary, R. (2018). Study of solar and wind energy using as water pump drive-land for agricultural irrigation. In *Proceedings—2018 4th international conference on science and technology, ICST 2018*, 2018–2021. https://doi.org/10.1109/ICSTC.2018.8528643
Saulat, H. (2017). *Renewable energy resources current status and barriers in their adaptation for Pakistan development of membranes view project waste management view project.* January 2015. http://scienceq.org/Journals/JBCE.php
Sayanthan, S., & Kannan, N. (2017). Renewable energy resource of Sri Lanka! A review. *International Journal of Environmental & Agriculture Research (IJOEAR), 3*(4), 80–85.
Sayigh, A. (2012). *Renewable energy: The only solution chairman of world renewable energy congress & director general of WREN* (Vol. 1(3), pp. 83–86).
Sciences, V. (2010). *Solar energy applications for agriculture Chikaire, J. Nnadi, F. N., Nwakwasi, R. N., Anyoha, N. O., Aja, O. O., Onoh, P. A., & Nwachukwu C. A. 2*(September), 58–62.
Seetharaman, Moorthy, K., Patwa, N., Saravanan, & Gupta, Y. (2019). Breaking barriers in deployment of renewable energy. *Heliyon, 5*(1), e01166. https://doi.org/10.1016/j.heliyon.2019.e01166
Shinde, V. B., & Wandre, S. S. (2015). Solar photovoltaic water pumping system for irrigation: A review. *African Journal of Agricultural Research, 10*(22), 2267–2273.
Singh, S., Singh, P. P., & Dhaliwal, S. S. (2004). Multi-shelf portable solar dryer. *Renewable Energy, 29*(5), 753–765. https://doi.org/10.1016/j.renene.2003.09.010
Sri Lanka Survey Department. (2019). *Mapping manual 2019.*
Stokes, L. C. (2013). The politics of renewable energy policies: The case of feed-in tariffs in Ontario, Canada. *Energy Policy, 56*, 490–500. https://doi.org/10.1016/j.enpol.2013.01.009
The International Hydropower Association. (2021). *Hydropower status report 2021.*
Thibbotuwawa, M. (2017). *Achieving food security: A plausible reality or a pipedream for Sri Lanka?* Talkling Economics. https://www.ips.lk/talkingeconomics/2017/08/23/achieving-food-security-a-plausible-reality-or-a-pipedream-for-sri-lanka/
Tisdell, C. A. (2014). Sustainable agriculture: An introduction. In *Handbook of sustainable development* (2nd ed., pp. 517–531). https://doi.org/10.4337/9781782544708.00046
Tiwari, S., & Tiwari, G. N. (2016). Thermal analysis of photovoltaic-thermal (PVT) single slope roof integrated greenhouse solar dryer. *Solar Energy, 138*, 128–136. https://doi.org/10.1016/j.solener.2016.09.014
Training Within Industry. (2018). *What are the advantages and disadvantages of geothermal energy?* Training Within Industry. https://www.twi-global.com/technical-knowledge/faqs/geothermal-energy/pros-and-cons
Tyagi, V. V., Rahim, N. A. A., Rahim, N. A., & Selvaraj, J. A. L. (2013). Progress in solar PV technology: Research and achievement. *Renewable and Sustainable Energy Reviews, 20*, 443–461. https://doi.org/10.1016/j.rser.2012.09.028
U.S. Department of Commerce. (2021). *Sri Lanka- country commercial guide.* International Trade Administration.
U.S. Energy Information Administration. (2021). *Geothermal explained.* Independent Statistics and Analysis. https://www.eia.gov/energyexplained/geothermal/
Vithanage, I. (2019). Changing Energy Scenario of Sri Lanka. In *IEEJ* (Issue August).
Wijayatunga, P., & Kolantharaj, J. (2021). *Powering Sri Lanka's clean energy transition.* Asian Development Bank. https://www.adb.org/news/op-ed/powering-sri-lanka-s-clean-energy-transition-priyantha-wijayatunga-and-jaimes-kolantharaj

Xu, J., Ni, T., & Zheng, B. (2015). Hydropower development trends from a technological paradigm perspective. *Energy Conversion and Management, 90*, 195–206. https://doi.org/10.1016/j.enconman.2014.11.016

Yan, J., Yang, Y., Campana, P. E., & He, J. (2019). City-level analysis of subsidy-free solar photovoltaic electricity price, profits and grid parity in China. *Nature Energy, 4*(August), 709–717. https://doi.org/10.1038/s41560-019-0441-z

Young, M., & Vilhauer, R. (2003). Sri Lanka wind farm analysis and site selection assistance. In *National renewable energy laboratory* (Issue August).

Zhao, Z. Y., Chang, R. D., & Chen, Y. L. (2016). What hinder the further development of wind power in China?-A socio-technical barrier study. *Energy Policy, 88*, 465–476. https://doi.org/10.1016/j.enpol.2015.11.004

Zhu, Y., Ding, Z., & Gong, Y. (2008). Advanced agricultural irrigation system applying wind power generation. In *2008 IEEE energy 2030 conference, ENERGY 2008* (pp. 2–3). https://doi.org/10.1109/ENERGY.2008.4781005

Correction to: Mitigation of the Effects of Climate Change on Agriculture Through the Adoption of Precision Agriculture Technologies

Muharrem Keskin and Yunus Emre Sekerli

Correction to:
Chapter "Mitigation of the Effects of Climate Change on Agriculture Through the Adoption of Precision Agriculture Technologies" in: M. Behnassi et al. (eds.), *Climate-Smart and Resilient Food Systems and Security*, https://doi.org/10.1007/978-3-031-65968-3_20

In the original version of Chapter 20, Figures 1 and 2 have been replaced. The chapter and the book have been updated with the changes.

The updated version of this chapter can be found at
https://doi.org/10.1007/978-3-031-65968-3_20

M. Behnassi et al. (eds.), *Climate-Smart and Resilient Food Systems and Security*,
https://doi.org/10.1007/978-3-031-65968-3_24

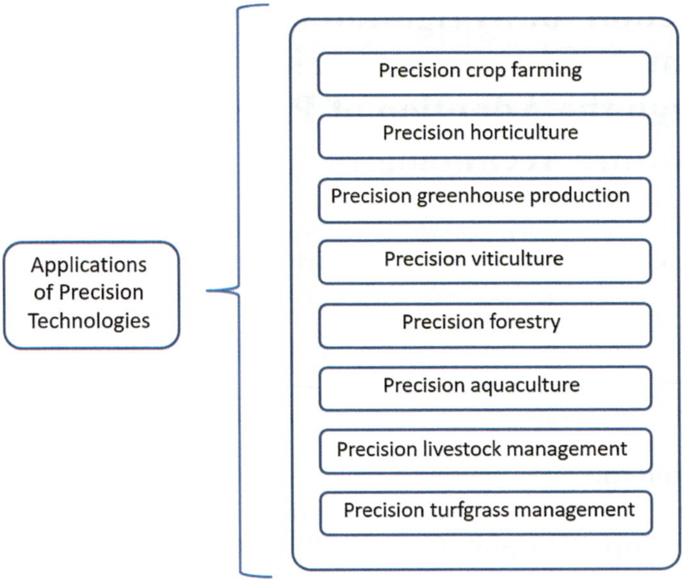

Fig. 1 Application of precision technologies to various production systems. *Source* Developed by the authors

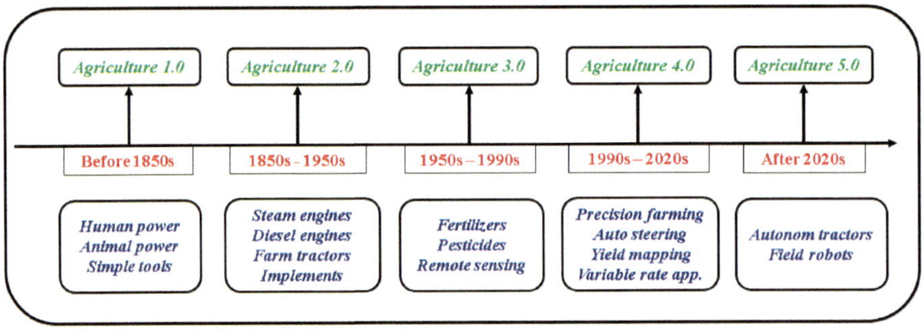

Fig. 2 Historical development stages of agricultural technologies. *Source* Developed by the authors

Postface: A Call to Action for Climate-Smart Food Systems

Achieving the second SDG of Zero Hunger hinges on a complete transformation of our food and agriculture systems. The specter of climate change, with its erratic weather patterns, extreme weather events, and biodiversity loss, exposes the unsustainable and vulnerable character of conventional agriculture and threatens the future of global food security. In response, stakeholders across the agricultural landscape are engaging in a paradigm shift embracing promising integrated approaches with the potential to help manage emerging challenges.

This book, *Climate-Smart and Resilient Food Systems and Security*, rises to this challenge by embodying and advocating for a holistic approach. The book's contributors and editors, thought leaders in their respective fields, delve into the transformative potential of Climate-Smart Agriculture (CSA). They explore how CSA can revolutionize food production by integrating scientific research, technological innovation, and robust policy development enabling well-crafted policy frameworks.

This book goes beyond mere potential. It addresses some of the most pressing challenges faced by vulnerable populations around the globe. These include the devastating impact of the COVID-19 pandemic on Africa's food security and the water management risks plaguing many developing countries. Recognizing the uneven impact of climate change across regions, the book emphasizes the need for tailored mitigation and adaptation strategies, particularly for communities in the Global South.

While the book acknowledges the challenges, it also brims with hope as the authors shine a light on potential and promising solutions. As climate change intensifies the precarity of rural communities, the book shines a light on subsistence farmers and smallholders—often overlooked yet significant contributors to global food production. Using concrete and compelling case studies, the authors brought to the forefront of the climate-agriculture discussion these declining forms of farming and propose initiatives to empower subsistence farmers and smallholders as vulnerable groups, recognizing their crucial role in building sustainable and resilient food systems.

M. Behnassi et al. (eds.), *Climate-Smart and Resilient Food Systems and Security*, https://doi.org/10.1007/978-3-031-65968-3

The book supports a balanced approach, emphasizing the importance of both policy changes, and research and technological advancements. It underscores the importance of collaborative efforts between policymakers and researchers in building resilient food systems. Therefore, the book offers valuable insights on CSA-friendly policy frameworks, smart breeding principles, the vital contribution of agroforestry, and the integration of technologies like drone applications and remote sensing in CSA practices.

Climate-Smart and Resilient Food Systems and Security, which assembled a collection of thought-provoking work, stands as a beacon, urging the global community to act. The book offers a wealth of practical solutions that can be implemented now. As the book convincingly emphasizes, CSA has the potential to not only address environmental challenges but also to transform agri-food systems, making them more socially equitable, inclusive, and economically viable.

The development of this book would not have been possible without the dedication of the outstanding editorial team, the thought-provoking contributions of each author, and the fruitful collaboration of reviewers. A special acknowledgment goes to everyone who contributes to the process, making possible such a project that is adding another valuable resource to the ever-growing body of knowledge on CSA. Working with Springer Nature has provided a trusted platform for this critical dialogue between all contributors on the future of our food systems.

This postface is not an ending, but a beginning of other initiatives. It is a call to action for policymakers, researchers, farmers, and the global community at large. Let us embrace the transformative potential of Climate-Smart Agriculture and work together to build a future where food security, sustainability, and resilience are a reality for all.

Aisha Baig
Mirza B. Baig
Mohamed Behnassi

Biographical Notes of Authors

Adnan, *Shahzada*

Dr. Adnan obtained a Ph.D. in Meteorology from COMSATS University Islamabad, Pakistan. He joined Pakistan Meteorological Department (PMD) in the year 2005 and is currently working as Deputy Director/Sr. Meteorologist. During his tenure, he served in Climate Data Processing Centre, National Agrometeorological Centre, National Weather Forecast Centre, and National Drought Monitoring and Early Warning Centre of PMD. He served as a PMD technical team member for providing a close liaison with the National Command and Operation Centre (NCOC), Pakistan, during the COVID-19 pandemic, with a focus on the prediction of coronavirus projection and expected peak in Pakistan. Dr. Adnan's research work successfully predicted major COVID-19 surges (2020–2022) well before time and the same was presented as well as communicated to the NCOC which helped to devise plans of action to curb the coronavirus spread and its impact. His research contributions were appreciated by NCOC, and his research work was uploaded by the World Health Organization and the National Library of Medicine (USA). Dr. Adnan developed a District Drought Hazard Map of Pakistan, which was subsequently adopted by the UN World Food Program and the National Disaster Management Authority (NDMA), Pakistan, for the development of the First Food Security Map of Pakistan using integrated context analysis. Besides this, this map helped in policy development and food security management.

Dr. Adnan is a prolific researcher who has so far authored 35 articles which have been published in high-impact, peer-reviewed international journals. These articles cover topics like drought, aridity, extreme weather events, agrometeorology, climate change, and weather-health relationships. Notably, his research is freely available to the scientific community and researchers on Research Gate, accumulating over 88,237 reads and 1298 citations across the globe in the past ten years. His name is enlisted in the directory of Productive Scientists of Pakistan 2017 of the Pakistan Council for Science and Technology, Ministry of Science and Technology. Moreover, he received Best Research Paper Award (2016) from Higher Education Commission Pakistan and three Research Productivity awards (2015–2017) from COMSATS University Islamabad, Islamabad.

Dr. Adnan has also assisted/co-supervised numerous Ph.D., M.S., and B.S. scholars at various universities of Pakistan. He has also provided peer reviews for over 100 international research articles on climate change and its impacts, which appeared in prestigious journals like Springer, Elsevier, Taylor & Francis, Science Direct, Wiley, and, notably, the Royal Meteorological Society and the American Meteorological Society.

Ahmad, *Maqsood*

Dr. Ahmad is currently serving as an Assistant Professor at Balochistan University of Information Technology, Management Sciences (BUITEMS), located in Quetta, Pakistan. He holds a Ph.D. in Environmental Science from GC University, Lahore, Pakistan, and has pursued research on environmental and industrial monitoring, and biodegradation of organic pollutants. Dr. Ahmad has made significant contributions to the fields of environmental science and public health.

Ahmed, *Nazeer*

Professor Ahmed has a blend of experience in the Agriculture Department Balochistan and Balochistan University of Information Technology, Engineering and Management Sciences (BUITEMS), Quetta. At BUITEMS, besides teaching and research, Prof. Ahmed held various administrative positions from Chairperson of a department to Pro Vice Chancellor. He obtained his Ph.D. from Wurzburg University, Germany. The main theme of his research work in his Ph.D. was to explore the early signaling events involved in plant–microbe interaction, particularly plant interaction with mutualistic bacteria. The current focus of his research is biodiversity conservation. Utilizing state-of-the-art DNA barcoding and metabarcoding approaches, he is attempting to catalog the flora and fauna of Pakistan through collaborative endeavors with the related scientific community of Pakistan.

Alaoui, *Si Bennasseur*

Dr. Alaoui is an Emeritus Professor at Hassan II Agronomic and Veterinary Sciences Institute, Rabat, Morocco. Originally from a peasant family, he completed his primary studies in Lahri, his native village, and his secondary studies in Khenifra city. In 1982, he obtained the Diploma of "Advanced Agronomy" (DAA) from the National Institute of Agronomy, Paris-Grignon, France, and the Diploma of State Engineer in Agronomy from the IAV Hassan II in 1983. In 1993, he obtained a Doctorate in Agronomic Sciences from IAV Hassan II and a Ph.D. in Weed Science from the University of Minnesota (USA). He is the author of several scientific and technical articles and the book entitled "Les Filières Bio: Enjeux et Promesses pour le Maroc et l'Afrique", for which was awarded the First Prize of the 11th edition of the Grand Prix Hassan II for Innovation and Research in the agricultural field for 2019. His last book was "Crop Rotations and Green Manure for Better Soils and More Sustainable Agricultural Production". He is a consultant to several national and international organizations including FAO, IFAD, GIZ, ISESCO, and MAMDA.

Ashfaq, *Muhammad*

Dr. Ashfaq is currently working as Professor/Subject Specialist, Faculty of Social Sciences and Humanities, MNS University of Agriculture, Multan-Pakistan. He completed his graduation in 1985 and earned an M.S. degree in Agricultural Economics in 1988. He won the scholarship in 1990–91 from USAID for training in Agricultural Marketing and Sales, Colorado State University, USA. Later on, he won a scholarship from the World Bank for higher studies in Australia and went to University of New England, Armidale, Australia, in 1995 and successfully completed his Ph.D. degree. He carried out his doctoral-level research on the key issue of agricultural policy in Pakistan. He also won a prestigious fellowship and went to University of Laval, Canada, for his post-doctorate studies. Dr. Ashfaq has rendered services as a consultant to various national and international organizations. He is also acting as Ph.D./M.S. supervisor for ICDD, Kassel University, Germany. He has also won the TWAS Associateship Award in 2011 and Australian Alumni Excellence Award in 2013 for Innovation and Research. He has supervised 12 Ph.D. students and more than 150 M.S./M.Phil./M.Sc. students. He is the author/co-author of 5 books, 12 chapters, and more than 100 research papers published in peer-reviewed journals. He is also working as an editor/section editor of different journals. He has completed 30 research projects as PI/Co-PI which were funded by national and international organizations.

Baig, *Aisha*

Ms. Baig holds a Bachelor of Arts in International Development from the University of Guelph, Ontario, Canada. In her undergraduate studies, she took a particular interest in topics related to food, agriculture, and resource economics. One of the most transformative experiences of her studies was the opportunity to complete a field course at the Instituto Nacional de Ciencias Agricolas (INCA) in San José de las Lajas, Cuba. In this course, Aisha worked directly with Cuban farmers and witnessed first-hand the massive potential of agroecology to create more sustainable and resilient food systems. In December 2022, Aisha earned a Master of Arts degree in International Affairs from Carleton University with a specialization in International Organizations and Global Public Policy. Her research and studies focused on addressing complex global issues through international cooperation, including the ways in which diverse international actors can collectively pursue adaptation and mitigation strategies to protect global public goods. Presently, Aisha is completing an internship with an international development consultancy firm based in Turkey, gaining practical experience to complement her academic knowledge. She has an upcoming role in Tanzania, where she will work with women entrepreneurs to advance the goals of gender equality and sustainable economic development. Aisha has a passion for working with vulnerable individuals and finds immense satisfaction in connecting with diverse communities. She is driven by the desire to address systemic issues and contribute to a more equitable, and inclusive global society. In her spare time, Aisha enjoys volunteering, traveling, and reading—to continuously expand her knowledge of other cultures and global affairs. She is particularly interested in content created by authors from the Global South, as they offer a unique perspective on challenges faced by communities worldwide. Through her own writing, Aisha hopes to inspire others' interest in sustainable development and encourage them to adopt a global lens when analyzing resource shortages and the climate crisis.

Baig, *Saima*

Baig is a Chartered Environmentalist from the Society for the Environment, UK (that operates under a Royal Charter), a member of the Institute of Environmental Management and Assessment (IEMA), UK, a member of the International Union for Conservation of Nature (IUCN) Commission on Ecosystem Management, and a member of IUCN Commission on Environmental, Economic and Social Policy. She has a Master of Environmental Management (Yale School of the Environment, Connecticut, USA), in addition to a Master of Business Administration and Bachelor of Science degrees. She has over twenty years of experience working in areas of sustainable development, environmental services, and climate change. Having previously worked throughout Asia, primarily with IUCN as a trainer, researcher, project manager, and head of program, she is now an environmental consultant and has published several studies on environmental and biodiversity conservation and climate change in Asia, Africa, and Latin America.

Bareerah, *Fatima*

Bareerah is a Water Resources Engineer with 11 years of field, policy, and research experience. Since 2008, she has worked in many areas focusing on water resources management. Staring from the corporate sector to the Federal government at the planning and development level, she is presently with a full-time engagement with Pakistan Council of Research in Water Resources. Her areas of functionalities include

Sustainable Development Goals, Integrated Water Resources Management Process, application and management of smart technologies in the field, Karez Cultural Landscape of Balochistan, Artificial Groundwater Recharge, and Rainwater Harvesting. She is an advocate of bottom-up policy and innovation development for water resources development, management, and use.

Bazmi, *Aqeel Ahmed*

Dr. Bazmi is currently working as Professor at the COMSATS University Islamabad-Lahore Campus. He holds a Ph.D. from UTM, Malaysia, a B.E. from NED University of Engineering and Technology, Karachi, and an M.A. Economics from University of Karachi, Pakistan. His research interests include the areas of Climate Change, Energy Conservation, Energy Policy, and Sustainable Development.

Bhutto, *Abdul Waheed*

Dr. Bhutto is currently working as Professor at the Dawood University of Engineering and Technology, Karachi, Pakistan. His research interests include the areas of Climate Change, Energy Conservation, Energy Policy, and Sustainable Development.

Butt, *Zahid Ali*

Dr. Butt holds a Ph.D. in Botany and has a special interest in environmental microbes and their uses for the betterment of humankind. He is presently working as Associate Professor at the Department of Botany, Government College Women University, Sialkot, Pakistan.

Choudhary, *Madhu*

Dr. Choudhary is mainly involved in assessing microbial (fungi and bacteria) diversity, soil enzymatic changes, and microbial biomass C and N under different CSA-based cereal systems. She also investigated soil flora and fauna (earthworms, arthropods, etc.) with different CSA practices under different crops and cropping systems.

Culas, *Richard*

Dr. Culas is a Senior Lecturer in Agricultural Economics/Agribusiness at the School of Agricultural, Environmental and Veterinary Sciences, Charles Sturt University, Australia. He has twenty years of experience in teaching economics and agribusiness subjects, and the supervision of higher degree research and post-graduate projects. During his career, Dr. Culas has served as an external examiner for Ph.D. theses, an expert reviewer for Qatar National Research Fund, and peer-reviewer for more than twenty-five international journals. He chaired contributed sessions for AARES conferences and served on several committees at Charles Sturt. Richard's areas of research *include agriproducts trade, groundwater management, food security, agricultural land use, and deforestation from Australian and international contexts.* He has worked in funded projects worth about A$7.5m by sources such as Soil CRC, ACIAR, Hort. Innovation, The Crawford Fund, and Charles Sturt. He has presented his research at international and national conferences and published in high-impact international journals. He has authored or co-authored over one hundred publications. Richard was a team member of the Strategic Research Area "Improving Rural Livelihoods and Environments in Developing Countries", in the Institute for Land Water and Society, Charles Sturt, which has received DVC's Award for Research Excellence in 2016. He also received an Emerald Literati Award "Outstanding Reviewer", from Emerald Publishing in 2018, for his professional contributions.

Dagar, *Jagdish Chandra*

Dr. Dagar is well recognized for his research in agroforestry, reclamation ecology, and management of salt-affected and waterlogged soils, judicious utilization of saline waters in agriculture, and ecology of halophytes including mangroves, biodiversity, and climate change.

Daoui, *Khalid*

Dr. Daoui obtained his Diploma in Agronomic Engineering from the National School of Agriculture, Morocco, in 1993. He occupied the post of Assistant Coordinator of the Maghreb Faba Bean Research Network from 1993 to 1998. The main objective of this project was the establishment of a research network on food legumes among Maghreb countries (Morocco, Algeria, and Tunisia). In 1999, he joined the National Institute of Agricultural Research (INRA) as a researcher. He conducted research on phosphorus use efficiency by faba bean. In 2007, he obtained his Ph.D. from the Catholic University of Louvain, Belgium. The subject of his Ph.D. was "Research of strategies for improving phosphorus use efficiency in Vicia faba L. conducted under rainfed conditions in Morocco". Now, he occupies the position of Senior Scientist at the Regional Center of Agricultural Research in Meknes, Morocco. The main subjects of interest are agroforestry, food legume agronomy, and nutrient use efficiency.

Diallo, *Thierno Malick*

Dr. Diallo is an Economist and a Lecturer, Gaston Berger University of Saint-Louis, Senegal. His research centers around labor economics, gender, poverty, and rural economics. He has been involved in research collaborations with the Partnership for Economic Policy (PEP), the International Institute of Tropical Agriculture (IITA), the International Fund for Agricultural Development (IFAD), and the African Economic Research Consortium (AERC).

Farhan, *Muhammad*

Dr. Farhan is an accomplished academic and researcher, currently serving as an Associate Professor at Government College University (GCU) in Lahore, Pakistan. His doctoral research focused on the vital areas of bioremediation and biodegradation of organic pollutants, reflecting a keen interest in sustainable solutions for environmental challenges.

Haruna, *Usman*

Dr. Haruna is a Professor of Agricultural Economics and former Dean, Faculty of General Studies, Federal University, Dutse, Jigawa State, Nigeria. He was also the former Provost of the Jigawa State College of Agriculture, Hadejia, Nigeria. He has been involved in teaching both undergraduate and postgraduate courses in Agricultural Economics since 1997. He is also involved in several research activities. His major areas of research include farm management, Sustainable Agricultural Production Systems (SAPS), agricultural marketing, agricultural policy, and development economics. Furthermore, he published more than 100 academic articles in reputable national and international journals. Professor Haruna also supervised more than 50 theses/dissertations of M.Sc. and Ph.D. students to completion, and he is a member of six academic associations. Currently, he is the President of the Africa Farm Management Association.

Jain, *Rahul*

Dr. Jain serves as an Assistant Professor in the Department of Mechanical Engineering at the University Departments, RTU, Kota. He earned his Bachelor's degree in Production and Industrial Engineering from the Rajasthan Technical University, Kota, in 2012. Subsequently, he pursued a Master's degree in Manufacturing System Engineering at MNIT Jaipur, completing the program in 2014. He later completed his Ph.D. from MNIT Jaipur in the field of industrial engineering. His research interests span ergonomics, data analytics, posture biomechanics, and industrial design. He has contributed to the field by publishing over 50 research papers in various international journals and conferences. In addition, he has provided guidance to one Ph.D. candidate and two postgraduate students in the field of industrial engineering.

Jat, *Hanuman Sahay*

Dr. Jat is working on conservation agriculture (CA) and climate-smart agriculture (CSA) to address the natural resource degradation in the north-western IGP. Much of his focus is on cereal systems' resilience to sustain agri-food systems. He also served CIMMYT under the Sustainable Intensification Program. Currently engaged in how salt behaves with CSA practices in salt-affected ecologies and on out-scaling of smart management practices in the region.

Jemziya, *Fathima*

Ms. Jemziya is a dedicated agricultural and food science professional with a passion for addressing the challenges of climate change in the context of Sri Lankan agriculture. Holding degrees in Agriculture and Food Science & Technology, Ms. Jemziya has demonstrated a profound commitment to advancing sustainable practices in the field. With a solid academic foundation, she has also actively contributed to the scientific community by authoring and co-authoring several research papers that have been published in renowned journals, addressing critical issues related to postharvest handling, food security, and climate change adaptation strategies in agriculture. Her work in this field not only showcases her academic prowess but also underscores her genuine concern for the environment and the welfare of the agricultural community in Sri Lanka. Her commitment to advancing knowledge and practices in the agriculture and food science domains is a testament to her potential as a future leader in this critical field.

Kamran, *Ashraf*

Mr. Kamran is currently working as Lecturer at the Department of Food Sciences, Government College University Faisalabad, Sahiwal Campus. He has more than 2 years of teaching and research experience, which includes working as Lecturer in Comsats University, Islamabad, Sahiwal Campus, and University of Montgomery, Sahiwal, Pakistan. He also served as Secretary Press & Public Relations in Pakistan Society of Food Scientists & Technologists. Kamran research background is linked with Multigrain Products and their Nutritional Aspects. Major areas of his research are the antioxidant potential of multigrain products and their beneficial aspects and sustainability issues in environment and agriculture with a major focus on beneficial elements for heavy metal toxicity in cereals and vegetables. Nowadays, he is a Ph.D. scholar at the East China University of Science and Technology, Shanghai, China.

Kan, *Mustafa*

Dr. Kan is working as Associate Professor at the Department of Agricultural Economics, Agricultural Faculty, at Kirsehir Ahi Evran University, Kirsehir-Turkiye. He has about 20 years of teaching and research experience. His main areas of expertise are rural development, rural tourism, local foods, geographical indication systems, climate change, and agricultural support policies. He managed and carried out many national and international projects and published his work in peer-reviewed journals.

Karim, *Sadia*

Sadia Karim is currently working as Assistant Professor at the Dawood University of Engineering and Technology, Karachi, Pakistan. His research interests include the areas of Sustainable Development.

Keskin, *Muharrem*

Dr. Keskin is a Professor at the Department of Biosystems Engineering of Hatay Mustafa Kemal University, Antakya, Hatay, Turkey. He is also a Visiting Scholar for online teaching at Konya Food and Agriculture University located in Konya, Turkiy. He has research and teaching tasks in the areas of precision agriculture, agricultural machinery, spectroscopy, and occupational safety in agriculture. He received an M.S. and a Ph.D. degree from the Department of Biosystems Engineering of Clemson University located in Clemson, SC, USA. He is the Head of the Agricultural Machinery Systems Department and Precision Agriculture Technologies Research Lab at Hatay Mustafa Kemal University. He has more than 20 years of experience in teaching and research on precision agriculture technologies.

Komal, *Nayab*

Ms. Komal is a young, hardworking, and dedicated Environmentalist. She has completed her M. Phil. from Department of Environmental Sciences, The University of Lahore, Pakistan, in 2021. She has experience in working with project delivery teams, environmental engineering firms, and in the role of agroforestry in climate change mitigation. She published research articles in national and international journals. She is interested in further exploring the role of forests especially agroforests and plantations in climate change mitigation by using some models and advanced techniques.

Mohamed Najim, *Mohamed Mujithab*

Professor Dr. M. M. Mohamed Najim earned his B.Sc. Honors in Agriculture from University of Peradeniya, Sri Lanka, in 1994, M. Eng. in Irrigation Engineering and Management from Asian Institute of Technology, Thailand, in 2000, and Ph.D. in Water Resources Engineering from University Putra Malaysia in 2004. He has 29 years of experience as a University academic. He served as the Vice-Chancellor of South Eastern University of Sri Lanka from June 2015 to August 2021. He has served as the General Secretary of Sri Lanka Association for the Advancement of Science from 2012–2015. He was awarded many prestigious awards including Presidential Award for Scientific Research. He has supervised 5 Ph.D. research students, one M.Phil. student, and many M.Sc. students. He is a life member of many professional associations. He has published 63 full research papers out of which 28 are published in peer-reviewed indexed journals, 17 in peer-reviewed journals, and 18 in conference proceedings. He has also published more than 75 short scientific communications. He has contributed to publishing 9 scholarly book chapters.

Mrabet, *Rachid*

Dr. Mrabet holds a Ph.D. in Agronomy from Colorado State University, 1997, and an M.Sc. in Soil Physics from University of Minnesota/IAV Hassan II, 1989. He is a Distinguished Research Director (INRA Morocco) with 33 years of research, expertise, and advisory in agronomy, soil science, environment quality, conservation agriculture, climate change, and carbon farming as well as biophysical modeling. Dr. Mrabet authored 21 books, 77 chapters, and 90 peer-reviewed papers. He served as an IPCC review editor for Special Report on 1.5°C, a Coordinating Lead Author (CLA) for AR6 Group III Chap. 7 on AFOLU, a CLA for MedECC First Assessment Report (MAR1) Chapters on Food systems, and currently CLA for UNEP GEO-7. He is also the Deputy President of WASWAC, Panelist at International Conservation Agriculture Advisory Panel for Africa, and Consultant for several international organizations (FAO, GIZ, ICARDA, UNECA, CIHEAM, OADA, ACT, and COP22). Dr. Mrabet received Distinguished Research Award from WASWAC, the King Hassan II National Prize, the 2020 North South Prize of the Council of Europe, and 2022 Gulbenkian Prize for Humanity as CLA for MedECC and IPCC. He coordinated/participated in more than 60 national and international projects and serves as Associate Editor for JSWC, STR, ISWCR, Editor in Chief of Afrimed AJ, and reviewer and special Guest Editor for several international journals.

Mufeezah, *Ahsan*

Mufeezah is an Environmentalist and Geospatial Analyst with 6 years of experience as researcher and trainer. Since 2012, she has delivered health and safety training in various institutions focusing on water-borne diseases. From 2016 to date, she has been part of various research teams in Pakistan Council of Research in Water Resources (PCRWR). As part of the water resources management research team, she has worked closely with communities for qualitative research on water issues and gender, water quality assessment, and geospatial analysis for hotspot identification of water pollution risk areas. Other functionalities include training on water quality analysis and mapping, water risk assessment, artificial groundwater recharge and rainwater harvesting, and groundwater regulatory framework development. She is a strong believer and supporter of efficient water use technologies as well as water resource management to achieve Sustainable Development Goals.

Muneeb, *M. Musthafa*

Dr. Muneeb is currently working as a Senior Lecturer at the Biosystems Department of South Eastern University of Sri Lanka for the past five years. He has contributed scientifically in broad areas of ecology, biodiversity, conservation science, animal science, and genetics. He is supervising several postgraduate and undergraduate students. He received his Ph.D. from the University of Malaya, Malaysia in 2018, with excellence in ecology and biodiversity. He received International Graduate Research Assistant Scheme (iGRAS) from the University of Malaya. Before that, he worked as a researcher at Ege University, Turkey, and worked at the Chinese Academy of Agricultural Sciences (CAAS). He completed his M.Sc. in Animal Science from King Saud University, Saudi Arabia, while serving as a researcher at the Animal Biotechnology lab. He earned his bachelor's degree in agriculture from the University of Peradeniya, Sri Lanka, in 2007. Dr. Muneeb has published over 55 peer-reviewed and indexed publications in various areas including Ecology, Biodiversity, Genetic Diversity, Environmental Conservation, Food Science and Technology, etc. Dr. Muneeb has published three book chapters and participated in 21 international conferences. He has been involved in different journals as a reviewer. He is a member of several IUCN Commissions, the Global Mountain Biodiversity Assessment (GMBA), the Society for Open, Reliable, and Transparent Ecology and Evolutionary Biology (SORTEE), and the Entomological Society of Malaysia (ENTOMA). He contributed to the strategic plan preparation (2024–2028) of the South Eastern University of Sri Lanka in May 2023. He has been instrumental in establishing the Sri Lankan Journal of Technology and served as the editor-in-chief from 2020 to 2023. Dr. Muneeb has been appointed as the Curriculum Committee Chairman of the Faculty of Technology, South Eastern University of Sri Lanka since January 2021. He is serving on several international and local academic committees.

Narmilan, *Amarasingam*

Narmilan is presently a Lecturer at the Department of Biosystems Technology of the South Eastern University of Sri Lanka. He is currently a Ph.D. candidate at the School of Electrical Engineering and Robotics, Faculty of Engineering of the Queensland University of Technology. He completed double bachelor's degrees, receiving a Bachelor's in Agricultural Science (BSc Agri. (Hons.)) at the Eastern University, Sri Lanka, and a Bachelor's in Information Technology (BIT) at the University of Colombo, Sri Lanka. He holds an M.Sc. in Agriculture from the Eastern University, (mostly silvopastoral), forest ecology, ecology and vegetation, environmental education, long-distance learning, and M.Sc. programs. She has taught courses on agroforestry, forest soils, forest ecology, and forest vegetation. She has over 130 publications in scientific journals, conference proceedings, and books. She acts as a scientific editor in two international scientific journals and many as a reviewer. Since 1993, she has been a member of the Honorary Scientific (Honorary Society) American Company "Xi Sigma Pi".

Parihar, *Chittar Mal*

Dr. Parihar is notable for handling the soil quality aspects (physical and chemical) of different crops and cropping systems. He has been deeply involved in experimental design, data interpretation, and broader contextualization of experimental results toward achieving sustainable intensification at scale.

Poonia, *Tanuja*

Dr. Poonia is mainly involved in the agronomic management of crops and cropping systems at research platform (RP) as well as at farmers' fields for demonstration and testing of individual technological interventions like the use of Happy Seeder with full residue retention in rice, direct seeded rice (DSR), crop diversification, nutrient management using Green Seeker and Nutrient Expert Software tool, and performance of different cultivars under CA-based management system.

Qureshi, *Rahmatullah*

Dr. Qureshi is a Tenured Professor and Chairman of the Department of Botany at Pir Mehr Ali Shah Arid Agriculture University, Rawalpindi, Pakistan. He is a national plant biodiversity specialist and works on plant diversity, conservation issues, and ecosystem services in various ecological zones of Pakistan. He has contributed 250 publications, including four international books and 8 book chapters on ecosystem services, medicinal plants, plant biodiversity, vegetation ordination, etc. He has earned 12 projects and recently worked as a pioneer Project Director on the PSDP Project for Hemp Cultivation in Pakistan. His cumulative impact factor is 230 with over 5480 citations, an H-Index of 38, and an i10-Index of 122. He has supervised 15 Ph.D. and 66 M. Phil. theses. He has participated in various national and international conferences/symposia and delivered 42 talks as a Guest/Invited/Keynote speaker. He has also participated in various Talk shows and briefings on Television as an expert. He is editor of three research journals of international repute and Editor-in-Chief of the Journal of Botanical Research and Applications. In the scientific community, he is a renowned person and remained an elective member of the Executive Council of the Pakistan Botanical Society, where he worked as Vice-President of Punjab Province from 2011 to 2017. He is a member of various technical and curriculum committees of different universities, etc. He won five research productivity awards granted by the Pakistan Council for Science & Technology. Realizing his contribution toward plant biodiversity and related issues, he has been nominated by the government of Pakistan as country representative in the Multidisciplinary Expert Panel (MEP) for the Intergovernmental Science-Policy Platform on Biodiversity and

Ecosystem Services (IPBES), Bonn, Germany. For the last two years, he has been awarded Best Research Professor by the University.

Rana, *Shahjahan Shabbir Ahmed*

Professor Dr. Rana is a distinguished expert in the fields of horticulture and biotechnology, with a career marked by groundbreaking research, academic leadership, and a commitment to advancing sustainable agriculture practices. He has diverse experience in the Lasbela University of Agriculture Water & Marine Sciences (LUAWMS), Agriculture Department Balochistan and Balochistan University of Information Technology, Engineering and Management Sciences (BUITEMS), Quetta. He has earned his Ph.D. from Northwest A&F University, China. He has demonstrated an application of biotechnological tools in horticultural commodities to extend their shelf life without exposure to reefer conditions. Dr. Rana has demonstrated exceptional leadership in academia, holding various key positions from Chairperson to Deputy Director of various departments of the University. He has been actively involved in curriculum development, teaching Chinese, and fostering an environment that encourages innovative thinking and research.

Raza, *Syed Ali*

Dr. Raza obtained his Ph.D. in chemistry from Government College University Lahore Pakistan. He is currently serving as Research Officer in Water Resources Zone erstwhile Directorate of Land Reclamation Irrigation Department Government of the Punjab, Pakistan. He has 15 years of experience in research on natural products and crop, water, and soil relationships.

Sani, *Imran Ali*

Dr. Sani is an accomplished Assistant Professor at the Department of Biotechnology, Balochistan University of Information Technology, Engineering, and Management Sciences (BUITEMS) in Quetta, Pakistan. He has dedicated his career to the fields of Entomology, Plant Biotechnology, and DNA technology.

Sekerli, *Yunus Emre*

Dr. Sekerli is an Assistant Professor at the Department of Biosystems Engineering of Hatay Mustafa Kemal University, Antakya, Hatay, Turkey. He completed his Ph.D. research in the same department on "the prediction of nutrient content of turfgrass using optical sensors". He has teaching and research responsibilities mainly in the area of Biosystems Engineering. In 2014, he had a summer training period at Warsaw University of Life Sciences on precision agriculture applications in growing winter wheat. He has published research papers on proximal sensing, non-destructive quality evaluation of agricultural and food products, precision agriculture, food processing, and food quality. He is also a member of Precision Agriculture Technology Research Lab at Hatay Mustafa Kemal University.

Sharma, *Parbodh Chander*

Dr. Sharma has mainly worked on crop physiology, marker-assisted selection, and mutation for crop improvement under salt stress conditions in India. He is also involved in the sustainable intensification of cereal-based systems in the IGP through Conservation Agriculture-based management systems.

Syed, *Ale Raza Shah*

Mr. Syed is a Doctoral Scholar at the School of Economics & Finance at Xi'an Jiaotong University, Xi'an, P.R. China. He completed his Master's in Energy Economics from the Government College University Faisalabad, Pakistan. His research has appeared in various journals such as Renewable Sustainable Energy Reviews, Energy, Renewable Energy, Applied Energy, Ecological Indicators, Resources Policy, and Sustainability. His research interests are related to Energy Economics, Development Economics, and Environmental Economics.

Syed, *Asif Ali Naqvi*

Dr. Syed is Assistant Professor at the Department of Economics, Government College University, Faisalabad, Pakistan. Syed holds a Ph.D. degree in Agricultural Economics from University of Agriculture Faisalabad, Pakistan. His key areas of research include Impact Assessment, Environmental Economics, and Energy Markets. He has published papers in the Journal of Cleaner Production, Sustainability, Renewable Energy, Environmental Science, and Pollution Research, and Pak Journal of Agri Sciences.

Thariq, *MG Mohamed*

Dr. Thariq serves as a Senior Lecturer at the Department of Biosystems Technology of the South Eastern University of Sri Lanka. Dr. Thariq, before joining the university in 2016, worked as Production Manager for the State Timber Corporation of Sri Lanka. He completed a Bachelor Degree in Agriculture from the University of Peradeniya, Sri Lanka, an M.Sc. in Forestry and Environmental Management from the University of Sri Jayewardenepura, Sri Lanka, and a Ph.D. in Ergonomics from the University of Moratuwa, Sri Lanka. He has more than 30 research publications in indexed and peer-reviewed journals and in research conferences. His research interests include ergonomics in agriculture, climate-resilient animal production, animal nutrition, and goat farming. He has supervised 30 undergraduate research projects. He served as the Dean of the Faculty of Technology of the South Eastern University of Sri Lanka. Presently, he is an Associate Editor of the Sri Lanka Journal of Technology published by the Faculty of Technology of the South Eastern University of Sri Lanka.

Udayanga, *Lahiru*

Dr. Udayanga earned his B.Sc. (Special) in Environmental Conservation and Management in 2005, followed by a Ph.D. in Geo-Informatics and Medical Entomology in 2019 from University of Kelaniya, Sri Lanka. From 2018 onwards, he is serving as an Academic at the Wayamba University of Sri Lanka. He has been awarded many prestigious awards including the Presidential Award for Scientific Research. He conducts research in the fields of environmental modeling, climate change, medical entomology, and remote sensing. Up to now, he has published 68 full research papers out of which, 34 have been published in peer-reviewed science citation indexed journals, along with 13 manuscripts published in peer-reviewed journals, and 21 as conference proceedings. Further, he has published 79 short scientific communications and contributed to 5 scholarly book chapters. He is also a life member of many professional associations.

ul Hasan, *Faizan*

ul Hasan is a civil engineer with a specialization in Water Resources Management. He has more than 20 years of experience in the water sector particularly research and development activities on different issues of water supply and treatment, irrigation, and water/soil conservation in various areas of Pakistan. During his professional career, he worked for the integrated management of water resources, particularly surface and groundwater resources in the irrigated and dry (rainfed) areas. More specifically, he worked on ICT technologies for smart water management, urban and rural water supply management, water quality monitoring, water treatment, climate change, community education and mobilization, water conservation technologies, human water security for underprivileged areas, and renewable energy options. Presently, he is a Ph.D. researcher at the Center for Environmental Governance, University of Canberra, Australia.

Umar, *Mohammed Bashir*

Dr. Umar is currently a Senior Lecturer in Agricultural Extension at the Federal University Gashua, Yobe State, Nigeria. He took appointment in 1991 as an Assistant Lecturer with the then State Polytechnic and has been involved in teaching and research in Agricultural Extension and Rural Sociology. Dr. Umar Attended Ahmadu Bello University, Zaria, Nigeria, where he obtained a Bachelor's degree in Agriculture in 1991. Later, in 2006, he obtained a Master's degree in Agricultural Extension and Rural Sociology from the same university. In 2016, he obtained a Ph.D. in the same field from the same university. His major areas of research include the impact of Hadejia Valley Irrigation Project on crop productivity and poverty reduction. Dr. Umar has several publications in national and international journals and conferences. He joined the services of the University in 2019 and his current interests include extension communication, poverty reduction, climate change, and sustainable development.

uz Zaman, *Qamar*

Dr. uz Zaman is a young environmentalist working as an Assistant Professor in the Department of Environmental Sciences at the University of Lahore (UOL). He obtained his B.Sc. (Hons.), M.Sc. (Hons.), and Ph.D. in (Plant Sciences/Agronomy) from University of Agriculture, Faisalabad, Punjab, Pakistan. He has a good record of publications and has been working on different projects to put his proper share in developing advanced technologies for students. Major areas of his research are